**Pharmacokinetics and Pharmacodynamics of Biotech Drugs**

*Edited by*
*Bernd Meibohm*

*Related Titles*

M.S. Chorghade (Ed.)

**Drug Discovery and Development**

**2 Volumes**

2006. ISBN 0-471-39846-2

J. Knäblein (Ed.)

**Modern Biopharmaceuticals**

**4 Volumes**

**Design, Development and Optimization**

2005. ISBN 3-527-31184-X

M. Schleef (Ed.)

**DNA-Pharmaceuticals**

**Formulation and Delivery in Gene Therapy, DNA Vaccination and Immunotherapy**

2005. ISBN 3-527-31187-4

O. Kayser, R.H. Müller (Eds.)

**Pharmaceutical Biotechnology**

**Drug Discovery and Clinical Applications**

2004. ISBN 3-527-30554-8

R.J.Y. Ho, M. Gibaldi

**Biotechnology and Biopharmaceuticals**

**Transforming Proteins and Genes into Drugs**

2003. ISBN 0-471-20690-3

# Pharmacokinetics and Pharmacodynamics of Biotech Drugs

## Principles and Case Studies in Drug Development

*Edited by*
*Bernd Meibohm*

WILEY-VCH

WILEY-VCH Verlag GmbH & Co. KGaA

**The Editor**

***Prof. Dr. Bernd Meibohm***
University of Tennessee
Health Science Center
Department of Pharmaceutical Sciences
College of Pharmacy
874 Union Ave. Suite 5p
Memphis, TN 38163
USA

1st Reprint 2008



**Library of Congress Card No.:** applied for

**British Library Cataloguing-in-Publication Data:**
A catalogue record for this book is available from the British Library

**Bibliographic information published by the Deutsche Nationalbibliothek**
The Deutsche Nationalbibliothek lists this publication in the Deutsche Nationalbibliografie; detailed bibliographic data are available in the Internet at http://dnb.d-nb.de.



**Composition** ProSatz Unger, Weinheim
**Printing** Strauss GmbH, Mörlenbach
**Bookbinding** Litges & Dopf Buchbinderei GmbH, Heppenheim

Printed in the Federal Republic of Germany
Printed on acid-free paper

**ISBN-13:** 978-3-527-31408-9
**ISBN-10:** 3-527-31408-3

# Foreword

Pharmacokinetics and pharmacodynamics (PK/PD) have become essential disciplines in drug research and development. Rational use of PK/PD allows for better decision making and streamlines dose optimization. In the past, PK/PD concepts have been primarily been applied to the development of small drug molecules. However, in recent years more and more drug candidates come form the field of biotechnology and are larger molecules. Pharmacokinetics and Pharmacodynamics of Biotech Drugs gives an excellent overview of the state of the art of applying PK/PD concepts to large molecules.

After a comprehensive introduction, the basic PK/PD properties of peptides, monoclonal antibodies, antisense oligonucleotides and gene delivery vectors are reviewed. In the second part, the book covers a number of challenges and opportunities in this field such as bioanalytical methods, bioequivalence and pulmonary delivery. The text finishes with a detailed presentation of some real-life examples and case studies which should be of particular interest to the reader and integrate many of the concepts presented earlier in the text.

The book was written by a group of international expert scientists in the field. It is well-structured and easy to follow. The book is of great value for everybody working in this area.

*Hartmut Derendorf, Ph.D.*
Distinguished Professor and Chairman
Department of Pharmaceutics
University of Florida

*Pharmacokinetics and Pharmacodynamics of Biotech Drugs: Principles and Case Studies in Drug Development.* Edited by Bernd Meibohm

ISBN: 3-527-31408-3

# Preface

In recent years, biotechnologically-derived drugs (biotech drugs) including proteins, peptides, monoclonal antibodies and antibody fragments, as well as antisense oligonucleotides and DNA preparations for gene therapy, have been a major focus of research and development (R&D) efforts in the pharmaceutical industry, and biotech drugs constitute already a sizable fraction of the medications used in clinical practice.

Pharmacokinetic (PK) and pharmacodynamic (PD) concepts impact every stage of the drug development process starting from lead optimization to the design of Phase III pivotal trials. PK and PK/PD evaluations are widely considered cornerstones in the development of new drug products and are usually deeply embedded in the discovery and development plan. The widespread application of PK/PD concepts in all phases of drug development has repeatedly been promoted by industry, academia, and regulatory authorities, most recently through FDA's Critical Path to New Medical Products initiative and the concept of integrated model-based drug development.

An understanding of PK and PD and the related dose-concentration-effect relationship is crucial to any drug – including biotech products – since it lays the foundation for dosing regimen design and rational clinical application. While general PK and PD principles are just as applicable to biotech agents as they are to traditional small molecule drugs, PK and PK/PD analyses of biotech agents frequently pose extra challenges related to factors such as their similarity to endogenous molecules and/or nutrients and their immunogenicity.

This textbook provides a comprehensive overview on the PK and PD of biotech-derived drug products, highlights the specific requirements and challenges related to PK and PK/PD evaluations of these compounds and provides examples of their application in preclinical and clinical drug development. The impetus for this project originated from the notion that at the time of its initiation there was no comprehensive publication on the market that specifically addressed this topic.

Following a short introduction, the book is structured into three sections: The 'Basics' section discusses individually the pharmacokinetics of peptides, monoclonal antibodies, antisense oligonucleotides and gene delivery vectors. The subsequent 'Challenges and Opportunities' section includes more detailed considerations on selected topics, including technical challenges such as bioanalytical

*Pharmacokinetics and Pharmacodynamics of Biotech Drugs: Principles and Case Studies in Drug Development.* Edited by Bernd Meibohm

ISBN: 3-527-31408-3

methodologies, noncompartmental data analysis and exposure-response assessments. It furthermore discusses biopharmaceutical challenges as exemplified by the delivery of oligonucleotides and of peptides and proteins to the lung, and provides insights into the opportunities provided by chemical modification of biotech drugs and the regulatory challenges related to follow-on biologics. The third and final section provides examples for the 'Integration of Pharmacokinetic and Pharmacodynamic Concepts into the Biotech Drug Development Plan', including the preclinical and early clinical development of tasidotin, and the clinical development programs for cetuximab and pegfilgrastim.

The book addresses an audience with basic knowledge in clinical pharmacology, PK and PD, and clinical drug development. It is intended as a resource for graduate students, postdocs, and junior scientists, but also for those more experienced pharmaceutical scientists that have no experience in the PK and PD evaluation of biotech drugs and wish to gain knowledge in this area.

I would like to express my gratitude to all contributors of this project for providing their unique array of expertise to this book project which allowed us to compile a wide variety of viewpoints relevant to the PK and PK/PD evaluation of biotech drugs and derived products. In addition, I would like to thank Dr. Romy Kirsten and Dr. Andrea Pillmann at Wiley-VCH for their assistance in producing this book and Ms. Faith Barcroft for her invaluable text editing.

Finally I would like to dedicate this book to my family for their patience, encouragement and support during this project.

Memphis, Summer 2006 *Bernd Meibohm*

# Contents



*Pharmacokinetics and Pharmacodynamics of Biotech Drugs: Principles and Case Studies in Drug Development.* Edited by Bernd Meibohm

ISBN: 3-527-31408-3

# List of Contributors

***Larry Arthaud***
Department of Pharmacokinetics
Genzyme Corporation
4545 Horizon Hill Blvd.
San Antonio, TX 78229
USA

***Jeffrey S. Barrett***
Laboratory for Applied PK/PD
The Children's Hospital of Philadelphia
Abramson Research Center, Room 916 H
34th Street and Civic Center Blvd.
Philadelphia, PA 19104
USA

***Peter L. Bonate***
Department of Pharmacokinetics
Genzyme Corporation
4545 Horizon Hill Blvd.
San Antonio, TX 78229
USA

***Paolo Caliceti***
Department of Pharmaceutical Sciences
School of Pharmacy
University of Padua
Via Francesco Marzolo, 5
35131 Padua
Italy

***Kun Cheng***
Department of Pharmaceutical Sciences
University of Tennessee Health Science Center
26 S. Dunlap Street
Memphis, TN 38163
USA

***Floyd E. Fox***
Clinical Pharmacology
ImClone Systems Inc.
59 ImClone Drive
Somerville, NJ 08876
USA

***Richard S. Geary***
Pharmacokinetics and Drug Metabolism
Isis Pharmaceuticals
1896 Rutherford Road
Carlsbad, CA 92008
USA

***Gregory E. Hardee***
Quality Assurance
Isis Pharmaceuticals
1896 Rutherford Road
Carlsbad, CA 92008
USA

*Pharmacokinetics and Pharmacodynamics of Biotech Drugs: Principles and Case Studies in Drug Development*. Edited by Bernd Meibohm

ISBN: 3-527-31408-3

***Jeffrey R. Hughes***
Pharmaceutical Sciences
College of Pharmacy
University of Florida
Box 100494, JHMHC
Gainesville, FL 32610
USA

***Charlotte Kloft***
Department of Clinical Pharmacy
Faculty of Pharmacy
Martin-Luther-Universität
Halle-Wittenberg
Wolfgang-Langenbeck-Strasse 4
06120 Halle
Germany

***Katharina Kuester***
Department of Clinical Pharmacy
Institute of Pharmacy
Freie Universität Berlin
Kelchstrasse 31
12169 Berlin
Germany

***Andreas Kovar***
Clinical Pharmacology
and Pharmacokinetics
Merck KGaA
Frankfurter Strasse 250
64293 Darmstadt
Germany

***Jean W. Lee***
Pharmacokinetics and Drug
Metabolism
Protein Labs
Amgen Inc.
One Amgen Center Drive
Thousand Oaks, CA 91320
USA

***Arthur A. Levin***
Isis Pharmaceuticals
1896 Rutherford Road
Carlsbad, CA 92008
USA

***Ram I. Mahato***
Department of Pharmaceutical
Sciences
College of Pharmacy
University of Tennessee Health
Science Center
26 S. Dunlap Street
Memphis, TN 38163
USA

***Bernd Meibohm***
Department of Pharmaceutical
Sciences
College of Pharmacy
University of Tennessee Health
Science Center
874 Union Avenue, Suite 5p
Memphis, TN 38163
USA

***Martin Meyer***
Department of Pharmaceutics
College of Pharmacy
Box 100494, JHMHC
University of Florida
Gainesville, FL 32610
USA

***Arno Nolting***
Clinical Pharmacology and
Pharmacokinetics
Merck KGaA
Frankfurter Strasse 250
64293 Darmstadt
Germany

***Gururaj Rao***
Department of Pharmaceutics
College of Pharmacy
Box 100494, JHMHC
University of Florida
Gainesville, FL 32610
USA

***Ke Ren***
Department of Pharmaceutics
College of Pharmacy
Box 100494, JHMHC
University of Florida
Gainesville, FL 32610
USA

***Lorin K. Roskos***
Pharmacokinetics and Toxicology
Amgen Inc.
6701 Kaiser Drive
Fremont, CA 94555
USA

***Katherine Stephenson***
Department of Pharmacokinetics
Genzyme Corporation
4545 Horizon Hill Blvd.
San Antonio, TX 78229
USA

***Arthur B. Straughn***
Department of Pharmaceutical Sciences
College of Pharmacy
University of Tennessee Health Science Center
874 Union Avenue, Suite 5p
Memphis, TN 38163
USA

***Mohammad Tabrizi***
Pharmacokinetics and Toxicology
Amgen Inc.
6701 Kaiser Drive
Fremont, CA 94555
USA

***Lisa Tang***
Department of Pharmaceutical Sciences
College of Pharmacy
University of Tennessee Health Science Center
874 Union Avenue, Room 105p
Memphis, TN 38163
USA

***Lloyd G. Tillman***
Pharmaceutical Development
Isis Pharmaceuticals
1896 Rutherford Road
Carlsbad, CA 92008
USA

***Francesco M. Veronese***
Department of Pharmaceutical Sciences
School of Pharmacy
University of Padua
Via Francesco Marzolo, 5
35131 Padua
Italy

***Bing-Bing Yang***
Pharmacokinetics and Drug Metabolism
Amgen Inc.
One Amgen Center Drive
Thousand Oaks, CA 91320
USA

***Rosie Z. Yu***
Department of Pharmacokinetics
Isis Pharmaceuticals
1896 Rutherford Road
Carlsbad, CA 92008
USA

# Part I
# Introduction

*Pharmacokinetics and Pharmacodynamics of Biotech Drugs: Principles and Case Studies in Drug Development.* Edited by Bernd Meibohm

ISBN: 3-527-31408-3

# 1
# The Role of Pharmacokinetics and Pharmacodynamics in the Development of Biotech Drugs

*Bernd Meibohm*

## 1.1
## Introduction

During the past two decades, advances in biotechnology have triggered the development of numerous new drug products. This group of so-called biotech drugs is a subset of the therapeutic group of biologics. Therapeutic biologic products, or biologics, are defined by the U.S. Food and Drug Administration (FDA) as any virus, therapeutic serum, toxin, antitoxin, or analogous product applicable to the prevention, treatment or cure of diseases or injuries of man. Biologics are a subset of drug products distinguished by their manufacturing process. While classical drugs are synthesized via a chemical process, biologics are manufactured utilizing biological processes and are typically derived from living material – human, plant, animal, or microorganism. Biotech drugs can be considered as those biologics that are manufactured using biotechnology-based production processes.

The similarity in the drug development and evaluation process for biotech drugs and conventional, chemically synthesized drugs has recently been acknowledged in the FDA's 2003 decision to transfer certain product oversight responsibilities from the Center for Biologics Evaluation and Research (CBER) to the Center for Drug Evaluation and Research (CDER). The biologics for which oversight was transferred include monoclonal antibodies for *in vivo* use, proteins intended for therapeutic use, including cytokines (e.g., interferons), enzymes (e.g., thrombolytics), growth factors, and other novel proteins that are derived from plants, animals, or microorganisms, including recombinant versions of these products, and other non-vaccine and non-allergenic therapeutic immunotherapies. Classical biologics such as blood, blood components and vaccines remain under the regulatory authority of the CBER. Even under this new structure, however, the biologic products transferred to the CDER will continue to be regulated as licensed biologics – that is, a Biologic License Application (BLA) must be submitted to obtain marketing authorization as compared to a New Drug Application (NDA) which is used for traditional, chemically manufactured drug products.

*Pharmacokinetics and Pharmacodynamics of Biotech Drugs: Principles and Case Studies in Drug Development*. Edited by Bernd Meibohm

ISBN: 3-527-31408-3

For the purpose of this book, biotech drugs include not only therapeutically used peptides and proteins, including monoclonal antibodies, but also oligonucleotides and DNA preparations for gene therapy. Although oligonucleotides are, due to their chemically defined production process, classified by the FDA as classical drugs requiring an NDA prior to marketing authorization, and DNA preparations for gene therapy are regulated by the CBER, they are both included in the class of biotech drugs as their therapeutic application relies heavily on the principles of molecular biology and they are considered by analysts as biotech compounds.

## 1.2
## Biotech Drugs and the Pharmaceutical Industry

In parallel with the development of the discipline of biotechnology during the past two decades, an increasing fraction of pharmaceutical R&D has been devoted to biotechnology-derived drug products. It has been estimated that more than 250 million patients have benefited from already approved biotechnology medicines to treat or prevent heart attacks, stroke, multiple sclerosis, leukemia, hepatitis, rheumatoid arthritis, breast cancer, diabetes, congestive heart failure, kidney cancer, cystic fibrosis and other diseases [1]. This number is expected to increase significantly with the introduction of new biotech drugs into the marketplace. According to a survey by the Pharmaceutical Research and Manufacturers of America (PhRMA) in 2004, 324 biotechnology medicines were in development for almost 150 diseases. These include 154 medicines for cancer, 43 for infectious diseases, 26 for autoimmune diseases, and 17 for AIDS/HIV and related conditions. These potential medicines – all of which were at the time of the survey either in human clinical trials or under review by the FDA – will enlarge the list of 108 biotechnology medicines already approved and available to patients (Fig. 1.1) [1].

Biotech and genomic companies currently perform almost one-fifth of all pharmaceutical R&D, and this figure is set to double during the next 10 years [2]. It has been suggested that over half of all the New Active Substances developed during the next 10–15 years will result from research into antibodies alone. Biotechnology products accounted for more than 35% of the 37 New Active Substances that were launched in 2001 [2]. This success in drug development is underlined by the fact that several biotech drugs have achieved blockbuster status, earning more than US$ 1 billion in annual sales, including Epoetin-α (Epogen/Procrit/Eprex), interferon-α2b (IntronA, PEG-Intron/Rebetron combination therapy), and filgrastim (Neupogen) [3].

Since the development of biotech drugs generally rests on a fundamental understanding of the related disease, their clinical development has also proven to be more successful than for conventional, chemically derived small-molecule drugs. Only 8% of the new chemical entities that entered the clinical phases of drug development between 1996 and 1998 reached the market, compared to 34% of biotech drugs (Fig. 1.2). This means that biologics have, at the time of their first-in-

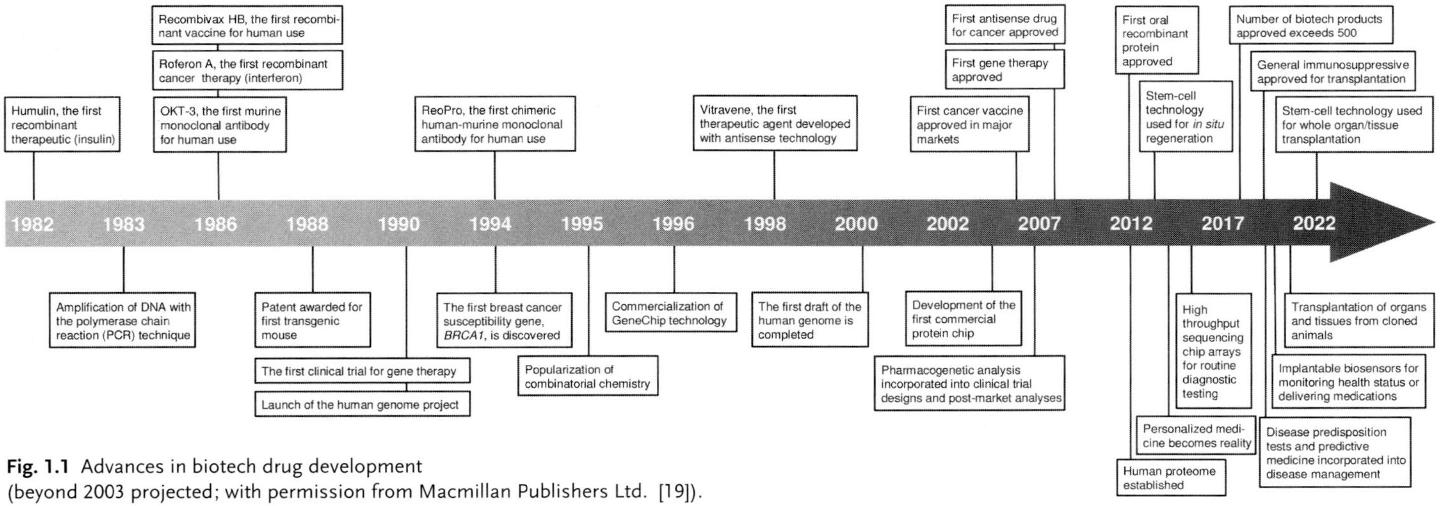
1982
1983
1986
1988
1990
1994
1995
1996
1998
2000
2002
2007
2012
2017
2022
Humulin, the first recombinant therapeutic (insulin)
Amplification of DNA with the polymerase chain reaction (PCR) technique
Recombivax HB, the first recombinant vaccine for human use
Roferon A, the first recombinant cancer therapy (interferon)
OKT-3, the first murine monoclonal antibody for human use
Patent awarded for first transgenic mouse
The first clinical trial for gene therapy
Launch of the human genome project
ReoPro, the first chimeric human-murine monoclonal antibody for human use
The first breast cancer susceptibility gene, BRCA1, is discovered
Popularization of combinatorial chemistry
Commercialization of GeneChip technology
Vitravene, the first therapeutic agent developed with antisense technology
The first draft of the human genome is completed
First cancer vaccine approved in major markets
Development of the first commercial protein chip
Pharmacogenetic analysis incorporated into clinical trial designs and post-market analyses
First antisense drug for cancer approved
First gene therapy approved
First oral recombinant protein approved
Stem-cell technology used for in situ regeneration
Human proteome established
Personalized medicine becomes reality
High throughput sequencing chip arrays for routine diagnostic testing
Number of biotech products approved exceeds 500
General immunosuppressive approved for transplantation
Stem-cell technology used for whole organ/tissue transplantation
Transplantation of organs and tissues from cloned animals
Implantable biosensors for monitoring health status or delivering medications
Disease predisposition tests and predictive medicine incorporated into disease management

**Fig. 1.1** Advances in biotech drug development
(beyond 2003 projected; with permission from Macmillan Publishers Ltd. [19]).

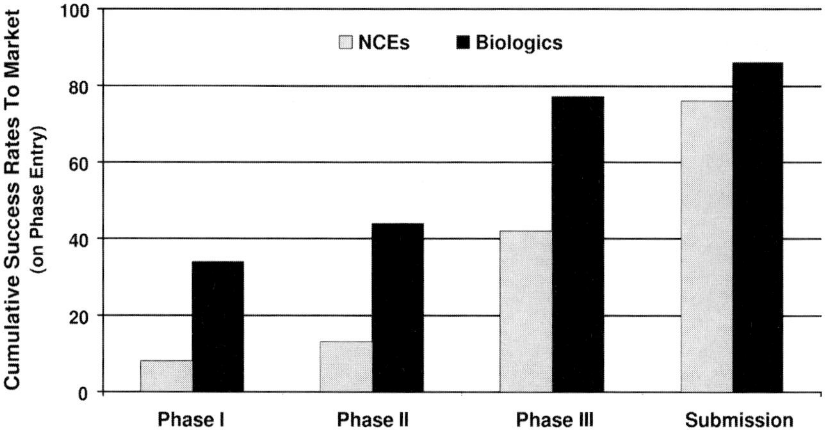


**Fig. 1.2** Success rates for clinical drug development are much higher for biologics than for traditional, chemically defined drug compounds. NCE: New chemical entity (modified from [2]).

man studies, a fourfold greater chance than traditional, chemically defined drugs of making it into the marketplace. Thus, greater use of biologics will likely reduce the attrition rate at every stage of the clinical drug development process [2]. Based on these facts, it can be predicted that biotech drugs will play a major – if not dominant – role in the drug development arena of the next decades.

## 1.3
## Pharmacokinetics and Pharmacodynamics in Drug Development

The general paradigm of clinical pharmacology is that administration of a dose or the dosing regimen of a drug results in defined drug concentrations in various body compartments and fluids. These are, in turn, the driving force for the drug's desired and undesired effects on the human body that collectively constitute the drug's efficacy and safety profile. Based on this paradigm, the basis for the pharmacotherapeutic use of biotech drugs is similar to that of small molecules – a defined relationship between the intensity of the therapeutic effect and the amount of drug in the body or, more specifically, the drug concentration at its site of action (i.e., an exposure–response relationship). The relationship between the administered dose of a drug, the resulting concentrations in body fluids and the intensity of produced outcome may be either simple or complex, and thus obvious or hidden. However, if no simple relationship is obvious, it would be misleading to conclude *a priori* that no relationship exists at all rather than that it is not readily apparent [4, 5].

The dose–concentration–effect relationship is defined by the pharmacokinetic (PK) and pharmacodynamic (PD) characteristics of a drug. Pharmacokinetics comprises all processes that contribute to the time course of drug concentrations in various body fluids, generally blood or plasma – that is, all processes affecting drug absorption, distribution, metabolism, and excretion. In contrast, pharmacodynamics characterizes the effect intensity and/or toxicity resulting from certain drug concentrations at the assumed effect site. When simplified, pharmacokinetics characterizes "what the body does to the drug", whereas pharmacodynamics assesses "what the drug does to the body" [6]. Combination of both pharmacological disciplines by integrated PK/PD modeling allows a continuous description of the effect–time course resulting directly from the administration of a certain dose (Fig. 1.3) [4, 5].

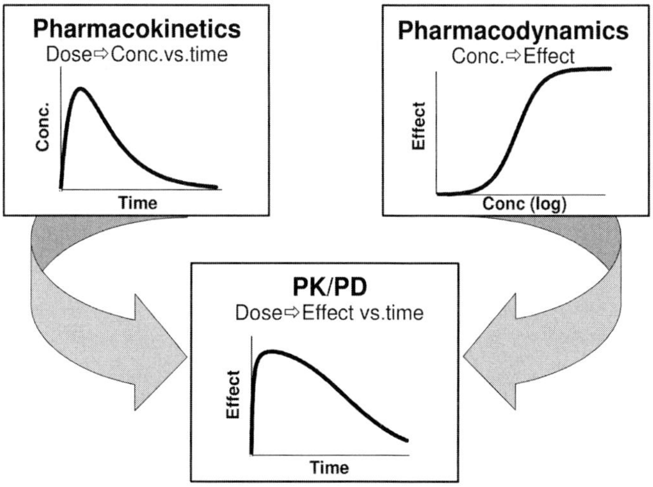


**Fig. 1.3** Pharmacokinetic/pharmacodynamic (PK/PD) modeling as combination of the classic pharmacological disciplines pharmacokinetics and pharmacodynamics (from [5]).

The increased application and integration of PK/PD concepts in all stages of preclinical and clinical drug development is one potential tool to enhance the information gain and the efficiency of the decision-making process during drug development [7]. PK/PD analysis supports the identification and evaluation of drug response determinants, especially if mechanism-based modeling is applied [8]. PK/PD analysis also facilitates the application of modeling and simulation (M&S) techniques in drug development, and allows predictive simulations of effect intensity-time courses for optimizing future development steps.

Pharmaceutical drug development has traditionally been performed in sequential phases, preclinical as well as clinical Phases I to III, in order to answer the two basic questions – which compound should be selected for development, and how it should be dosed. This information-gathering process has recently been characterized as two successive learning-confirming cycles (Fig. 1.4) [8, 9]. The first cycle (traditional Phases I and IIa) comprises learning – in healthy subjects – what dose is tolerated and confirming that this dose has some measurable bene-

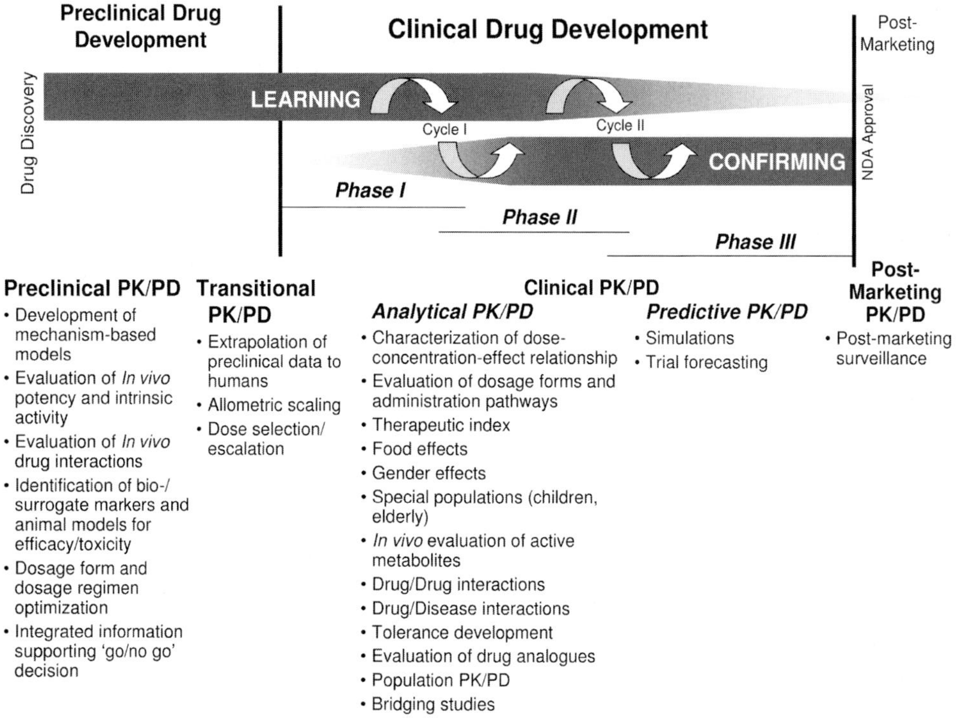


**Fig. 1.4** Examples of the application of pharmacokinetic/pharmacodynamic (PK/PD) concepts in preclinical and clinical drug development (from [10]).

fits in the targeted patients. An affirmative answer at this first cycle provides the justification for a larger and more costly second learn–confirm cycle (Phases IIb and III), where the learning step is focused on how to use the drug in representative patients for maximizing its benefit/risk ratio, while the confirming step is aimed at demonstrating an acceptable benefit/risk ratio in a large patient population. It has repeatedly been suggested to leave the sequential approach of preclinical/clinical phases and to streamline drug development by combining preclinical and early clinical development as parallel, exploratory endeavors and to expand the learning process to all phases of drug development. Such a strategy might provide a deeper understanding of the drug's action prior to taking it further in development. This will ensure that the limited resources available in drug development are allocated to the most promising drug candidates [10].

For several years, the widespread application of PK/PD concepts in all phases of drug development has repeatedly been promoted by industry, academia, and regulatory authorities [11–16]. Rigorous implementation of PK/PD concepts in drug product development provides a rationale, scientifically based framework for efficient decision-making regarding the selection of potential drug candidates, for maximum information gain from the performed experiments and studies, and for conducting fewer, more focused clinical trials with improved efficiency and cost-effectiveness [2, 10]. Examples of applications of PK/PD in drug development are provided in Figure 1.4.

## 1.4 PK and PK/PD Pitfalls for Biotech Drugs

Pharmacokinetic and pharmacodynamic principles are equally applicable to conventional small-molecule drugs and biotech drugs such as peptides, proteins, and oligonucleotides. Since biotech drugs are frequently identical or similar to endogenous substances, however, they often exhibit unique pharmacokinetic and pharmacodynamic properties that are different from traditional small-molecule drugs and resemble more those of endogenous macromolecules.

The distribution and metabolism of protein-based biotech drugs, for example, generally follows the mechanisms of endogenous and nutritional proteins. This includes, for example, unspecific proteolysis as a major elimination pathway for proteins rather than oxidative hepatic metabolism typical for the majority of small-molecule drugs. As a consequence, drug interactions studies focused on cytochrome P-450 enzymes do not usually need to be performed for protein-based biotech drugs [17].

Due to their structural similarity as polypeptides, it is generally much easier for peptide-based biotech drugs to predict how they will be distributed, metabolized and eliminated, and they typically have much faster development cycles. As the handling of peptides is relatively well preserved between different mammalian species, this also implies that knowledge generated in pharmacokinetic studies in animals can be extrapolated to predict the situation in humans with a relatively high reliability. Thus, allometric scaling is usually much more successful for biotech drugs than for traditional small-molecule compounds [17].

Another pharmacokinetic feature frequently observed for biotech drugs, but only rarely seen for traditional small-molecule drugs, is target-mediated drug disposition [18]. In this case, interaction of the drug with its pharmacological target is not reversible, but initiates the elimination of the drug, for example through intracellular metabolism after internalization of a drug–receptor complex. If the number of pharmacological target molecules is in the same magnitude or larger than the number of drug molecules, drug elimination via interaction with the pharmacological target may constitute a substantial fraction of the overall elimination clearance of the drug. In this case, pharmacokinetics and pharmacodynamics are no longer independent processes, but become inseparable and bidirectionally interdependent, in contrast to being unidirectionally interdependent as is the case if drug concentrations determined by pharmacokinetics are the driving force of drug effect via the concentration–effect relationship described by pharmacodynamics.

Target-mediated drug disposition is often associated with nonlinearity in the pharmacokinetics of the affected drug, as the elimination pathway mediated via interaction with the pharmacological target is frequently saturated at therapeutic concentrations. The consequence is an over-proportional increase in systemic exposure with increasing dose once this elimination pathway becomes saturated.

As mentioned earlier, one of the reasons for the success of biotech compounds in drug development is the fact that the biological approach rests on a fundamental understanding of the disease at the molecular level [2, 19]. Nevertheless, nonlinear pharmacokinetics, target-mediated disposition, as well as their metabolic handling may not only pose extra challenges, but also provide opportunities during the preclinical and clinical development of biotech drugs that are different from small-molecule drug candidates and may require additional resources and unique expertise. Some of the associated challenges, pitfalls and opportunities will be addressed in the subsequent chapters of this textbook.

## 1.5
## Regulatory Guidance

Regulatory guidance documents supporting the drug development process with regard to pharmacokinetics and PK/PD evaluations have, in general, a similar relevance for biotech drugs as they have for traditional, chemically defined small-molecule compounds. These include for example the exposure–response guidance document from 2003 [20] and the population pharmacokinetics guidance document from 1999 [20] as issued by the FDA, and the ICH E4 guideline on "Dose Response Information To Support Drug Registration" [21] of the International Conference on Harmonization of Technical Requirements for Registration of Pharmaceuticals for Human Use (ICH).

Additional guidance documents, however, have been issued that address the specific needs of and requirements for biotech drugs. Besides specific guidance documents for biologics with regard to chemical characterization, stability and manufacturing, there are also documents affecting clinical pharmacology evaluations. The ICH S6 guideline on "Preclinical Safety Evaluation of Biotechnology-Derived Pharmaceuticals", for example, addresses among other topics preclinical pharmacokinetics and pharmacodynamics as well as exposure–response and drug metabolism studies [22].

## 1.6
## Future

Biotech drug development is frequently described as occurring in waves. The initial wave occurred following the introduction of recombinant human insulin in 1982, and comprised recombinant endogenous proteins. The introduction of multiple monoclonal antibodies into pharmacotherapy constitutes the second wave of innovation. Further waves are expected to occur during the second part of this decade, with the widespread introduction of humanized or fully human monoclonal antibodies and the marketing approval of antibody-derived products such as antibody fragments and fusion proteins [23].

The further development and promotion of model-based drug development through the FDA's critical path initiative is likely to further foster the application of PK/PD concepts and M&S in the development of biotech drug compounds. This move is further bolstered by initiatives of the pharmaceutical industry as well as professional organizations in pharmaceutical sciences and clinical pharmacology to embrace a more holistic and integrated approach to quantitative methods in drug development stretching from drug discovery to post-marketing surveillance, with PK/PD-based M&S at its core.

In the near future, one of the major challenges for the biotech industry, as well as for the regulatory authorities, will be the introduction of biogeneric drugs, also termed biosimilar drugs or "follow-on" biologics. This group of drugs is the biotech analogue of generic drug products for traditional small-molecule compounds. Biogenerics have been defined as new protein drug products that are pharmaceutically and therapeutically equivalent to a reference product, for example an innovator product after expiration of its patent protection.

The activity of those biotech drugs that are macromolecules is often dependent upon their conformation, which is based on the secondary, tertiary and sometimes quaternary structure. As conformational changes can result in changes of the drug's activity profile, it is crucial to ensure that the essential conformational features are maintained in biogeneric compared to innovator products. This situation is further complicated by the fact that the manufacturing conditions are largely determining the final product, and that changes in this process may already result in activity changes, for example conformational changes or changes in the glycosylation pattern. In addition, there is often only limited knowledge available as to which features are essential for the biotech drug's *in-vivo* effectiveness [24]. Although the extensive discussions on biogeneric products are still ongoing, the European Medicines Agency (EMEA) has recently issued a "Guideline on Similar Biological Medicinal Products" [25]. In this document, the EMEA argues that the standard generic approach for the establishment of bioequivalence applied to chemically derived small-molecule drugs is not appropriate for biologics or biotechnology-derived drug products due to their complexity. Instead, a "similar biological medicinal products" approach, based on a comparability exercise, must be applied. The issue of biogenerics is discussed in detail in Chapter 8.

Biotech drugs, including peptides, proteins and antibodies, oligonucleotides and DNA, are projected to cover a substantial market share in the healthcare systems of the future. It will be crucial for their widespread application in pharmacotherapy, however, that their respective drug development programs are successfully completed in a rapid, cost-efficient and goal-oriented manner. Model-based drug development utilizing pharmacokinetic and pharmacodynamic concepts including exposure–response correlations has repeatedly been promoted by industry, academia, and regulatory authorities for all preclinical and clinical phases of drug development, and is believed to result in a scientifically driven, evidence-based, more focused and accelerated drug product development process [10]. Thus, PK/PD concepts are likely to continue expanding their role as a cornerstone in the successful development of biotech drug products in the future.

## 1.7 References


1 Pharmaceutical Research and Manufacturers of America. **2004**. *Medicines in Development – Biotechnology – Survey.* Washington, DC.

2 Arlington, S., S. Barnett, S. Hughes, and J. Palo. **2002**. *Pharma 2010: The Threshold to Innovation.* IBM Business Consulting Services, Somers, NY.

3 Berg, C., R. Nassr, and K. Pang. **2002**. The evolution of biotech. *Nat. Rev. Drug Discov.* 1:845–846.

4 Meibohm, B., and H. Derendorf. **1997**. Basic concepts of pharmacokinetic/pharmacodynamic (PK/PD) modelling. *Int. J. Clin. Pharmacol. Ther.* 35:401–413.

5 Derendorf, H., and B. Meibohm. **1999**. Modeling of pharmacokinetic/pharmacodynamic (PK/PD) relationships: concepts and perspectives. *Pharm. Res.* 16:176–185.

6 Holford, N.H., and L.B. Sheiner. **1982**. Kinetics of pharmacologic response. *Pharmacol. Ther.* 16:143–166.

7 Curtis, G., W. Colburn, G. Heath, T. Lenehan, and T. Kotschwar. **2000**. Faster Drug Development. *Appl. Clin. Trials* 9:52–55.

8 Sheiner, L.B., and J. Wakefield. **1999**. Population modelling in drug development. *Stat. Methods Med. Res.* 8:183–193.

9 Sheiner, L.B. **1997**. Learning versus confirming in clinical drug development. *Clin. Pharmacol. Ther.* 61:275–291.

10 Meibohm, B., and H. Derendorf. **2002**. Pharmacokinetic/pharmacodynamic studies in drug product development. *J. Pharm. Sci.* 91:18–31.

11 Steimer, J.L., M.E. Ebelin, and J. Van Bree. **1993**. Pharmacokinetic and pharmacodynamic data and models in clinical trials. *Eur. J. Drug Metab. Pharmacokinet.* 18:61–76.

12 Sheiner, L.B., and J.L. Steimer. **2000**. Pharmacokinetic/pharmacodynamic modeling in drug development. *Annu. Rev. Pharmacol. Toxicol.* 40:67–95.

13 Peck, C.C., W.H. Barr, L.Z. Benet, J. Collins, R.E. Desjardins, D.E. Furst, J.G. Harter, G. Levy, T. Ludden, J.H. Rodman, et al. **1994**. Opportunities for integration of pharmacokinetics, pharmacodynamics, and toxicokinetics in rational drug development. *J. Clin. Pharmacol.* 34:111–119.

14 Machado, S.G., R. Miller, and C. Hu. **1999**. A regulatory perspective on pharmacokinetic/pharmacodynamic modelling. *Stat. Methods Med. Res.* 8:217–245.

15 Lesko, L.J., M. Rowland, C.C. Peck, and T.F. Blaschke. **2000**. Optimizing the science of drug development: Opportunities for better candidate selection and accelerated evaluation in humans. *J. Clin. Pharmacol.* 40:803–814.

16 Breimer, D.D., and M. Danhof. **1997**. Relevance of the application of pharmacokinetic-pharmacodynamic modelling concepts in drug development. The 'wooden shoe' paradigm. *Clin. Pharmacokinet.* 32:259–267.

17 Tang, L., A.M. Persky, G. Hochhaus, and B. Meibohm. **2004**. Pharmacokinetic aspects of biotechnology products. *J. Pharm. Sci.* 93:2184–2204.

18 Levy, G. **1994**. Pharmacologic target-mediated drug disposition. *Clin. Pharmacol. Ther.* 56:248–252.

19 Nagle, T., C. Berg, R. Nassr, and K. Pang. **2003**. The further evolution of biotech. *Nat. Rev. Drug Discov.* 2:75–79.

20 CDER/FDA. **2003**. *Exposure Response Relationships: Guidance for Industry.* US Department of Health and Human Services, Food and Drug Administration, Center for Drug Evaluation and Research, Rockville.

**21** International Conference on Harmonization. **1994**. *ICH E4 – Dose-Response Information to Support Drug Registration*. European Agency for the Evaluation of Medicinal Products, London.

**22** International Conference on Harmonization. **1997**. *ICH S6 – Preclinical Safety Evaluation of Biotechnology-Derived Pharmaceuticals*. European Agency for the Evaluation of Medicinal Products, London.

**23** Reichert, J., and A. Pavolu. **2004**. Monoclonal antibodies market. *Nat. Rev. Drug Discov.* 3 : 383–384.

**24** Griffiths, S. **2004**. Betting on biogenerics. *Nat. Rev. Drug Discov.* 3 : 197–198.

**25** EMEA. **2005**. *Guideline on Similar Biological Medicinal Products*. European Agency for the Evaluation of Medicinal Products, London.

# Part II
# The Basics

*Pharmacokinetics and Pharmacodynamics of Biotech Drugs: Principles and Case Studies in Drug Development.* Edited by Bernd Meibohm

ISBN: 3-527-31408-3

# 2
# Pharmacokinetics of Peptides and Proteins

*Lisa Tang and Bernd Meibohm*

## 2.1
## Introduction

Although small numbers of peptide- and protein-based therapeutics have long been used in medical practice (e.g., calcitonin or glucagon), it has been the advances in biotechnology and their application in drug development that have propelled peptides and proteins from niche products to mainstream therapeutics. Recombinant human insulin, which was approved in 1982, was the first of these biotechnologically derived drug products, and many more have followed during the past 25 years. Today, peptide- and protein-based drug products constitute a sizable fraction of all clinically used medications. Nevertheless, there are distinct differences between these drug products and small molecule-based therapeutics that may require the use of different technologies and methodological approaches and experimental designs during their preclinical and clinical development, including pharmacokinetic and exposure–response assessments.

Pharmacokinetic and exposure–response concepts impact every stage of the drug development process, starting from lead optimization to the design of Phase III pivotal trials [1]. An understanding of the concentration–effect relationship is crucial to any drug – including peptides and proteins – as it lays the foundation for dosing regimen design and rational clinical application. General pharmacokinetic principles are equally applicable to protein- and peptide-based drugs as they are to traditional small-molecule drugs. This includes pharmacokinetic-related recommendations for drug development, such as the recently published exposure–response guidance document of the U.S. Food and Drug Administration (FDA) and the ICH E4 guideline of the International Conference on Harmonization [2, 3].

The assessment and interpretation of the pharmacokinetics of peptides and proteins, however, frequently poses extra challenges, and requires additional resources compared to small-molecule drug candidates. One such challenge arises from the fact that most peptide- and protein-based drugs are identical or similar to endogenous molecules, and must be identified and quantified next to a myriad

*Pharmacokinetics and Pharmacodynamics of Biotech Drugs: Principles and Case Studies in Drug Development.* Edited by Bernd Meibohm

ISBN: 3-527-31408-3

of structurally similar molecules. The resulting bioanalytical challenges are discussed in Chapter 6. The following sections of this chapter provide a general discussion on the pharmacokinetics of peptide- and protein-based therapeutics. Additionally, they outline some of the associated challenges and obstacles in the drug development process and illustrate them with examples of approved and experimental drugs. An overview of the basic pharmacokinetic parameters of select FDA-approved protein- and peptide-based drugs is provided as reference in Table 2.1.

## 2.2
## Administration Pathways

Peptides and proteins, unlike conventional small-molecule drugs, are generally not therapeutically active upon oral administration [4–6]. The lack of systemic bioavailability is mainly caused by two factors: high gastrointestinal enzyme activity, and low permeability through the gastrointestinal mucosa. In fact, the substantial peptidase and protease activity in the gastrointestinal tract makes it the most efficient body compartment for peptide and protein metabolism. Furthermore, the gastrointestinal mucosa presents a major absorption barrier for water-soluble macromolecules such as peptides and proteins [4]. Absorption may be, at least for peptides, further impeded through presystemic metabolism by the functional system of cytochrome P450 3A and p-glycoprotein (P-gp) [7–9].

Due to the lack of activity after oral administration for most peptides and proteins, administration by injection or infusion – that is, by intravenous (IV), subcutaneous (SC), or intramuscular (IM) administration – is frequently the preferred route of delivery for these drug products. In addition, other non-oral administration pathways have been utilized, including nasal, buccal, rectal, vaginal, transdermal, ocular, or pulmonary drug delivery. Some of these delivery pathways will be discussed in the following sections in the order of the increasing biopharmaceutic challenges to obtain adequate systemic exposure.

### 2.2.1
### Administration by Injection or Infusion

Injectable administration of peptides and proteins offers the advantage of circumventing presystemic degradation, thereby achieving the highest concentration in the biological system. Examples of FDA-approved proteins given by the IV route include the tissue plasminogen activator (t-PA) analogues alteplase and tenecteplase, the recombinant human erythropoietin epoetin-α, and the granulocyte colony-stimulating factor filgrastim. However, IV administration as either a bolus dose or constant rate infusion may not always provide the desired concentration–time profile depending on the biological activity of the product, and IM or SC injections may be more appropriate alternatives. For example, luteinizing hormone-releasing hormone (LH-RH) in bursts stimulates the release of follicle-stimulating

**Table 2.1** Pharmacokinetic parameters of select FDA-approved protein/peptide drugs as reported in the prescribing information.

| **Generic name/** ***Trade name*** | **Class** | **Manufacturer** | **Route** | **Bioavailability** | **Clearance**[a)] | **Volume of distribution**[b)] | **Half-life** |
|---|---|---|---|---|---|---|---|
| Abarelix/*Plenaxis* | LHRH antagonist | Praecis | IM | | 208 ± 48 L/day | 4040 ± 1607 L | 13.2 ± 3.2 days |
| Agalsidase/*Fabrazyme* | α-Galactosidase | Genzyme | IV | – | Nonlinear PK | | 45–102 min |
| Aldesleukin/*Proleukin* | Antineoplastic (IL-2) | Chiron | IV | – | 268 mL/min | | 85 min |
| Alefacept/*Amevive* | Immunosuppressant | Biogen | IV, IM | 63% (IM) | 0.25 mL/h/kg | 94 mL/kg | 270 h |
| Alteplase/*Activase* | Thrombolytic | Genentech | IV | – | 380–570 mL/min | PLV | <5 min |
| Anakinra/*Kineret* | Antirheumatic (IL-1Ra) | Amgen | SC | 95% | | | 4–6 h |
| Asparaginase/*Elspar* | Antineoplastic | Merck | IV, IM | | | 70–80% PLV | 8–30 h (IV)<br>39–49 h (IM) |
| Cetrorelix/*Cetrotide* | LHRH antagonist | Serono | SC | 85% | 1.28 mL/min/kg | 1.16 L/kg | 62.8 (38.2–108) h |
| Cyclosporine | Immunosuppressant | | | | | | |
| *Neoral* | | Novartis | PO | 10–89% | 5–7 mL/min/kg | 3–5 L/kg | 8.4 (5–18) h |
| *Sandimmune* | | Novartis | IV, PO | 30 (PO) | | | 19 (10–27) h |
| Darbepoetin-α/*Aranesp* | Anti-anemic | Amgen | IV, SC | 37 (30–50)% | | | 21 h (IV)<br>49 (24–72) h (SC) |
| Denileukin diftitox/ *Ontak* | Antineoplastic | Ligand | IV | – | 1.5–2.0 mL/min/kg | 0.06–0.08 L/kg | 70–80 min |
| Desmopressin/ *DDAVP* | Antidiuretic | Aventis | PO, IN | 3.2% (IN)<br>0.16%(PO) | | | 75.5 min (IN)<br>1.5–2.5 h (PO) |
| Drotrecogin-α/*Xigris* | Activated protein C | Eli Lilly | IV | – | 40 L/h | | |
| Epoetin-α/ *Epogen, Procrit* | Anti-anemic | Amgen, Ortho | IV, SC | | | | 4–13 h (IV)<br>16.3 ± 3.0 h (SC) |

Table 2.1 (continued)

| Generic name/ *Trade name* | Class | Manufacturer | Route | Bioavailability | Clearance[a] | Volume of distribution[b] | Half-life |
|---|---|---|---|---|---|---|---|
| Eptifibatide/*Integrilin* | GPIIb/IIIa inhibitor | Millennium | IV | – | 55–58 mL/h/kg | | 2.5 h |
| Etanercept/*Enbrel* | Antirheumatic | Amgen | SC | | 89 mL/h | | 115 (98–300) h |
| Exenatide/*Byetta* | Incretin mimetic | Amylin | SC | | 9.1 L/h | 28.3 L | 2.4 h |
| Follitropin/*Gonal-f* | Follicle-stimulating hormone | Serono | SC | 66 ± 39% | 0.7 ± 0.2 L/h | 10 ± 3 L | 32 h |
| Follitropin beta/ *Follistim AQ* | Follicle-stimulating hormone | Organon | SC | 77.8% | 0.01 L/h/kg | 8 L | 33.4 h |
| Ganirelix/*Antagon* | LHRH antagonist | Organon | SC | 91% | 2.4 ± 0.2 L/h | 43.7 ± 11.4 L | 12.8 ± 4.3 h |
| Glucagon/*Glucagon* | rh Glucagon | Eli Lilly | IV, IM, SC | | 13.5 mL/min/kg | 0.25 L/kg | 8–18 min (IV) |
| Goserelin/*Zoladex* | LHRH agonist | AstraZeneca | SC | | 110.5 ± 47.5 mL/min (men)<br>163.9 ± 71.0 mL/min (women) | 44.1 ± 13.6 L (men)<br>20.3 ± 4.1 L (women) | 4.2 ± 1.1 h (men)<br>2.3 ± 0.6 h (women) |
| Insulins | | | | | | | |
| Lispro *(Humalog)* | Insulin analogue | Eli Lilly | SC | 55–77% | | 0.26–0.36 L/kg | 60 min (IV) |
| Aspart *(Novolog)* | Insulin analogue | Novartis | SC | 55–77% | 1.22 L/h/kg | | 81 min (SC) |
| rh *(Humulin R)* | | Eli Lilly | SC | 55–77% | | 0.26–0.36 L/kg | 90 min (IV) |
| Glargine *(Lantus)* | Insulin analogue | Aventis | SC | | | | 11 h ($T_{25\%}$) |
| Inhaled rh *(Exubera)* | | Pfizer | IH | | Assumed to be identical to rh insulin | | |
| Interferon β-1b/ *Betaseron* | Biological response modifier | Berlex/Chiron | SC | 50% | 9.4–28.9 mL/min/kg | 0.25–2.88 L/kg | 8 min – 4.3 h (IV) |
| Interferon γ-1b/ *Actimmune* | Immunomodulator | InterMune | SC | >89% | 1.4 L/min | | 38 min (IV)<br>2.9 h (IM)<br>5.9 h (SC) |

**Table 2.1** (continued)

| Generic name/ *Trade name* | Class | Manufacturer | Route | Bioavailability | Clearance[a)] | Volume of distribution[b)] | Half-life |
|---|---|---|---|---|---|---|---|
| Leuprolide/*Eligard* | LHRH agonist | Atrix | SC | – | 8.36 L/h | 27 L | 3 h |
| Laronidase/*Aldurazyme* | Lysosomal hydrolase | Genzyme | IV | – | 1.7–2.7 mL/min/kg | 0.24–0.6 L/kg | 1.5–3.6 h |
| Octreotide/*Sandostatin* | Somatostatin analogue | Novartis | IV, SC | 100% | 7–10 L/h | 13.6 L | 1.7–1.9 h |
| Oprelvekin/*Neumega* | Thrombopoietic stimulant (IL-11) | Wyeth | SC | >80% | | | 6.9 ± 1.7 h |
| Pegaspargase/*Oncaspar* | Antineoplastic | Enzon | IM, IV | | | | 5.69 ± 3.25 days |
| Pegfilgrastim/*Neulasta* | Hematopoietic stimulant | Amgen | SC | | Nonlinear PK | | 15–80 h |
| Peginterferon α-2a/ *Pegasys* | Biological response modifier | Roche | SC | 84% | 94 mL/h | | 80 (50–140) h |
| Pramlintide/*Symlin* | Amylin analogue | Amylin | SC | 30–40% | 1 L/min | 15–27 L | 48 min |
| Reteplase/*Retavase* | Thrombolytic | Centocor | IV | – | 250–450 mL/min | | 13–16 min |
| Sargramostim/*Leukine* | Hematopoietic stimulant | Berlex | IV, SC | | 431 $mL/min/m^2$ (IV)<br>549 $mL/min/m^2$ (SC) | | 60 min (IV)<br>162 min (SC) |
| Tenecteplase/*TNKase* | Thrombolytic | Genentech | IV | – | 99–119 mL/min | PLV | 90–130 min |
| Teriparatide/*Forteo* | Parathyroid hormone | Eli Lilly | SC | 95% | 94 L/h (men)<br>62 L/h (women) | 0.12 L/kg (IV) | 5 min (IV)<br>1 h (SC) |
| Triptorelin/*Trelstar* | LHRH agonist | Pfizer | IM | 83% | 212 ± 32 mL/min | 30–33 L | 2.81 ± 1.21 h |
| Urokinase/*Abbokinase* | Thrombolytic | Abbott | IV | – | | 11.5 L | 12.6 ± 6.2 min |

**a)** PLV plasma volume (40 mL/kg); rh = recombinant human; IV = intravenous; IM = intramuscular; IN = intranasal; SC = subcutaneous.

**b)** Clearance and volume of distribution terms for drugs not administered by IV usually reflect clearance and volume terms divided by bioavailability (i.e., CL/F or V/F).

hormone (FSH) and luteinizing hormone (LH), whereas a continuous baseline level will suppress the release of these hormones [10]. To avoid the high peaks from an IV administration of leuprorelin, an LH-RH agonist, a long-acting monthly depot injection of the drug is approved for the treatment of prostate cancer and endometriosis [11]. A recent study comparing SC versus IV administration of epoetin-$\alpha$ in patients receiving hemodialysis reports that the SC route can maintain the hematocrit in a desired target range with a lower average weekly dose of epoetin-$\alpha$ compared to IV [12].

The drawbacks of SC and IM injections include potentially decreased bioavailability that is secondary to variables such as local blood flow, injection trauma, protein degradation at the site of injection, and limitations of uptake into the systemic circulation related to effective capillary pore size and diffusion. The bioavailability of numerous peptides and proteins is, for example, markedly reduced after SC or IM administration compared to their IV administration. The pharmacokinetically derived apparent absorption rate constant is thus the combination of absorption into the systemic circulation and presystemic degradation at the absorption site. The true absorption rate constant $k_a$ can then be calculated as:

$$k_a = F \cdot k_{app}$$

where $F$ is the bioavailability compared to IV administration. A rapid apparent absorption rate constant $k_{app}$ can thus be the result of a slow absorption and a fast presystemic degradation – that is, a low systemic bioavailability [13].

Several approved peptides and proteins, including insulin, enfuvirtide [14], and the recently approved pramlintide [15], are administered as SC injections. Following a SC injection, peptide and protein therapeutics may enter the systemic circulation either via blood capillaries or through lymphatic vessels [16]. In general, macromolecules larger than 16 kDa are predominantly absorbed into the lymphatic system, whereas those under 1 kDa are mostly absorbed into the blood circulation. Studies with recombinant human interferon $\alpha$-2a (rhIFN $\alpha$-2a) indicate that, following SC administration, high concentrations of the recombinant protein are found in the lymphatic system, which drains into regional lymph nodes [17]. Furthermore, there appears to be a linear relationship between the molecular weight of the protein and the proportion of the dose absorbed by the lymphatics (Fig. 2.1) [18]. This is of particular importance for those agents for which the therapeutic targets are lymphoid cells (i.e., interferons and interleukins) [17]. Clinical studies show that palliative low-to-intermediate-dose SC recombinant interleukin-2 (rIL-2) in combination with rhIFN $\alpha$-2a can be administered to patients in the ambulatory setting with efficacy and safety profiles comparable to the most aggressive IV rIL-2 protocol against metastatic renal cell cancer [19].

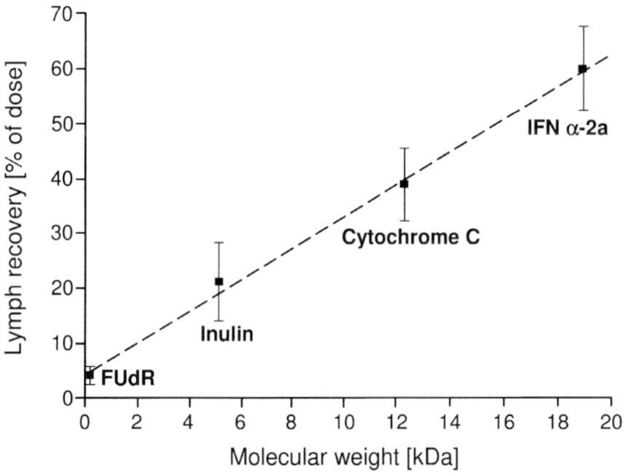


**Fig. 2.1** Correlation between molecular weight and cumulative recovery (mean ± SD) of rINF α-2a (MW 19 kDa), cytochrome C (MW 12.3 kDa), inulin (MW 5.2 kDa) and 5-fluoro-2'-deoxyuridine (FUDR) (MW 246 Da) in the efferent lymph from the right popliteal lymph node following subcutaneous administration into the lower part of the right hind leg in sheep (n = 3). The linear regression line has a correlation coefficient r = 0.998 (p <0.01) (from [18]).

### 2.2.2
### Inhalational Administration

Inhalational delivery of peptides and proteins offers the advantage of ease of administration, the presence of a large surface area (75 $m^2$) available for absorption, high vascularity of the administration site, and bypass of hepatic first-pass metabolism. Disadvantages of inhalation delivery include the presence of certain proteases in the lung, potential local side effects of the inhaled agents on the lung tissues (i. e., growth factors and cytokines), and molecular weight limitations (Fig. 2.2) [20, 21].

The success of inhaled peptide and protein drugs can be exemplified by inhaled recombinant human insulin products, with Exubera® being the first approved product (2006), and several others in clinical development. Inhaled insulin offers the advantages of ease of administration and rapid onset with a shorter duration of action for tighter postprandial glucose control as compared to subcutaneously administered regular insulin [22]. A study involving 26 patients with type II diabetes mellitus showed that inhalation insulin treatment for 3 months significantly improved glycemic control when compared to baseline, as assessed by hemoglobin $A_{1c}$ levels [23]. Another clinical study with 249 type II diabetic subjects showed comparable efficacy between inhaled insulin and the conventional SC insulin regimen [24]. These clinical trials demonstrate that the non-invasive administration of inhaled insulin offers similar efficacy, but better patient compliance than subcutaneously administered insulin.

Dornase-α, which is indicated for the treatment of cystic fibrosis, is another example of a protein drug successfully administered through the inhalation route.

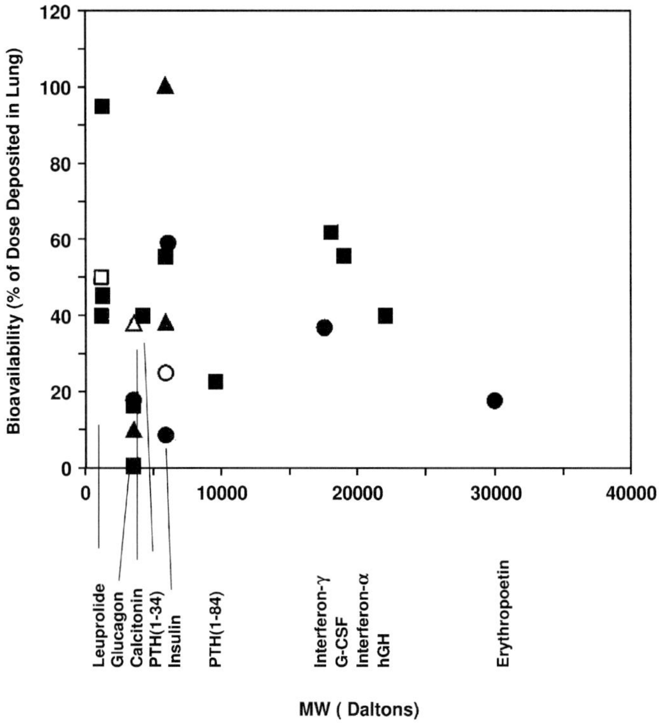


**Fig. 2.2** Plasma bioavailability of therapeutic peptides versus molecular weight (MW) after pulmonary administration. Bioavailability is expressed as percentage of the dose deposited in the lungs relative to subcutaneous administration in humans (open symbols) and various animal species (solid symbols; square and circle: rodents; triangle: monkey). PTH: parathyroid hormone; GH: growth hormone (from [20]).

In a multi-center, two-year clinical study, inhaled dornase-α was shown to significantly improve lung function and reduce the risk of respiratory exacerbations in pediatric cystic fibrosis patients [25].

### 2.2.3
### Intranasal Administration

Similar to the inhalation route, intranasal administration of peptides and proteins offers the advantages of ease of administration, delivery to a surface area rich in its vascular and lymphatic network, and the bypassing of hepatic first-pass metabolism [26]. Intranasal absorption of a variety of peptide and protein drugs including calcitonin, oxytocin, LH-RH, growth hormone, interferons, and even vaccines has been extensively investigated over the past decade. In general, polypeptides with a molecular weight up to 2000 Da have been found to be pharmacologically active via the intranasal route. In contrast, pharmaceutical peptides with molecular weights of 2000–6000 Da (i. e., insulin, calcitonin, and LH-RH) require the addition of absorption enhancers in order to reach adequate bioavailability [10]. Limitations that may preclude the use of the intranasal route of administration include high variability in absorption associated with the site of deposition, the type of delivery system, changes in mucus secretion and mucociliary clearance, as well as the presence of allergy, hay fever, or the common cold in the target population [27].

Intranasal administration has recently been proposed as a means to deliver protein therapeutics directly into the central nervous system (CNS), thereby bypassing the blood–brain barrier [28]. In particular, one study reported achieving a higher concentration of $^{125}$I-labeled recombinant human insulin-like growth factor-I ($^{125}$I-rhIGF-I) in the CNS after intranasal administration rather than IV administration [29]. The effectiveness of intranasal delivery of IGF-1 for the treatment of stroke has been reported in a more recent study through a rat model. The investigators reported that 150 µg of intranasal IGF-1 could effectively reduce induced infarct size and neurologic deficits as compared to controls [29].

### 2.2.4 Transdermal Administration

Transdermal drug delivery offers the advantages of bypassing metabolic and chemical degradation in the gastrointestinal tract, as well as first-pass metabolism by the liver. Methods frequently used to facilitate transdermal delivery include sonophoration and iontophoresis. Both methodologies increase skin permeability to ionic compounds – sonophoration by applying low-frequency ultrasound, and iontophoresis by applying a low-level electric current. Therapeutic doses of insulin, interferon-$\gamma$, and epoetin-$\alpha$ have all been successfully delivered transdermally via sonophoresis [30]. Additionally, transdermal iontophoresis has been applied in delivery of a host of proteins and peptides including leuprolide [31], insulin [32], growth hormone-releasing factor [33], calcitonin [34], and parathyroid hormone [35].

### 2.2.5 Peroral Administration

Most peptides and proteins are currently formulated as parenteral formulations because of their poor oral bioavailability. Nevertheless, oral delivery of peptides and proteins would be the preferred route of administration if bioavailability issues could be overcome, as it offers the advantages of convenient, pain-free administration. Although various factors such as permeability, chemical and metabolic stability and gastrointestinal transit time can affect the rate and extent of absorption of orally administered peptides and proteins, molecular size is generally considered the ultimate obstacle [36].

Several promising strategies have emerged from the intensive recent research efforts into the oral delivery of peptides and proteins [6, 36, 37]. Absorption enhancers may be used either to temporarily disrupt the intestinal barrier so that drug penetration is increased, or to serve as transport carriers for the protein via complex formation. The oral co-administration of parathyroid hormone, an 84-amino acid protein, with N-[38]amino caprylic acid, a transport carrier, resulted in positive bioactivity as demonstrated in a rodent model of osteoporosis [38]. While parathyroid hormone has no oral bioavailability when administered alone, co-administration of this absorption enhancer resulted in 2.1% oral bioavailability relative to SC administration in monkeys [38]. Increasing intestinal paracellular absorption was demonstrated in a study involving insulin and immunoglobulins co-administered with zonula occulens toxin (Zot), another permeation enhancer. In animal models, Zot reversibly increased the intestinal absorption of both insulin and immunoglobulins in a time-dependent manner [5].

Encapsulation in microparticles or nanoparticles may be used to shield peptides and proteins from enzymatic degradation. These solid particles may be taken up via endocytosis by the intestinal cells or passage through paracellular tight junctions. Due to their stability in the gastrointestinal tract, they appear more favorable for oral delivery than liposomes. In particular, the gut-associated lymphoid tissue (GALT) organized in Peyer's patches has been suggested as a useful oral delivery target for encapsulated peptides and proteins. Peyer's patches cover approximately 25% of the gastrointestinal mucosal surface area and are characteristically high in phagocytotic activity with limited lysosomal activity. More importantly, protein and peptide delivery through GALT offers the advantage of bypassing hepatic first-pass metabolism [6]. The concept has successfully been demonstrated by oral delivery of glucagon-like peptide-1 (GLP-1) in PLGA microspheres to diabetic mice. Mice treated with the microsphere preparation had indeed a lower glycemic response to oral glucose challenge than mice treated with GLP-1 without encapsulation into microspheres [39].

Because Peyer's patches contain a large number of IgA-committed cells that can be stimulated by antigens absorbed through membranous (M) cells, they have also been targeted for oral vaccine delivery [40]. While significant gastrointestinal degradation complicates oral vaccine delivery, recent studies have shown successful chitosan microparticle absorption in oral vaccine delivery through Peyer's patches using ovalbumin as a model vaccine [41].

Other strategies of oral peptide and protein delivery include amino acid backbone modification, alternate formulation design, chemical conjugation to improve their resistance to degradation, and inhibition of enzymatic degradation by co-administration of protease inhibitors [6,42]. However, novel approaches to improve oral protein and peptide drug delivery may not always be ideal. For example, the use of absorption enhancers including EDTA, bile salts, and surfactants can actually cause disaggregation of insulin and increase its rate of degradation. Thus, despite numerous approaches, development of orally administered proteins and peptides still poses a major challenge and remains a key area in drug delivery research [6].

## 2.3 Administration Route and Immunogenicity

Antibody formation is a frequently observed phenomenon during chronic dosing with protein drugs, especially for those derived from animal proteins. The presence of antibodies can obliterate the biological activity of a protein drug. In addition, protein–antibody complexation can also modify the distribution, metabolism and excretion (i. e., the pharmacokinetic profile) of the protein drug. Elimination can either be increased or decreased, with faster elimination of the complex occurring if the reticuloendothelial system is stimulated. Elimination is slowed down if the antibody–drug complex forms a depot for the protein drug. This effect would prolong the drug's therapeutic activity, which might be beneficial if the complex formation does not decrease therapeutic activity [43].

Immunogenicity may be affected by the route of administration. Extravascular injection has been shown to stimulate antibody formation more than IV application, but this is most likely due to the increased immunogenicity of protein aggregates and precipitates formed at the injection site [44]. A recent study investigated the effect of the route of administration of INF-β preparations on inducing anti-INF-β antibodies in multiple sclerosis patients. The results indicate that IM injections appear less immunogenic compared to SC injections, resulting in both a lower serum level of anti-INF-β antibodies as well as a delay in their appearance [45].

The immunogenicity of proteins can be reduced by chemical modifications of the molecule, for example by conjugation with polyethylene glycol (PEG), a process known as PEGylation [46, 47]. PEGylation can shield antigenic determinants on the protein drug from detection by the immune system through steric hindrance [48]. This concept was successfully applied to overcome the high rate of allergic reactions towards L-asparaginase, resulting in pegaspargase [49]. More details of the effect of chemical modifications of peptides and proteins, including PEGylation, are discussed in Chapter 11.

Other major advantages of PEGylation include its ability to manipulate the pharmacokinetics and physico-chemical properties of the protein drug. Conjugation of protein drugs with PEG chains increases their hydrodynamic volume, which in turn can result in a reduced renal clearance, restricted biodistribution, and prolonged residence time. PEGylation can also protect against proteolytic degradation and increase drug solubility [47]. Pegfilgrastim is the PEGylated version of the granulocyte colony-stimulating factor filgrastim, which is administered for the management of chemotherapy-induced neutropenia. PEGylation minimizes filgrastim's renal clearance by glomerular filtration, thereby making neutrophil-mediated clearance the predominant route of elimination. Thus, PEGylation of filgrastim results in so-called "self-regulating pharmacokinetics", since pegfilgrastim has a reduced clearance and thus prolonged half-life and more sustained duration of action in a neutropenic setting because few mature neutrophils are available to mediate its elimination [50]. The pharmacokinetics of pegfilgrastim are discussed in detail in Chapter 15.

## 2.4 Distribution

Whole-body distribution studies are essential for classical small-molecule drugs in order to exclude any tissue accumulation of potentially toxic metabolites. This problem does not exist for protein drugs, where the catabolic degradation products (amino acids) are recycled in the endogenous amino acid pool. Therefore, biodistribution studies for peptides and proteins are performed primarily to assess targeting to specific tissues as well as to identify the major elimination organs [4].

The volume of distribution of a peptide or protein drug is determined largely by its physico-chemical properties (e. g., charge, lipophilicity), protein binding, and dependency on active transport processes. Due to their large size – and therefore limited mobility through biomembranes – most therapeutic proteins have small volumes of distribution, typically limited to the volumes of the extracellular space [26, 51].

After IV application, peptides and proteins usually follow a biexponential plasma concentration–time profile that can best be described by a two-compartment pharmacokinetic model [13]. The central compartment in this model represents primarily the vascular space and the interstitial space of well-perfused organs with permeable capillary walls, especially liver and kidneys, while the peripheral compartment comprises the interstitial space of poorly perfused tissues such as skin and (inactive) muscle [4].

In general, the volume of distribution of the central compartment ($V_c$), in which peptides and proteins initially distribute after an IV administration, is typically equal to or slightly larger than the plasma volume of 3–8 L (approximate body water volumes for a 70-kg person: interstitial 12 L, intracellular 27 L, intravascular 3 L). Furthermore, the steady-state volume of distribution ($V_{SS}$) is usually no more than twice the initial volume of distribution, or approximately 14–20 L [13, 37, 43]. This distribution pattern has been described for the somatostatin analogue octreotide ($V_c$ 5.2–10.2 L; $V_{SS}$ 18–30 L), and t-PA analogue tenecteplase ($V_c$ 4.2–6.3 L; $V_{SS}$ 6.1–9.9 L) [52]. Epoetin-α also has a volume of distribution estimated to be close to the plasma volume at 0.0558 L/kg after an IV administration to healthy volunteers [53]. Similarly, $V_{SS}$ for darbepoetin-α has been reported as 0.0621 L/kg after an IV administration in patients undergoing dialysis [54], and distribution of thrombopoietin has also been reported to be limited to the plasma volume (~3 L) [55].

Active tissue uptake and binding to intra- and extravascular proteins, however, can substantially increase the volume of distribution of peptide and protein drugs, as for example observed with atrial natriuretic peptide (ANP) [56].

There is a tendency for $V_{SS}$ and $V_c$ to correlate one with another, which implies that the volume of distribution is predominantly determined by distribution in the vascular and interstitial space as well as unspecific protein binding in these distribution spaces. The distribution rate is inversely correlated with molecular size and is similar to that of inert polysaccharides, suggesting that passive diffusion through aqueous channels is the primary distribution mechanism [57].

The distribution, elimination and pharmacodynamics are, in contrast to conventional drugs, frequently interrelated for peptides and proteins. The generally low volume of distribution should not necessarily be interpreted as low tissue penetration. Receptor-mediated specific uptake into the target organ, as one mechanism, can result in therapeutically effective tissue concentrations despite a relatively small volume of distribution. Nartograstim, a recombinant derivative of the granulocyte-colony stimulating factor (G-CSF), for example, is characterized by a specific, dose-dependent and saturable tissue uptake into the target organ bone marrow, presumably via receptor-mediated endocytosis [58].

Another factor that can influence the distribution of therapeutic peptides and proteins is binding to endogenous protein structures. Physiologically active endogenous peptides and proteins frequently interact with specific binding proteins involved in their transport and regulation.

A wide range of protein drugs, including growth hormone [59], recombinant human DNases used as mucolytics in cystic fibrosis [60], and recombinant human vascular endothelial growth factor (rhVEGF) [61], have all been shown to associate with specific binding proteins. Protein binding not only affects whether the peptide or protein drug will exert any pharmacological activity, but on many occasions it may also have an inhibitory or stimulatory effect on the biological activity of the agent [59]. Recombinant cytokines, when injected into the bloodstream, may encounter various cytokine-binding proteins including soluble cytokine receptors and anti-cytokine antibodies. In either case, the binding protein may either prolong the cytokine circulation time by acting as a storage depot, or it may enhance the cytokine clearance [62]. Growth hormone, for example, has at least two binding proteins in plasma [63]; this protein binding substantially reduces growth hormone elimination with a tenfold smaller clearance of total compared to free growth hormone, but also decreases its activity via reduction of receptor interactions.

Apart from these specific bindings, peptides and proteins may also be non-specifically bound to plasma proteins. For example, metkephamid, a met-enkephalin analogue, was described as being 44–49% bound to albumin [64], while octreotide, a somatostatin analogue, is up to 65% bound to lipoproteins [27].

Aside from the physico-chemical properties and protein binding of peptides and proteins, site-specific and target-oriented receptor-mediated uptake can also influence biodistribution. In the case of rhVEGF, the administration of high doses of the protein results in nonlinear pharmacokinetics. This nonlinearity has been attributed to saturable binding, internalization, and degradation of VEGF mediated by high-affinity receptors that line the vasculature (see Section 2.5.5) [61].

## 2.5 Elimination

In general, peptides and protein drugs are almost exclusively eliminated by metabolism via the same catabolic pathways as endogenous or dietary proteins, resulting in amino acids that are reutilized in the endogenous amino acid pool for *de-novo* biosynthesis of structural or functional body proteins. This has, for example, been described for enfuvirtide, a 36-amino acid synthetic peptide used in the treatment of HIV-1 infection [65].

Non-metabolic elimination pathways such as renal or biliary excretion are generally negligible for most peptides and proteins, although biliary excretion has been described for some peptides and proteins such as immunoglobulin A and octreotide [27]. Clearance through biliary excretion has also been observed for the opioid peptides DPDPE [66], as well as the prodrug form of DADLE [67]. If biliary excretion of peptides and proteins occurs, it generally results in subsequent metabolism of these compounds in the gastrointestinal tract (see Section 2.5.2) [13].

General tendencies in the *in-vivo* elimination of proteins and peptides may often be predicted from their physiological function. Peptides, for example, frequently have hormone activity and usually have short elimination half-lives. This is desirable for a close regulation of their endogenous levels and thus function. In contrast, transport proteins such as albumin or α-1 acid glycoprotein have elimination half-lives of several days or weeks, which enables and ensures the continuous maintenance of necessary concentrations in the bloodstream.

The elimination of peptides and proteins can occur unspecifically almost everywhere in the body, or it can be limited to a specific organ or tissue. The locations of intensive peptide and protein metabolism include not only the liver, kidneys, and gastrointestinal tissue, but also the blood and other body tissues. The determining factors for rate and mechanism of protein and peptide clearance include molecular weight (Table 2.2) as well as a molecule's physico-chemical properties, including size, overall charge, lipophilicity, functional groups, glycosylation pattern, secondary and tertiary structure and propensity for particle aggregation [68]. The metabolism rate generally increases with decreasing molecular weight from

**Table 2.2** Molecular weight as major determinant of the elimination mechanisms of peptides and proteins. As indicated, mechanisms may overlap. Endocytosis may occur at any molecular weight range. (modified from [43]).

| Molecular weight [Da] | Elimination site(s) | Predominant elimination mechanism(s) | Major determinant |
|---|---|---|---|
| <500 | Blood, liver | Extracellular hydrolysis<br>Passive lipoid diffusion | Structure, lipophilicity |
| 500–1000 | Liver | Carrier-mediated uptake<br>Passive lipoid diffusion | Structure, lipophilicity |
| 1000–50 000 | Kidney | Glomerular filtration and subsequent degradation processes (see Fig. 2.2) | Molecular weight |
| 50 000–200 000 | Kidney, liver | Receptor-mediated endocytosis | Sugar, charge |
| 200 000–400 000 | | Opsonization | α2-macroglobulin, IgG |
| >400 000 | | Phagocytosis | Particle aggregation |

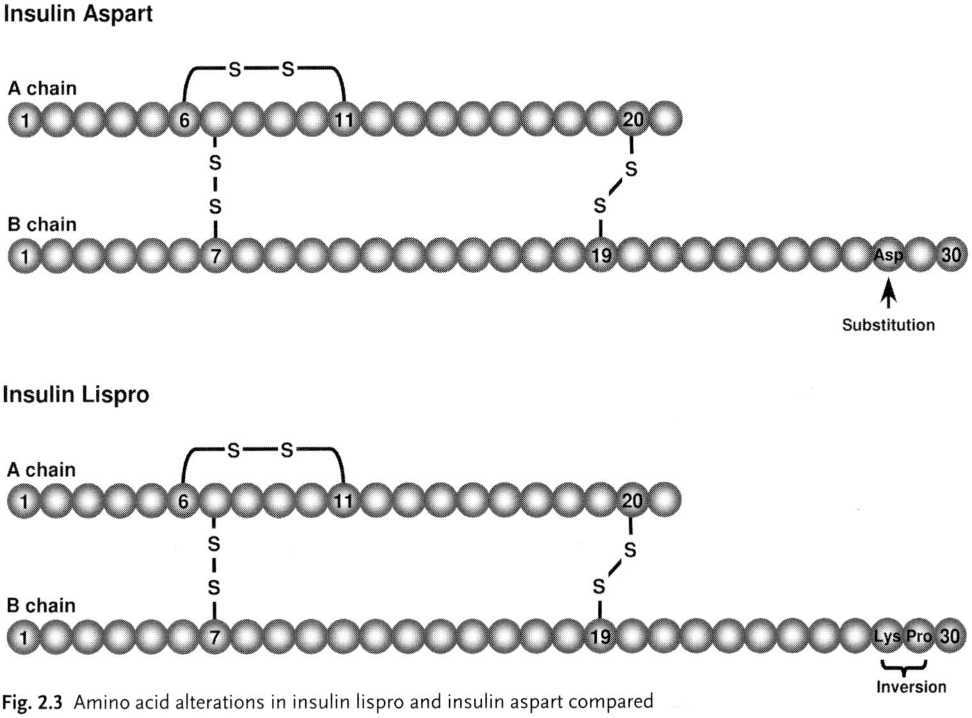


**Fig. 2.3** Amino acid alterations in insulin lispro and insulin aspart compared to regular human insulin (from [71]).

large to small proteins to peptides. Due to the unspecific degradation of numerous peptides and proteins in blood, clearance can exceed cardiac output – that is, >5 L/min blood clearance and >3 L/min for plasma clearance [13]. Investigations into the detailed metabolism of peptides and proteins are relatively difficult because of the myriad of molecule fragments that may be formed [69, 70].

A model example of the dependency of clearance on the physico-chemical properties of a protein is given by regular human insulin and its rapid-acting analogues insulin lispro and insulin aspart. The insulin analogues differ structurally from regular insulin through amino acid substitutions on the B chain, leading to conformational changes that result in a shift in binding to the C-terminal portion (Fig. 2.3). These structural alterations allow the rapid-acting insulin analogues to have an onset of action between 5 and 15 min, and an effective duration lasting at most 6 h as compared to regular human insulin with a much later onset of 30–60 min and a longer duration up to 8–10 h. These properties of the rapid-acting insulin analogues allow them to facilitate a much tighter control of postprandial hyperglycemia in diabetic patients compared to regular human insulin [71].

### 2.5.1
### Proteolysis

Proteolytic enzymes such as proteases and peptidases are ubiquitous throughout the body. Sites capable of extensive peptide and protein metabolism are not only limited to the liver, kidneys, and gastrointestinal tissue, but also include the blood and vascular endothelium as well as other organs and tissues. As proteases and peptidases are also located within cells, intracellular uptake is *per se* more an elimination rather than a distribution process [13]. While peptidases and proteases in the gastrointestinal tract and in lysosomes are relatively unspecific, soluble peptidases in the interstitial space and exopeptidases on the cell surface have a higher selectivity and determine the specific metabolism pattern of an organ. The proteolytic activity of subcutaneous tissue, for example, results in a partial loss of activity of SC compared to IV administered interferon-γ.

### 2.5.2
### Gastrointestinal Elimination

For orally administered peptides and proteins, the gastrointestinal tract is the major site of metabolism. Presystemic metabolism is the primary reason for their lack of oral bioavailability. Parenterally administered peptides and proteins, however, may also be metabolized in the intestinal mucosa following intestinal secretion. At least 20% of the degradation of endogenous albumin takes place in the gastrointestinal tract [13].

### 2.5.3
### Renal Elimination

For parenterally administered and endogenous peptides and proteins, the kidneys are the major elimination organ if the peptide/protein size is less than the glomerular filtration limit of ~60 kDa, though controversy persists with regard to glomerular filtration selectivity in terms of size, molecular conformation, and charge of the peptide or protein. The importance of the kidneys as elimination organ has been shown for interleukin-2, M-CSF and interferon-α [57, 63].

Complex mathematical models have been developed in order to calculate the sieving coefficient, or the average filtrate-to-plasma concentration ratio along the length of a representative capillary [72]. Glomerular size–selectivity studies have used dextran and Ficoll, a co-polymer of sucrose and epichlorohydrin, as test macromolecules over a wide range of molecular sizes [73]. Despite several existing models for size selectivity which provide basic frameworks in relating structure to actual functional properties of the glomerular barrier, future studies are still warranted to elucidate the effects of shape and deformability, as well as the steric hindrance of protein and peptide molecules [72–75]. In addition to size–selectivity, glomerular charge–selectivity has also been observed where anionic polymers pass through the capillary wall less readily than neutral polymers, which in turn

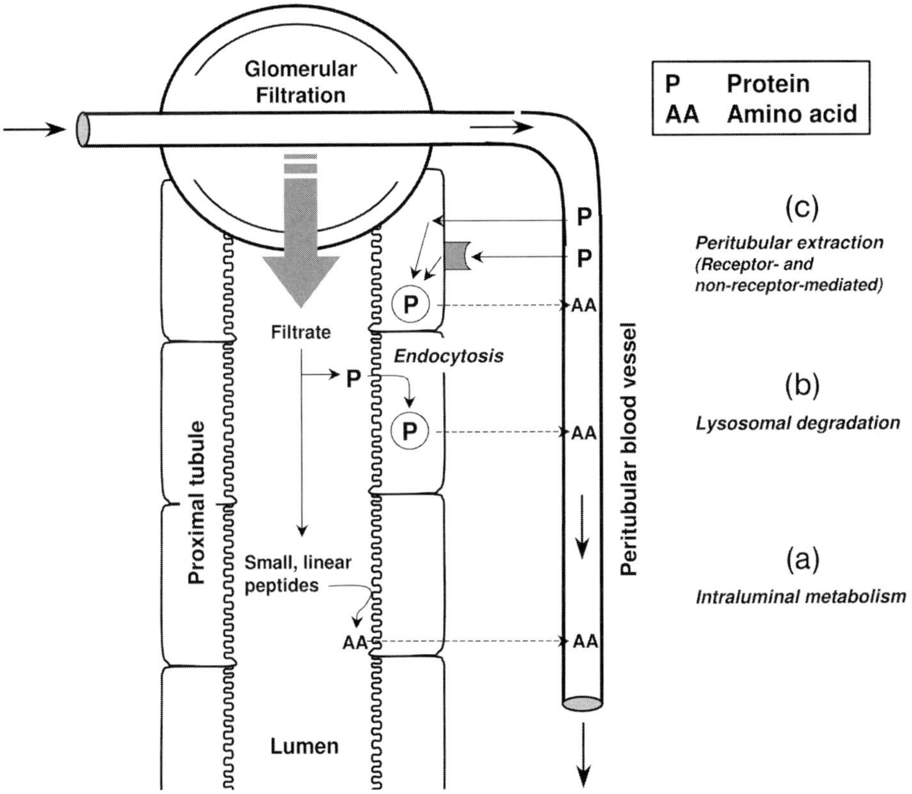


**Fig. 2.4** Renal elimination processes of peptides and proteins: Glomerular filtration followed by either (a) intraluminal metabolism or (b) tubular reabsorption with intracellular lysosomal metabolism, and (c) peritubular extraction with intracellular lysosomal metabolism (modified from [76]).

pass through less readily than cationic polymers. Many of the charge–selectivity studies have utilized combinations of dextran (neutral)/dextran sulfate (anionic) and neutral horseradish peroxidase (nHPR)/anionic horseradish peroxidase (aHPR) [75].

Renal metabolism of peptides and small proteins is mediated through three highly effective processes (Fig. 2.4). Consequently, only minuscule amounts of intact protein are detectable in the urine.

The first mechanism involves the glomerular filtration of larger, complex peptides and proteins, followed by reabsorption into endocytic vesicles in the proximal tubule and subsequent hydrolysis into small peptide fragments and amino acids [76]. This mechanism of elimination has been described for IL-2 [77], IL-11 [78], growth hormone [79], and insulin [80].

The second mechanism entails glomerular filtration followed by intraluminal metabolism, predominantly by exopeptidases in the luminal brush border membrane of the proximal tubules. The resulting peptide fragments and amino acids are reabsorbed into the systemic circulation. This route of disposition applies to small linear peptides such as angiotensin I and II, bradykinin, glucagon, and LH-RH [81, 82]. Recent studies implicate the proton-driven peptide transporters PEPT1 and PEPT2 as the main route of cellular uptake of small peptides and peptide-like drugs from the glomerular filtrates [83]. These high-affinity transport proteins seem to exhibit selective uptake of di- and tripeptides, which implicates their role in renal amino acid homeostasis [84].

For both mechanisms, glomerular filtration is the dominant, rate-limiting step, as subsequent degradation processes are not saturable under physiologic conditions [13, 76]. Hence, the renal contribution to the overall elimination of peptides and proteins is reduced if the metabolic activity for these proteins is high in other body regions, and it becomes negligible in the presence of unspecific degradation throughout the body. In contrast, the contribution to total clearance approaches 100% if the metabolic activity is low in other tissues, or if distribution is limited. For recombinant IL-10, for instance, elimination correlates closely with glomerular filtration rate, making dosage adjustments necessary in patients with impaired renal function [85].

The third mechanism is peritubular extraction of peptides and proteins from postglomerular capillaries and intracellular metabolism. Experiments using iodinated growth hormone ($^{125}$I-rGH) have demonstrated that, while reabsorption into endocytic vesicles at the proximal tubule is still the dominant route of disposition, a small percentage of the hormone may be extracted from the peritubular capillaries [79, 86]. Peritubular transport of proteins and peptides from the basolateral membrane has also been shown for insulin [87] and the mycotoxin ochratoxin A [88].

### 2.5.4
### Hepatic Elimination

Aside from gastrointestinal and renal metabolism, the liver may also contribute substantially to the metabolism of peptide and protein drugs. Proteolysis usually starts with endopeptidases that attack in the middle part of the protein, and the resulting oligopeptides are then further degraded by exopeptidases. The ultimate metabolites of proteins, amino acids and dipeptides, are finally reutilized in the endogenous amino acid pool. The rate of hepatic metabolism is largely dependent on specific amino acid sequences in the protein.

Substrates for hepatic metabolism include insulin, glucagon, and t-PAs [89,90]. For insulin, an acidic endopeptidase (termed "endosomal acidic insulinase") appears to mediate internalized insulin proteolysis at a number of sites [91]. Specifically, the endosomal activity results from cathepsin D, an aspartic acid protease [92]. Similarly, proteolysis of glucagon has also been attributed to membrane-bound forms of cathepsins B and D [93].

An important first step in the hepatic metabolism of proteins and peptides is uptake into the hepatocytes. Small peptides may cross the hepatocyte membrane via passive diffusion if they have sufficient hydrophobicity. Uptake of larger proteins, such as t-PA (65 kDa), is facilitated via receptor-mediated transport processes. Radio-iodinated t-PA ($^{125}$I-t-PA) studies implicate the role of mannose and asialoglycoprotein receptors in the liver for facilitating t-PA uptake and clearance [90]. Additionally, evidence also points to another hepatic membrane receptor – the low-density lipoprotein receptor-related protein – for contributing to overall t-PA clearance [90, 94].

### 2.5.5
### Receptor-Mediated Endocytosis

For conventional small-molecule drugs, receptor binding is usually negligible compared to the total amount of drug in the body, and rarely affects their pharmacokinetic profile. In contrast, a substantial fraction of a peptide and protein dose can be bound to receptors. This binding can lead to elimination through receptor-mediated uptake and subsequent intracellular metabolism. The endocytosis process is not limited to hepatocytes, but can also occur in other cells, including the therapeutic target cells. The binding and subsequent degradation via interaction with these generally high-affinity, low-capacity binding sites is a typical example for a pharmacologic target-mediated drug disposition, where binding to the pharmacodynamic target structure affects drug disposition [95]. Since the number of receptors is limited, drug binding and uptake can usually be saturated within therapeutic concentrations, or more specifically at relatively low molar ratios between the protein drug and the receptor. As a consequence, the pharmacokinetics of these drugs frequently does not follow the rule of superposition, i. e., clearance, and potentially other pharmacokinetic parameters are dose-dependent. Thus, receptor-mediated elimination constitutes a major source for the nonlinear pharmacokinetic behavior of numerous peptide and protein drugs, resulting in a lack of dose proportionality [4].

For example, M-CSF, in addition to linear renal elimination, undergoes a nonlinear elimination pathway that follows Michaelis–Menten kinetics and is linked to a receptor-mediated uptake into macrophages. At low concentrations, M-CSF follows linear pharmacokinetics, whereas at high concentrations the nonrenal elimination pathways are saturated, and this results in nonlinear pharmacokinetic behavior (Fig. 2.5) [96, 97]. Similarly, pharmacokinetic analysis of the *in-vivo* tissue distribution of the G-CSF derivative nartograstim revealed a nonlinear clearance process by the bone marrow and spleen with increasing doses of nartograstim [58]. Further studies with nartograstim suggested that nonlinearity in the early-phase bone marrow clearance might be due to the down-regulation of G-CSF receptors on the cell surface [98]. For recombinant human vascular endothelial growth factor (rhVEGF), a mechanism-based target-mediated drug distribution model had to be developed in order to accurately describe the drug's nonlinear pharmacokinetics in patients with coronary artery disease [61]. Nonlinear elimination of rhVEGF has been shown to be caused by binding to saturable high-affinity receptors followed by internalization and degradation.

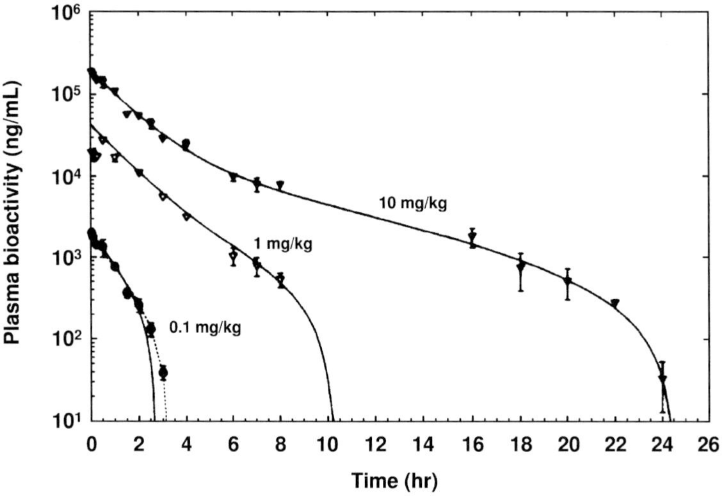


**Fig. 2.5** Nonlinear pharmacokinetics of M-CSF, presented as measured and modeled plasma concentration–time curves (mean ± SE) after intravenous injection of 0.1 mg/kg (n = 5), 1.0 mg/kg (n = 3), and 10 mg/kg (n = 8) in rats (from [97]).

## 2.6 Interspecies Scaling

Peptides and proteins exhibit distinct species specificity with regard to structure and activity. Peptides and proteins with identical physiological function may have different amino acid sequences in different species, and may have no activity or be even immunogenic if used in a different species.

The extrapolation of animal data to predict pharmacokinetic parameters by allometric scaling is an often-used tool in drug development, with multiple approaches available at variable success rates [99–101]. In the most frequently used approach, pharmacokinetic parameters between different species are related via body weight using a power function:

$$P = a \cdot W^b$$

where $P$ is the pharmacokinetic parameter scaled, $W$ is the body weight (in kg), $a$ is the allometric coefficient, and $b$ is the allometric exponent. $a$ and $b$ are specific constants for each parameter of a compound. General tendencies for the allometric exponent are 0.75 for rate constants (i. e., clearance, elimination rate constant), 1 for volumes of distribution, and 0.25 for half-lives.

For most traditional small-molecule drugs, allometric scaling is often imprecise, especially if hepatic metabolism is a major elimination pathway and/or if there are interspecies differences in metabolism. For peptides and proteins, however, allometric scaling has frequently proven to be much more precise and reliable, probably because of the similarity in handling peptides and proteins between different mammalian species [4, 43]. Interspecies scaling can lay the foundation for determining dosing as it relates to efficacy and toxicity. Clearance and volume of distribution of numerous therapeutically used proteins such as growth hormone or t-PA follow a well-defined, weight-dependent physiologic relationship between laboratory animals and humans [85, 102]. This information often provides the basis for quantitative predictions for toxicology and dose-ranging studies based on preclinical findings.

In a study investigating the allometric relationships of pharmacokinetic parameters for five therapeutic proteins, the allometric equations for clearance and volumes of distribution, however, were found to be different for each protein [102]. This variability was attributed to possible species specificity and immune-mediated clearance mechanisms. Species specificity refers to the inherent differences in structure and activity across species. Minute differences in the amino acid sequence may render an agent inactive when administered to foreign species, and may even generate an immunogenic response. Immunogenicity has been clearly demonstrated in a study with the tumor necrosis factor receptor-immunoglobulin fusion protein lenercept. This all-human sequence protein elicits an immune response in laboratory animals which ultimately results in the rapid clearance of the protein [103].

To further complicate the matter, studies have shown that the extent of glycosylation and/or sialylation of a protein molecule also exhibits species specificity that may affect drug efficacy, clearance, and immunogenicity. This is of particular importance if the production of human proteins is performed using bacterial cells [63]. While epoetin-$\alpha$, interferon-$\alpha$, and follicle-stimulating hormone are naturally glycosylated, others, such as insulin and growth hormone, are not. This requires careful selection of the animal model to assure an adequately intended pharmacological effect. Despite such differences, allometric scaling can still form the basis of estimating effective regimens for initial human studies. This is especially significant given that materials are often in limited supply for preclinical and early clinical development, where dose optimization is crucial [102].

## 2.7 Conclusions

Peptide and protein drugs are subject to the same general principles of pharmacokinetics and exposure–response correlations as conventional small-molecule drugs. Due to their similarity to protein nutrients and/or especially regulatory endogenous peptides and proteins, however, numerous caveats and pitfalls related to bioanalytics and pharmacokinetics must be considered and addressed during their drug development process. Furthermore, pharmacokinetic/pharmacodynamic correlations are frequently complicated due to the close interaction of peptide and protein drugs with endogenous substances and receptors, as well as regulatory feedback mechanisms. Additional investigations and resources are necessary to overcome some of these difficulties in order to ensure a rapid and successful drug development process. Nevertheless, pharmacokinetic evaluations provide a cornerstone in the development of dosage regimens for a rational, scientifically based clinical application of peptide and protein therapeutics that ultimately will ensure the success of this class of compounds in applied pharmacotherapy.

## 2.8 References


1 Meibohm, B., and H. Derendorf. **2002**. Pharmacokinetic/pharmacodynamic studies in drug product development. *J. Pharm. Sci.* 91:18–31.

2 CDER/FDA. **2003**. Exposure Response Relationships: Guidance for Industry. US Department of Health and Human Services, Food and Drug Administration, Center for Drug Evaluation and Research, Rockville.

3 International Conference on Harmonization. **1994**. ICH E4 – Dose-response information to support drug registration. European Agency for the Evaluation of Medicinal Products, London.

4 Tang, L., A.M. Persky, G. Hochhaus, and B. Meibohm. **2004**. Pharmacokinetic aspects of biotechnology products. *J. Pharm. Sci.* 93:2184–2204.

5 Fasano, A. **1998**. Novel approaches for oral delivery of macromolecules. *J. Pharm. Sci.* 87:1351–1356.

6 Mahato, R.I., A.S. Narang, L. Thoma, and D.D. Miller. **2003**. Emerging trends in oral delivery of peptide and protein drugs. *Crit. Rev. Ther. Drug Carrier Syst.* 20:153–214.

7 Lan, L.B., J.T. Dalton, and E.G. Schuetz. **2000**. Mdr1 limits CYP3A metabolism in vivo. *Mol. Pharmacol.* 58:863–869.

8 Meibohm, B., I. Beierle, and H. Derendorf. **2002**. How important are gender differences in pharmacokinetics? *Clin. Pharmacokinet.* 41:329–342.

9 Wacher, V.J., J.A. Silverman, Y. Zhang, and L.Z. Benet. **1998**. Role of P-glycoprotein and cytochrome P450 3A in limiting oral absorption of peptides and peptidomimetics. *J. Pharm. Sci.* 87:1322–1330.

10 Handelsman, D.J., and R.S. Swerdloff. **1986**. Pharmacokinetics of gonadotropin-releasing hormone and its analogs. *Endocr. Rev.* 7:95–105.

11 Periti, P., T. Mazzei, and E. Mini. **2002**. Clinical pharmacokinetics of depot leuprorelin. *Clin. Pharmacokinet.* 41:485–504.

12 Kaufman, J.S., D.J. Reda, C.L. Fye, D.S. Goldfarb, W.G. Henderson, J.G. Kleinman, and C.A. Vaamonde. **1998**. Subcutaneous compared with intravenous epoetin in patients receiving hemodialysis. Department of Veterans Affairs Cooperative Study Group on Erythropoietin in Hemodialysis Patients. *N. Engl. J. Med.* 339:578–583.

13 Colburn, W. **1991**. Peptide, peptoid, and protein pharmacokinetics/pharmacodynamics. In: P. Garzone, W. Colburn, and M. Mokotoff (Eds.), *Peptides, peptoids, and proteins*. Harvey Whitney Books, Cincinnati, OH, pp. 94–115.

14 Fuzeon. **2004**. Prescribing Information. Roche Pharmaceuticals, Nutley, NJ.

15 Symlin. **2005**. Prescribing information. Amylin Pharmaceuticals, San Diego, CA.

16 Porter, C.J., and S.A. Charman. **2000**. Lymphatic transport of proteins after subcutaneous administration. *J. Pharm. Sci.* 89:297–310.

17 Supersaxo, A., W. Hein, H. Gallati, and H. Steffen. **1988**. Recombinant human interferon-α-2a: delivery to lymphoid tissue by selected modes of application. *Pharm. Res.* 5:472–476.

18 Supersaxo, A., W.R. Hein, and H. Steffen. **1990**. Effect of molecular weight on the lymphatic absorption of water-soluble compounds following subcutaneous administration. *Pharm. Res.* 7:167–169.

**19** Schomburg, A., H. Kirchner, and J. Atzpodien. **1993**. Renal, metabolic, and hemodynamic side-effects of interleukin-2 and/or interferon-α: evidence of a risk/benefit advantage of subcutaneous therapy. *J. Cancer Res. Clin. Oncol.* 119:745–755.

**20** Laube, B.L. **2001**. Treating diabetes with aerosolized insulin. *Chest* 120: 99S–106S.

**21** Cleland, J.L., A. Daugherty, and R. Mrsny. **2001**. Emerging protein delivery methods. *Curr. Opin. Biotechnol.* 12:212–219.

**22** Patton, J.S., J.G. Bukar, and M.A. Eldon. **2004**. Clinical pharmacokinetics and pharmacodynamics of inhaled insulin. *Clin. Pharmacokinet.* 43:781–801.

**23** Cefalu, W.T., J.S. Skyler, I.A. Kourides, W.H. Landschulz, C.C. Balagtas, S. Cheng, and R.A. Gelfand. **2001**. Inhaled human insulin treatment in patients with type 2 diabetes mellitus. *Ann. Intern. Med.* 134:203–207.

**24** Hollander, P.A., L. Blonde, R. Rowe, A.E. Mehta, J.L. Milburn, K.S. Hershon, J.L. Chiasson, and S.R. Levin. **2004**. Efficacy and safety of inhaled insulin (exubera) compared with subcutaneous insulin therapy in patients with type 2 diabetes: results of a 6-month, randomized, comparative trial. *Diabetes Care* 27:2356–2362.

**25** Quan, J.M., H.A. Tiddens, J.P. Sy, S.G. McKenzie, M.D. Montgomery, P.J. Robinson, M.E. Wohl, and M.W. Konstan. **2001**. A two-year randomized, placebo-controlled trial of dornase-α in young patients with cystic fibrosis with mild lung function abnormalities. *J. Pediatr.* 139:813–820.

**26** Zito, S.W. **1997**. Pharmaceutical biotechnology: a programmed text. Technomic, Lancaster, PA.

**27** Chanson, P., J. Timsit, and A.G. Harris. **1993**. Clinical pharmacokinetics of octreotide. Therapeutic applications in patients with pituitary tumours. *Clin. Pharmacokinet.* 25:375–391.

**28** Lawrence, D. **2002**. Intranasal delivery could be used to administer drugs directly to the brain. *Lancet* 359:1674.

**29** Liu, X.F., J.R. Fawcett, R.G. Thorne, T.A. DeFor, and W.H. Frey, II. **2001**. Intranasal administration of insulin-like growth factor-I bypasses the blood-brain barrier and protects against focal cerebral ischemic damage. *J. Neurol. Sci.* 187: 91–97.

**30** Mitragotri, S., D. Blankschtein, and R. Langer. **1995**. Ultrasound-mediated transdermal protein delivery. *Science* 269:850–853.

**31** Kanikkannan, N. **2002**. Iontophoresis-based transdermal delivery systems. *BioDrugs* 16:339–347.

**32** Pillai, O., and R. Panchagnula. **2003**. Transdermal iontophoresis of insulin. V. Effect of terpenes. *J. Control. Release* 88:287–296.

**33** Kumar, S., H. Char, S. Patel, D. Piemontese, K. Iqbal, A.W. Malick, E. Neugroschel, and C.R. Behl. **1992**. Effect of iontophoresis on in vitro skin permeation of an analogue of growth hormone releasing factor in the hairless guinea pig model. *J. Pharm. Sci.* 81:635–639.

**34** Chang, S.L., G.A. Hofmann, L. Zhang, L.J. Deftos, and A.K. Banga. **2000**. Transdermal iontophoretic delivery of salmon calcitonin. *Int. J. Pharm.* 200:107–113.

**35** Suzuki, Y., K. Iga, S. Yanai, Y. Matsumoto, M. Kawase, T. Fukuda, H. Adachi, N. Higo, and Y. Ogawa. **2001**. Iontophoretic pulsatile transdermal delivery of human parathyroid hormone (1–34). *J. Pharm. Pharmacol.* 53:1227–1234.

**36** Widera, A., K.J. Kim, E.D. Crandall, and W.C. Shen. **2003**. Transcytosis of GCSF-transferrin across rat alveolar epithelial cell monolayers. *Pharm. Res.* 20:1231–1238.

**37** Kageyama, S., H. Yamamoto, H. Nakazawa, J. Matsushita, T. Kouyama, A. Gonsho, Y. Ikeda, and R. Yoshimoto. **2002**. Pharmacokinetics and pharmacodynamics of AJW200, a humanized monoclonal antibody to von Willebrand factor, in monkeys. *Arterioscler. Thromb. Vasc. Biol.* 22:187–192.

**38** Leone-Bay, A., M. Sato, D. Paton, A.H. Hunt, D. Sarubbi, M. Carozza, J. Chou, J. McDonough, and R.A. Baughman. **2001**. Oral delivery of biologically active parathyroid hormone. *Pharm. Res.* 18:964–970.

**39** Joseph, J.W., J. Kalitsky, S. St-Pierre, and P.L. Brubaker. **2000**. Oral delivery of glucagon-like peptide-1 in a modified polymer preparation normalizes basal glycaemia in diabetic db/db mice. *Diabetologia* 43 : 1319–1328.

**40** O'Hagan, D.T. **1992**. Oral delivery of vaccines. Formulation and clinical pharmacokinetic considerations. *Clin. Pharmacokinet.* 22 : 1–10.

**41** Van Der Lubben, I.M., F.A. Konings, G. Borchard, J.C. Verhoef, and H.E. Junginger. **2001**. In vivo uptake of chitosan microparticles by murine Peyer's patches: visualization studies using confocal laser scanning microscopy and immunohistochemistry. *J. Drug Target.* 9 : 39–47.

**42** Pauletti, G.M., S. Gangwar, T.J. Siahaan, A. Jeffrey, and R.T. Borchardt. **1997**. Improvement of oral peptide bioavailability: Peptidomimetics and prodrug strategies. *Adv. Drug Deliv. Rev.* 27: 235–256.

**43** Braeckman, R. **1997**. Pharmacokinetics and pharmacodynamics of peptide and protein drugs. In: D. Crommelin, and R. Sindelar (Eds.), *Pharmaceutical Biotechnology*. Harwood Academic Publishers, Amsterdam, pp. 101–122.

**44** Working, P., and P. Cossum. **1991**. Clinical and preclinical studies with recombinant human proteins: effect of antibody production. In: P. Garzone, W. Colburn, and M. Mokotoff (Eds.), *Peptides, peptoids, and proteins*. Harvey Whitney Books, Cincinnati, OH, pp. 158–168.

**45** Perini, P., A. Facchinetti, P. Bulian, A.R. Massaro, D.D. Pascalis, A. Bertolotto, G. Biasi, and P. Gallo. **2001**. Interferon-beta (INF-beta) antibodies in interferon-beta1a- and interferon-beta1b-treated multiple sclerosis patients. Prevalence, kinetics, cross-reactivity, and factors enhancing interferon-beta immunogenicity in vivo. *Eur. Cytokine Netw.* 12: 56–61.

**46** Harris, J.M., N.E. Martin, and M. Modi. **2001**. Pegylation: a novel process for modifying pharmacokinetics. *Clin. Pharmacokinet.* 40 : 539–551.

**47** Molineux, G. **2003**. Pegylation: engineering improved biopharmaceuticals for oncology. *Pharmacotherapy* 23 : 3S–8 S.

**48** Walsh, S., A. Shah, and J. Mond. **2003**. Improved pharmacokinetics and reduced antibody reactivity of lysostaphin conjugated to polyethylene glycol. *Antimicrob. Agents Chemother.* 47 : 554–558.

**49** Graham, M.L. **2003**. Pegaspargase: a review of clinical studies. *Adv. Drug Deliv. Rev.* 55 : 1293–1302.

**50** Zamboni, W.C. **2003**. Pharmacokinetics of pegfilgrastim. *Pharmacotherapy* 23 : 9S–14 S.

**51** Reilly, R.M., J. Sandhu, T.M. Alvarez-Diez, S. Gallinger, J. Kirsh, and H. Stern. **1995**. Problems of delivery of monoclonal antibodies. Pharmaceutical and pharmacokinetic solutions. *Clin. Pharmacokinet.* 28 : 126–142.

**52** Tanswell, P., N. Modi, D. Combs, and T. Danays. **2002**. Pharmacokinetics and pharmacodynamics of tenecteplase in fibrinolytic therapy of acute myocardial infarction. *Clin. Pharmacokinet.* 41 : 1229–1245.

**53** Ramakrishnan, R., W.K. Cheung, M.C. Wacholtz, N. Minton, and W.J. Jusko. **2004**. Pharmacokinetic and pharmacodynamic modeling of recombinant human erythropoietin after single and multiple doses in healthy volunteers. *J. Clin. Pharmacol.* 44 : 991–1002.

**54** Allon, M., K. Kleinman, M. Walczyk, C. Kaupke, L. Messer-Mann, K. Olson, A.C. Heatherington, and B.J. Maroni. **2002**. Pharmacokinetics and pharmacodynamics of darbepoetin-$\alpha$ and epoetin in patients undergoing dialysis. *Clin. Pharmacol. Ther.* 72 : 546–555.

**55** Jin, F., and W. Krzyzanski. **2004**. Pharmacokinetic model of target-mediated disposition of thrombopoietin. *AAPS PharmSci* 6:E9.

**56** Tan, A.C., F.G. Russel, T. Thien, and T.J. Benraad. **1993**. Atrial natriuretic peptide. An overview of clinical pharmacology and pharmacokinetics. *Clin. Pharmacokinet.* 24 : 28–45.

**57** McMartin, C. **1992**. Pharmacokinetics of peptides and proteins: opportunities and challenges. *Adv. Drug Res.* 22 : 39–106.

**58** Kuwabara, T., T. Uchimura, H. Kobayashi, S. Kobayashi, and Y. Sugiyama. **1995**. Receptor-mediated clearance of G-CSF derivative nartograstim in bone marrow of rats. *Am. J. Physiol.* 269:E1–E9.

**59** Toon, S. **1996**. The relevance of pharmacokinetics in the development of biotechnology products. *Eur. J. Drug Metab. Pharmacokinet.* 21:93–103.

**60** Mohler, M., J. Cook, D. Lewis, J. Moore, D. Sinicropi, A. Championsmith, B. Ferraiolo, and J. Mordenti. **1993**. Altered pharmacokinetics of recombinant human deoxyribonuclease in rats due to the presence of a binding protein. *Drug Metab. Dispos.* 21:71–75.

**61** Eppler, S.M., D.L. Combs, T.D. Henry, J.J. Lopez, S.G. Ellis, J.H. Yi, B.H. Annex, E.R. McCluskey, and T.F. Zioncheck. **2002**. A target-mediated model to describe the pharmacokinetics and hemodynamic effects of recombinant human vascular endothelial growth factor in humans. *Clin. Pharmacol. Ther.* 72:20–32.

**62** Piscitelli, S.C., W.G. Reiss, W.D. Figg, and W.P. Petros. **1997**. Pharmacokinetic studies with recombinant cytokines. Scientific issues and practical considerations. *Clin. Pharmacokinet.* 32:368–381.

**63** Wills, R.J., and B.L. Ferraiolo. **1992**. The role of pharmacokinetics in the development of biotechnologically derived agents. *Clin. Pharmacokinet.* 23:406–414.

**64** Taki, Y., T. Sakane, T. Nadai, H. Sezaki, G.L. Amidon, P. Langguth, and S. Yamashita. **1998**. First-pass metabolism of peptide drugs in rat perfused liver. *J. Pharm. Pharmacol.* 50:1013–1018.

**65** Patel, I.H., X. Zhang, K. Nieforth, M. Salgo, and N. Buss. **2005**. Pharmacokinetics, pharmacodynamics and drug interaction potential of enfuvirtide. *Clin. Pharmacokinet.* 44:175–186.

**66** Hoffmaster, K.A., M.J. Zamek-Gliszczynski, G.M. Pollack, and K.L. Brouwer. **2005**. Multiple transport systems mediate the hepatic uptake and biliary excretion of the metabolically stable opioid peptide [D-penicillamine2,5]enkephalin. *Drug Metab. Dispos.* 33:287–293.

**67** Yang, J.Z., W. Chen, and R.T. Borchardt. **2002**. In vitro stability and in vivo pharmacokinetic studies of a model opioid peptide, H-Tyr-D-Ala-Gly-Phe-D-Leu-OH (DADLE), and its cyclic prodrugs. *J. Pharmacol. Exp. Ther.* 303:840–848.

**68** Meijer, D., and K. Ziegler. **1993**. *Biological Barriers to Protein Delivery.* Plenum Press, New York, pp. 339–408.

**69** Brugos, B., and G. Hochhaus. **2004**. Metabolism of dynorphin A(1–13). *Pharmazie* 59:339–343.

**70** Muller, S., A. Hutson, V. Arya, and G. Hochhaus. **1999**. Assessment of complex peptide degradation pathways via structured multicompartmental modeling approaches: the metabolism of dynorphin A1–13 and related fragments in human plasma [In Process Citation]. *J. Pharm. Sci.* 88:938–944.

**71** Hirsch, I.B. **2005**. Insulin analogues. *N. Engl. J. Med.* 352:174–183.

**72** Edwards, A., B.S. Daniels, and W.M. Deen. **1999**. Ultrastructural model for size selectivity in glomerular filtration. *Am. J. Physiol.* 276:F892–902.

**73** Oliver, J.D., 3rd, S. Anderson, J.L. Troy, B.M. Brenner, and W.H. Deen. **1992**. Determination of glomerular size-selectivity in the normal rat with Ficoll. *J. Am. Soc. Nephrol.* 3:214–228.

**74** Venturoli, D., and B. Rippe. **2005**. Ficoll and dextran vs. globular proteins as probes for testing glomerular permselectivity: effects of molecular size, shape, charge, and deformability. *Am. J. Physiol. Renal Physiol.* 288:F605–613.

**75** Deen, W.M., M.J. Lazzara, and B.D. Myers. **2001**. Structural determinants of glomerular permeability. *Am. J. Physiol. Renal Physiol.* 281:F579–596.

**76** Maack, T., C. Park, and M. Camargo. **1985**. Renal filtration, transport and metabolism of proteins. In: D. Seldin, and G. Giebisch (Eds.), *The Kidney.* Raven Press, New York, pp. 1773–1803.

**77** Anderson, P.M., and M.A. Sorenson. **1994**. Effects of route and formulation on clinical pharmacokinetics of interleukin-2. *Clin. Pharmacokinet.* 27:19–31.

**78** Takagi, A., H. Masuda, Y. Takakura, and M. Hashida. **1995**. Disposition characteristics of recombinant human interleu-

kin-11 after a bolus intravenous administration in mice. *J. Pharmacol. Exp. Ther.* 275:537–543.
**79** Johnson, V., and T. Maack. **1977**. Renal extraction, filtration, absorption, and catabolism of growth hormone. *Am. J. Physiol.* 233:F185–196.
**80** Rabkin, R., M.P. Ryan, and W.C. Duckworth. **1984**. The renal metabolism of insulin. *Diabetologia* 27:351–357.
**81** Carone, F.A., and D.R. Peterson. **1980**. Hydrolysis and transport of small peptides by the proximal tubule. *Am. J. Physiol.* 238:F151–158.
**82** Carone, F.A., D.R. Peterson, and G. Flouret. **1982**. Renal tubular processing of small peptide hormones. *J. Lab. Clin. Med.* 100:1–14.
**83** Inui, K., T. Terada, S. Masuda, and H. Saito. **2000**. Physiological and pharmacological implications of peptide transporters, PEPT1 and PEPT2. *Nephrol. Dial. Transplant.* 15 Suppl 6:11–13.
**84** Daniel, H., and M. Herget. **1997**. Cellular and molecular mechanisms of renal peptide transport. *Am. J. Physiol.* 273:F1–8.
**85** Andersen, S., L. Lambrecht, S. Swan, D. Cutler, E. Radwanski, M. Affrime, and J. Garaud. **1999**. Disposition of recombinant human interleukin-10 in subjects with various degrees of renal function. *J. Clin. Pharmacol.* 39:1015–1020.
**86** Krogsgaard Thomsen, M., C. Friis, B. Sehested Hansen, P. Johansen, C. Eschen, J. Nowak, and K. Poulsen. **1994**. Studies on the renal kinetics of growth hormone (GH) and on the GH receptor and related effects in animals. *J. Pediatr. Endocrinol.* 7:93–105.
**87** Nielsen, S., J.T. Nielsen, and E.I. Christensen. **1987**. Luminal and basolateral uptake of insulin in isolated, perfused, proximal tubules. *Am. J. Physiol.* 253:F857–867.
**88** Groves, C.E., M. Morales, and S.H. Wright. **1998**. Peritubular transport of ochratoxin A in rabbit renal proximal tubules. *J. Pharmacol. Exp. Ther.* 284:943–948.
**89** Authier, F., B.I. Posner, and J.J. Bergeron. **1996**. Endosomal proteolysis of internalized proteins. *FEBS Lett.* 389:55–60.
**90** Smedsrod, B., and M. Einarsson. **1990**. Clearance of tissue plasminogen activator by mannose and galactose receptors in the liver. *Thromb. Haemost.* 63:60–66.
**91** Authier, F., G.M. Danielsen, M. Kouach, G. Briand, and G. Chauvet. **2001**. Identification of insulin domains important for binding to and degradation by endosomal acidic insulinase. *Endocrinology* 142:276–289.
**92** Authier, F., M. Metioui, S. Fabrega, M. Kouach, and G. Briand. **2002**. Endosomal proteolysis of internalized insulin at the C-terminal region of the B chain by cathepsin D. *J. Biol. Chem.* 277:9437–9446.
**93** Authier, F., J.S. Mort, A.W. Bell, B.I. Posner, and J.J. Bergeron. **1995**. Proteolysis of glucagon within hepatic endosomes by membrane-associated cathepsins B and D. *J. Biol. Chem.* 270:15798–15807.
**94** Bu, G., S. Williams, D.K. Strickland, and A.L. Schwartz. **1992**. Low density lipoprotein receptor-related protein/alpha 2-macroglobulin receptor is an hepatic receptor for tissue-type plasminogen activator. *Proc. Natl. Acad. Sci. USA* 89:7427–7431.
**95** Levy, G. **1994**. Mechanism-based pharmacodynamic modeling. *Clin. Pharmacol. Ther.* 56:356–358.
**96** Bartocci, A., D.S. Mastrogiannis, G. Migliorati, R.J. Stockert, A.W. Wolkoff, and E.R. Stanley. 1987. Macrophages specifically regulate the concentration of their own growth factor in the circulation. *Proc. Natl. Acad. Sci. USA* 84:6179–6183.
**97** Bauer, R.J., J.A. Gibbons, D.P. Bell, Z.P. Luo, and J.D. Young. **1994**. Nonlinear pharmacokinetics of recombinant human macrophage colony- stimulating factor (M-CSF) in rats. *J. Pharmacol. Exp. Ther.* 268:152–158.
**98** Kuwabara, T., T. Uchimura, K. Takai, H. Kobayashi, S. Kobayashi, and Y. Sugiyama. **1995**. Saturable uptake of a recombinant human granulocyte colony-stimulating factor derivative, nartograstim, by the bone marrow and spleen of rats in vivo. *J. Pharmacol. Exp. Ther.* 273:1114–1122.

**99** Mahmood, I., and J.D. Balian. **1999**. The pharmacokinetic principles behind scaling from preclinical results to phase I protocols. *Clin. Pharmacokinet.* 36 : 1–11.

**100** Mahmood, I. **2002**. Interspecies scaling: predicting oral clearance in humans. *Am. J. Ther.* 9 : 35–42.

**101** Boxenbaum, H. **1982**. Interspecies scaling, allometry, physiological time, and the ground plan of pharmacokinetics. *J. Pharmacokinet. Biopharm.* 10 : 201–227.

**102** Mordenti, J., S.A. Chen, J.A. Moore, B.L. Ferraiolo, and J.D. Green. **1991**. Interspecies scaling of clearance and volume of distribution data for five therapeutic proteins. *Pharm. Res.* 8 : 1351–1359.

**103** Richter, W.F., H. Gallati, and C.D. Schiller. **1999**. Animal pharmacokinetics of the tumor necrosis factor receptor-immunoglobulin fusion protein lenercept and their extrapolation to humans. *Drug Metab. Dispos.* 27 : 21–25.

# 3
# Pharmacokinetics of Monoclonal Antibodies

*Katharina Kuester and Charlotte Kloft*

## 3.1
## Introduction

The foundations for the generation of novel pharmaceuticals, the monoclonal antibodies (mAbs), were laid approximately 30 years ago. Due to their ability to selectively hit a specific target, the mAbs can be regarded as very closely fitting the concept first proposed by Paul Ehrlich, a century ago, of a "magic bullet" to treat medical disorders.

The first therapeutic mAbs were developed and entered preclinical and clinical trials during the late 1970s. Subsequently, during the early 1980s, recombinant DNA and expression technologies were applied to the antibody research field, while during the later part of that decade different expression systems (bacterial, yeast, and mammalian cells) became available as novel production and engineering tools. Later still, during the 1990s, these techniques allowed not only the production of large quantities of the drug substance but also the design and engineering of a wide range of antibody structures.

Following the success of recombinant proteins such as insulin, therapeutic mAbs today represent the second wave of innovation created by the biotechnology industry during the past 20 years. The recent success of a number of new mAb therapies, for example rituximab (Rituxan®) and infliximab (Remicade®), suggests a resurgence of the biotech industry for the coming years. For serious chronic diseases such as cancer or rheumatoid arthritis, mAb therapy has indeed proven its clinical efficacy.

Natalizumab (Tysabri®) was regarded as a highly promising new therapeutic principle for the treatment of multiple sclerosis, but its withdrawal from the market in 2004 was a clear setback. The voluntary decision to discontinue the sale of natalizumab was taken by its manufacturers following the report in the SENTINEL trial of two cases of a rare but usually fatal demyelinating disorder, the so-called progressive multifocal leukoencephalopathy [1]. Although the mechanisms of and reasons for this adverse event are not yet fully understood, they are most likely attributable to some special characteristics of this mAb (such as the IgG4 isotype; see Section 3.3.1.1) rather than to a drug class effect.

*Pharmacokinetics and Pharmacodynamics of Biotech Drugs: Principles and Case Studies in Drug Development.* Edited by Bernd Meibohm

ISBN: 3-527-31408-3

Despite this setback, the clinical and financial prospect for mAbs remains impressive and highly favorable:

- In 2002, mAbs represented ~16% of the value of the total global biotech industry, with a rising tendency.
- At present, between one-third and one-half of potential biotech drugs in human clinical trials or under development are mAbs.
- To date, approximately 20 mAbs have been approved for various indications.

Monoclonal antibodies have a significant potential as therapeutic agents because of their ability to bind to specific structures as targets. This principle of "targeted therapy" results in high clinical efficacy whilst minimizing adverse reactions, and thus increases mAb tolerability and use.

For a successful preclinical and clinical development, approval and therapeutic application of mAbs, a thorough understanding of the pharmacokinetic characteristics of these compounds is crucial at every stage. This chapter addresses the pharmacokinetic characteristics of mAbs within the context of the principle of "targeted therapy". In the first two sections, the basic principles of the immune system and the role of physiological antibodies will be presented. A thorough description of the various types of physiological and therapeutic antibodies with the different effector functions and modes of action will follow. Next, the prerequisites of mAb therapy will be provided. The next three sections will provide detailed insights into the pharmacokinetic characteristics of antibodies and discuss various approaches to the analysis of pharmacokinetic data for mAbs. Finally, the pharmacokinetics and their relationship to pharmacodynamics will be discussed both for approved therapeutic mAbs and for some of those under development.

## 3.2 The Human Immune System

The first examples of immunotherapy with serum emerged at the end of the nineteenth century, since which time different types of immunotherapy with various classes of antibodies have been developed that will, in the future, generate exciting new pharmaceuticals. To correctly classify and understand the therapeutic principle of antibodies, the constitution of the immune system of the human body will briefly be presented in the following sections.

In order to recognize and protect the human body against foreign substances or invading microorganisms, the immune system has developed several defense mechanisms, the ultimate goal being to eliminate these potentially harmful interferences from the body. In general, the entire immune system is divided into two main branches – humoral and cellular. An overview of the immune system based on this subdivision is provided in Table 3.1.

**Table 3.1** An overview of the human immune system.

| | Different immune response types | |
|---|---|---|
| | **Specific** | **Nonspecific** |
| Humoral | Antibodies | Complement system |
| Cellular | T lymphocytes, APC, (B lymphocytes) | Macrophages, natural killer cells, granulocytes, monocytes |

APC: antigen-presenting cell.

### 3.2.1 The Cellular Immune Response

The cellular immune system or cell-mediated immunity involves different cell types which also synthesize and release a number of soluble factors or mediators (cytokines). It is further subdivided into two components:

- Specific cellular immunity is a type of (more specialized) *acquired* immunity which is based primarily on T lymphocytes. Several subpopulations of T lymphocytes are differentiated in the *thymus*, where one subset has identical structures on the cell surface to recognize a specific set of antigen. Each subpopulation performs different effector functions. The dominant T-cell type involved is the so-called cytotoxic T lymphocyte. In addition, T-helper cells are important for enhancing the immune response. By synthesizing and secreting various cytokines such as interferon $\gamma$ (INF-$\gamma$ or interleukin 2 (IL-2), T-lymphocytes also have an important role in controlling other major parts of the immune system.
- Non-specific cellular immunity: All other immune cell types such as granulocytes, various cells of the mononuclear phagocytosis system (macrophages/monocytes, dendritic cells) or natural killer cells form part of the (less specialized) *innate* immunity.

The major part of this section will deal with *humoral immunity*, which is mediated by the release of antibodies, and the response of the complement system. In the following section, attention will be focused on the role and structure of the different types of antibodies.

### 3.2.2 The Humoral Immune Response

Humoral immunity can also be divided into *nonspecific* and *specific* components:

- Nonspecific humoral immunity is mediated via the complement system. After activation of the complement system a cascade is initiated in which more than 15 individual glycoprotein compounds in the plasma react with each other in a predetermined manner. The result includes the attraction and stimulation of immune cells (e.g., macrophages), the initiation of phagocytosis, or the lysis of cells that are targeted by the complement system.

- Specific humoral immunity is activated by the contact of an antigen with the immune system, with the immune response being mediated by B lymphocytes that mature in the bone marrow and express immunoglobulins on their surface. At one to two weeks after the initial contact with the antigen, immunologically active B lymphocytes begin to secrete antibodies; these are strictly defined products that contain a complementary region towards the binding part of the antigen (antigenic determinant). The principle of the interlocking interaction between an antibody and an antigen can be compared to a key that matches a lock.

## 3.3 Physiological Antibodies

Under physiological conditions, the immune system is capable of generating its own antibodies against invading material with antigenic determinants. These antibodies are called immunoglobulins (Ig), and are glycoproteins produced by B lymphocytes. As part of the specific humoral immune system they are secreted into the blood or lymph system to identify and neutralize foreign invading objects such as microorganisms (bacteria, parasites, or viruses) or their products, or other non-endogenous substances and objects.

### 3.3.1 Classes of Antibodies

Antibodies can be classified according to the GADME system based on their configuration and function. The five different *classes* (also referred to as *isotypes*) are presented, along with their function, in Table 3.2. Also integrated into this overview are the molecular mass, half-life and the proportion of each class. In the following section, emphasis will be placed on the kinetic aspects of the isotypes.

#### 3.3.1.1 Immunoglobulin G

Immunoglobulin G (IgG) represents the most important class of immunoglobulins, with a serum concentration of approximately 12 mg/mL. IgG molecules have the lowest molecular mass and are the predominant antibody for immunochemistry. Due to differences in chemical structure (see Section 3.3.2) and biological function among IgGs, four subclasses (also referred to as *subisotypes*) can be distinguished. The proportion of IgG1, IgG2, IgG3 and IgG4 is 70 : 20 : 7 : 3. Furthermore, there are also several subtypes for certain subclasses such as IgG1a and IgG1b. IgG has the longest half-life among the different isotypes (21 days), with the exception of the subclass IgG3 (only 7 days). The numerous biological func-

**Table 3.2** Immunoglobulins.

| Class | Sub-classes | Function | Molecular mass [kDa] | Proportion of total Ig [%] | $t_{1/2}$ [days] |
|---|---|---|---|---|---|
| IgG | IgG1 | Main Ig in blood and | 150 | 75 | 21 |
| | IgG2 | extravascular region. | | | 21 |
| | IgG3 | Binds to antigen | | | 7 |
| | IgG4 | and toxins | | | 21 |
| IgA | IgA1 | Main Ig in seromucous | Monomer: 160 | 15 | 6 |
| | IgA2 | excretion. | Dimer: 390 | | 6 |
| | | Surface protection | Secretory dimer: 385[a)] | | |
| IgD | IgD | Mainly in humans. On B lymphocytes | 180 | 0.5 | 3 |
| IgM | IgM | First Ig on B lymphocytes. Favors agglutination | 970 | 7 | 10 |
| IgE | IgE | Main role in allergies. Surface protection | 190 | 0.002 | 2 |

**a)** Heterogeneous distribution; also larger polymers.

tions of IgG and modes of action will be described in Section 3.5. Two kinetic properties of IgG shall be emphasized:

- Kinetics of the immune response by antibodies: Following the first exposure of an individual to an unknown antigen, the first occurrence of specific IgG can be detected approximately three weeks later (primary response). If these particular antigens appear in the body for a second time, however, IgG will be produced much more rapidly and in larger amounts to eliminate the antigen (secondary response).
- Distribution: The IgG class (except subclass IgG2) is the only class of immunoglobulins capable of crossing the placental barrier. Since a fetus is unable to synthesize immunoglobulins during the first three to four months of pregnancy, the initial immunity of the fetus is of maternal origin. The protection by the transferred IgG, however, is lost three months after birth due to catabolism of the maternal molecules.

#### 3.3.1.2 **Immunoglobulins A, D, M, and E**

- IgA accounts for 15% of all immunoglobulins. The approximate concentration in serum is low (1.8 mg/mL), but higher in secretions where it is always present in the dimeric form. IgA is specialized in the defense against antigens on the surface of human mucosal membranes, for example in the intestine or in the nose. A large number of exogenous pathogens, including parasites, are neutralized by preventing them from penetrating epithelial cells.

- IgD is found on B lymphocytes during certain stages of maturation, and is jointly responsible for their activation. The serum concentration is low (0–0.04 mg/mL).
- IgM forms pentamers, and is the first class of immunoglobulins to be produced by the immune system if an unknown antigen enters the body, or following immunization (approximately one to two weeks). After the primary reaction, the concentration decreases due to the cessation of production (in contrast to the increase in IgG).
- IgE has the lowest percentage of total Ig at only 0.002%, and the serum concentration approximates only 0.00002 mg/mL. This isotype is derived from adenoid tissue and then transported into the blood. In spite of its small quantities, IgE is very important in, and is responsible for, about 90% of allergic reactions. The impact of this immunoglobulin is to trigger the release of vasodilators for an inflammatory response, for example histamine. IgE may also protect external mucosal surfaces by promoting inflammation and enabling IgG, complement proteins and leucocytes to enter the tissue.

In general, immunoglobulins – as proteins with hydrophilic and glycosylated moieties – have a high molecular weight of approximately 150–200 kDa. The only exception is the tentameric IgM, which has a larger molecular mass of 970 kDa.

### 3.3.2
### Chemical Structure of Antibodies

Physiological antibodies have as a monomer a typical common chemical structure consisting of several structural elements, two identical heavy and long chains (H-chains), and two identical light and short chains (L-chains). These chains are held together by a number of disulfide bridges, and may be glycosylated. The L-chain consists of 220 amino acids, and appears in two different configurations, the λ- and κ-chains. Any individual of a species produces both types of L-chain. The ratio of κ- to λ-chains varies with the species; for example, humans have 60% and mice 95% κ-chains. In any Ig molecule, however, the L-chains are always either both κ-chains or both λ-chains. The type of H-chain of about 450–550 amino acids determines the GADME classification referred to in Section 3.1.1. There are five different types of H-chains: type γ in the IgG molecule; type α in the IgA molecule; type δ in the IgD molecule; type μ in the IgM molecule; and type ε in the IgE molecule. Thus, the different classes and subclasses of human Ig differ in the structure of the heavy chain, the number and localization of the disulfide bridges, and the glycosylation pattern.

As shown in Fig. 3.1, the shape of antibodies resembles the letter "Y" which is due to the disulfide bridges between the two H-chains and between each H-chain with one L-chain (IgG). These endings tend to crystallize, and for this reason the

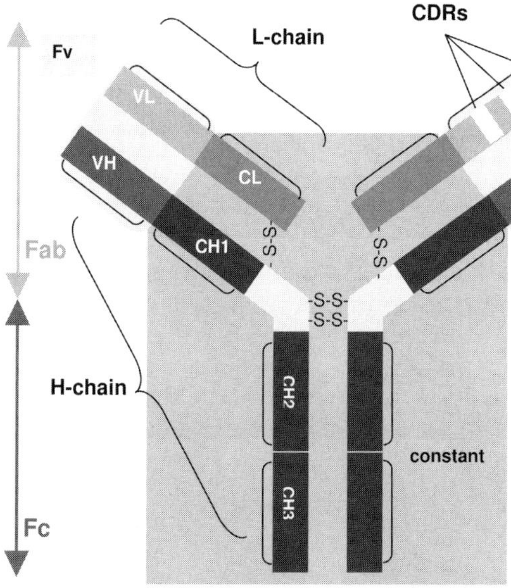


**Fig. 3.1** The structure of antibodies. H-chain: heavy chain, consisting of VH, CH1, CH2 and CH3; L-chain: light chain, consisting of VL and CL; VH: variable part of H-chain; VL: variable part of L-chain; CL: constant part of L-chain; CH1–3: constant parts 1, 2 or 3 of H-chain; -S-S-: disulfide bridge; Fc: crystallizable fragment; Fab: antigen-binding fragment; Fv: variable part of Fab; CDR: complementarity-determining regions, special hypervariable sequences of amino acids.

region is referred to as "fragment crystallizable" (Fc). The Fc region is responsible and important for: (1) the binding to the Fc receptor of cytotoxic immune cells (see Section 3.5); (2) the activation of the complement system (see Section 3.5.2.1); (3) the passage of the placenta (see Section 3.3.1.1); and (4) the catabolism of an antibody (see Section 3.8).

The antigen-binding fragment (Fab) is located where H- and L-chains are linked by both covalent bindings (disulfide bridge) and noncovalent bindings. At the end of this part there is a region in both L-chains which is variable, the so-called Fv region. Within the Fv region special hypervariable sequences of amino acids, the complementarity-determining regions (CDR), are responsible for the huge differentiation of antibodies as they vary in each immunoglobulin in both length and sequence. The steric structure of the summary of only six CDRs builds a contact surface, the counterpart to the epitope, also referred to as the antigenic determinant, which binds to the antibody. The area between these CDRs, called the framework region (FR), provides the basis for the stability of the frame. Based on these structural features, antibodies show very specific recognition of highly diverse antigenic determinants, despite their generally uniform structure.

## 3.4 Therapeutic Antibodies

Therapeutic antibodies must be distinguished as mAbs (and derived products) and polyclonal antibodies when discussing their different pharmacokinetic properties and therapeutic applications. mAbs are already – and will in the future – be much more important in drug development and applied pharmacotherapy due to their favorable properties and higher clinical success rates compared to polyclonal antibodies. Thus, although both types of antibody will be discussed in the following sections, most emphasis will be placed on mAbs.

### 3.4.1 Therapeutic Polyclonal Antibodies

Polyclonal antibodies are a mixture of antibodies produced from different B-lymphocyte lines, and are accordingly directed against various antigens. Typically, polyclonal antibodies are obtained by immunizing animals with a chosen antigen and subsequently harvesting immunoglobulin fractions from their serum. Rodents (rabbits > mice > rats) are the most commonly used laboratory animals for the production of polyclonal antibodies. For the production of larger amounts of antibodies, however, goats, sheep and horses are frequently used.

Physiologically, polyclonal antibodies are formed in the body following an active immunization. After the injection of a chosen antigen, which is usually bound to a carrier to increase the molecular mass for recognition as foreign by the immune system, the antigenic determinants of the antigen are recognized by the surface immunoglobulins on B lymphocytes. This triggers the differentiation of the B lymphocytes into plasma cells, which subsequently results in the production of immunoglobulin molecules specific for that antigen. Initially, low-affinity IgM class antibodies are secreted, but during the course of the immune response – and particularly if repeated injections of the antigen are administered – IgG, IgA and IgE class antibodies with higher affinity are generated. Due to the different B-lymphocyte cell lines involved in this process, the antibodies produced are characterized by an inconsistent composition and thus vary in their specificity as well as affinity towards an epitope.

The advantage of polyclonal antibodies lies in a higher tolerance towards minor changes of the antigen, as a slightly modified antigen will most probably also be recognized. Polyclonal antibodies as therapeutics are also used if the specific epitope is not established due to practical or financial reasons.

Polyclonal antibodies however, also have some drawbacks. Due to the non-human molecular heterogeneity, unspecific reactions are likely to occur and may cause a large variety of adverse reactions. In addition, the dose to be administered to target a specific antigen is relatively high compared to that for mAbs; this is due to the heterogeneity in specificity and affinity. Furthermore, the immune system will produce anti-antibodies to attack the non-human structures on polyclonal therapeutic antibodies, thereby potentially leading to serious hypersensitivity reactions.

The antithymocyte globulin (ATG) Fresenius® S, produced in rabbits, is an example of a polyclonal antibody mixture used in immunosuppressive therapy in combination with other immunosuppressive agents (e.g., cyclosporine, prednisone) after organ transplantation.

## 3.4.2 Therapeutic mAbs

The first report of the treatment of a tumor patient with a mAb was published in 1980 [2]. The principle of the production of mAbs can be traced back to the Nobel prize (Medicine 1984)-winning research of César Milstein and Georges Köhler [3]. The first extracted mAbs were produced using a method called the hybridoma technique, and emanated from murine B cells. The biotechnological steps involved in the hybridoma technique are initiated by the immunization of a mouse with a specific antigen (Fig. 3.2). The subsequent immune reaction results in an increase of B lymphocytes that produce and secrete antibodies. These antibodies react with the antigen and accumulate in the spleen. After removal of the spleen, the B lymphocytes are isolated and an antibody-producing B lymphocyte is fused with a malignant myeloma cell (an "immortal" cancer cell capable of replicating indefinitely). The result is a so-called "hybridoma cell" or cell line, which produces only a single type of antibody targeted against a particular antigen. The major advantage of the fusion of the two cell types (myeloma cell and B lymphocyte) is the unification of their original characteristics, the capability of the myeloma cell for unlimited growth, and the ability of the B-lymphocyte to produce a specific antibody.

For the production of mAbs, the cell line with the best binding to the targeted epitope of the antigen is chosen from several engineered hybridoma cell lines. The obtained species of antibodies is referred to as "mAbs" because they derive from one original B lymphocyte and thus they are all identical (clones).

### 3.4.2.1 Murine mAbs

Murine mAbs are antibodies of murine origin. The first murine mAb to be approved for clinical use was muromonab-CD3 (Orthoclone OKT3®), an mAb with CD3 on T cells as the molecular target, and used in organ transplantation.

The therapeutic application of antibodies of non-human origin may cause problems because murine mAbs are recognized by the human immune system as extrinsic substances. In general, the first administration of 100% murine mAbs is well tolerated. However, it can induce the production of specific anti-antibodies by the human body, the so-called human anti-murine antibodies (HAMAs), against the murine mAbs. Repeated administration of 100% murine mAbs may cause an immune response with influenza-like symptoms and even severe states of shock.

New technology developed during the 1980s allowed the development of the following antibody species.

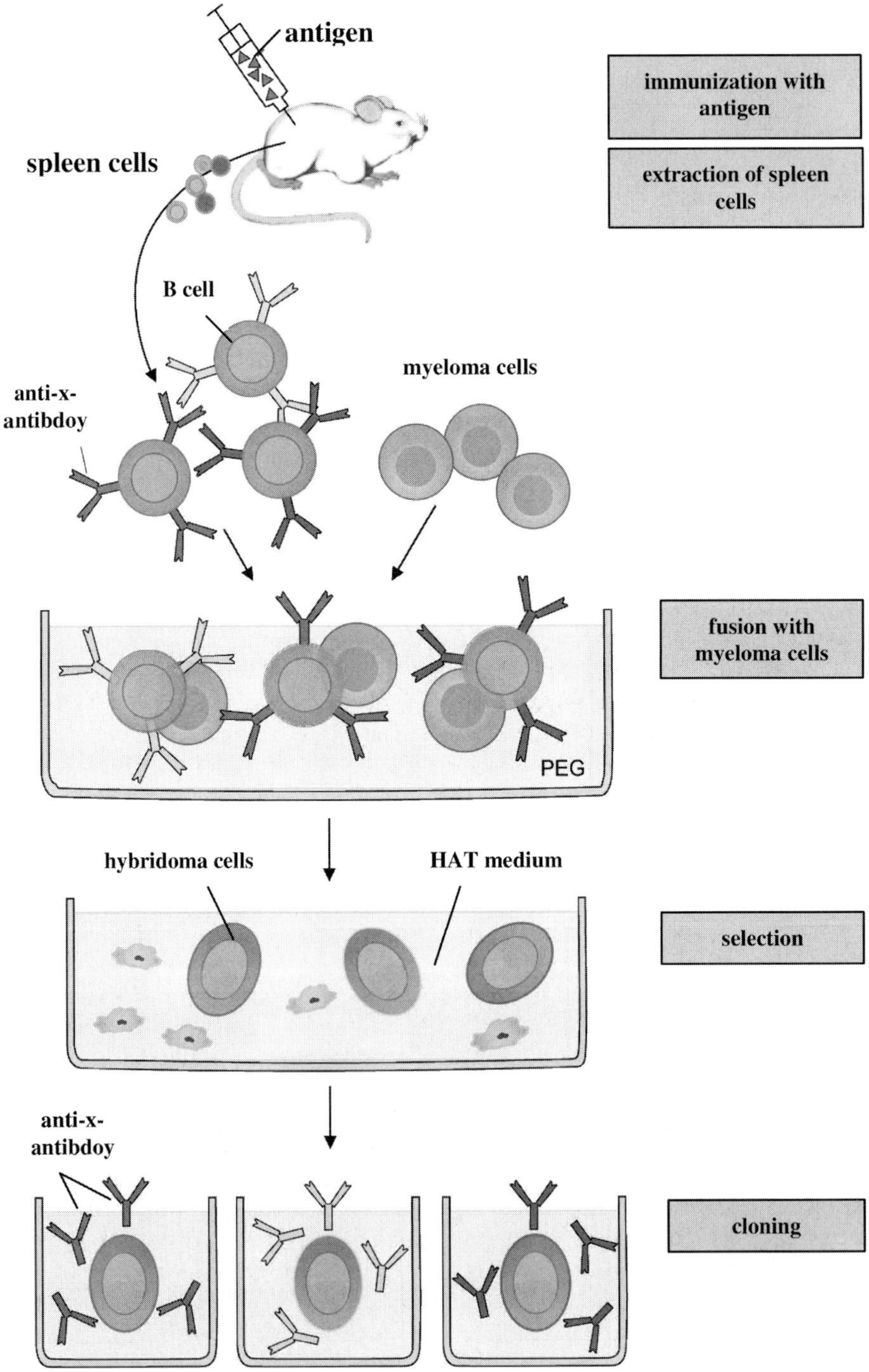


**Fig. 3.2** (legend see p. 55)

#### 3.4.2.2 **Chimeric mAbs**

The introduction of recombinant gene technology made it possible to overcome, or at least minimize, some of the limitations of murine mAbs by conferring the Fv region to another antibody. Transfer of the Fv region of a murine antibody to the constant regions of a human antibody achieved a significant decrease in the murine fraction to approximately one-third, resulting in so-called chimeric mAbs. The term "chimeric" can be traced back to Greek mythology, where the Chimera was a creature with both human and animal proportions. In spite of the extensive reduction of the murine part, immune reactions are still observed for chimeric mAbs, though to a smaller extent than for murine mAbs. In particular, the remaining framework regions in the molecule are responsible for the immune response. The antibodies produced against the chimeric mAb are referred to as human anti-chimeric antibodies (HACAs).

#### 3.4.2.3 **Humanized mAbs**

Further progress was achieved by the humanization of murine antibodies. In humanized mAbs, the murine fraction of 5–10% consists only of the murine CDRs and, if necessary, some exclusive parts of the framework region. The human immune system can also produce antibodies against these humanized antibodies, so-called human anti-human antibodies (HAHAs).

The extent of antibody formation against therapeutic mAbs is an adverse reaction that depends heavily on several factors, including the type of mAb, dosage regimen, and route of administration. It ranges from <1% up to 60% of the treated patient population. Therefore, close monitoring of the development of HAHAs, HACAs or HAMAs under therapy is obligatory. These anti-antibodies can also influence the pharmacokinetics, and especially the clearance, of therapeutic mAbs (see Section 3.9.3).

#### 3.4.2.4 **Human mAbs**

More recent developments using cDNA libraries of B lymphocytes and phage display technology have made it possible to produce complete human mAbs. The

◀ **Fig. 3.2** Hybridoma technique for the production of monoclonal antibodies. Spleen (milt) cells, which have been taken from mice (being immunized with an antigen X) contain anti-X-antibody-producing B cells. These cells are fused with myeloma cells in the presence of polyethylene glycol (PEG) and then taken to the HAT (hypoxanthine-aminopterin-thymidine) medium. HAT will induce death to myeloma cells because of the absence of the enzyme hypoxanthine-guanine-phosphoribosyl-transferase (HGPRT). Hybridoma cells, however, possess this enzyme due to the originating spleen cells and will survive in the HAT medium selectively. The growing hybridoma cells will be tested for their ability to produce antibodies and will accordingly be cloned. As a result, each cell clone secretes a specific antibody, the monoclonal antibody against the antigen X. (Modified from Vollmar, A., Dingermann, T. 2005. *Immunologie*. Wissenschaftliche Verlagsgesellschaft mbH, Stuttgart, with permission).

first licensed human mAb was adalimumab (Humira®), which is directed against tumor necrosis factor-α (TNF-α). Adalimumab received marketing approval by the U.S. Food and Drug Administration (FDA) in December 2002. Despite its nature as being completely human, with some *in-vitro* modification after production to increase the affinity to the target, reports have been made describing the occurrence of HAHAs.

Examples of the four different types of mAbs, together with the year of their first approval and manufacturer, are listed in Table 3.3. The different types of mAbs, together with the appropriate suffixes in the compound names, their murine fraction and their degree of immunogenicity, are shown in Fig. 3.3.

**Table 3.3** An overview of monoclonal antibodies.

| Class of antibody | Suffix (or prefix) | Example compound | Approval year | Manufacturer |
|---|---|---|---|---|
| Murine | 'muro-...ab' | Muromonab-CD3 | 1986 (first approval) | Ortho Biotech |
| Chimeric | '-ximab' | Rituximab | 1997 | Biogen Idec/ Genentech |
| Humanized | '-xumab, -zumab' | Alemtuzumab | 2001 | Genzyme |
| Human | '-mumab' | Adalimumab | 2002 | Abbott |

#### 3.4.2.5 **Further Species of mAbs**

In addition to these four types of mAbs, several further species of mAbs have been developed by new engineering technologies, predominantly during the past decade. The production of these mAbs is based on the ability to generate human (or at least human-like) binding sites from a number of starting positions and to link them genetically or chemically to a wide variety of effector elements (see Section 3.5).

##### 3.4.2.5.1 **Primatized mAbs**

These antibodies are genetically engineered from cynomolgus macaque monkey and human components, and are structurally indistinguishable from human antibodies. They may, therefore, be less likely to cause adverse reactions in humans compared to humanized mAbs, making them potentially suitable for long-term treatment in chronic diseases [4]. In December 2004, there was a notice that lumiliximab, a primatized anti-CD23 macaque/human chimeric antibody that inhibits the production of IgE antibodies, was under development for the treatment of allergic conditions (Biogen Idec).

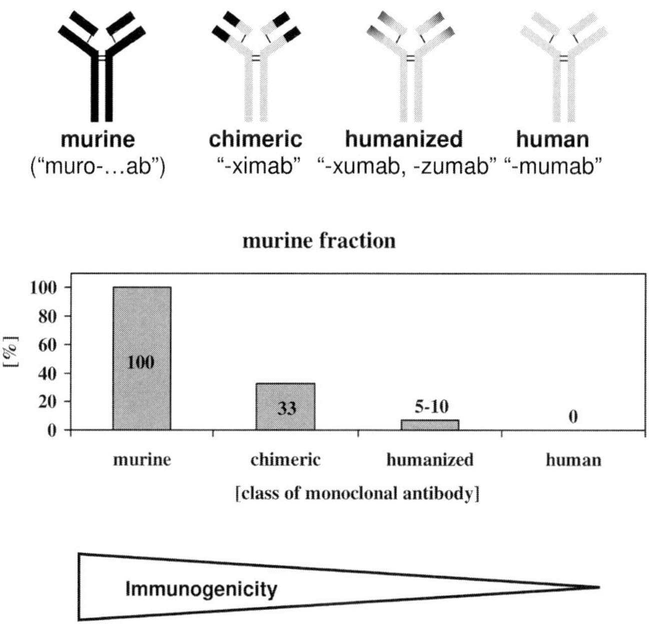


**Fig. 3.3** Classes of monoclonal antibodies. Upper panel: The different classes of monoclonal antibodies (murine, chimeric, humanized and human, with typical suffixes in the compound names). Lower panel: Murine fractions of different classes are illustrated, with declining immunogenicity of the decreasing murine fractions.

#### 3.4.2.5.2 **Bispecific mAbs**

Bispecificity means that one binding part of the mAb is responsible for specific targeting (for example, a structure solely available on pathogenic cells), and the other part for binding to a specific receptor on immune cells to enhance or recruit additional effector mechanisms. In other bispecific antibodies the second binding site can also be specific for toxins such as saponin to enhance killing of the targeted cell [5]. Bispecific antibodies can be produced in three different ways: (1) by linkage of two Fab regions (chemically); (2) by linkage of two Fab regions via leucine zippers (genetically); or (3) by fusion of two hybridoma cells resulting in a hybrid-hybridoma (also called a quadroma).

#### 3.4.2.5.3 **Monoclonal Intrabodies**

One very new and powerful area of research is that of "monoclonal intrabodies" (also called intracellular antibodies), which represent a new class of targeting molecules with potential use in gene therapy. Due to their totally different targets, namely inside a cell, and thus different effector functions, they may have a high potential in the treatment of human diseases [6].

The engineered single chain antibodies consist of one heavy chain (only variable domain), which is linked via a peptide chain to one light chain (also only variable domain). Therefore, intrabodies mainly occur in the single chain fragment (scFv) format engineered by conjugated neighboring genes encoding the heavy chain and light chain variable regions. Yet, the affinity of the parent antibody is still preserved. Thus, despite the smaller size of the polypeptide (1/5 to 1/6) with a molecular mass of ~30 kDa, monoclonal intrabodies maintain the complete recognition ability but not always total functional ability [7].

Monoclonal intrabodies show certain advantages over intact immunoglobulin molecules, especially with respect to pharmacokinetic properties. Recognition of these antibody fragments by the human immune system should also be minimized. In addition, most scFv antibodies can nowadays be expressed in prokaryotic systems in a convenient procedure. Due to the high specificity and affinity that can be achieved in intrabodies, they represent very effective biologicals against intracellular or viral targets. This principle has already been used for targets such as structural, regulatory and enzymatic proteins of the human immunodeficiency virus (HIV-1). However, there are many significant problems to overcome before intrabodies will actually be used in the treatment of human diseases such as HIV or cancer.

In an attempt to increase efficacy in therapy, the production of bivalent or multivalent species by linking subunits (non-covalently, covalently, or by disulfide bonds) has been proposed. For example, a bispecific single-chain antibody fragment has been investigated for malaria therapy. This is a combination of two scFv, one directed against the CD3 molecule on human T lymphocytes, and the other against an epitope of a surface protein of *Plasmodium falciparum*, the parasite which causes malaria [8].

In summary, only a few therapeutic polyclonal antibodies are still marketed today. The main advantages of mAbs are their high specificity towards the target and the capability of an unlimited production of these homogeneous biological molecules. In future, new antibody or antibody-derived pharmaceuticals will be developed and can be expected to have favorable efficacy, a lack of immunogenicity, and appropriate pharmacokinetics such that they can be used to treat intracellular medical disorders.

## 3.5 Effector Functions and Modes of Action of Antibodies

Antibodies display several different effector functions and modes of action as part of their function in the human immune system.

### 3.5.1 Biological Effector Functions of mAbs

The following list provides an overview of the variety of possible defense mechanisms of the immune system after an antigen–antibody reaction has taken place:

- Antibody-dependent cellular cytotoxicity by natural killer (NK) cells.
- Complement-dependent cytotoxicity (CDC).
- Neutralization of exotoxins and viruses.
- Prevention of bacterial adherence to host cells.
- Membrane attack complex (MAC) resulting in cytolysis.
- Agglutination of microorganisms.
- Immobilization of bacteria and protozoa.
- Opsonization.

The Fc and Fab regions of the antibodies are responsible for different mechanisms in the immune response. Fc enhances CDC, ADCC and opsonization (phagocytosis and cytolysis). Independent from the Fc region are the following direct mechanisms of therapeutic mAbs: blockage of ligands or ligand interaction via receptor modulation and subsequent signal transduction.

### 3.5.2 Modes of Action of mAbs

In the following, the four most important and best-understood effector functions/modes of action of therapeutic mAbs or antibody-derived products (e. g., antibody fragments) will be discussed, and will focus on cancer therapy. The different modes of action of mAbs in cancer therapy are depicted in Fig. 3.4. As this topic is not entirely within the scope of this chapter, the reader is referred to textbooks on immunology for more details on the modes of action of mAbs.

#### 3.5.2.1 Antibody-Dependent Cellular Cytotoxicity (ADCC)

One effector function/mode of action of mAbs is the so-called "antibody-dependent cellular cytotoxicity". mAbs are able to activate this very important function of the human immune system. After several steps, immune cells will finally kill target cells, for example cancer cells. In general, ADCC is induced by unconjugated (also called "naked") mAbs or antibody fragments. They can either induce apoptosis, negative growth signals or indirectly activate host defense mechanisms [5].

The chronological sequence of ADCC is as follows [9]:
- The Fab region of the mAb reacts with its antigen on the cell surface.
- Fc receptors on host NK cells and monocytes are activated by the antigen–antibody interaction and are bound to the Fc region of the mAbs.
- The resulting complex triggers a cytolytic response and induces apoptosis.
- In addition, macrophages are able to opsonize antibody-bound cells.

With the modification of mAbs using a new technology (the Potelligent™ Technology), binding affinity for Fc receptors was reported to be enhanced and thus ADCC activity increased. As a result, this type of mAb would be expected to achieve a higher tumor cell killing activity than conventional antibodies.

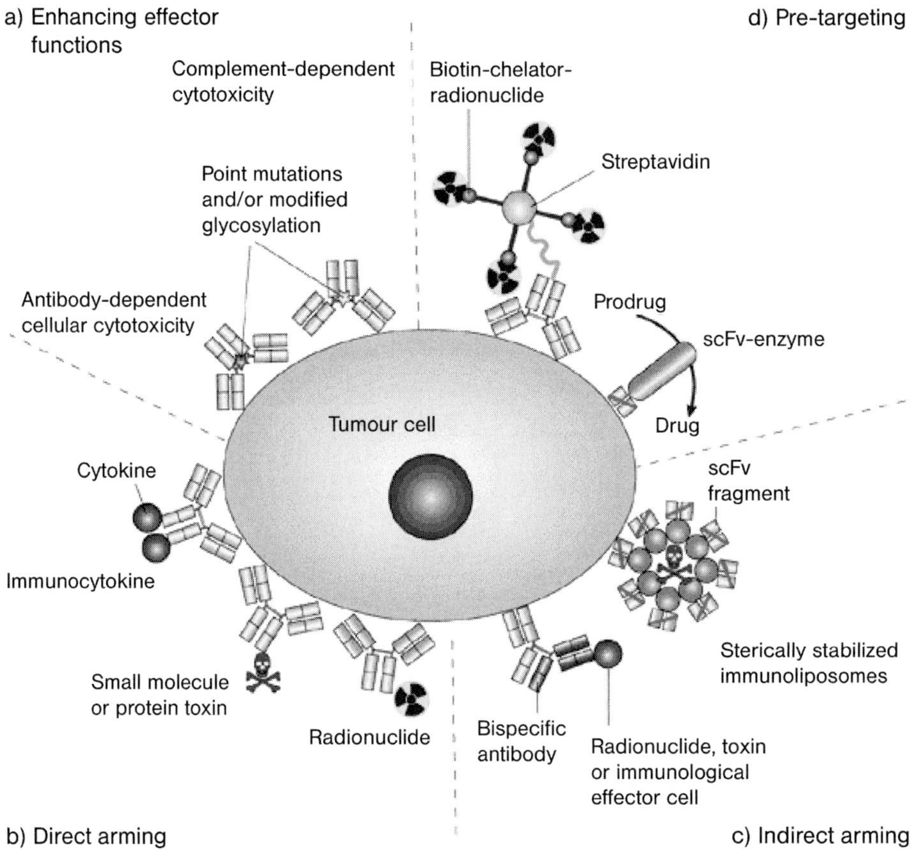


**Fig. 3.4** Modes of action of monoclonal antibodies in cancer therapy. scFv: single-chain Fv fragment. (Reproduced from Carter, P. 2001. Improving the efficacy of antibody-based cancer therapies. *Nat. Rev. Cancer* 1 : 118–129).

#### 3.5.2.2 **Complement-Dependent Cytotoxicity**

Another effector function/mode of action of mAbs is the so-called "complement-dependent cytotoxicity" or "complement-mediated cell death". Activation of the complement system can lead to lysis of the antigen-presenting cell, or can induce inflammation reactions aimed at eliminating these cells efficiently. The successive steps of the complement activation can be summarized in a simplified way [9]:

- The glycoprotein C1q of the complement system binds to the Fc regions of two antibodies which must have previously formed a Fc dimer.
- The glycoprotein iC3 b is produced.
- The cytolytic membrane attack complex (MAC) is formed.
- In addition, ADCC is enhanced by binding of iC3 b to complement receptor 3 on the surface of immune cells.

It must be noted that not all mAbs of the IgG1 or IgG3 isotypes are capable of activating CDC, most likely because in some cases the density of the antigen may be too low to support the formation of Fc dimers.

Clearly, one prerequisite for ADCC or CDC is an intact immune system. Hence, unconjugated mAbs are unsuitable for the treatment of immunocompromised patients. In contrast to classical cytotoxic agents, tumor cells will also be attacked in the $G_0$ phase by the antibody-induced ADCC and CDC mechanisms, this being a major advantage of mAbs in antineoplastic therapy.

#### 3.5.2.3 Blockage of Interaction between (Patho)Physiological Substance and Antigen

Apart from the recruitment of the body's own effector mechanism of the immune system, "naked" mAbs or antibody fragments show further diverse mechanisms of action. As the antigen often represents a receptor or an enzyme, the interaction with its (patho)physiological ligand or substrate is blocked by the antigen–antibody interaction. By this mechanism, therapeutic mAbs are able to disrupt (or at least perturb) downstream reactions that may be followed by down-regulation of the antigen, modification of cell growth, inhibition of angiogenesis, or induction of apoptosis. For example, the humanized "naked" mAb matuzumab which is directed against the epidermal growth factor receptor (EGFR) shows promising results in clinical studies. The mode of action is neither ADCC nor CDC, but rather blocking the binding of ligands such as EGF to the receptor, thereby impairing downstream signaling [10].

#### 3.5.2.4 Conjugated Unlabeled mAbs

Conjugated mAbs combine the specific targeting of a mAb with another beneficial principle within one substance. For oncologic applications, for example, mAbs are conjugated with a cytotoxic drug. The conjugated therapeutic mAbs selectively bind to the targeted cells with the antigen on the surface (e.g., tumor cells), and the conjugated cytotoxic substance is only delivered to these cells. The major advantage of this "targeted therapy" lies in a reduced systemic toxicity compared to conventional therapy with a cytotoxic agent.

In clinical studies, conjugates with methotrexate, doxorubicin, derivatives of ricin, and *Staphylococcus* enterotoxin A or immunomodulatory cytokines have been investigated [11]. It has also been shown that it is favorable for therapy to conjugate a cytotoxin such as *Pseudomonas* exotoxin to a mAb which is targeted against a specific antigen at the surface of tumor cells [11]. In 2000, the FDA approved gemtuzumab ozogamicin (Mylotarg®), a calicheamicin derivative conjugated to a mAb for the treatment of CD33+ acute myeloid leukemia (AML).

#### 3.5.2.5 Radioactively Labeled mAbs

Another mode of action is the targeted delivery of a defined dose of radiation by a mAb, which is used almost exclusively in tumor therapy. One common radiation source used is $^{131}I$ (iodine-131 isotope), which emits beta radiation. Other beta-emitting nuclides include $^{90}Y$, $^{77}Lu$, $^{67}Cu$, $^{186}Re$, and $^{188}Re$. Alpha-emitting nuclides such as $^{212}Bi$, $^{213}Bi$ and $^{211}At$ are also of interest because of their high-level radiation (>6000 keV). Due to their small range of activity they act very effectively, and with particular selectivity at the tumor target.

The isotope $^{131}I$ has several advantages: it is readily available, cost-effective, chemically easily bound to proteins, and sufficiently stable with a radioactive half-life of approximately 8 days. Furthermore, the concomitantly emitted gamma radiation can be used for imaging methods, although it may present safety concerns to both the patient and the environment. Radiation hazards from undirected high-level gamma radiation might occur, depending on the $^{131}I$-activity administered. In the USA, the Nuclear Regulatory Commission's guideline on the release of patients undergoing radioimmunotherapy with $^{131}I$ permits the use of this therapy in an outpatient setting.

Compared to $^{131}I$ (610 keV), the beta-emitting nuclide $^{90}Y$ (2280 keV) has a higher radiation energy and a wider range of activity in tissue (12 mm versus 2 mm, respectively); this results in a very good therapeutic response, especially in the irradiation of large tumors. In addition, there is also a difference in the stability of the radioisotope to the antibody that influences the pharmacokinetics and pharmacodynamics of these compounds. The degradation of $^{131}I$ is more rapid than that of $^{90}Y$, which leads to a longer residence time at the target site and delivery of a larger radiation dose of $^{90}Y$ compared to $^{131}I$. Since $^{90}Y$ is a pure beta-emitter it is not possible to observe the distribution of the labeled mAb in the human body by imaging techniques and thus to ensure successful tumor targeting. In pharmacotherapy with ibritumomab tiuxetan (Zevalin), this deficiency is overcome by using the following approach. Prior to the start of therapy with $^{90}Y$-labeled ibritumomab tiuxetan, a low dose of the $^{111}In$-labeled molecule is administered. In this way, the tumor-specific binding of ibritumomab can be visualized by imaging methods (gamma-emitting $^{111}In$) before regular treatment with the $^{90}Y$-labeled mAb is started.

## 3.6 Prerequisites for mAb Therapy

The successful application of mAbs in the treatment of diseases means that several requirements need to be fulfilled, including four major components: the patient; the antibody; the target cell; and the antigen.

### 3.6.1 The Patient

For unconjugated mAbs in particular, it is a fundamental necessity for successful treatment that the immune system of the patient is completely intact. In particular, ADCC and CDC must operate efficiently as the primary effector functions.

### 3.6.2
### The Antibody

For successful therapy, high specificity and high affinity of the mAb towards the complementary antigen are needed, so that only the targeted site is affected by the therapy. Cross-reactivity – that is, the reaction of the therapeutic antibody with a second or more antigens – must be limited to a minimum. Early attempts at applying mAbs to patients suffered major drawbacks due to the development of HAMA (see Section 3.4.2.1). In this case, murine mAbs are rapidly blocked in their binding abilities to the specific antigen and removed from the circulation. Thus, high serum stability is now considered a prerequisite for the development of new mAbs. Current and future developments aim at humanizing chimeric mAbs or developing fully human mAbs if multiple dosing in chronic treatment is needed [9].

### 3.6.3
### The Target Cell

The target of a mAb must be as "unique" as possible for the disease to be treated. If the target resides on the surface of a cell, a high expression rate is important. In these cases, or if the target is a soluble substance such as TNF-$\alpha$, the antigen must be selectively expressed, or at least overexpressed at the target site. In cancer therapy the tumor-specific antigen should be (over)expressed only on the malignant cells/tissue, but not on benign or healthy cells/tissue. An example is the fibroblast activation protein (FAP) of activated tumor stroma fibroblasts. This antigen is a 95-kDa glycoprotein with proteolytic activity enhancing tumor cell invasion and metastasis as well as angiogenesis. Activated tumor stromal fibroblasts consistently induce the formation of this cell-surface molecule. It is expressed in >90% of malignant epithelial tumors, but rarely occurs in healthy tissue. Thus, FAP of activated tumor stroma fibroblasts presents a promising antigenic target in carcinoma patients [9, 12]. If the expression of the antigen is not exclusive in the target cells/tissue, adverse reactions are more likely. In some cases, the antigen is only (over)expressed in a limited number of patients. The humanized mAb trastuzumab for example is directed against a receptor (her-2/neu) that is only overexpressed in 20–30% of the entire patient population. Hence, treatment is only successful and thus adequate in the antigen-positive patients.

### 3.6.4
### The Antigen

Another requirement for successful mAb therapy concerns the antigen itself. It is essential that the antigen presents a homogeneous target population and does not show heterogeneity. Hence, all antigen molecules should be recognized as the target by the antibody. This requirement is very often fulfilled and is highly advantageous in oncology as malignant cells themselves are often heterogeneous and display genetic variability. Mutations in the target structure are highly undesirable.

## 3.7
## Issues in the Bioanalysis of Antibodies

Pharmacokinetic data analysis requires determination of the analyte in various body fluids. In the case of therapeutic antibodies, serum is the most common matrix to be analyzed. For a critical interpretation of pharmacokinetic data the chosen bioanalytical methods must be considered. The most frequently used for mAbs include enzyme-linked immunosorbent assay (ELISA), capillary electrophoresis (CE)/polyacrylamide gel electrophoresis (PAGE), fluorescence-activated cell sorting (FACS), and surface plasmon resonance (SPR). The challenges and limitations of bioanalytical methods used for the analysis of mAb concentrations are discussed in detail in Chapter 6.

A common analytical problem in immunoassays is that of cross-reactivity, which is the unspecific binding of the therapeutic antibodies determined in the assay with a second (or more) antigen(s). These interfering bindings must be kept to a minimum. Among other factors, specificity of the enzyme-labeled anti-Fc and anti-Fab antibodies used for detection is of major importance. In general, each assay design bears its special drawbacks that must be carefully evaluated for its intended use.

Another challenge may occur from the samples to be analyzed. If human anti-idiotype antibodies, HAxAs (HAMAs, HACAs or HAHAs) against the therapeutic mAb have been developed in the patient, almost all assays may unfortunately give unreliable results of the concentration of therapeutic antibody. If the HAxAs bind to the Fc region of the therapeutic antibody and do not interfere with the antigen or with the detection antibody in the respective ELISA set-up, it should not matter if HAxAs are present. However, if the HAxAs bind to the Fab region and interfere with the antigen binding, the results will be useless because the measured values do not correlate with the actual concentration of the therapeutic antibody.

Typically, the relationship in the assay between the detection signal and the concentration of the mAb is nonlinear. A simple regression analysis is, therefore, not possible. Usually, a third-to-fifth degree polynomial is used to relate the detection signal to the concentration. Because of this relationship, a typical sample dilution is not possible and careful evaluations are necessary if the concentration of the antibody is above the upper limit of quantification of the assay. As a consequence, the assay should ideally cover the whole concentration range of all samples measured.

Compared to other bioanalytical methods such as high-performance liquid chromatography (HPLC), the methods used to quantitate mAbs often display less precision and a higher between-day variability. In choosing a bioanalytical method it must also be considered that some assays measure the unbound fraction, the bound fraction, or both. When using FACS, only the fraction of the therapeutic antibody that is bound to its antigen on the cells is counted. In contrast, ELISA measures only the unbound fraction in serum that can react with the offered antigen.

In summary, because of the differences between assays, careful consideration of which assay has been applied is highly recommended when evaluating the concentration data of mAbs in pharmacokinetic analyses.

## 3.8 Catabolism of Antibodies

Before discussing the pharmacokinetic characteristics of therapeutic mAbs, the present section will focus on the catabolism of physiological antibodies. This process is also highly relevant for therapeutic mAbs.

The rate of catabolic elimination of intact antibodies is variable and depends on the immunoglobulin class/isotype. The catabolic rate constant and resulting half-lives were determined for the four different subclasses of IgG: IgG1, IgG2 and IgG4, with 93% of the total IgG immunoglobulins, show a very long half-life of 18–21 days. The other subclass, IgG3, and the two larger classes, IgM and IgA, have substantially shorter half-lives of 7, 5 and 5–6 days, respectively.

### 3.8.1 Proteolytic Degradation

Although the mechanisms involved are not yet fully understood, one elimination process of IgG as proteins is a slow proteolytic degradation; this takes place predominantly in hepatic and reticuloendothelial cells.

### 3.8.2 Neonatal Fc Receptor (Fc-Rn)

One factor responsible for the extraordinarily long half-lives of the three IgG subclasses was identified a few years ago [13], after having been postulated several decades earlier by Brambell et al. [14]. In the regulation of IgG homeostasis, these authors proposed a protective role of a "factor" to which immunoglobulins were bound. After isolation and cloning of the "factor", experiments revealed that the "factor" not only influenced catabolism/elimination of antibodies but also mediated their absorption in the gastrointestinal tract of neonates (see Section 3.9.1). Due to this role, it was named the Fc receptor of neonates (Fc-Rn). Further investigations showed that Fc-Rn only binds IgG class antibodies, since in Fc-Rn knockout mice the catabolism rate constant of IgG was significantly increased (10- to 13-fold) whereas those of the other immunoglobulins remained unchanged.

The protective function against catabolism is illustrated in Fig. 3.5. The IgG enters a cell by receptor-mediated endocytosis (pinocytosis), and is bound to Fc-Rn. As the binding sites are limited at physiological concentrations, not all IgG molecules will be bound to the membrane-associated receptor by the Fc region. After cellular uptake, the intracellular vesicle (phagosome) fuses with a lysosome.

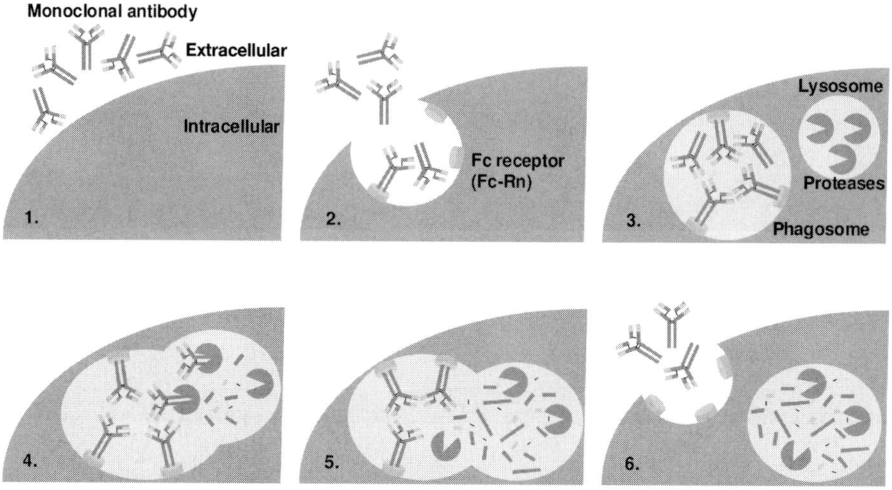


**Fig. 3.5** Schematic flow-chart of antibody degradation by proteases combined with the protection by Fc-Rn in phagosomes. (1) IgG molecules are transported to the extracellular space (= interstitial fluid). (2) IgG molecules enter a cell by receptor-mediated endocytosis where the Fc region of the antibody is bound to the limited binding sites of Fc-Rn. (3) The intracellular vesicle (phagosome) fuses with a lysosome containing proteases. (4) Proteases degrade the unbound antibody molecules. (5) The degraded products reside in the lysosome part. (6) The intact IgG molecules bound to Fc-Rn are transported to the cell membrane where they can be returned to the extracellular space.

The resultant decrease in pH value leads to an increase in binding affinity to the protective Fc-Rn, and to degradation by proteases of unbound antibody molecules. The degradation products will reside in the lysosome, while the intact IgG molecules bound to Fc-Rn will be transported to the cell membrane and can be returned intact to the extracellular space. The Fc-Rn can transport the bound antibody bidirectionally to the apical and basolateral membrane, delivering it to the interstitial fluid or the systemic circulation. As a result, the residence time of the intact antibody in the body is prolonged.

Further experiments revealed that the higher the serum IgG concentration, the shorter the half-life (Fig. 3.6). At lower concentrations, IgG is bound to phagosome membrane-associated receptors and protected from proteolysis (Fig. 3.7, left column). At high concentrations, however, phagosome Fc receptors are saturated much more rapidly, and more unbound IgG molecules are digested; this results in a higher apparent catabolic rate and hence a shorter half-life (Fig. 3.7, right column). Therefore, the protective role of Fc-Rn on IgG homeostasis fits very well with the hypothesis by Brambell et al. The impact of serum IgG concentrations on IgG catabolism was demonstrated in several animal studies. The administration of a high dose of purified human IgG (~2 g/kg) to mice resulted in a large decrease (>60%) in baseline mouse IgG1 and IgG3 serum concentrations. Similar results were reported for the effect of a large dose of human IgG (2 g/kg) in rats.

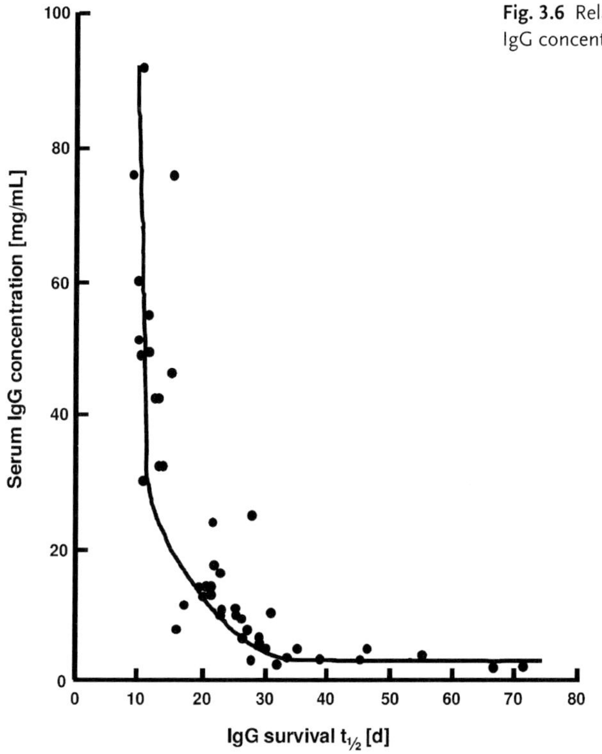


**Fig. 3.6** Relationship between serum IgG concentration and its half-life.

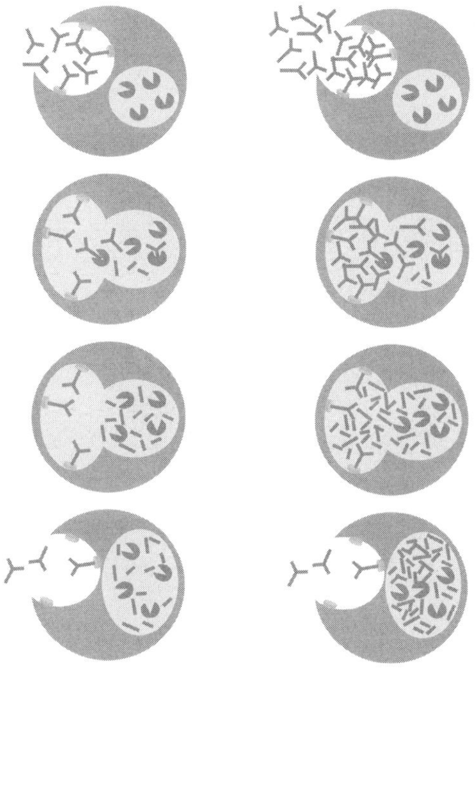

**Fig. 3.7** Protective influence of Fc-Rn at low (left column) and high (right column) IgG concentrations (not IgG3).

## 3.9 Pharmacokinetic Characteristics of mAbs

Compared to small-molecule drugs, therapeutic mAbs display different pharmacokinetic characteristics, including nonlinear pharmacokinetic behavior. As the majority of therapeutic mAbs present IgG (or more specially IgG1) molecules, the emphasis will be placed on this isotype, although the characteristics of newer types of molecule such as antibody fragments will also be included.

### 3.9.1 Absorption

Due to their high molecular mass (and other reasons), the vast majority of mAbs that have been approved or are currently in clinical development are administered by intravenous (IV) infusion. This route allows the total dose to be available in the circulation, as F (the systemically available fraction of the dose) is, by definition, 1. In consequence, maximum concentrations in serum are rapidly observed, and are higher compared to those achieved by other routes. Therefore, adverse reactions after IV administration occur more often but are generally reversible. In addition, IV infusions represent the most inconvenient (they often require hospitalization) as well as time- and cost-consuming means of administration. Consequently, extravascular routes have been chosen as alternatives, including subcutaneous administration (SC; e.g., adalimumab, efalizumab) and intramuscular administration (IM; e.g., palivizumab) (Table 3.4).

**Table 3.4** Approved dosing regimens and pharmacokinetic parameters related to absorption/bioavailability for therapeutic monoclonal antibodies.

| Monoclonal antibody | Dose[a] | Dose unit | Dosing interval [h] | Route of administration | $C_{max}$[b] [mg/L] | $t_{max}$[c] [h] | F[d] [%] |
|---|---|---|---|---|---|---|---|
| Abciximab | 0.25/ 0.125 | mg/kg µg/kg | Individual | IV bolus/ IV infusion | n.a. | 0.17 | 100 |
| Adalimumab | 40 | mg | 336 | SC | 4.7 ± 1.6 | 131 ± 56 | 64 |
| Alefacept | 7.5/15[e] | mg | 168 | IV/IM | 0.96 ± 0.26/ 0.36 ± 0.19 | 2.8 ± 1.9/ 86 ± 60 | 100/ 63 |
| Alemtuzumab | 3/10/30 | mg | 24[f] | IV | 0.5–8[g] | 2 | 100 |
| Basiliximab | 20 | mg | 96 | IV | 7.1 ± 5.1 | 0.5 | 100 |
| Bevacizumab | 5 | mg/kg | 336 | IV | n.a. | n.a. | 100 |
| Cetuximab | 400/250 | $mg/m^2$ | 168 | IV | 185 ± 55 | 2 | 100 |
| Daclizumab | 1 | mg/kg | 336 | IV | 32 ± 22 | 0.25 | 100 |

**Table 3.4** (continued)

| Monoclonal antibody | Dose[a)] | Dose unit | Dosing interval [h] | Route of administration | $C_{max}$[b)] [mg/L] | $t_{max}$[c)] [h] | F[d)] [%] |
|---|---|---|---|---|---|---|---|
| Efalizumab | 0.7/1.0 | mg/kg | 168 | SC | 12 ± 8 | 24–48 | 50 |
| Etanercept | 25 | mg | 84 | SC | 1.7 ± 0.7 | 48 | 76 |
| Gemtuzumab ozogamicin | 9 | mg/m$^2$ | h) | IV | 3.1 ± 2.1 | n.a. | 100 |
| Ibritumomab tiuxetan | Individual | mCi/kg | i) | IV | n.a. | 0.4 | 100 |
| Infliximab | 3 | mg/kg | 336/672/ 1344[j)] | IV | 77 | 2 | 100 |
| Muromonab | 5 | mg | 24 | IV | 0.9[k)] | <0.01 | 100 |
| Omalizumab | 150–375 | mg | l) | SC | n.a. | n.a. | 62 |
| Palivizumab | 15 | mg/kg | 672 | IM | 40 | 720 | n.a. |
| Rituximab | 375 | mg/m$^2$ | 168 | IV | 206 ± 60 | m) | 100 |
| Tositumomab | n) | mg | n) | IV | n.a. | n.a. | 100 |
| Trastuzumab | 4/2 | mg/kg | 168/504 | IV | 110 | 1.5 | 100 |

n.a.: not available, not applicable.

**a)** If initial dosing regimen differs from maintenance dosing regimen, the loading dose and dosing interval will be listed first, followed by the maintenance dose and dosing interval.
**b)** Maximum concentration after single dose.
**c)** Time of maximum concentration after single dose.
**d)** Systemically available fraction (bioavailability) after single dose.
**e)** 7.5 mg (IV) or 15 mg (IM) once weekly.
**f)** First three days daily (1$^{st}$ day: 3 mg, 2$^{nd}$ day: 10 mg, 3$^{rd}$ day: 30 mg); afterwards twice-thrice weekly (30 mg).
**g)** After first 30 mg dose (= third dose).
**h)** Second dose after 2 weeks.
**i)** Day 1: 250 mg/m$^2$ rituximab; day 8: rituximab 250 mg/m$^2$, then 0.3–0.4 mCi/kg (max. 32 mCi) ibritumomab-tiuxetan.
**j)** Second dose after 2 weeks, 3$^{rd}$ dose 4 weeks later, the following doses with an interval of 8 weeks.
**k)** After the 3$^{rd}$ dose (not after single dose).
**l)** Every 2 or 4 weeks.
**m)** After end of infusion with a rate of 50 mg/h for the first 30 min; thereafter an increase by 50 mg/h is possible, with a maximum rate of 400 mg/h.
**n)** Two steps with sequential infusions each: 1. Dosimetric step: 2 × 225 mg + 35 mg tositumomab + $^{131}$I tositumomab. 2. Therapeutic step (7–14 days later): same amounts, but higher radioactivity.

The mechanism of absorption after SC or IM administration is thought to occur via the lymphatic system. The mAbs enter the lymphatic system by convective flow of interstitial fluid into the porous lymphatic vessels. The molecular mass cut-off of these pores is >100-fold the molecular mass of mAbs. From the lymphatic vessels, the mAbs are transported unidirectionally into the venous system. As the flow rate of the lymphatic system is relatively low, mAbs are absorbed over a long time period after administration. The resulting time of maximum concentration ($t_{max}$) is much later (typically 1–8 days), and the systemically available fraction (F) is equal or lower (typically 0.5–1.0) compared to the IV administration of mAbs. For example, SC injection of 40 mg adalimumab results in a $t_{max}$ of approximately 5 days, and F is approximately 64%.

Both routes, however, have one major limitation – that the injection volume is limited. The upper limit of volume not causing any discomfort in patients (mainly pain) is 2.0–2.5 mL for SC injection, and 4.5–5.0 mL for IM injection. With respect to the solubility and doses of the mAbs that are to be given, these small injection volumes very often present a problem. The solubility of IgG usually is in the range of 100 mg/mL, so that the maximum SC or IM doses possible range from 200 to 250 mg, or 450 to 500 mg, respectively.

In addition to these three parenteral routes, oral administration has been investigated and applied, albeit to a much smaller extent. The reasons for this become evident from the chemical structure of classical mAbs:

- A high molecular mass of ~150–1000 kDa and high polarity prevent, or at least disfavor, permeation by diffusion across the epithelial cells along the gastrointestinal tract.
- The acidic environment of the stomach encourages denaturation of the glycoprotein mAbs.
- As glycoproteins, mAbs are substrates of proteases in the gastrointestinal tract. Since the number of proteases in the gastrointestinal tract is limited, proteolytic enzyme degradation is saturable, leading to less degradation if higher doses are administered.

It has been shown that antibodies can reach the systemic circulation after oral administration, but only to a very small extent. The antibodies pass the intestinal epithelium not by passive transcellular but by receptor-mediated transcellular or paracellular transport. The Fc part of the antibody is responsible for the saturable receptor-mediated transport, especially IgG in breast-fed neonates. As the receptor is found primarily in the gastrointestinal tract of neonates, it was called Fc-Rn (Fc receptor neonatal). Apart from this location, Fc-Rn has also been discovered in other tissues such as the liver. Its role will be further discussed in Section 3.9.3.

The pharmacokinetic parameters of approved mAbs with regard to absorption, maximum concentration ($C_{max}$) and $t_{max}$ after single dose administration, systemically available fractions (F), and the commonly used dosing regimens and routes of administration are summarized in Table 3.4.

### 3.9.2 Distribution

In general, the distribution of classical mAbs in the body is poor. Limiting factors are, in particular, the high molecular mass and the hydrophilicity/polarity of the molecules. Nevertheless, mAbs are able to reach targets outside the systemic circulation. Distribution throughout the body has been visualized for conjugated mAbs using imaging techniques such as single photon emission computed tomography (SPECT).

#### 3.9.2.1 Transport

Permeation of mAbs across the cells or tissues is accomplished by transcellular or paracellular transport, involving the processes of diffusion, convection, and cellular uptake. Due to their physico-chemical properties, the extent of passive diffusion of classical mAbs across cell membranes in transcellular transport is minimal. Convection as the transport of molecules within a fluid movement is the major means of paracellular passage. The driving forces of the moving fluid containing mAbs from: (1) the blood to the interstitial space of tissue; or (2) the interstitial space to the blood via the lymphatic system, are gradients in hydrostatic pressure and/or osmotic pressure. In addition, the size and nature of the paracellular pores determine the rate and extent of paracellular transport. The pores of the lymphatic system are larger than those in the vascular endothelium. Convection is also affected by tortuosity, which is a measure of hindrance posed to the diffusion process, and defined as the additional distance a molecule must travel in a particular human fluid (i. e., *in vivo*) compared to an aqueous solution (i. e., *in vitro*).

Cellular uptake of mAbs takes place via endocytosis and can be either receptor-mediated, or non-receptor-mediated. Endocytosis is an absorptive process of large and polar molecules such as mAbs, and involves the formation of intracellular vesicles from parts of the cell membrane. The process can be divided in three different subtypes:

- Phagocytosis (literally "cell eating"): the cell membrane is folded around large molecules outside the cell such as mAbs to envelop them. A vesicle called a "phagosome" is formed as a subcellular compartment by internalization. The phagosome may merge with other intracellular vesicles such as lysosomes, which contain proteolytic enzymes. In this way, the included mAbs can be degraded.

- Pinocytosis (literally "cell drinking"): by invagination, the cell membrane forms a "pit" that is filled with interstitial fluid. All molecules (including mAbs) present in the interstitial fluid will also move into the pit, which becomes detached from the cell membrane to form an intracellular vesicle.

- Receptor-mediated endocytosis is a form of pinocytosis that is actively initiated. The trigger is the binding of a large extracellular molecule (e.g., a mAb) to a receptor on the surface of the cell membrane. Analogous to pinocytosis, a “pit” is formed around the bound antibody and subsequently internalized to form an intracellular vesicle.

In all three cases the contents of the intracellular vesicles will finally be released into the cytoplasm, or transported to the cell membrane to be released into the extravascular space. Receptor-mediated endocytosis is the predominant mechanism of cellular uptake for mAbs. As mentioned previously (see Sections 3.9.1 and 3.9.3), Fc-Rn is present in a large variety of cells and is very often involved in this process. In addition, the antigen–antibody complex via Fab can also undergo “receptor“-mediated endocytosis. The impact of this internalization process on the pharmacokinetics of mAbs will be discussed later.

#### 3.9.2.2 **Volume of Distribution**

Generally, the concentration of mAbs in the interstitial fluid is rather small, with reported serum:tissue ratios in biopsy samples ranging from 2 to 10. However, as this sampling technique normally uses homogenization during preparation, only a mixed concentration across all different cell types and fluids of the entire tissue is provided, and the ratios will most likely be an overestimation of the real situation.

Generally, the estimated volumes of distribution are small and relatively homogeneous (Table 3.5). mAbs initially distribute into a restricted central volume ($V_c$) of 3–5 L, which in humans approximates the serum volume. In order to evaluate the value in patients with respect to physiology, it must be kept in mind that, especially in tumor patients, the cellular fraction in blood and the hematocrit may be decreased and hence the serum volume will be increased.

The steady-state volume of distribution ($V_{SS}$, the sum of central and peripheral distribution volume) of approximately 5–10 L (0.06–0.2 L/kg) suggests a limited distribution outside serum, which is consistent with the behavior of endogenous IgG immunoglobulins. As expected for intact mAbs as macromolecules of ≥150 kDa, transport through the endothelial capillaries occurs only to a small extent. If a compartmental or physiological approach in pharmacokinetic modeling is chosen, the low values for the estimate of the peripheral volume and intercompartmental clearance, Q, also indicate that distribution is limited. A typical Q-value of 20–50 mL/h might be a good indicator for extravascular transport when compared to albumin (60 kDa protein), where 60% of the protein is located outside the serum. As can be expected for intact mAbs as macromolecules of 150 kDa or more, transport through the endothelial capillaries occurs to a smaller extent.

With the development of special types of mAbs such as monoclonal intrabodies, antibody fragments with considerably lower molecular mass (see Section 3.4.2.5), enhanced tissue penetration is expected. Studies conducted *in vitro* or in animals revealed that small-sized antibody fragments can penetrate tissues more easily, might potentially cross the blood–brain barrier, and can be delivered locally to the

**Table 3.5** Pharmacokinetic disposition parameters of therapeutic monoclonal antibodies.

| Monoclonal antibody | Volume of distribution ($V_{ss}$) [L] | Clearance (CL) [mL/h] | Half-life ($t_{1/2}$) [days] |
|---|---|---|---|
| Abciximab | 0.07[a)] | 183 ± 72[b)] | 30[c)] |
| Adalimumab | 5.0–6.0 | 11–15 | 14 |
| Alefacept | 0.094[a)] | 0.25[d)] | 11.25 |
| Alemtuzumab | 9 | n.a. | 8 |
| Basiliximab | 8.6 ± 4.1 | 75 | 7.2 ± 3.2 |
| Bevacizumab | 2.9 | 231 | 13–21 |
| Cetuximab | 4.0 | 35–140 | 2.9–4.2 |
| Daclizumab | 5.9 | 15 | 11.25–38.3 |
| Efalizumab | 0.110/0.058[e)] | 24 ± 19 | 5.5–10.5 |
| Etanercept | 7.6 | 120 | 2.9 |
| Gemtuzumab ozogamicin | 20 | 336 ± 51.5 | 66.2 ± 34.9 |
| Ibritumomab tiuxetan | n.a. | n.a. | 1.2 |
| Infliximab | 3.0–4.1 | n.a. | 8–9.5 |
| Muromonab | n.a. | n.a. | 0.75 |
| Omalizumab | 0.096 ± 0.035[a)] | 3.5 ± 1.7[f)] | 22 ± 8.7 |
| Palivizumab | n.a. | n.a. | 9.5–27 |
| Rituximab | n.a. | 38.2 ± 18.2 | 9.4 |
| Tositumomab | n.a. | 30.2–260.8[g)] | 1.17–4.79 |
| Trastuzumab | 2.95 | 16–41 | 2.85–10 |

n.a.: not available
**a)** L/kg
**b)** mL/min
**c)** min
**d)** mL/h/kg
**e)** 0.11 L/kg after administration of 0.03 mg/kg; 0.058 L/kg after administration of 10 mg/kg
**f)** mL/kg/day
**g)** mg/h (unit according to FDA Product Approval Documentation – Labeling document)

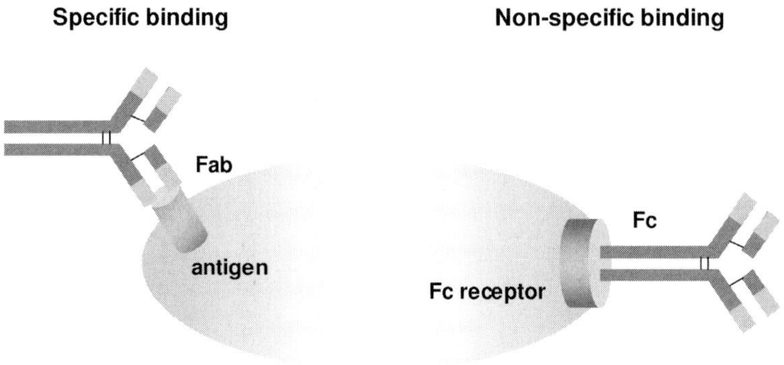


**Fig. 3.8** Two types of monoclonal antibody binding. Left: Fab-directed binding. Right: Fc-directed binding.

lung through inhalation. With a dimension of approximately 30 kDa, monoclonal intrabodies as single-chain Fv recombinant proteins show a higher extravascular distribution and thus, for example in oncologic applications, a better tumor penetration. Some mono- or bivalent Fab fragments have shown excellent tumor penetration and good efficacy in preclinical and clinical cancer trials, though the tumor residence time is much shorter than for complete mAbs.

#### 3.9.2.3 **Types of Binding**

The vast majority of mAbs undergoes almost irreversible binding, with affinity constants in the range of to $10^{10}$ to $10^{11}$ M. As binding is usually related to distribution, the two different types of binding that must be distinguished will be presented here. As shown in Fig. 3.8, the nonspecific binding occurs via the Fc region, and the resulting effector functions of the bound antibody have been described previously (see Section 3.5). Specific binding is mediated by the Fab region. Here, three different cases of complex equilibria must be considered, depending upon the type of antigen that is targeted:

(A) The antigen presents a soluble substance, such as TNF-$\alpha$ (e. g., infliximab).

(B) The antigen is located on the surface of a (activated) cell and presents a receptor or an enzyme. The antigen *is not shed* to the systemic circulation.

(C) The antigen has the same properties as in case B, but *is shed* to the systemic circulation, such as the her-2/neu receptor (e. g., trastuzumab).

The complexity of the competitive situation increases from case A to C (Fig. 3.9). In cases A and B, both the mAb and the physiological receptor or physiological ligand/enzyme merely compete for the same binding site, the antigen. In case C, however, the antigen occurs at different sites (on the cell surface and in the systemic circulation), so that there are in total four binding equilibria. The mAb and the physiological ligand/enzyme can bind to the soluble antigen as well as to the

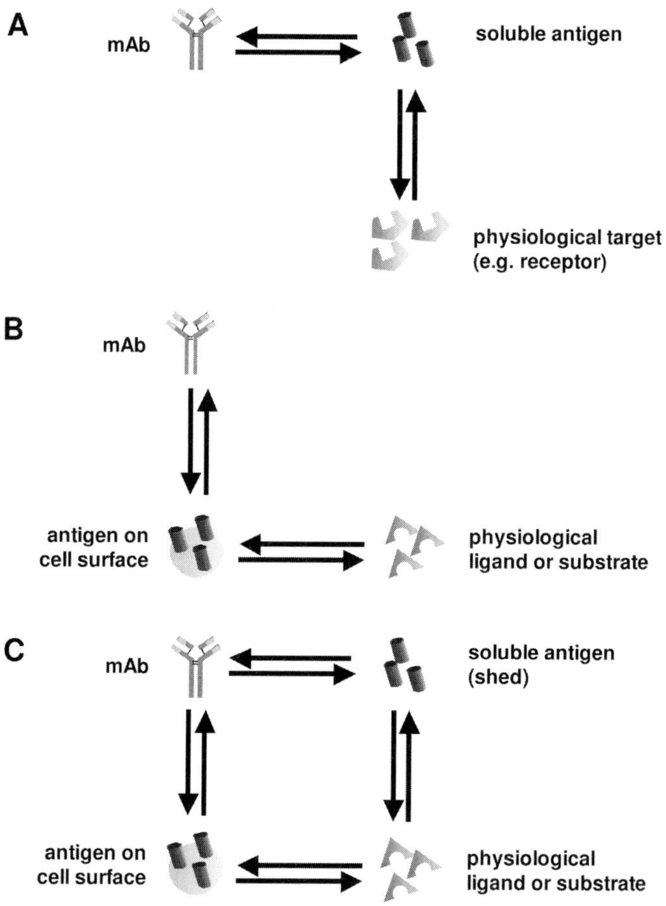


**Fig. 3.9** Binding equilibria for different types of antigen. (A) The antigen is a soluble substance, for example TNF-α (B) The antigen is a receptor or enzyme on a cell surface, but is *not* shed. (C) The antigen is a receptor or enzyme on a cell surface and is shed, for example her-2/neu or CD20.

antigen on the cell surface. The overall equilibrium depends on the concentrations of the mAb, the concentration of the physiological ligand/enzyme, the expression rate of the antigen on the cell surface, and the extent of shedding of the antigen. A high expression rate on (activated) cells results in a high density of the antigen that might also lead to issues of steric hindrance. In general, the mAb preferentially binds to soluble antigen, whereas the physiological ligand/enzyme favors the antigen on the cell surface.

## 3.9.3 Elimination

### 3.9.3.1 Clearance

As glomerular filtration has an approximate molecular size limit of 20–30 kDa, mAbs do not undergo filtration in the kidneys due to their relatively large size. The situation is different, however, for low molecular-mass antibody fragments, which can be filtered. Tubular secretion has not been reported to occur to any significant extent for mAbs, and peptides/small proteins are readily reabsorbed in the proximal or distal tubule of the nephron (potentially also mediated by the neonatal Fc receptor, Fc-Rn), or are even metabolized. Thus, renal elimination in total is uncommon or low for mAbs. Biliary excretion of mAbs has been reported only for IgA molecules, and only to a very small extent. Therefore, total clearance (CL) does usually not comprise renal or biliary clearance.

Across mAbs, clearance estimates range from about 11 to almost 400 mL/h (except for abciximab: ~11 000 mL/h), and display a much higher variability among mAbs compared to volume of distribution (see Table 3.5). A data trend towards nonlinearity in clearance can be found even in simpler approaches of modeling (see Section 3.10). Therapeutic mAbs show different elimination pathways, including catabolic proteolysis, irreversible binding to antigen, and irreversible binding by anti-idiotype antibodies.

### 3.9.3.2 Proteolysis

MAbs undergo proteolytic degradation similar to that of physiological immunoglobulins. The protective mechanism of Fc-Rn against catabolism is mainly responsible for the long terminal half-lives of the therapeutic antibodies if the IgG molecule is complete and does not belong to the IgG3 subclass. The half-lives of mAbs range from several days to weeks (Table 3.5). The shorter half-lives of the approved therapeutic mAbs compared to physiological immunoglobulin (IgG1) might be attributable to a less-efficient binding to the Fc-Rn receptor and lower protection against proteolysis (see Fig. 3.5). It has been reported that the different glycosylation pattern of therapeutic and physiological antibodies might contribute to this observation. The only approved Fab antibody fragment, abciximab, lacking the Fc part responsible for binding to Fc-Rn is characterized by an unusually short half-life of 30 min. Thus, the elimination half-lives of IgG antibody-derived therapeutics are directly related to the binding affinity to the Fc-Rn. Genetically modified mice that lack the expression of Fc-Rn receptors demonstrated antibody hypercatabolism and faster elimination. The development of etanercept, for example, as a fusion protein can be seen as a strategy to attach an "instable" protein to the Fc region for stabilization against proteolytic degradation. Attempts have also been made to increase the binding affinity of molecules to the Fc-Rn receptor, thereby enhancing the protective mechanism. So far, however, no licensed products have evolved.

Another reason for the different half-lives among mAbs lies in the origin of the therapeutic antibodies. Degradation will be more rapid for more nonhuman antibodies. There is a clear relationship between the human fraction of the mAb and the half-life. The lower the nonhuman fraction, the longer the half-life – that is, the order of half-life is murine (few days) < chimeric < humanized ≤ human (few weeks) mAbs. Fc-Rn has been reported to have high affinity to human and rabbit Fc regions, but low or no affinity to mouse, rat, or horse immunoglobulins.

Antibody fragments, such as the chimeric Fab fragment abciximab, and other fragments (in particular single-chain fragments under development) usually show a very short circulation period *in vivo*; this is advantageous in acute conditions, but limits the use for chronic diseases. By conjugation with polyethylene glycol (PEG) or additional glycosylation, however, it has been shown that proteolysis can be diminished for these compounds, which suggests that the half-life can be "tuned" as required to address acute as well as chronic diseases (for details, see Chapter 11).

#### 3.9.3.3 **Binding to Antigen**

Binding of mAbs not only affects distribution (similar to the situation for small-molecule drugs) but also reflects another means of elimination. Binding of the Fab region to the antigen with high affinity must be regarded as almost irreversible. The antigen–antibody complex, if located on the surface of a cell, will be internalized and subsequently degraded.

#### 3.9.3.4 **Binding to Anti-Idiotype Antibodies**

A third elimination pathway occurs if anti-idiotype antibodies are formed as an immune response of the human body to the administration of mAbs. Following repeated administration, anti-idiotype antibodies are usually observed after one to two weeks, with the extent of the adverse reaction strongly depending on several factors:

- Type of mAb: the lower the extent of "humanization", the more anti-idiotype antibodies are likely to be formed.
- Dosage regimen: single doses rarely evoke a strong immunological response, whereas multiple dosing often results in anti-idiotype antibody formation.
- Route of administration: SC injection has a higher incidence of forming anti-idiotype antibodies than IM or IV administration.
- Patient's genetics: patients with autoimmune diseases or prior anti-antibody production are more likely to produce an immune response.

The formed anti-idiotype antibodies almost irreversibly bind the therapeutic antibodies and therefore, by neutralization, eliminate them from the body. In total, the formation of anti-idiotype antibodies will alter the pharmacokinetics and consequently the pharmacodynamic effect (loss of effectiveness) of affected mAbs.

#### 3.9.3.5 Drug Interaction Studies

mAb drugs do not undergo classical metabolic reactions involving the superfamily of cytochrome P450 (CYP) isoenzymes. Hence, co-medication with CYP isoenzyme substrates, inducers or inhibitors is not expected to result in clinically relevant pharmacokinetic drug–drug interactions. The clinical development of mAbs therefore usually does not include *in-vitro* or *in-vivo* drug–drug interaction studies with CYP substrates, inducers or inhibitors.

### 3.9.4
### Comparison of Pharmacokinetics of mAbs and Traditional Small-Molecule Drugs

Compared to small-molecule drugs (as chemically defined entities), therapeutic mAbs (as biologics) display several different, partly unique pharmacokinetic characteristics; for a summary, see Table 3.6. The tissue penetration of small-molecule drugs is often much more extensive compared to that of the macromolecular mAbs. Binding of small-molecule drugs to macromolecules is usually reversible, whereas the binding of mAbs to the antigens is practically irreversible and can significantly contribute to their elimination. With regard to degradation, the majority of small-molecule drugs is metabolized predominantly by hepatic enzymes, whereas mAbs as glycoproteins undergo ubiquitous proteolytic degradation. At therapeutic doses, small-molecule drugs rarely show nonlinear pharmacokinetics, which in contrast is seen very frequently for mAbs at therapeutic doses. For both small-molecule and mAb drugs, the unbound molecules – that is, the nonbound, free fraction of drug – are considered to exert the pharmacodynamic effects. Addi-

**Table 3.6** Comparison of pharmacokinetic characteristics between traditional small-molecule drugs and therapeutic monoclonal antibodies.

| Characteristic | Small-molecule drugs | Monoclonal antibodies |
|---|---|---|
| Tissue penetration | Often good | Usually poor |
| Binding | Usually implies distribution | Usually implies clearance |
| Degradation | Metabolic degradation | Proteolytic degradation |
| Renal clearance | Often important | Uncommon |
| Unbound concentration | Considered to exert effect | Considered to exert effect + may cause immunogenicity |
| Pharmacokinetics | Usually linear<br>Usually independent from dynamics | Often nonlinear<br>Often dependent on pharmacodynamics and HAxA |

HAxA: anti-idiotype antibodies against therapeutic monoclonal antibodies, such as HAMA (human anti-murine antibody), HACA (human anti-chimeric antibody), and HAHA (human anti-human antibody).

tionally, the unbound concentration of mAbs induces also the production of anti-antibodies that cause negative immunological reactions.

## 3.10 Pharmacokinetic Modeling of mAbs

Although, during the early applications of therapeutic mAbs, pharmacokinetic modeling was rarely applied, a variety of analytical techniques has been used over the years to characterize the pharmacokinetics of this class of compounds. The application and information derived from three different methods of noncompartmental analysis, individual compartmental analysis, and population analysis will be discussed in the following sections.

### 3.10.1 Noncompartmental Pharmacokinetic Analysis

In pharmaceutical research and drug development, noncompartmental analysis is normally the first and standard approach used to analyze pharmacokinetic data. The aim is to characterize the disposition of the drug in each individual, based on available concentration–time data. The assessment of pharmacokinetic parameters relies on a minimum set of assumptions, namely that drug elimination occurs exclusively from the sampling compartment, and that the drug follows linear pharmacokinetics; that is, drug disposition is characterized by first-order processes (see Chapter 7). Calculations of pharmacokinetic parameters with this approach are usually based on statistical moments, namely the area under the concentration–time profile (area under the zero moment curve, AUC) and the area under the first moment curve (AUMC), as well as the terminal elimination rate constant ($\lambda_z$) for extrapolation of AUC and AUMC beyond the measured data. Other pharmacokinetic parameters such as half-life ($t_{1/2}$), clearance (CL), and volume of distribution (V) can then be derived.

An assumption concerning the number of compartments is, by nature, not required. For reliable results and precise parameter estimates, however, a relatively large number of data points per individual are required. Phase 1 studies of mAbs usually provide sufficient data for a noncompartmental analysis, but the assumption of linear pharmacokinetics is not valid for most mAbs. This prerequisite, however, was frequently neglected during the early years of therapeutic mAb development, and an overall estimate for CL, for example, was frequently reported in the literature. In dose-escalating studies, however, the concentration–time plots of the raw data clearly indicate that the slope of the terminal phase is not parallel for the different doses, but increases with increasing dose (Fig. 3.10). As a result, the listing of different clearance values for different doses can be found. For example, the clearance of trastuzumab was reported to be 88.3 mL/h for a 10-mg dose, 34.3 mL/h for a 50-mg dose, 25.0 mL/h for a 100-mg dose, 19.0 mL/h for a 250-mg dose, and 16.7 mL/h for a 300-mg dose.

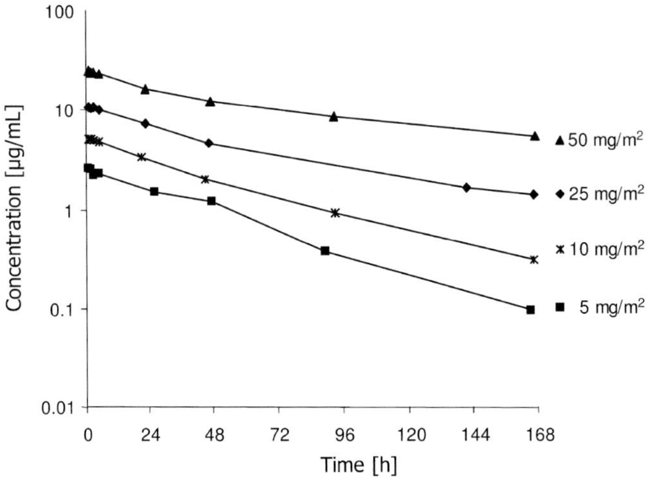


**Fig. 3.10** Concentration–time data for different doses of sibrotuzumab.

Despite these limitations, even today noncompartmental analysis approaches are sometimes the only way in which pharmacokinetic data of mAbs are analyzed. Especially for the mechanistic understanding of the behavior of mAbs in the body, a noncompartmental analysis cannot be recommended.

### 3.10.2 Individual Compartmental Pharmacokinetic Analysis

In contrast to noncompartmental analysis, in compartmental analysis a decision on the number of compartments must be made. For mAbs, the standard compartment model is illustrated in Fig. 3.11. It comprises two compartments, the central and peripheral compartment, with volumes V1 and V2, respectively. Both compartments exchange antibody molecules with specific first-order rate constants. The input into (if IV infusion) and elimination from the central compartment are zero-order and first-order processes, respectively. Hence, this disposition model characterizes linear pharmacokinetics. For each compartment a differential equation describing the change in antibody amount per time can be established. For

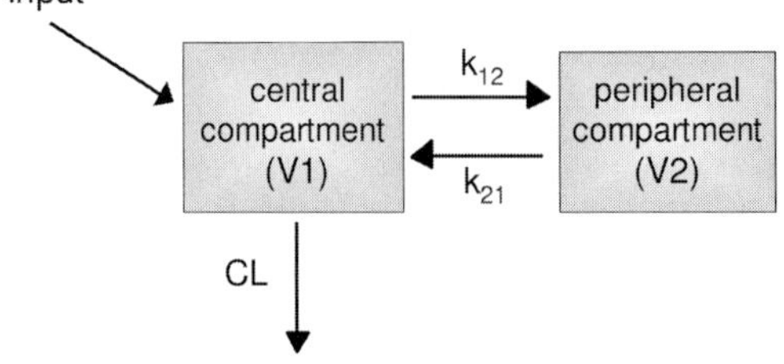


**Fig. 3.11** Schematic depiction of the simple two-compartment pharmacokinetic model with an input, transport between two compartments, and elimination from the central compartment.

the simple two-compartment model depicted in Fig. 3.11 the integrated forms of these differential equations are available to calculate individually the pharmacokinetic parameter values and the resulting model-predicted concentration–time profile. However, due to the nonlinear disposition of mAbs this linear disposition model is often inadequate to describe the concentration–time data of mAbs. To overcome this model misspecification, more sophisticated models with nonlinear components needed to be developed to adequately represent the complex elimination pathways of mAbs (as discussed in Section 3.9.3). Since nonlinearity in pharmacokinetics is difficult to quantify in individual subjects, population data analysis techniques were identified as a more valuable approach for providing a deeper understanding of the pharmacokinetic behavior of mAbs.

### 3.10.3 Population Pharmacokinetic Analysis

Although the fundamental research for population analysis techniques was carried out almost 30 years ago, modern computing power has allowed a more sophisticated development of the technique and the transfer to applied research. The hallmark characteristic of population analysis is the simultaneous evaluation of all pharmacokinetic data of all individuals available. Pooling of data from different studies is, therefore, possible and also desirable to obtain a dataset representing the entire target population with a large variety of different study designs, dosing regimens and patient characteristics. The obtained datasets can be unbalanced with regard to the number of data points contributed by each individual subject (Fig. 3.12). In a typical dataset there are individuals with several data points (connected by a thin line), but in the extreme only one or two data points per individual may be available. In the simultaneous evaluation, all concentration–time data contribute to the description of the typical concentration–time course of the population, represented by the thick line in Fig. 3.12. Apart from the typical phar-

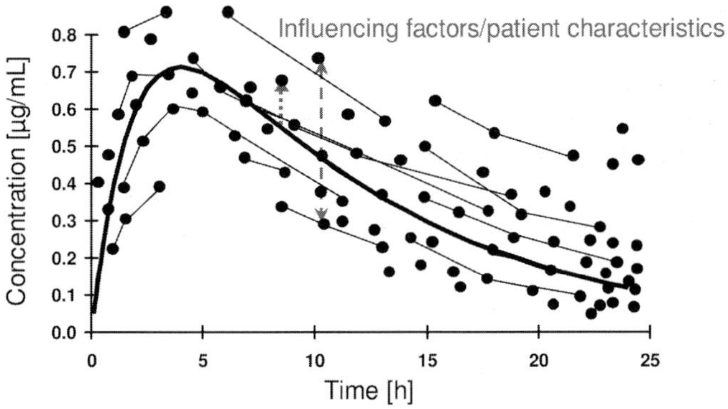


**Fig. 3.12** Data set for a population pharmacokinetic data analysis. Individual data points from one individual are connected by thin lines. The thick line represents the population time course.

macokinetic parameters to describe the typical time course, variability parameters are also estimated in a population pharmacokinetics analysis describing for example the deviation of single data points from the typical time course. The technique is applied as a valuable tool in drug development for multiple purposes, including methods to explore rational dosing regimens, to identify subgroups of special populations and characterize their pharmacokinetics, and to generate population parameters (typical disposition and variability) necessary to perform therapeutic drug monitoring by Bayesian estimation.

Various methods are available to estimate population parameters, but today the nonlinear mixed effects modeling approach is the most common one employed. Population analyses have been performed for mAbs such as basiliximab, daclizumab and trastuzumab, as well as several others in development, including clenoliximab and sibrotuzumab. Population pharmacokinetic models comprise three submodels: the structural; the statistical; and covariate submodels (Fig. 3.13). Their development and impact for mAbs will be discussed in the following section.

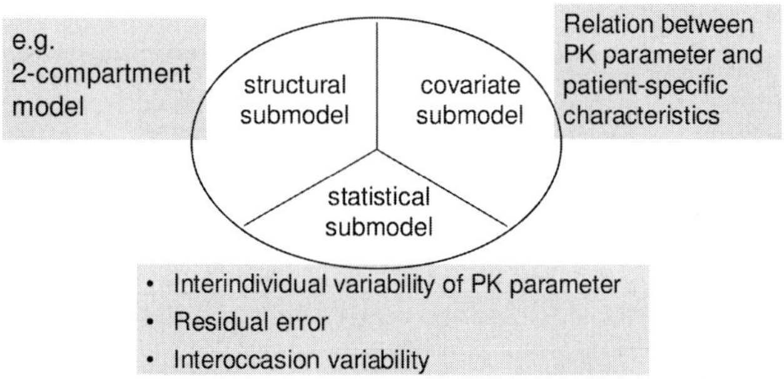


**Fig. 3.13** Components of a population pharmacokinetic model.

#### 3.10.3.1 **Structural Submodel**

The structural submodel describes the central tendency of the time course of the antibody concentrations as a function of the estimated typical pharmacokinetic parameters and independent variables such as the dosing regimen and time. As described in Section 3.9.3, mAbs exhibit several parallel elimination pathways. A population structural submodel to mechanistically cover these aspects is depicted schematically in Fig. 3.14. The principal element in this more sophisticated model is the incorporation of a second elimination pathway as a nonlinear process (Michaelis–Menten kinetics) into the structural model with the additional parameters $V_{max}$, the maximum elimination rate, and $k_m$, the concentration at which the elimination rate is 50% of the maximum value. The addition of this second nonlinear elimination process from the peripheral compartment to the linear clearance process usually significantly improves the fit of the model to the data. Total clearance is the sum of both clearance parts. The dependence of total clearance on mAb concentrations is illustrated in Fig. 3.15, using population estimates of the linear ($CL_L$) and nonlinear clearance ($CL_{NL}$) components. At low concentra-

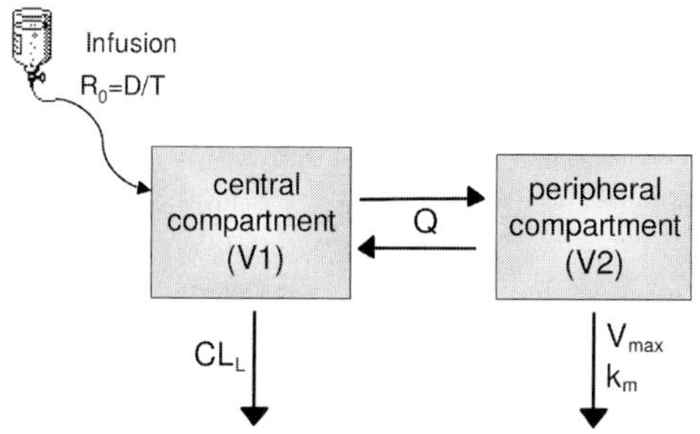


**Fig. 3.14** Schematic depiction of a two-compartment pharmacokinetic model typical for the pharmacokinetics of mAbs, with an input, transport between two compartments and two elimination pathways, one from each compartment.

tions relative to $k_m$, nonlinear clearance typically is much higher than the clearance for the linear pathway, and total clearance is almost independent of serum concentration. The elimination capacity via the linear fraction of clearance is generally low. As concentration increases relative to $k_m$, the nonlinear pathway becomes saturated – that is, total clearance becomes highly dependent on serum concentration. At concentrations much higher than $k_m$, $CL_L$ dominates the total clearance and the influence of the nonlinear clearance pathway becomes negligible.

Mechanistically, the linear clearance pathway represents most probably the slow and nonspecific proteolytic degradation of mAbs in hepatic endothelial and reticu-

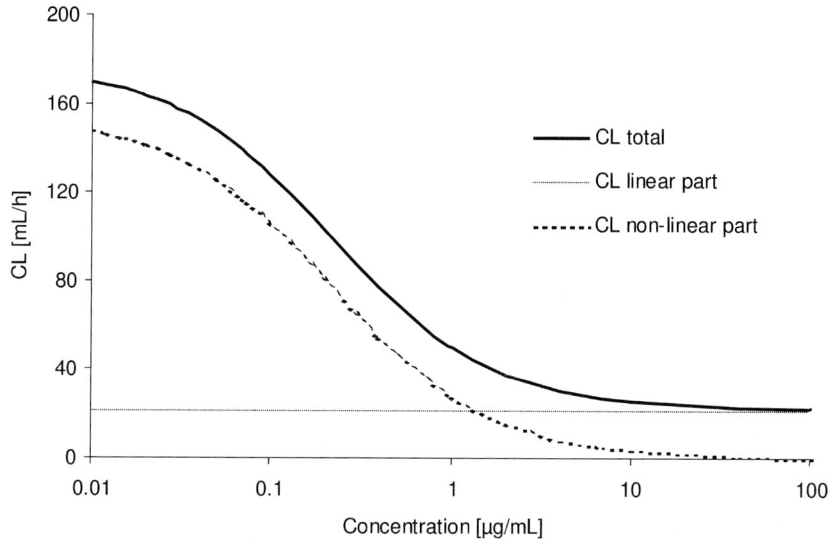


**Fig. 3.15** Dependence of clearance on concentration in the pharmacokinetic model shown in Fig. 3.14.

loendothelial cells. At higher antibody concentrations, where only the linear part of the CL determines total clearance, a terminal half-life of approximately one to two weeks can usually be derived from the population parameter estimates.

The nonlinear elimination route is usually operative at low serum concentrations relative to $k_m$, and results in a convex-upward terminal serum concentration–time profile with a limiting terminal half-life of approximately a few days (Fig. 3.16). The typical values for $k_m$ and $V_{max}$ suggest a low capacity, high-affinity saturable elimination process. This Michaelis–Menten process was hypothesized to be attributable to the saturable binding to Fc-Rn (see Section 3.8.2). Saturable binding has been reported to occur at high IgG1 concentrations (hyperglobulinemia). While the physiological range lies between 5 and 11 mg/mL in adults, however, the maximum concentrations at the end of a mAb infusion do generally not exceed 0.7–0.9 mg/mL but are typically much lower for most antibodies (see Table 3.4). These figures suggest that a saturation of the Fc-Rn is not responsible for the nonlinear pattern.

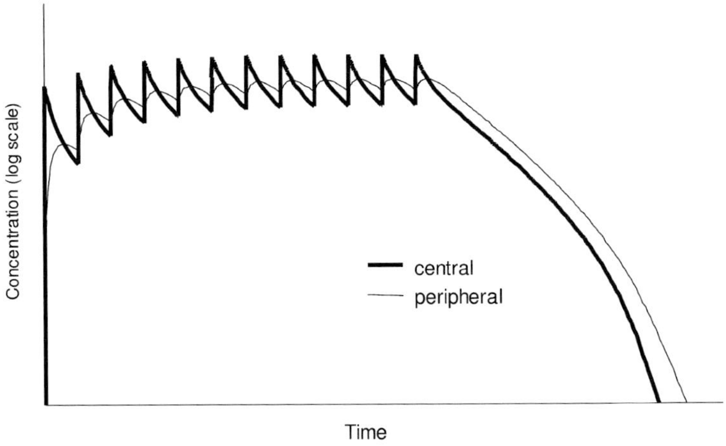


**Fig. 3.16** Convex-upward terminal concentration–time profile for a mAb with two elimination pathways, one linear and one non-linear route.

Furthermore, the nonlinear pathway is not effective at high but at low concentrations of the mAb (Fig. 3.15). This suggests that the nonlinear behavior is presumably attributable to specific interactions with the antigen – that is, a target-mediated elimination pathway. One potential mechanism for the nonlinearity might be steric hindrance for binding or limited access at high antibody concentrations, or complete binding of available antigen. Another potential source of pharmacokinetic nonlinearity that has been observed for the T-cell antigen of clenoliximab, might be up-modulation or diminishing down-modulation of the antigen at decreasing antibody concentrations. In the case of antigens located on the surface of soluble cells, the amount of antigen and/or of interacting therapeutic antibody might be quantified by FACS analysis and incorporated into the modeling. Overall, there are still gaps in the understanding of the entire clearance mechanism of mAbs.

#### 3.10.3.2 **Statistical Submodel**

The statistical submodel characterizes the pharmacokinetic variability of the mAb and includes the influence of random – that is, not quantifiable or uncontrollable factors. If multiple doses of the antibody are administered, then three hierarchical components of random variability can be defined: inter-individual variability; inter-occasional variability; and residual variability. Inter-individual variability quantifies the unexplained difference of the pharmacokinetic parameters between individuals. If data are available from different administrations to one patient, inter-occasional variability can be estimated as random variation of a pharmacokinetic parameter (for example, CL) between the different administration periods. For mAbs, this was first introduced in sibrotuzumab data analysis. In order to individualize therapy based on concentration measurements, it is a prerequisite that inter-occasional variability (variability within one patient at multiple administrations) is lower than inter-individual variability (variability between patients). Residual variability accounts for model misspecification, errors in documentation of the dosage regimen or blood sampling time points, assay variability, and other sources of error.

#### 3.10.3.3 **Covariate Submodel**

Covariate analysis aims at identifying patient- or study-specific characteristics with a significant influence on pharmacokinetic parameters, and at quantifying their impact. Covariates include demographic data such as age, gender or weight, laboratory parameters such as organ function, disease-related data such as severity, co-illnesses, or concomitant medication, and patient status such as genetic or smoking status. Significant and clinically relevant covariates should reduce inter-individual variability in pharmacokinetic parameters and finally allow for a more appropriate dose selection according to the needs of the individual patient. So far, covariate analysis has only been performed for a very limited number of mAbs. In many cases, for example sibrotuzumab [13], body weight was found significantly to affect pharmacokinetic parameters, leading to marked differences in the model-predicted concentration–time profiles after multiple dosing of the same dose in patients with low and high body weights (Fig. 3.17).

In the case of trastuzumab, where shed antigen is found in the circulation, high antigen concentrations resulted in lower trough trastuzumab concentrations. This effect was modeled using only a linear clearance term and the shed antigen baseline concentration as covariate.

Common demographic covariates other than weight, such as age, and liver or kidney function, do not seem to alter the pharmacokinetics of mAbs, and should not be considered for dosage adjustments in the clinical use of mAbs. Other measures of body size – including body surface area (BSA), which is traditionally used

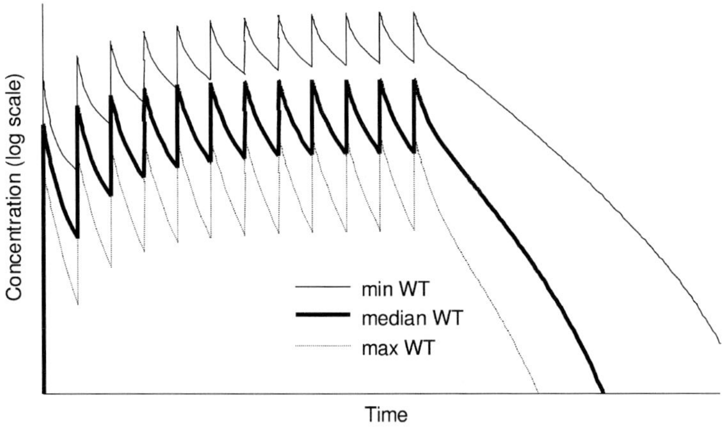


**Fig. 3.17** Weight-dependent concentration–time profile following the same fixed-dose dosing regimen to three patients with minimum, median, and maximum body weight (WT, range ~100 kg) of the studied patient population.

for dosing in oncology – were found to be inferior to body weight in explaining inter-patient variability. Consequently, the dosing of mAbs based on BSA cannot be recommended.

In conclusion, robust population pharmacokinetic models may contribute to the mechanistic understanding of the fate of mAbs in the body. Factors that may influence the pharmacokinetics of mAbs should be investigated for their potential influence on dosage regimen design in clinical trials and therapeutic use.

## 3.11
## Pharmacodynamics of mAbs

Compared to polyclonal antibodies, mAbs display molecular homogeneity and significantly higher specificity leading to increased *in-vivo* activity. For example, 100–170 mg serum containing polyclonal antibodies against the tetanus toxin are necessary to achieve the same effect as 0.7 mg of a respective mAb. This section will provide an overview on therapeutic antibodies which have either been approved or are in clinical development, and will classify them according to different pharmacodynamically relevant properties.

In total, 19 mAbs have been approved by the FDA and/or EMEA since 1986. The approval year, as well as the proprietary name and indication for use, are listed-αbetically in Table 3.7. A comprehensive summary of the structure-related pharmacodynamic characteristics of the approved mAbs is provided in Table 3.8. All 19 approved mAbs belong to the IgG class, with 15 to IgG1, three to IgG2, and one to the IgG4 subclasses. With regard to the origin and extent of humanization, there are three murine, five chimeric, eight humanized and three human (one human and two chimeric proteins) mAbs available on the market. There has been a clear tendency during the past few years towards the development of more human-like mAbs.

With respect to the different modes of actions and effector function (as discussed in Section 3.5), there are:

- 12 intact unconjugated mAbs: adalimumab, alemtuzumab, basiliximab, cetuximab, daclizumab, efalizumab, infliximab, muromonab, omalizumab, palivizumab, rituximab, trastuzumab.
- Three intact conjugated unlabeled mAbs: alefacept (+ protein), etanercept (+ protein), gemtuzumab ozogamicin (+ toxin).
- Two intact radiolabeled mAbs: ibritumomab tiuxetan ($^{90}$Y and $^{111}$In), tositumomab ($^{131}$I).
- One mAb fragment: abciximab.

Three different pharmacodynamic principles of action can be distinguished for mAbs, comprising lysis or apoptotic activity, coating activity, and inactivating activity (see Section 3.5.2), depending on the type of antigen and the antigen–antibody interaction.

In the first two cases, the ideally disease-specific antigen is located on the surface of a cell (see Table 3.8). The majority of mAbs (abciximab, alefacept, alemtuzumab, basiliximab, cetuximab, daclizumab, efalizumab, gemtuzumab ozogamicin, ibritumomab tiuxetan, muromonab, palivizumab, rituximab, tositumomab, and trastuzumab) exert their pharmacodynamic activity by Fc-mediated ADCC/CDC activation or delivery of toxic substance to the targeted cell [15]. Cell death will be the ultimate result. In some cases "only" a coating of the antigen-binding site by the antibody takes place, which will lead to a "down-regulation" of the antigen. As more knowledge is acquired, it is being recognized that by additional blockage of downstream signaling, apoptosis might also finally be induced.

Four mAbs (adalimumab, bevacizumab, infliximab, omalimumab) are directed against soluble substances as antigens (cytokine, growth factor, immunoglobulin). The pharmacodynamic principle is to block the antigen-binding activity to its receptor and thereby inactivate the target antigen function. The binding of the therapeutic antibody to the target antigen finally leads to agglutination and neutralization of the antigen. The question on the degree of occupancy for a therapeutic effect has not been satisfactorily addressed in the literature.

Efficacious mAbs are used in a variety of therapeutic indications such as cancer, rheumatoid arthritis, Crohn's disease, psoriasis, organ transplantation, asthma, infectious diseases, and cardiovascular diseases, while for other diseases research and development is currently ongoing. The current three major therapeutic areas include oncology (eight mAbs against solid tumors and lymphoma/leukemia), inflammatory diseases (five mAbs), and the immunosuppression/prophylaxis or treatment of organ rejection in transplantation (three mAbs). The reader is referred to Chapter 12 for a detailed discussion on the exposure–response relationships and pharmacodynamics of therapeutically administered mAbs.

**Table 3.7** Indication and approval of therapeutic monoclonal antibodies.

| Monoclonal antibody | Brand name | Indication for use | Year of approval[a] |
|---|---|---|---|
| Abciximab | ReoPro | Unstable angina (percutaneous coronary intervention) | 1994 |
| Adalimumab | Humira | Rheumatoid arthritis | 2002 |
| Alefacept | Amevive | Chronic plaque psoriasis | 2003 (not by EMEA) |
| Alemtuzumab | Campath | B-cell chronic lymphocytic leukemia (B-CLL) | 2001 |
| Basiliximab | Simulect | Treatment of the acute rejection of kidney, heart and liver transplants | 1998 |
| Bevacizumab | Avastin | Metastasized colorectal carcinoma | 2004 |
| Cetuximab | Erbitux | Metastasized colorectal carcinoma (EGFR-positive) | 2004 |
| Daclizumab | Zenapax | Prophylaxis of acute rejection after allogenic kidney transplantation | 1997 |
| Efalizumab | Raptiva | Plaque psoriasis | 2003 |
| Etanercept | Enbrel | Rheumatoid arthritis; psoriatic arthritis | 1998 |
| Gemtuzumab ozogamicin | Mylotarg | Relapsed acute myeloid leukemia | 2000 (not by EMEA)[b] |
| Ibritumomab tiuxetan | Zevalin | CD20-positive follicular non-Hodgkin's lymphoma | 2002 |
| Infliximab | Remicade | Rheumatoid arthritis; Morbus Crohn; ankylosing spondylitis | 1998 |
| Muromonab | Orthoclone OKT 3 | Prophylaxis of rejection after heart transplantation; immunosuppression | 1986/1992 |
| Omalizumab | Xolair | Asthma (positive skin test or *in-vitro* reactivity to a perennial aeroallergen) | 2003 (not by EMEA)[c] |
| Palivizumab | Synagis | Respiratory tract infections caused by respiratory syncytial virus | 1998 |
| Rituximab | Rituxan/ MabThera· | Follicular lymphoma (grade III- IV); CD20-positive non-Hodgkin's lymphoma | 1997 |
| Tositumomab | Bexxar | CD20-positive, follicular, non-Hodgkin's lymphoma | 2003 (not by EMEA)[d] |
| Trastuzumab | Herceptin | Metastasized breast cancer | 1998 |

**a)** Approved by FDA and EMEA unless stated otherwise.
**b)** European Commission granted orphan designation (October 2000).
**c)** Committee for Medicinal Products for Human Use (CHMP) recommended to grant a marketing authorization (July 2005).
**d)** European Commission granted orphan designation (February 2003).

**Table 3.8** Characterization of approved therapeutic monoclonal antibodies.

| Monoclonal antibody | Class[a] | Subclass | Antigen | Type of antigen | Physiological ligand/ substrate/receptor |
|---|---|---|---|---|---|
| Abciximab | Chimeric[b] | IgG1 | GPIIb/IIIa (CD41)[c] | Receptor | Fibrinogen, von Willebrand factor, other adhesive molecules |
| Adalimumab | Human | IgG1 | TNF-α[d] | Soluble substance | TNF-α receptor[d] |
| Alefacept | Human[e] | IgG2-Fc | CD2 | Surface marker | LFA-3[f] |
| Alemtuzumab | Humanized (rat CDR) | IgG1 | CD52 | Surface marker | – |
| Basiliximab | Chimeric | IgG1 | IL-2R (CD25)[g] | Receptor | IL-2[h] |
| Bevacizumab | Humanized | IgG1 | VEGF[i] | Soluble substance | VEGF receptor |
| Cetuximab | Chimeric | IgG1 | EGFR[j] | Receptor | EGF, TNF-α TNF-β[k] |
| Daclizumab | Humanized | IgG1 | IL-2R (CD25)[g] | Receptor | IL-2[h] |
| Efalizumab | Humanized | IgG1 | CD11a | Receptor | LFA-1[l] |
| Etanercept | Human[m] | IgG1-Fc | TNF-α/TNF-β receptor[d] | Receptor | TNF-α TNF-β[d] |
| Gemtuzumab ozogamicin | Humanized[n] | IgG4 | CD33 | Surface marker | Host cell |
| Ibritumomab tiuxetan | Murine[o] | IgG1 | CD20 | Channel constituent | Calcium |
| Infliximab | Chimeric | IgG1 | TNF-α[d] | Soluble substance | TNF-α receptor[d] |
| Muromonab | Murine | IgG2a | CD3ε | Part of T-cell receptor complex | Peptides presented by MHC on APC[p] |
| Omalizumab | Humanized | IgG1 | IgE | Soluble substance | FcεRI[q] |
| Palivizumab | Humanized | IgG1 | RSV F protein[r] | Surface protein | Host cell |
| Rituximab | Chimeric | IgG1 | CD20 | Channel constituent | Calcium |
| Tositumomab | Murine[s] | IgG2a | CD20 | Channel constituent | Calcium |
| Trastuzumab | Humanized | IgG1 | her2/neu[t] | Receptor | EGF, TNF-α[u] |

Footnotes see page 90.

**Footnotes to Table 3.8**

a) If not stated otherwise: non-human part is of murine origin.
b) Chimeric Fab fragment comprised of murine variable fraction and human constant region.
c) Glycoprotein IIb/IIIa.
d) Tumor necrosis factor.
e) Fusion protein (also called chimeric protein): two molecules of human leukocyte function antigen-3 (LFA-3) conjugated to human Fc region.
f) Leukocyte function antigen-3.
g) Interleukin 2 receptor.
h) Interleukin 2.
i) Vascular endothelial growth factor.
j) Epidermal growth factor receptor.
k) Epidermal growth factor receptor, tumor necrosis factor.
l) Leukocyte function antigen-1.
m) Fusion protein (also called chimeric protein): two molecules of human tumor necrosis factor receptor 2 (TNFR2/p75) conjugated to human Fc region.
n) Conjugated with calicheamicin (bacterial toxin).
o) Conjugated with $^{111}$In or $^{90}$Y.
p) MHC on APC: major histocompatibility complex on antigen-presenting cells.
q) Fcε receptor type I.
r) Respiratory syncytial virus fusion protein.
s) Conjugated with $^{131}$I.
t) Human epidermal growth factor receptor 2.
u) Epidermal growth factor, tumor necrosis factor: other ligands (e.g., PICK1) may also potentially bind.

## 3.12 Conclusions

The current global mAb pipeline comprises approximately 140 drugs in development (with approximately 75 in clinical development), and is prepared to deliver as many as 13 new mAbs between 2005 and 2008. Despite problems and limitations in different arenas with regard to production and preclinical/clinical development (e.g., high cost and adverse reactions), continued effort is being undertaken in the development of mAb drug products. In 2003, sales of mAbs reached approximately US$ 5 billion, this being a 27% increase compared to 2001.

Important questions that remain to be addressed in the future include how to identify new targets for mAbs, how to increase the effectiveness and specificity of new mAbs, how to reduce the side effects of mAb therapy, how to validate specific assays, how to identify which targets are active in which group of patients (responders versus nonresponders), and how to predict the responsiveness. For an effective pharmacotherapy with mAbs, however, those factors which impact upon the pharmacokinetics and pharmacodynamics of antibody-based therapeutics must be understood. They are also relevant for optimizing the drug development process, for increasing its probability of success, and for the optimal use of mAbs in clinical applications. An exciting future with diversity in novel targets is foreseen for the therapeutic application of mAbs in many acute and chronic diseases.

## Acknowledgments

The authors acknowledge Ms. Sabrina Albrecht, Ms. Stephanie Mendow, and Ms. Uta Zschoeckner for their technical help in these studies.

## 3.13 References

**1** Langer-Gould, A., S.W. Atlas, A.J. Green, A.W. Bollen, and D. Pelletier. **2005**. Progressive multifocal leukoencephalopathy in a patient treated with natalizumab. *N. Engl. J. Med.* 353 : 375–381.

**2** Nadler, L.M., P. Stashenko, R. Hardy, W.D. Kaplan, L.N. Button, D.W. Kufe, K.H. Antman, and S.F. Schlossman. **1980**. Serotherapy of a patient with a monoclonal antibody directed against a human lymphoma-associated antigen. *Cancer Res.* 40 : 3147–3154.

**3** Kohler, G., and C. Milstein. **1975**. Continuous cultures of fused cells secreting antibody of predefined specificity. *Nature* 256 : 495–497.

**4** Schopf, R.E. **2001**. Idec-114 (Idec). *Curr. Opin. Investig. Drugs* 2 : 635–638.

**5** Reff, M.E., K. Hariharan, and G. Braslawsky. **2002**. Future of monoclonal antibodies in the treatment of hematologic malignancies. *Cancer Control* 9 : 152–166.

**6** Lobato, M.N., and T.H. Rabbitts. **2004**. Intracellular antibodies as specific reagents for functional ablation: future therapeutic molecules. *Curr. Mol. Med.* 4 : 519–528.

**7** Rondon, I.J., and W.A. Marasco. **1997**. Intracellular antibodies (intrabodies) for gene therapy of infectious diseases. *Annu. Rev. Microbiol.* 51 : 257–283.

**8** Molnar, E., J. Prechl, A. Isaak, and A. Erdei. **2003**. Targeting with scFv: immune modulation by complement receptor specific constructs. *J. Mol. Recognit.* 16 : 318–323.

**9** Lin, M.Z., M.A. Teitell, and G.J. Schiller. **2005**. The evolution of antibodies into versatile tumor-targeting agents. *Clin. Cancer Res.* 11 : 129–138.

**10** Weiner, L.M., and P. Carter. **2003**. The rollercoaster ride to anti-cancer antibodies. *Nat. Biotechnol.* 21 : 510–511.

**11** Pai, L.H., R. Wittes, A. Setser, M.C. Willingham, and I. Pastan. **1996**. Treatment of advanced solid tumors with immunotoxin LMB-1: an antibody linked to *Pseudomonas* exotoxin. *Nat. Med.* 2 : 350–353.

**12** Kloft, C., E.U. Graefe, P. Tanswell, A.M. Scott, R. Hofheinz, A. Amelsberg, and M.O. Karlsson. **2004**. Population pharmacokinetics of sibrotuzumab, a novel therapeutic monoclonal antibody, in cancer patients. *Invest. New Drugs* 22 : 39–52.

**13** Ghetie, V., S. Popov, J. Borvak, C. Radu, D. Matesoi, C. Medesan, R.J. Ober, and E.S. Ward. **1997**. Increasing the serum persistence of an IgG fragment by random mutagenesis. *Nat. Biotechnol.* 15 : 637–640.

**14** Brambell, F.W.R., W.A. Hemmings, and I.G. Morris. **1964**. A theoretical model of γ-globulin catabolism. *Nature* 203 : 1352–1355.

**15** Carter, P. **2001**. Improving the efficacy of antibody-based cancer therapies. *Nat. Rev. Cancer* 1 : 118–129.

# 4
# Pharmacokinetics and Pharmacodynamics of Antisense Oligonucleotides

*Rosie Z. Yu, Richard S. Geary, and Arthur A. Levin*

## 4.1
## Introduction

Decreasing gene expression by selectively blocking mRNA translation can be accomplished using single-strand antisense molecules or, as more recently demonstrated, by double-strand RNA constructs now known as RNAi. Antisense compounds are short synthetic oligonucleotides, usually between 15 and 25 nucleotides in length, designed to hybridize to RNA through Watson–Crick base pairing. Upon binding the target RNA, the oligonucleotide prevents translation of the encoded protein product in a sequence-specific manner. Since the rules for Watson–Crick base pairing are well characterized [1], antisense oligonucleotides represent, in principle, a simple method for the rational design of drugs. In practice, the exploitation of antisense oligonucleotide technology for therapies has presented a unique set of challenges, some of which relate to their pharmacokinetic behavior.

The oligonucleotide chemistry that has advanced furthest in clinical development is that of single-strand phosphorothioate oligodeoxynucleotides. Phosphorothioate (PS) oligodeoxynucleotides differ from native DNA only in the substitution of one non-bridging oxygen with sulfur in the phosphodiester bridge connecting one base to another. Thus, a 20-base PS oligodeoxynucleotide would contain 19 sulfur atoms, one for each internucleotide bridge (Fig. 4.1 a). The addition of sulfur serves to stabilize the oligonucleotide to nuclease digestion and increases nonspecific plasma protein binding, likely due to the increased charge density of the sulfur backbone. Thus, this simple chemical modification to natural DNA imparts these small oligonucleotides with favorable pharmacokinetic properties, including prolonged residence time in tissues and cells, improved tissue distribution, and reduced urinary excretion. Currently there is one antisense PS oligodeoxynucleotide product on the market, and numerous additional agents in clinical trials, several of which are in advanced stages of development (Table 4.1).

Significant resources have been applied towards the identification of chemical modifications that further improve upon the pharmacokinetic and pharmacodynamic properties of PS oligodeoxynucleotides (Table 4.2). The vast majority of

*Pharmacokinetics and Pharmacodynamics of Biotech Drugs: Principles and Case Studies in Drug Development.* Edited by Bernd Meibohm

ISBN: 3-527-31408-3

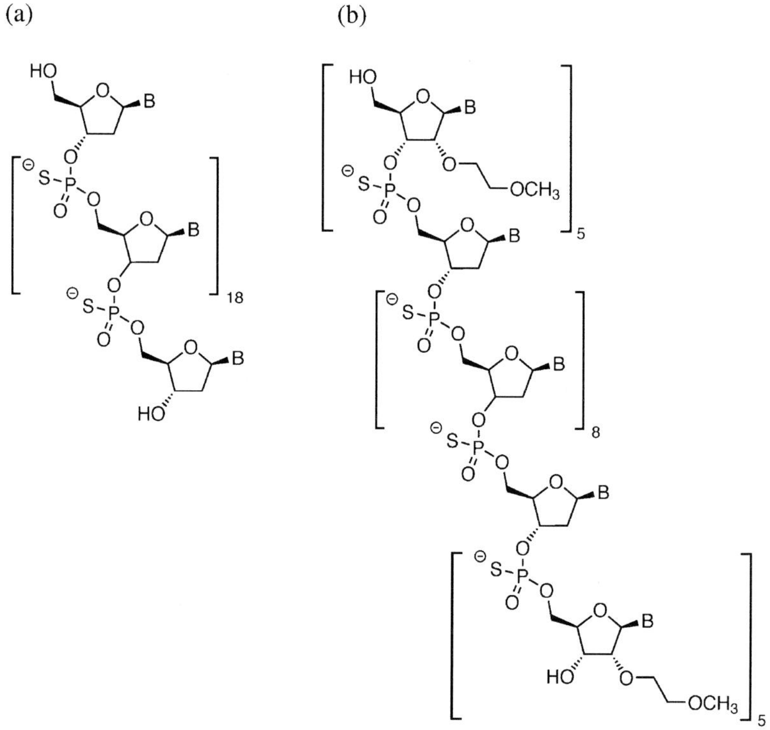


**Fig. 4.1** Representations of the chemical structure of first-generation oligonucleotides and 2′-MOE partially modified phosphorothioate oligonucleotides (second-generation oligonucleotides) (B = G, C, T, or A).

**Table 4.1** Antisense phosphorothioate oligodeoxynucleotides studied in human clinical trials.

| Oligonucleotide | Molecular target | Disease indication | Route of administration | Sponsor |
|---|---|---|---|---|
| Vitravene (fomivirsen, ISIS 2922) | Human cytomegalovirus IE-2 gene | CMV retinitis | Intravitreal | Isis/Ciba Vision |
| GEM91 | HIV | HIV infection | Intravenous | Hybridon |
| Alicaforsen (ISIS 2302) | ICAM-1 | Ulcerative colitis | Topical Enema | Isis |
| Affinitac (ISIS 3521, LY900003) | PKC-α | Cancer | Intravenous | Lilly/Isis |
| ISIS 5132 | c-raf kinase | Cancer | Intravenous | Isis |
| ISIS 2503 | H-ras | Cancer | Intravenous | Isis |

**Table 4.1** (continued)

| Oligonucleotide | Molecular target | Disease indication | Route of administration | Sponsor |
|---|---|---|---|---|
| Genasense (G 3139) | BCL-2 | Cancer | Intravenous | Genta |
| ISIS 14803 | HepC | Hepatitis C infection | Intravenous | Isis |
| INX-3001 | c-myb | Cancer | Intravenous | Inex |
| INX 3280 | c-myc | Restenosis | Local stent | Inex |
| Veglin | VEGF | Cancer | Intravenous | VasGene |
| RX-0201 | Akt | Cancer | Intravenous | Rexahn |
| GPI-2A | HIV gag gene | HIV | Intravenous | Novopharm |

**Table 4.2** Antisense 2′-alkoxy modified phosphorothioate oligonucleotides currently with clinical trial experience.

| Oligonucleotide | Molecular target | Disease indication | Route of administration | Sponsor |
|---|---|---|---|---|
| ISIS 104838 | TNF-α | Rheumatoid arthritis | Subcutaneous | Isis |
| ISIS 113715 | PTP-1B | Diabetes | Subcutaneous | Isis |
| ISIS 301012 | ApoB-100 | HoFH/ HeFH | Subcutaneous/ Oral | Isis |
| ATL 1102 | VLA-4 | Multiple sclerosis | Intravenous | ATL/Isis |
| ATL 1101 | IGF-1R | Psoriasis | Topical | ATL/Isis |
| OGX-011 | Clusterin | Cancer | Intravenous | Oncogenex/ Isis |
| LY2181308 | Survivin | Cancer | Intravenous | Lilly/Isis |
| GEM132 | Human CMV | CMV infection | Intravenous | Hybridon |
| GEM 92 | HIV-1, *gag* | AIDS | Intravenous | Hybridon |
| GEM 231 | PKA, RIα | Cancer | Intravenous | Hybridon |
| MG 98 | DNA-methyl transferase | Cancer | Intravenous | Methyl-Gene/MGI Pharma |
| MBI 1121 | E1-HPV | Genital warts | Local injection | Micrologix |
| AEG 35156/ GEM 640 | XIAP | Cancer | Intravenous | Aegera |

these modified PS oligonucleotides have followed a model of mixing 2′-alkoxy modifications placed at the 3′- and 5′-termini of the oligonucleotides with deoxynucleotides in the intervening gap. This RNA/DNA construct has provided even greater biological stability to the molecule, higher binding affinity to its target mRNA while maintaining RNase H activity [2, 3] and decreasing general nonhybridization toxicities [4]. Within this second chemical class of so-called second-generation oligonucleotides, there exist at least two subclasses – the 2′-*O*-methyl and 2′-*O*-(methoxyethyl) PS oligodeoxynucleotides (Fig. 4.1 b). As expected, there is significantly more information regarding the pharmacokinetic properties of first-generation PS oligodeoxynucleotides available in the literature [1, 5–20] (to cite just a few). Proprietary 2′-alkoxy partially modified PS antisense oligonucleotides are at a relatively early stage in their development, and available data are less plentiful [21–26]. RNAi, although an excellent *in-vitro* tool, is – relatively speaking – in its medicinal chemistry infancy [27], with no compound yet developed for systemic use and no literature from which to glean its pharmacokinetic behavior *in vivo*. Therefore, in this chapter we will review the pharmacokinetics of single-strand antisense oligonucleotides. Pharmacokinetic/pharmacodynamic properties of primarily PS 2′-*O*-(methoxyethyl) (abbreviated as 2′-MOE) partially modified single-strand PS oligodeoxynucleotides are reviewed and summarized in this chapter.

## 4.2 Pharmacokinetics

Antisense oligonucleotides are administered parenterally in the majority of reported *in-vivo* studies: intravenous (IV), intraperitoneal (IP), or subcutaneous (SC).

Parenteral administration has been the preferred route because it allows complete systemic availability. Non-parenteral administration of antisense oligonucleotides is only made possible with the aid of novel formulations intended to overcome barriers to absorption (see Chapter 10). Similar to first-generation antisense oligonucleotides (ASOs), the pharmacokinetics of 2′-MOE partially modified ASOs are characterized by:

- A plasma concentration–time profile that is poly-phasic with rapid distribution half-life ($\leq 1$ h) and long elimination half-life reflecting slow elimination from the tissues.
- High binding to plasma proteins (>90% across species).
- Plasma clearance that is dominated by distribution into the tissues.
- Long tissue elimination half-life cleared by nuclease-mediated metabolism.
- Minor urinary or fecal excretion of the intact drug.

The major difference in pharmacokinetics between PS oligodeoxynucleotides and 2′-MOE partially modified ASOs is that 2′-MOE partially modified ASOs have improved metabolic stability to nucleases [28]. Therefore, tissue residence time for the 2′-MOE partially modified ASOs is much longer compared to the first-generation ASOs, and this is reflected in an approximate five-fold increase in elimination half-lives (10–30 days) in tissue over the first-generation ASOs [8, 14, 18, 22, 29]. The pharmacokinetic section of this chapter will focus on 2′-MOE partially modified ASOs.

### 4.2.1
### Plasma Pharmacokinetics Across Species

The pharmacokinetic profile of 2′-MOE partially modified ASOs was similar in mice, rats, dogs, monkeys, and humans in that the drug was cleared within hours from the plasma and distributed to the tissues. Following IV administration, the plasma concentration–time profiles of 2′-MOE partially modified ASO are polyphasic, characterized by a rapid distribution phase (half-lives of 30–80 min), followed by at least one additional much slower elimination phase with half-lives reported from 10 to 30 days. The recent development of ultrasensitive hybridization ELISA methods have made it possible to follow plasma concentrations for up to three months after dose administration, enabling the investigators to determine terminal plasma elimination half-lives [24, 26, 30, 31]. Representative plasma concentration–time profiles with the rapid distribution phase along the slow terminal elimination phase in monkeys and humans for a 2′-MOE partially modified ASO, ISIS 104838 are shown in Figure 4.2 [26].

The initial rapid distribution to tissues dominates plasma clearance (>90%), which is characterized by a near-complete distribution with a several-fold log decline of plasma ISIS 104838 concentrations over a 24-h period. The rapid and complete distribution phase is similar to that of the first-generation PS oligodeoxynucleotides. However, plasma concentrations of 2′-MOE partially modified ASOs

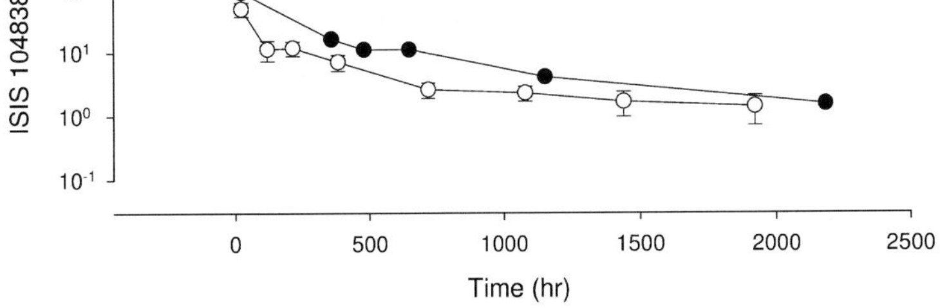


**Fig. 4.2** Polyphasic pharmacokinetic profiles of ISIS 104838 in plasma following 1-h IV infusion in monkey and human. ISIS 104838 concentrations in plasma were measured using a hybridization ELISA method (LLOQ = 0.7 ng/mL).

decline more slowly during the elimination phase. For example, the terminal elimination half-life for ISIS 104838 in monkey plasma is 27 days, which is consistent with the slow elimination of ISIS 104838 from monkey tissues, and reflects an equilibrium with oligonucleotide in tissue [26].

Similar to PS oligodeoxynucleotides, the pharmacokinetics of 2′-MOE partially modified ASOs is dose-dependent. Typically, the distribution half-life is seen to increase with increasing dose, while the area under the curve (AUC) increases were greater in proportion to the dose in mouse and monkey [26]. This observation suggests that there is a saturable component in the disposition of these compounds [26]. Consistent with a relatively rapid distribution from plasma, there is no evidence of measurable accumulation in peak plasma concentrations following repeat-dose administration on an every-other-day or every fourth-day schedule (data not shown). The nonlinear kinetics is characterized by slower apparent distribution half-lives at higher doses, and a slowing in overall plasma clearance at higher doses. The most likely explanation for the nonlinear kinetics is that the distribution to organs that take up a large fraction of the oligonucleotide dose (liver and kidney) is saturated as the dose increases [18, 19, 26, 29, 32]. Taken together, these data suggest that the nonlinear behavior observed in plasma clearance may be related to the saturation of distribution to the major organs of distribution.

Following SC injection, peak concentrations ($C_{max}$) occur at approximately 1 and 3 h after injection in mice and monkeys, respectively, and are several fold lower compared to IV administration. The dominant apparent plasma half-life for the distribution of ASO from plasma to tissue is approximately 30 min in mice, and 1–4 h in monkeys. The plasma half-life following SC administration is longer than that after IV injection, and is indicative of continued absorption from the injection site during the disposition phase. Consistent with the IV administration, peak plasma concentrations following SC injection increased in a dose-dependent manner in studies performed in mice and monkeys. In addition, the plasma clearance was attributed to a rapid distribution to the tissues and, to a much lesser degree, urinary excretion. The increase in AUC was greater in proportion to the dose in mouse and monkey [26]. Plasma absolute bioavailability following SC injection generally ranged between 70% and 100%. Plasma bioavailability may underestimate the ultimate complete tissue absorption, since animal studies have shown that the entire dose is ultimately distributed to tissues such that there is no difference between IV and SC administration with regard to end organ drug concentrations [26].

2′-MOE partially modified ASOs are also highly bound to plasma proteins, though not as extensively as PS oligodeoxynucleotides. At clinically relevant doses, more than 90% of ASO in plasma is bound to plasma proteins in mice, rats, monkeys, and humans [33]. For all species tested, there is little change in the whole plasma protein binding capacity over the concentration range from approximately 1 μg/mL to 75 μg/mL. A shift from lower-capacity proteins such as $\alpha_2$-macroglobulin and $\alpha_1$-acid glycoprotein to albumin most likely occurs when the concentration exceeds 10 μg/mL. Binding to plasma proteins has been shown to be both salt- and pH-dependent, which suggests that the binding is most likely a nonspecific electrostatic interaction. The binding affinity of ASO to $\alpha_2$-macroglobulin was found to be greater than that of ASO to albumin. The dissociation constants were measured as 3.1 μM ($K_{d1}$) and 11.9 μM ($K_{d2}$) for albumin, while the dissociation constants were 4 nM ($K_{d1}$) and 72 nM ($K_{d2}$) for $\alpha_2$-macroglobulin (measured using surface plasma resonance (SPR); internal data, unpublished). However, albumin appears to have a much greater binding capacity (>10-fold) to ASO than does $\alpha_2$-macroglobulin.

Protein binding may explain many other pharmacokinetic properties of this class of compounds. The high degree of protein binding in the circulation prevents any significant urinary excretion. Therefore, there is minimal glomerular filtration of the protein-bound oligonucleotide, and the excretion of intact compound in urine is a minor pathway in its clearance from the plasma [26].

Allometric scaling of plasma clearance estimated at doses ranging from 2 to 5 mg/kg across all species, from rat to human, shows a linear (slope of 0.98–1.01) relationship based on body weight alone (Fig. 4.3). Mouse appears to be the single "outlier", with five- to ten-fold more rapid weight-normalized plasma clearance. The faster plasma clearance in mice appears to be a function of a faster tissue distribution, due to a more rapid circulation to tissues, or larger liver and kidney volumes relative to body weight [34], or to a lower plasma protein-binding capacity in mice [33]. It is not clear which of these mechanisms is the major contributor, but all may play a role in the plasma clearance of ASOs. Nevertheless, similar to oligodeoxynucleotides, the pharmacokinetics of 2′-MOE partially modified ASO scales well across species, utilizing allometric correlation as a function of body weight.

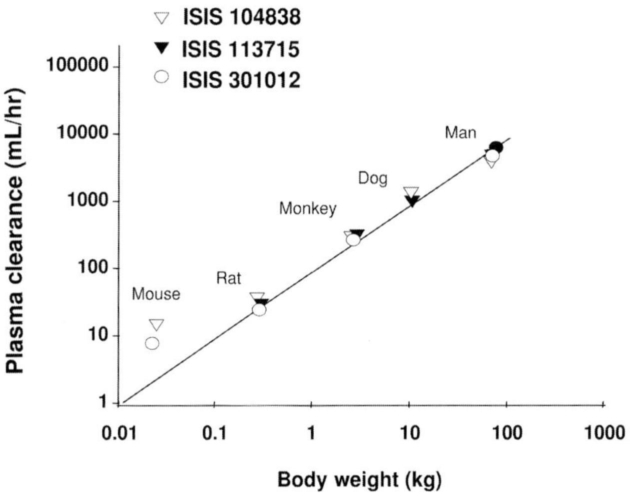


**Fig. 4.3** Allometric relationship of plasma clearance for ISIS 104838, ISIS 113715, and ISIS 301012. Each point represents the average of three to six individuals. Doses ranged from 2 to 5 mg/kg.

### 4.2.2
### Tissue Distribution

As a consequence of rapid plasma clearance during the distribution phase, 2′-MOE partially modified ASOs distribute both widely and extensively into the tissues in mice, rats, and monkeys, with peak concentrations generally achieved by 24–48 h post dose [26, 35]. The characteristics in tissue distribution are similar to those of first-generation ASOs [8, 11, 12, 18, 36, 37]. A representative total oligonucleotide tissue distribution at 24 h following a single 20 mg/kg dose in rats of [$^3$H]-ISIS 104838 and [$^3$H]-ISIS 113715, respectively, is shown in Figure 4.4. The highest concentrations of oligonucleotides in all species studied were found in kidney, liver, spleen, and lymph nodes, but oligonucleotides can be measured in almost every tissue, except brain, at 24 h after IV administration. Consistent with the pattern of distribution, liver and kidney were monitored closely for evidence of toxicity in mouse and monkey toxicology studies.

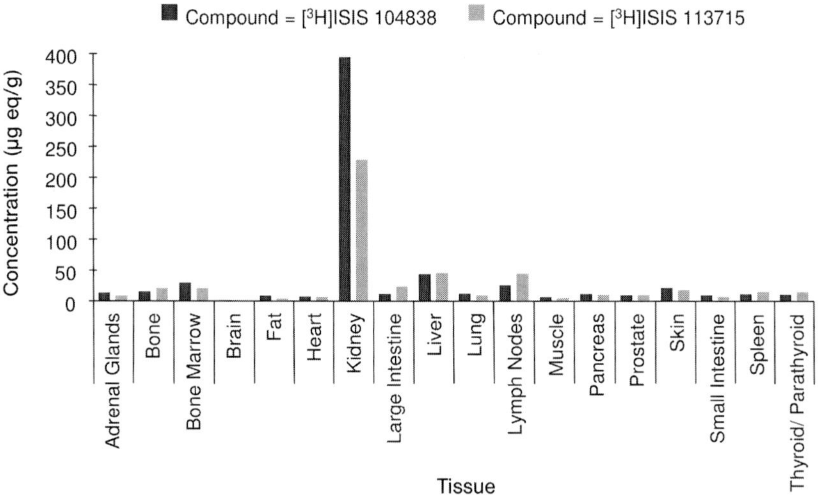


**Fig. 4.4** Tissue distribution of total radioactivity at 24 h after single IV bolus administration of 5 mg/kg [$^3$H]-labeled ISIS 104838 or ISIS 113715 to rats. Concentration values represent total radioactivity concentration in microgram equivalent/g (µg eq/g).

Concentrations of oligonucleotide in the liver and kidney of mice and monkeys after three months of treatment were dose-dependent, but were generally less than dose-proportional (especially in the case of mouse kidney), indicating a saturable process for tissue uptake and distribution. Tissue concentrations of oligonucleotides after four and 13 weeks of treatment were generally higher in monkeys than in mice at comparable dose levels [26]. In both mice and monkeys, it appeared that steady-state concentrations were almost attained in liver and kidney by four weeks with a loading regimen during the first week, because there were only slight increases over the next nine weeks. In line with the tissue half-life of 10 to 30 days, it could take two to three months to reach steady-state concentrations in tissues without a loading regimen.

2′-MOE partially modified ASOs in tissues are cleared slowly over the course of days, and tissue half-lives are approximately five-fold longer than for the first-generation ASOs [8, 14, 18, 22, 26, 29, 35]. The most rapid clearance from tissue was observed in mice, with half-lives of three to seven days [26]. Longer tissue half-lives were seen in monkeys, ranging from approximately 10 to 34 days, depending on the nucleotide sequence (Table 4.3). In rats administered the radiolabeled oligonucleotide, the radiolabel was cleared slowly, with half-lives of 18 to >30 days. The half-life observed in tissues appeared to be consistent with the long terminal plasma half-life. Indeed, the terminal elimination phase observed in the plasma of healthy human volunteers was similar to that seen in monkeys [24, 26]. Plasma concentrations of ISIS 104838 observed during the terminal elimination phase are shown to represent ISIS 104838 that is in equilibrium with tissue(s) (Fig. 4.5), and this provides a measure of the tissue elimination rate (based on nonclinical experience with measured tissue elimination over time). In addition, trough plasma concentrations can be used as a surrogate for predicting liver concentrations (Fig. 4.6). The terminal elimination phase observed in the plasma of healthy human volunteers was similar to that seen in monkeys (see Fig. 4.2) at a similar dose level. In summary, the pharmacokinetics of oligonucleotides in humans is predictable on the basis of pharmacokinetics measured in monkeys, and based on similarities in plasma pharmacokinetics, elimination half-life, and trough accumulation.

**Table 4.3** Elimination half-lives and exposure in liver and kidney across sequences for 2′-MOE partially modified antisense oligonucleotides in monkeys.

| | Half-life (days) | | Conc. (μg/g)[a] at Steady-state | |
|---|---|---|---|---|
| **Sequence** | **Liver** | **Kidney cortex** | **Liver** | **Kidney cortex** |
| ISIS 104838 | 13.0 | 17 | 201 | 436 |
| ISIS 113715 | 7.7 | 16 | 110 | 472 |
| ISIS 301012 | 34.0 | 33 | 126 | 228 |

**a)** Concentrations were normalized to 3 mg/kg per week.

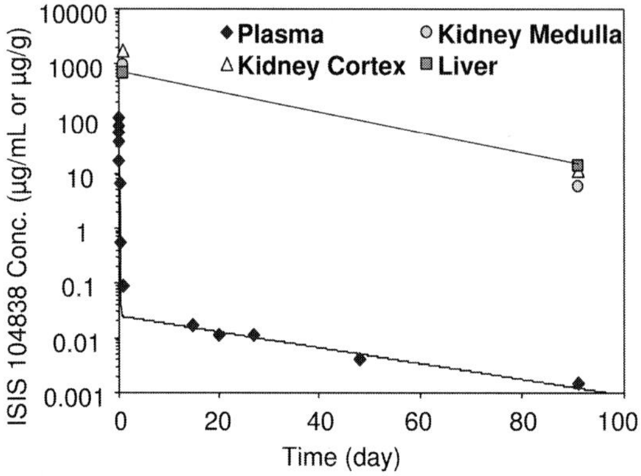


**Fig. 4.5** Correlation of plasma concentrations with tissue concentrations of ISIS 104838 following the final dose after 3 months' administration (10 mg/kg, administered every fourth day, via a 1-h IV infusion).

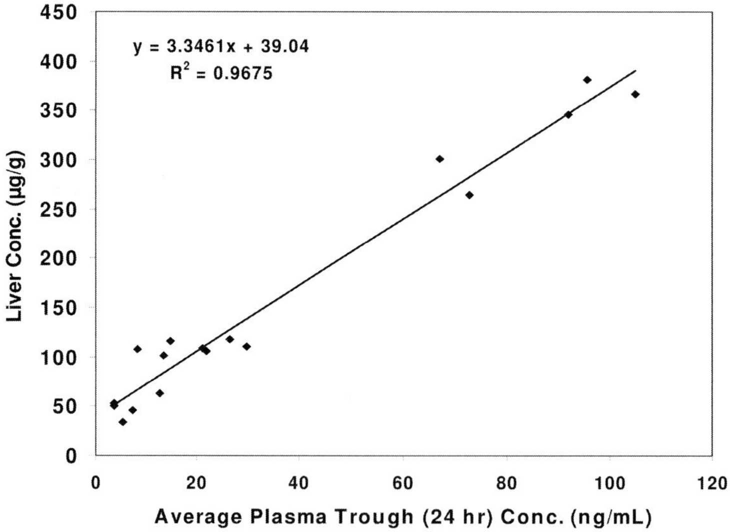


**Fig. 4.6** Correlation of plasma trough concentrations (24 h post dose) with liver concentrations of ISIS 113715 after 3 months' administration (1 to 10 mg/kg administered once weekly, via a 1-h IV infusion).

### 4.2.3 Metabolism

While distribution to tissues relies on the mechanism of plasma clearance, whole-body clearance is the result of metabolism and the excretion of low molecular-weight oligonucleotides. Ubiquitous nucleases are known to metabolize oligonucleotides, although the latter do not serve as substrates for P450 oxidative metabolism.

An HPLC-ES/MS method, which permits the separation of shorter oligonucleotide metabolites from the parent compound, has been used to identify oligonucleotide metabolites generated from tissues and urine following IV administration to rats of ISIS 104838 and ISIS 113715 [26, 35]. A proposed metabolic pathway for ISIS 113715 is illustrated in Fig. 4.7. The first step in the metabolism of oligonucleotides appears to be an endonuclease-mediated cleavage of the deoxynucleotide region in the center of the molecule; cleavage results in two shorter sequences, one with a 5′-MOE-protected terminus, and the other with a 3′-MOE-protected terminus. Subsequently, these endonuclease-cleaved molecules are subjected to exonucleolytic cleavage of the exposed deoxynucleotides to yield shorter oligonucleotide fragments and a mononucleotide or mononucleoside. The thiophosphate group on the nucleotide can be spontaneously oxidized to sulfate and phosphate, cleaved by a phosphatase, or excreted in urine as an entire thiophosphate mononucleotide. Monodeoxynucleotide and monodeoxynucleoside metabolites are identical to endogenous nucleotides, and are catabolized by the normal pathways

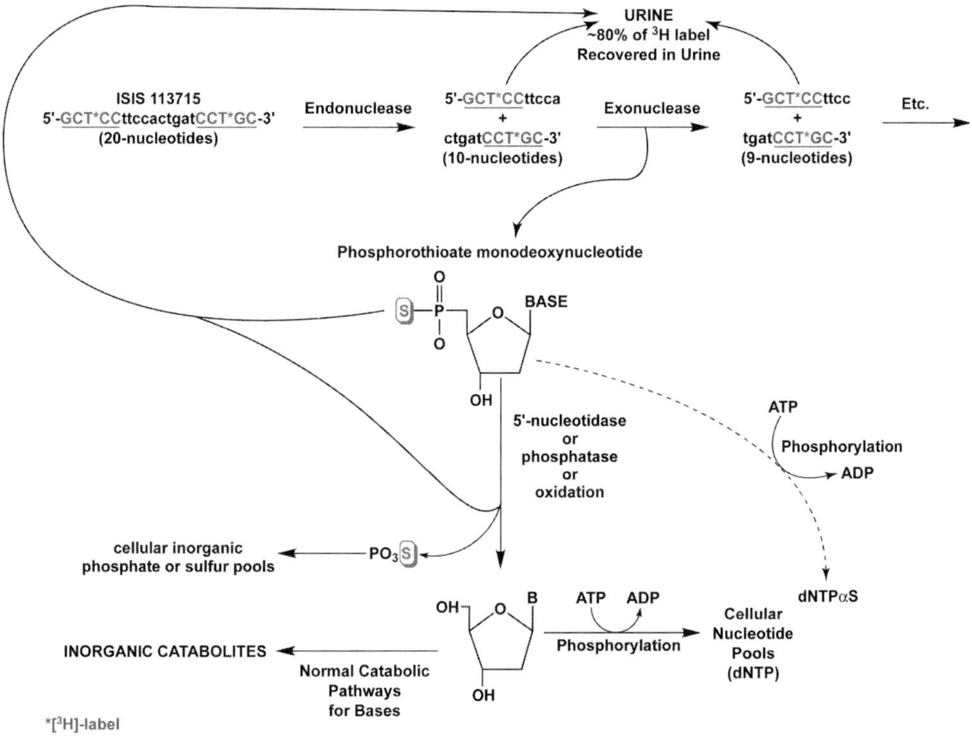


**Fig. 4.7** Metabolic pathways for the degradation of ISIS 104838. Upper-case letters indicate 2′-MOE modified nucleotides; lower-case letters indicate deoxynucleotides.

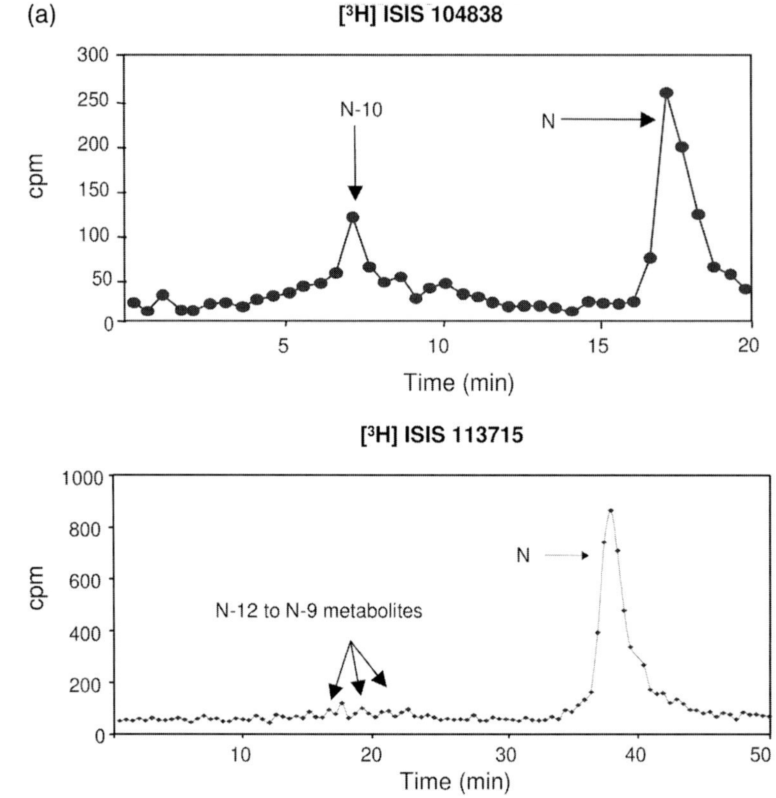


**Fig. 4.8** HPLC radiometric metabolite profiling of extracts of (a) plasma, (b) kidney, and (c) urine samples from rats treated with [$^3$H]-ISIS 104838 or [$^3$H]-ISIS 113715 [35].

for purines and pyrimidines. Metabolites consistent with endonuclease cleavage in the deoxy-portion of the parent oligonucleotides were observed in plasma, tissue and urine, as shown by HPLC with radiometric detection (Fig. 4.8, see also pp. 105 and 106). Although numerous chain-shortened oligonucleotide metabolites were identified in the tissues, the predominant oligonucleotide species was the parent compound at all time points analyzed. These data, when taken together, suggest that exonuclease metabolism of the intact molecule does not play a prominent role in the initial metabolism of the parent oligonucleotide. This metabolic pattern is consistent with the exonuclease-resistant 2′-MOE modifications placed at both the 3′- and 5′-ends of these 2′-MOE partially modified ASOs.

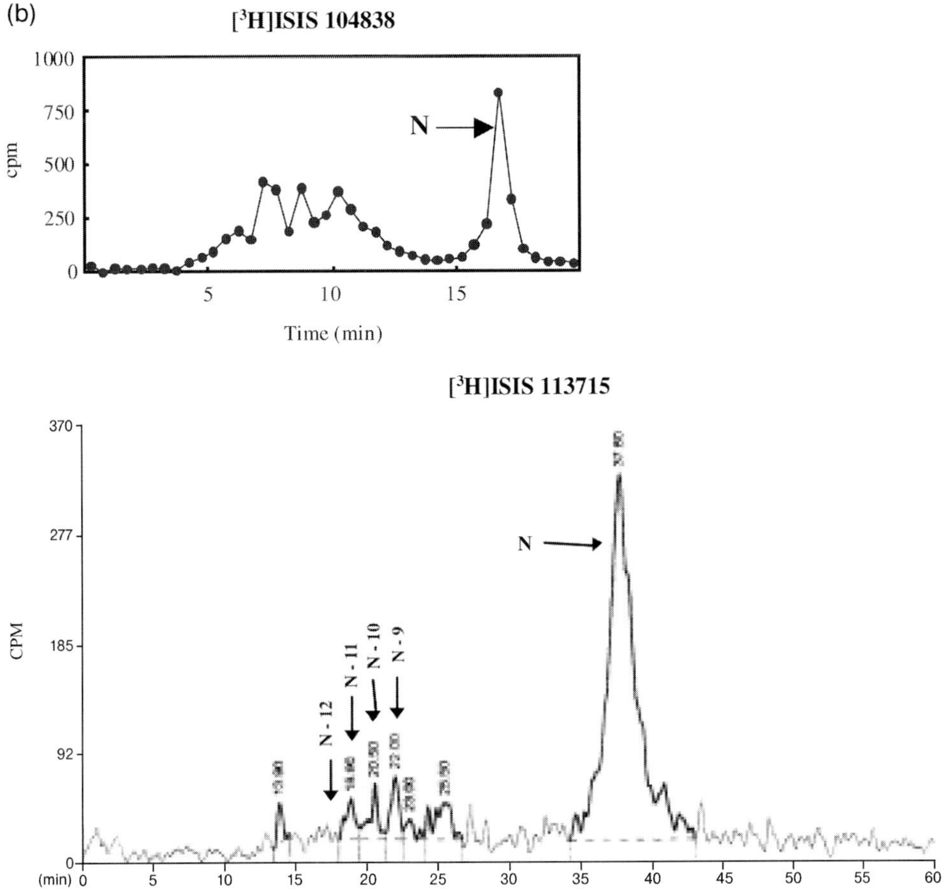


**Fig. 4.8b** (legend see p. 104)

## 4.2.4
## Elimination and Excretion

Similar to PS oligodeoxynucleotide, 2′-MOE partially modified ASOs are highly bound to plasma proteins, which prevents their glomerular filtration and limits urinary excretion (see Section 4.2.1). Therefore, urinary excretion is a minor elimination pathway for 2′-MOE partially modified ASOs across species (Table 4.4), including human [24, 26, 35]. Although remaining low, the percentage of urinary excretion increases with dose in all species, suggestive of a possible saturation of plasma protein binding at higher doses. Protein binding in plasma is relatively weak ($K_d$s in the μM range), and the binding sites of these hydrophilic drugs differ from the binding sites of low molecular-weight hydrophobic drugs. Therefore, few drug–drug interactions on the level of plasma protein binding are expected at clinically relevant concentrations.

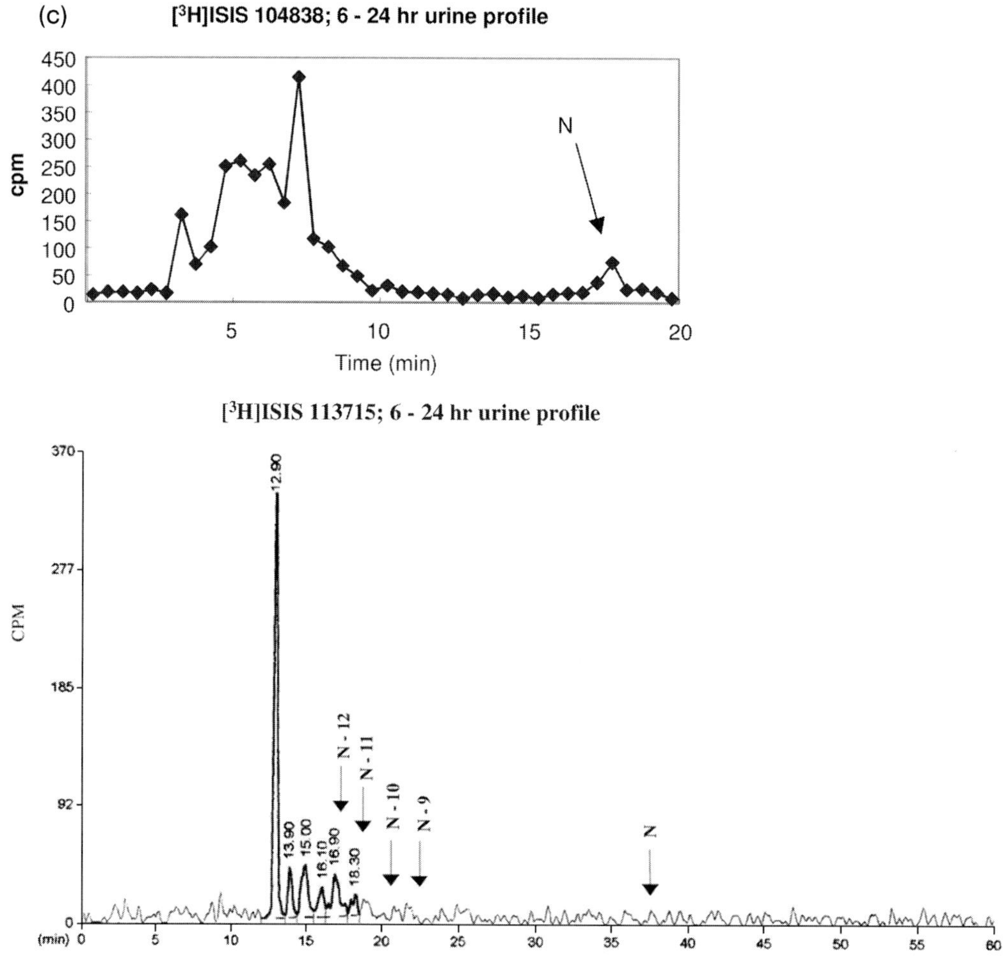


Note: N = parent compound

**Fig. 4.8c** (legend see p. 104)

As discussed in Section 4.2.3, the elimination of 2′-MOE partially modified ASOs is attributed to slow (but continuous) nuclease-mediated metabolism in tissue, followed by ultimate excretion of these shortened metabolites in the urine. This major elimination pathway for 2′-MOE partially modified ASOs is shown graphically in Fig. 4.9, where approximately 75% of the total radiolabeled dose is excreted in the urine by 90 days after a single dose (5 mg/kg) of [$^3$H]-ISIS 104838 to rats [26]. Another example of the ultimate elimination of this class of compound via slow metabolism in tissues, followed by the urinary excretion of metabolites, is the mass-balance study with [$^3$H]-labeled ISIS 113715. Here, between 8% and 37% of administered dose was excreted within the first 24 h, followed by a daily ur-

**Table 4.4** Twenty-four-hour urinary excretion following a single 1-h infusion in monkeys across sequences.

| Sequence | Dose | Route | Dose excreted as parent drug [%] | Dose excreted as total oligo [%] |
|---|---|---|---|---|
| ISIS 104838 | 3 mg/kg | 1-h IV | 0.1 | 1.1 |
| ISIS 113715 | 3 mg/kg[a)] | 1-h IV | 0.02 | 2.3 |
| ISIS 301012 | 4 mg/kg | 1-h IV | 1.3 | 2.3 |
| ISIS 112989 | 3 mg/kg | 1-h IV | 0.11 | 0.68 |

a) Urine samples collected following two doses of ISIS 113715.

inary excretion of ≤1–2% of the administered dose for the next 90 days; this was consistent with a slow tissue clearance. By 90 days post drug administration, the amount of radioactivity associated with the excreta accounted for the majority of the administered dose (>80%), with the larger proportion being recovered in the urine (ca. 70%) and the remainder in the feces [35]. It should be noted that the tritium radiolabel ([$^3$H]) was placed at the 5-C position of thymidine of the 2′-methoxyethyl-modified portion at the 3′-termini for ISIS 104838, and at both the 3′- and 5′-termini for ISIS 113715. The radiochemical purity of these compounds was >94%. Taken together, the metabolism and excretion data obtained from the rat mass-balance study indicate that, although tissue metabolism is slow, it is continuous and represents the primary route of whole-body elimination as oligonucleotide

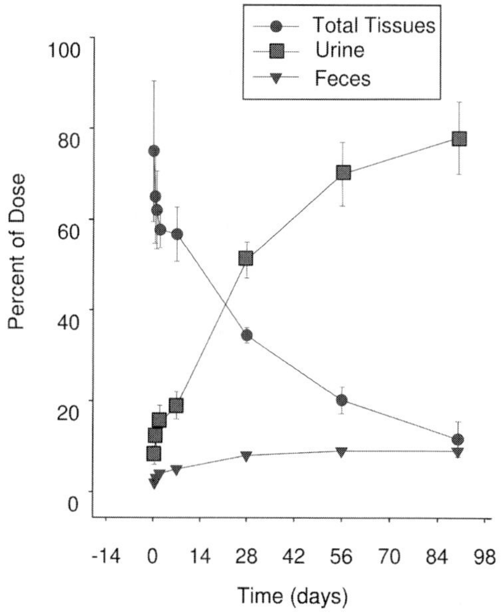


**Fig. 4.9** Mass-balance excretion of radiolabel residue associated with [$^3$H]-ISIS 104838 over 90 days after single 5 mg/kg IV bolus injection [26].

fragments are excreted from the body in urine. These data demonstrate the importance of metabolism and subsequent urinary excretion as the ultimate elimination pathway of the 2′-MOE partially modified ASOs.

## 4.3 Pharmacodynamics

The application of pharmacokinetic and pharmacodynamic analyses to guide and expedite drug development has received increasing interest during recent years [38]. Although these principles are well accepted and widely used for low molecular-weight drugs, only limited reports have been made to date of investigations into the pharmacokinetic and pharmacodynamic relationships of antisense therapeutic agents. A majority of studies with ASOs, both in laboratory animals and in the clinic, have characterized the pharmacology with respect to structure–activity relationships (i. e., sequence and mismatched sequence data) and dose–response relationships [39–42]. Hence, this section will review investigations into the correlation between oligonucleotide concentrations (exposure) in target organs or plasma (as a surrogate) and pharmacological effects.

The mechanism of action for antisense compounds is to inhibit gene expression sequence-specifically by hybridization to mRNA through Watson–Crick base-pair interactions; this is followed by degradation of the target mRNA through an RNase H-dependent terminating mechanism [43]. Consequently, the ASO prevents translation of the encoded protein product, or the disease-causing factor in a highly sequence-specific manner. Because of the unique mechanism of action of antisense therapeutics, investigations of the pharmacological effects of antisense oligonucleotide *in vivo* have focused primarily on:

- target mRNA reduction and the subsequent reduction in protein translation;
- downstream effects resulting from target protein reduction, which are dependent on the target studied; or
- clinical outcomes dependent upon disease indication.

Therefore, this review of pharmacokinetic/pharmacodynamics (PK/PD) correlation will include investigations between the effective concentrations at the target sites of antisense oligonucleotides with each of the pharmacological effects discussed above. Moreover, an establishment of the correlation between plasma equilibrium concentrations with concentrations at the target sites is pertinent, enabling plasma concentrations to be used as a surrogate in clinical studies to establish relationships between pharmacodynamics and pharmacokinetics.

Although the mechanisms of action for antisense oligodeoxynucleotide and 2′-MOE partially modified ASOs are the same, the 2′-MOE modifications provide increased affinity to the target mRNA while maintaining favorable RNase H activity and, therefore, enhanced potency and specificity over the first-generation ASOs [28, 40, 44, 45]. The pharmacodynamic section of this chapter will, once again, focus on 2′-MOE partially modified ASOs.

### 4.3.1
### Pharmacological Endpoint: Reduction of Target mRNA and Protein

Because the direct pharmacological response of antisense therapeutics is target mRNA reduction, establishment of the correlation of target organ concentration and target mRNA reduction is most widely used in studying the PK/PD relationships in animal models. Furthermore, any reduction in target mRNA levels correlates directly with a subsequent reduction in target protein levels. However, because target protein analysis relies on Western blotting and is considered semi-quantitative, protein levels are often used only on a confirmatory basis when assessing target reduction [46, 47]. Thus, any estimation of ASO pharmacodynamics is generally based on target mRNA expression.

Although ASOs are widely and extensively distributed to the tissues, the highest concentrations of oligonucleotides are always found in the kidney and liver (see Section 4.2.2). As such, liver targets are increasingly attractive due to the unique pattern of ASO biodistribution. Consequently, several PK/PD studies have been reported for 2′-MOE partially modified ASOs targeting the liver [46–49]. PK/PD relationships have also been demonstrated in the human synovium, using a 2′-MOE partially modified ASO, ISIS 104838, to target TNF-$\alpha$ in patients with rheumatoid arthritis [50], and also in the prostate gland, with a 2′-MOE partially modified ASO, OGX-011 targeting the clusterin mRNA in patients with prostatic cancer [42]. To date, however, no PK/PD studies targeting other tissues or organs with 2′-MOE partially modified ASOs have been reported.

One of the first reported applications of PK/PD modeling to characterize antisense PK/PD relationships was a study of a 2′-MOE partially modified ASO, ISIS 22023, targeting murine Fas mRNA expressed exclusively in hepatocytes [46, 47]. In this study, the PK/PD relationships of ISIS 22023 were characterized not only in the target organ (the mouse liver), but also in sub-organ levels within the liver. The results showed that the time course of mRNA reduction (pharmacological activity) correlated closely with the time course of ISIS 22023 concentrations in hepatocytes. A subsequent downstream effect was also demonstrated using protection from Fas antibody-induced fulminant hepatitis, consistent with target mRNA reduction and subsequent reduction in translation of Fas protein in the liver. Differential kinetics were observed in different cell types within the liver, where ISIS 22023 was cleared more rapidly from hepatocytes than from Kupffer and endothelial cells; hence, the kinetics of ISIS 22023 in the whole liver was not representative of the kinetics in the hepatocyte (Fig. 4.10). Moreover, the pharmacodynamics of ISIS 22023 correlated better with the pharmacokinetics in hepatocytes, supporting the concept that the presence of oligonucleotide in target cells results in a reduction in mRNA and, ultimately, in pharmacological activity. Using an integrated PK/PD model, the estimated $EC_{50}$ in hepatocytes for ISIS 22023 was $29.0 \pm 5.3$ µg/g liver. In summary, the results of this study demonstrated apparent differences in the patterns of uptake and elimination in different types of liver cell. The results also provide a comprehensive understanding of the kinetics of an antisense drug at the site of action, and show that the reductions in mRNA in-

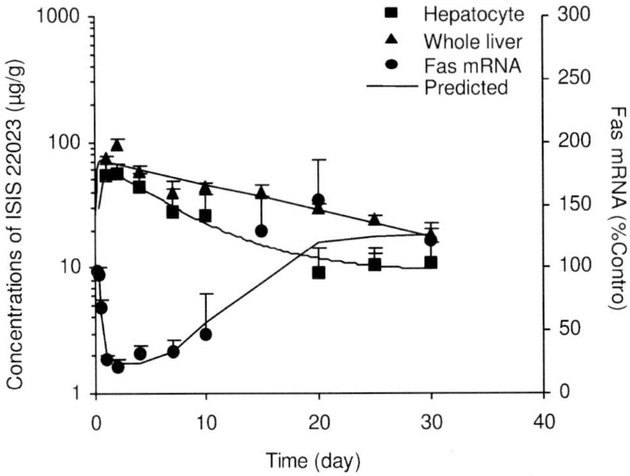


**Fig. 4.10** Pharmacokinetic/pharmacodynamic relationship of ISIS 22023 in mouse liver following SC administration of a single 50 mg/kg dose to mice. Symbols represent observed concentrations or Fas mRNA levels (error bars represent standard deviation for ISIS 22023 concentrations, or standard error for Fas mRNA levels; n = 3). Solid lines represent predicted ISIS 22023 concentrations or Fas mRNA levels using nonlinear regression.

duced by this antisense oligonucleotide correlate with its concentrations in the target cell. Based on the estimated PK/PD parameters from this study, a 24 mg/kg dose given once weekly should maintain liver concentrations of ISIS 22023 in excess of the $EC_{50}$, and Fas mRNA reduction greater than 50%, thus providing sufficient protection for mice subjected to an agonistic and normally lethal Fas antibody challenge. Indeed, results from a 20-week long-term study using the above simulation-based regimen showed reasonable correlation between model-predicted ISIS 22023 concentrations and Fas mRNA reduction with observed values (Fig. 4.11). Moreover, the targeted downstream effect, 100% protection, was achieved for up to 20 weeks in mice receiving ISIS 22023 treatment and subjected to agonistic Fas antibody challenge (R.Z. Yu, unpublished data).

An additional study which was performed with ISIS 116847, targeting mouse, rat, and human putative protein tyrosine phosphatase (PTEN) mRNA, showed that ASO concentrations in the target cell are critical for activity [48]; these data also indicated that dose regimen modifications can be used to optimize exposure. In this study, three treatment regimens were used to deliver the same total dose (200 mg/kg) over a four-week period: 25 mg/kg twice a week (BIW), 7.14 mg/kg daily (D), or 20 mg/kg daily as loading for the first five days, followed by 33 mg/kg once weekly maintenance (L/M). The results showed that whole-liver concentrations of ISIS 116847 were similar for the BIW regimen compared to the L/M regimen. However, hepatocyte concentrations of ISIS 116847 were doubled for the BIW regimen compared to the L/M regimen, suggesting more rapid clearance in hepatocytes than in the whole liver, since rats

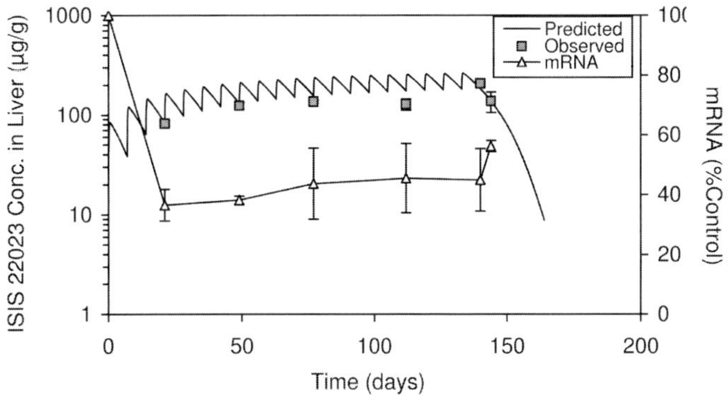


**Fig. 4.11** Correlations between observed and model-predicted ISIS 22023 concentrations and Fas mRNA expression in mouse liver after once-weekly SC administration of ISIS 22023 (24 mg/kg) for 20 weeks. Symbols represent observed concentrations or Fas mRNA levels (error bars represent standard deviation for ISIS 22023 concentrations, or standard error for Fas mRNA levels; n = 3). Solid lines represent predicted ISIS 22023 concentrations or Fas mRNA levels using nonlinear regression.

were treated two times less frequently with the L/M regimen compared to the BIW regimen. Subcellular trafficking differences across cell-types in the liver may also explain differences in activity.

Differences in ASO pharmacodynamics between species were noted in diabetic animal models [49]. In a PK/PD study with ISIS 116847, db/db mice and fa/fa rats were treated with doses ranging from 0 to 50 mg/g administered once weekly for four weeks. An inhibitory effect sigmoid $E_{max}$ model was used to fit the pharmacodynamic data. The liver $EC_{50}$ was estimated to be approximately 50 µg/g in mice, and 150 µg/g in rats. This species difference in $EC_{50}$ is likely related to species-specific sub-organ distribution in the liver, as observed previously with first-generation ASOs, where oligonucleotide concentrations in hepatocytes were two-fold higher in mice compared to rats at similar doses, and with the greatest differences being found in the nuclear and membrane fractions [51]. Further studies with other targets in different animal models are warranted in order to provide a better understanding of species similarities and differences in pharmacodynamics.

In the clinic, PK/PD relationships with the direct pharmacological effect, reduction of target mRNA and protein as the response measure, have been demonstrated in two studies with 2′-MOE partially modified ASOs [42, 50]. In a Phase IIa study in patients with rheumatoid arthritis, ISIS 104838 was targeted at the mRNA of TNF-α (a cytokine which is important in immune response and inflammation). The results showed that TNF-α mRNA and protein reduction in synovial tissue biopsies correlated positively with concentrations of ISIS 104838 in the synovial tissues [50]. Another investigation of PK/PD relationships was conducted in a Phase II dose-ranging study in prostate cancer patients treated with OGX-011, a 2′-MOE partially modified ASO targeting clusterin mRNA (weekly dose 320 to 640 mg) [42]. The clusterin gene encodes a cytoprotective chaperone protein that promotes cell survival and confers broad-spectrum resistance. Inhibition of clusterin expression in prostate cancer cells, as assessed by several methods such as quantitative real-time PCR and immunohistochemistry, occurred in a dose- and prostate drug concentration-dependent manner. The study results led to a recommended Phase II dose of OGX-011 of 640 mg administered once weekly, based on the optimal biologic activity in inhibiting clusterin expression and tolerability in humans at this dose.

The establishment of PK/PD relationships using plasma exposure has been of long-term interest in this field. However, PT oligonucleotides, including those with MOE-modified termini, are rapidly cleared from plasma within hours and distributed to the tissues. Hence, the elimination phase – which is in equilibrium with tissue levels – could not be quantitated with earlier assays [52–54]. However, with the development of highly sensitive and selective hybridization ELISA methods [24, 26, 30], characterization of the full pharmacokinetic behavior of antisense oligonucleotides is now possible. This includes monitoring the terminal elimination phase, which can be used to correlate plasma and tissue concentrations for PK/PD studies, and to establish exposure–response relationships using trough plasma levels. When the ELISA method first became available, a link-model was developed to correlate plasma kinetics with liver Fas mRNA reduction following a single IP dose of 50 mg/kg ISIS 22023 in mice (Fig. 4.12). The estimated half-life at the effect site was 4.3 days (unpublished data), which was shorter than that in the whole liver (18.8 days) [46]. These data were consistent with observations of differential sub-organ kinetics within the mouse liver, where ISIS 22023 is cleared more rapidly from hepatocytes (the effect site) than from the whole liver.

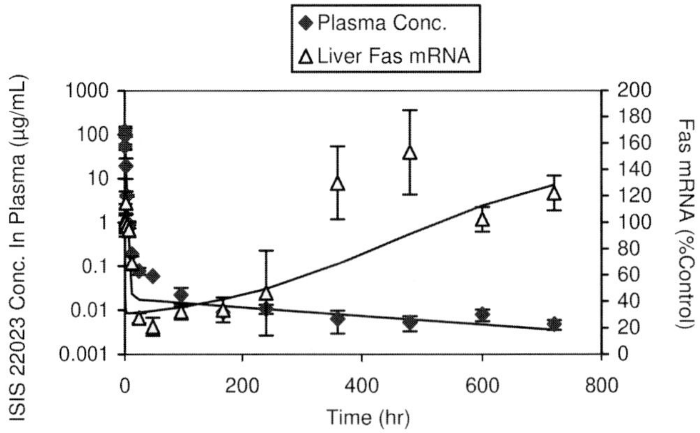


**Fig. 4.12** Pharmacokinetic/pharmacodynamic relationship of ISIS 22023 in mouse plasma with liver Fas mRNA reduction following subcutaneous administration of a single dose (50 mg/kg). Symbols represent observed concentrations or Fas mRNA levels (error bars represent standard deviation for ISIS 22023 concentrations, or standard error for Fas mRNA levels; n = 3). Solid lines represent predicted ISIS 22023 concentrations in plasma or Fas mRNA levels using nonlinear regression.

### 4.3.2
### Pharmacological Endpoint: Downstream Effects

The downstream effects, subsequent to the direct pharmacological effects of inhibition of target mRNA expression and protein translation, will depend on which gene the ASO targets. Some of the downstream effects are easily quantified with plasma biomarkers; for example, the measurement of plasma glucose or insulin for ASOs targeting metabolic diseases, or serum apoB-100 and LDL-cholesterol for ASOs targeting hyperlipidemia. However, some downstream effects cannot easily be quantified or have no assessable biomarkers; for example, ASOs targeting inflammatory conditions such as Crohn's disease or rheumatoid arthritis (RA). A semi-quantitative assessment scoring system is generally used when evaluating the downstream effect or clinical endpoint for inflammatory diseases (see Section 4.3.3).

To date, PK/PD relationships using downstream effects have not been reported in the literature. However, dose-dependent downstream effects have been demonstrated in numerous animal models and humans for a number of targets. For example, a significant dose-dependent decrease in plasma glucose and/or insulin was shown in diabetic mouse and rat models with ISIS 116847 targeting PTEN [39], with ISIS 113715 targeting protein-tyrosine phosphatase 1B (PTP1B) [55], and with several 2′-MOE partially modified ASOs targeting the glucagon receptor [56] or glucocorticoid receptor (GCCR) [57]. In a mouse model bearing a PC-3 tumor, a clusterin ASO demonstrated a dose-dependent reduction in PC-3 tumor volumes [40]. More recently, a mouse-specific 2′-MOE partially modified antisense compound, ISIS 147764, targeting apoB-100, has been shown to cause reductions in liver and serum apoB-100 levels in several mouse models of hyperlipidemia in both dose- and time-dependent manner [58]. Consistent with these findings, total cholesterol and LDL-cholesterol were also decreased accordingly. These findings, together with data from a Phase I trial with a human apoB-100 antisense drug [59], suggest that it might be possible not only to exploit this target for therapeutic activity but also, because of the easy readout of pharmacologic activity, to dissect out the details of the PK/PD relationship. Information gathered from these studies will clearly aid in the development of other antisense therapeutics.

### 4.3.3
### Relationship between ASO Pharmacokinetics and Clinical Outcome

In clinical studies, the clinical outcome or clinical endpoint are generally used as a measure of a drug's pharmacological effect. Depending on the disease indication, the clinical outcome can be either quantitative or semi-quantitative. Very few published studies have investigated the relationships between exposure and clinical outcome, or PK/PD relationships for ASOs. These investigations have included therapeutic areas of inflammatory diseases [24, 60], oncology [61, 62], and metabolic diseases including lipid-lowering agents [59, 63, 64]. In these studies, ASO was used either concomitantly with other marketed agents of different mechanisms, or as monotherapy.

Three approaches have been used to estimate pharmacodynamics using plasma concentrations as surrogate for target tissue exposure:

- Studying the correlation between clinical endpoint and total plasma exposure (plasma AUC or cumulative AUC).
- Estimation of plasma $EC_{50}$ by linking plasma kinetics to target dynamics and building a plasma PK/PD link model.
- Estimation of plasma $EC_{50}$ by an inhibitory effect $E_{max}$ model at equilibrium.

An example of the first approach is a population-based PK/PD study for a first-generation ASO, ISIS 2302, targeted to human intercellular adhesion molecule-1 (ICAM-1) mRNA for the treatment of Crohn's disease [60, 65]. A correlation between the plasma AUC and clinical endpoint (complete clinical remission rate) was established using a linear mixed effect model, and the exposure–response model was used to predict clinical outcome in a Phase III trial simulation [66]. Another example of the first approach is a Phase IIa study with ISIS 104838 targeting TNF-α mRNA [50]. Clinical outcome, defined as the response rate of achieving the American College of Rheumatology 20% improvement-criteria (achievement of ACR 20), was both dose- and exposure-dependent. The response was sustained for up to two months after treatment, consistent with the drug's long tissue half-lives.

A PK/PD relationship in humans has recently been reported for a novel hypolipidemic agent, ISIS 301012, a 2′-MOE partially modified antisense inhibitor of human apoB-100 [59, 64]. ApoB-100 is a structural component of LDL-cholesterol, and plays a key role in its metabolism and transport. Preliminary data from this study indicate that antisense inhibition of apoB-100 results in both dose- and concentration-dependent sustained reductions in serum levels of apoB-100, LDL-cholesterol and total cholesterol, but does not affect levels of HDL-cholesterol.

With the development of a sensitive hybridization-based ELISA bioanalytical assay [24, 30], it has been possible to measure the trough plasma concentrations and to define the terminal elimination kinetics of ASOs. In fact, plasma concentrations in the terminal elimination phase have been shown to represent ASO that is in equilibrium with tissue(s), and thus to provide a measure or surrogate of tissue exposure and elimination rate (based on nonclinical experience with measured tissue elimination over time) [31]. In addition, trough plasma concentrations can be used as a surrogate for predicting liver concentrations, and thus can be used as surrogate in estimating ASO pharmacodynamics in humans (see Fig. 4.6). Additional PK/PD analyses with linking plasma exposure at equilibrium to clinical endpoints are ongoing.

This chapter has described the pharmacokinetics of PT 2′-MOE partially modified ASO following parenteral administration, and has shown that ASOs in this chemical class distribute extensively to many tissue types, with prolonged half-lives. Moreover, exposure–response relationships have been established in both animal models and humans. However, there are a few tissues to which parenterally administered oligonucleotides do not distribute, or are distributed minimally; these include brain, muscle, eyes, and skin. Consequently, ASOs targeted to these tissues will require local delivery, as outlined in Chapter 10.

## 4.4 Summary

The pharmacokinetic characteristics of single-strand antisense PT oligonucleotides are governed by the molecule's chemical class, and are largely independent of sequence. The reliance of broad whole-body biodistribution of ASOs on plasma protein binding must not be underestimated. Alternative chemical modifications have, on occasion, failed to take this important property into consideration, and this has led to failures in distribution and loss of compound via rapid urinary excretion. Future gains are likely to be attributed to chemical modifications that balance plasma protein binding with optimal target binding.

Not surprisingly, the pharmacodynamics of ASOs are closely linked to target tissue and cell pharmacokinetics. Indeed, it is this property that demands careful consideration for the ultimate selection of an appropriate molecular target for this technology. If ASOs are to be rationally targeted, their use must take into consideration their intrinsic biodistribution, both to whole organs and to the various cell types within that organ. Particular attention in this chapter has been focused on the liver and, more specifically, upon the hepatocyte within the liver – an organ and cell type that is particularly well-served by the intrinsic biodistribution of ASOs.

The PK/PD properties of ASOs provide a framework for rapid drug development. Challenges to the promise of antisense therapeutics are not unlike those for any other new chemical entity, and they include proper application of the pharmacokinetic properties in order to select both target and delivery methods. The profound difference from the situation of small molecule discovery is the relative predictability of the pharmacokinetics, independent of target selection.

## 4.5 References


1 Watson, J., and F. Crick. **1953**. Molecular structure of nucleic acids: a structure for deoxyribose nucleic acid. *Nature* 171:737.

2 Lima, W.F., M. Venkatraman, and S.T. Crooke. **1997**. The influence of antisense oligonucleotide-induced RNA structure on E. coli RNase H1 activity. *J. Biol. Chem.* 272:18191–18199.

3 Manoharan, M. **1999**. 2′-Carbohydrate modifications in antisense oligonucleotide therapy: importance of conformation, configuration and conjugation. *Biochim. Biophys. Acta* 1489:117–130.

4 Henry, S.P., R.S. Geary, R. Yu, and A.A. Levin. **2001**. Drug properties of second-generation antisense oligonucleotides: how do they measure up to their predecessors? *Curr. Opin. Invest. Drugs* 2:1444–1449.

5 Geary, R.S., J.M. Leeds, S.P. Henry, D.K. Monteith, and A.A. Levin. **1997**. Antisense oligonucleotide inhibitors for the treatment of cancer: 1. Pharmacokinetic properties of phosphorothioate oligodeoxynucleotides. *Anti-Cancer Drug Design* 12:383–393.

6 Leeds, J.M., R.S. Geary, S.P. Henry, J. Glover, W. Shanahan, J. Fitchett, T. Burckin, L. Truong, and A.A. Levin. **1997**. Pharmacokinetic properties of phosphorothioate oligonucleotides. *Nucleosides Nucleotides* 16:1689–1693.

**7** Yu, R.Z., R.S. Geary, T. Ushiro-Watanabe, A.A. Levin, and S.L. Schoenfeld. **2001**. Pharmacokinetic properties in humans. In: S.T. Crooke (Ed.), *Antisense Drug Technology: Principles, Strategies, and Applications.* Marcel Dekker, Inc., New York, pp. 183–200.

**8** Agrawal, S., J. Temsamani, W. Galbraith, and J. Tang. **1995**. Pharmacokinetics of antisense oligonucleotides. *Clin. Pharmacokinet.* 28 : 7–16.

**9** Zhang, R., R.B. Diasio, Z. Lu, T. Liu, Z. Jiang, W.M. Galbraith, and S. Agrawal. **1995**. Pharmacokinetics and tissue distribution in rats of an oligodeoxynucleotide phosphorothioate (Gem 91) developed as a therapeutic agent for human immunodeficiency virus type-1. *Biochem. Pharmacol.* 49 : 929–939.

**10** Sands, H., L.J. Gorey-Feret, A.J. Cocuzza, F.W. Hobbs, D. Chidester, and G.L. Trainor. **1994**. Biodistribution and metabolism of internally $^{3}H$-labeled oligonucleotides. I. Comparison of a phosphodiester and a phosphorothioate. *Mol. Pharmacol.* 45 : 932–943.

**11** Grindel, J.M., T.J. Musick, Z. Jiang, A. Roskey, and S. Agrawal. **1998**. Pharmacokinetics and metabolism of an oligodeoxynucleotide phosphorothioate (GEM91) in cynomolgus monkeys following intravenous infusion. *Antisense Nucleic Acid Drug Dev.* 8 : 43–52.

**12** Cossum, P.A., H. Sasmor, D. Dellinger, L. Truong, L. Cummins, S.R. Owens, P.M. Markham, J.P. Shea, and S.T. Crooke. **1993**. Disposition of the $^{14}C$-labeled phosphorothioate oligonucleotide ISIS 2105 after intravenous administration to rats. *J. Pharmacol. Exp. Ther.* 267 : 1181–1190.

**13** Cossum, P.A., L. Troung, S.R. Owens, P.M. Markham, J.P. Shea, and S.T. Crooke. **1994**. Pharmacokinetics of a $^{14}C$-labeled phosphorothioate oligonucleotide, ISIS 2105, after intradermal administration to rats. *J. Pharmacol. Exp. Ther.* 269 : 89–94.

**14** Yu, R.Z., R.S. Geary, J.M. Leeds, T. Ushiro-Watanabe, M. Moore, J. Fitchett, J. Matson, T. Burckin, M.V. Templin, and A.A. Levin. **2001**. Comparison of pharmacokinetics and tissue disposition of an antisense phosphorothioate oligonucleotide targeting human Ha-*ras* mRNA in mouse and monkey. *J. Pharm. Sci.* 90 : 182–193.

**15** Glover, J.M., J.M. Leeds, T.G. Mant, D. Amin, D.L. Kisner, J.E. Zuckerman, R.S. Geary, A.A. Levin, and W.R. Shanahan, Jr. **1997**. Phase I safety and pharmacokinetic profile of an intercellular adhesion molecule-1 antisense oligodeoxynucleotide (ISIS 2302). *J. Pharmacol. Exp. Ther.* 282 : 1173–1180.

**16** Sereni, D., R. Tubiana, C. Lascoux, C. Katlama, O. Taulera, A. Bourque, A. Cohen, B. Dvorchik, R.R. Martin, C. Tournerie, A. Gouyette, and P.J. Schechter. **1999**. Pharmacokinetics and tolerability of intravenous trecovirsen (GEM 91), an antisense phosphorothioate oligonucleotide, in HIV-positive subjects. *J. Clin. Pharmacol.* 39 : 47–54.

**17** Leeds, J.M., and R.S. Geary. **1998**. Pharmacokinetic properties of phosphorothioate oligonucleotides in humans. In: S.T. Crooke (Ed.), *Antisense Research and Applications.* Springer, Heidelberg, pp. 217–231.

**18** Phillips, J.A., S.J. Craig, D. Bayley, R.A. Christian, R.S. Geary, and P.L. Nicklin. **1997**. Pharmacokinetics, metabolism and elimination of a 20-mer phosphorothioate oligodeoxynucleotide (CGP 69846A) after intravenous and subcutaneous administration. *Biochem. Pharmacol.* 54 : 657–668.

**19** Rifai, A., W. Byrsch, K. Fadden, J. Clark, and K.-H. Schlingensipen. **1996**. Clearance kinetics, biodistribution, and organ saturability of phosphorothioate oligodeoxynucleotides in mice. *Am. J. Pathol.* 149 : 717–725.

**20** Crooke, S.T., L.R. Grillone, A. Tendolkar, A. Garrett, M.J. Fratkin, J. Leeds, and W.H. Barr. **1994**. A pharmacokinetic evaluation of $^{14}C$-labeled afovirsen sodium in patients with genital warts. *Clin. Pharmacol. Ther.* 56 : 641–646.

**21** Geary, R.S., O. Khatsenko, K. Bunker, R. Crooke, M. Moore, T. Burckin, L. Truong, H. Sasmor, and A.A. Levin. **2001**. Absolute bioavailability of 2′-O-(2-methoxyethyl)-modified antisense oligonucleotides following intraduodenal instillation in rats. *J. Pharmacol. Exp. Ther.* 296 : 898–904.

**22** Geary, R.S., T. Ushiro-Watanabe, L. Truong, S.M. Freier, E.A. Lesnik, N.B. Sioufi, H. Sasmor, M. Manoharan, and A.A. Levin. **2001**. Pharmacokinetic properties of 2′-O-(2-methoxyethyl)-modified oligonucleotide analogs in rats. *J. Pharmacol. Exp. Ther.* 296 : 890–897.

**23** Zhang, R.W., R.P. Iyer, D. Yu, W.T. Tan, X.S. Zhang, Z.H. Lu, H. Zhao, and S. Agrawal. **1996**. Pharmacokinetics and tissue disposition of a chimeric oligodeoxynucleoside phosphorothioate in rats after intravenous administration. *J. Pharmacol. Exp. Ther.* 278 : 971–979.

**24** Sewell, L.K., R.S. Geary, B.F. Baker, J.M. Glover, T.G.K. Mant, R.Z. Yu, J.A. Tami, and A.F. Dorr. **2002**. Phase I trial of ISIS 104838, a 2′-methoxyethyl modified antisense oligonucleotide targeting tumor necrosis factor-alpha. *J. Pharmacol. Exp. Ther.* 303 : 1334–1343.

**25** Zhang, R., H. Wang, and S. Agrawal. **2005**. Novel antisense anti-MDM2 mixed-backbone oligonucleotides: Proof of principle, in vitro and in vivo activities, and mechanisms. *Curr. Cancer Drug Targets* 5 : 43–49.

**26** Geary, R.S., R.Z. Yu, T. Watanabe, S.P. Henry, G.E. Hardee, A. Chappell, J. Matson, H. Sasmor, L. Cummins, and A.A. Levin. **2003**. Pharmacokinetics of a tumor necrosis factor-alpha phosphorothioate 2′-O-(2-methoxyethyl) modified antisense oligonucleotide: comparison across species. *Drug Metab. Dispos.* 31 : 1419–1428.

**27** Manoharan, M. **2004**. RNA interference and chemically modified small interfering RNAs. *Curr. Opin. Chem. Biol.* 8 : 570–579.

**28** McKay, R.A., L.J. Miraglia, L.L. Cummins, S.R. Owens, H. Sasmor, and N.M. Dean. **1999**. Characterization of a potent and specific class of antisense oligonucleotide inhibitor of human protein kinase C-α expression. *J. Biol. Chem.* 274 : 1715–1722.

**29** Yu, R.Z., R.S. Geary, D.K. Monteith, J. Matson, L. Truong, J. Fitchett, and A.A. Levin. **2004**. Tissue disposition of a 2′-O-(2-methoxy) ethyl modified antisense oligonucleotides in monkeys. *J. Pharm. Sci.* 93 : 48–59.

**30** Yu, R.Z., B. Baer, A. Chappel, R.S. Geary, E. Chueng, and A.A. Levin. **2002**. Development of an ultrasensitive noncompetitive hybridization-ligation enzyme-linked immunosorbent assay for the determination of phosphorothioate oligodeoxynucleotide in plasma. *Anal. Biochem.* 304 : 19–25.

**31** Yu, R.Z., J. Matson, and R.S. Geary. **2001**. Terminal elimination rates for antisense oligonucleotides in plasma correlate with tissue clearance rates in mice and monkeys. In: Annual Meeting of American Association of Pharmaceutical Scientists, Denver, Colorado.

**32** Levin, A.A., R.S. Geary, J.M. Leeds, D.K. Monteith, R.Z. Yu, M.V. Templin, and S.P. Henry. **1998**. The pharmacokinetics and toxicity of phosphorothioate oligonucleotides. In: J.A. Thomas (Ed.), *Biotechnology and Safety Assessment.* Taylor & Francis, Philadelphia, PA, pp. 151–175.

**33** Watanabe, T.A., R.S. Geary, and A.A. Levin. **2004**. In vitro protein binding and drug interaction studies of an antisense oligonucleotide (ASO), Isis 113715, targeting human Ptb1 b mRNA. In: AAPS Annual Meeting and Exposition. American Association of Pharmaceutical Scientists, Baltimore, MD.

**34** Davies, B., and T. Morris. **1993**. Physiological parameters in laboratory animals and humans. *Pharm. Res.* 10 : 1093–1095.

**35** Grundy, J.S., R.Z. Yu, R.S. Geary, A.A. Levin, S. Zhao, A. Keyhani, and G. Piccirilli. **2002**. Comparative single-dose disposition of radiolabeled ISIS 104838 and ISIS 113715 in rat. In: Annual Meeting of Pharmaceutical Scientists, Toronto, Canada.

**36** Geary, R.S., J.M. Leeds, J. Fitchett, T. Burckin, L. Troung, C. Spainhour, M. Creek, and A.A. Levin. **1997**. Pharmacokinetics and metabolism in mice of a phosphorothioate oligonucleotide antisense inhibitor of C-raf-1 kinase expression. *Drug Metab. Dispos.* 25 : 1272–1281.

**37** Goodarzi, G., M. Watabe, and K. Watabe. **1992**. Organ distribution and stability of phosphorothioated oligodeoxyribonucleotides in mice. *Biopharm. Drug Dispos.* 13 : 221–227.

**38** Galluppi, G.R., M.C. Rogge, L.K. Roskos, L.J. Lesko, M.D. Green, D.W. Feigal, Jr., and C.C. Peck. **2001**. Integration of pharmacokinetic and pharmacodynamic studies in the discovery, development, and review of protein therapeutic agents: a conference report. *Clin. Pharmacol. Ther.* 69 : 387–399.

**39** Butler, M., R.A. McKay, I.J. Popoff, W.A. Gaarde, D. Witchell, S.F. Murray, N.M. Dean, S. Bhanot, and B.P. Monia. **2002**. Specific inhibition of PTEN expression reverses hyperglycemia in diabetic mice. *Diabetes* 51 : 1028–1034.

**40** Zellweger, T., H. Miyake, S. Cooper, K. Chi, B.S. Conklin, B.P. Monia, and M.E. Gleave. **2001**. Antitumor activity of antisense clusterin oligonucleotides is improved in vitro and in vivo by incorporation of 2′-O-(2-methoxy)ethyl chemistry. *J. Pharmacol. Exp. Ther.* 298 : 934–940.

**41** Yacyshyn, B.R., M.B. Bowen-Yacyshyn, L. Jewell, R.A. Rothlein, E. Mainolfi, J.A. Tami, C.F. Bennett, D.L. Kisner, and W.R. Shanahan. **1998**. A placebo-controlled trial of ICAM-1 antisense oligonucleotide in the treatment of steroid-dependent Crohn's disease. *Gastroenterology* 114 : 1133–1142.

**42** Chi, K.N., E. Eisenhauer, L. Fazli, E.C. Jones, S.L. Goldenberg, J. Powers, and M.E. Gleave. **2004**. A phase I pharmacokinetic and pharmacodynamic study of OGX-011, a 2′-methoxyethyl antisense to clusterin, in patients with localized prostate cancer prior to radical prostatectomy. In: 40th Annual Meeting of the American Society of Clinical Oncology, New Orleans, LA.

**43** Baker, B.F., and B.P. Monia. **1999**. Novel mechanisms for antisense-mediated regulation of gene expression. *Biochim. Biophys. Acta* 1489 : 3–18.

**44** Altmann, K.-H., D. Fabbro, N.M. Dean, T. Geiger, B.P. Monia, M. Muuler, and P. Nicklin. **1996**. Second-generation antisense oligonucleotides: Structure–activity relationships and the design of improved signal transduction inhibitors. *Biochem. Soc. Trans.* 24 : 630–637.

**45** Baker, B.F., S.S. Lot, T.P. Condon, S. Cheng-Flournoy, E.A. Lesnik, H.M. Sasmor, and C.F. Bennett. **1997**. 2′-O-(2-Methoxy)ethyl-modified anti-intercellular adhesion molecule 1 (ICAM-1) oligonucleotides selectively increase the ICAM-1 mRNA level and inhibit formation of the ICAM-1 translation initiation complex in human umbilical vein endothelial cells. *J. Biol. Chem.* 272 : 11994–12000.

**46** Yu, R.Z., H. Zhang, R.S. Geary, M. Graham, L. Masarjian, K. Lemonidis, R. Crooke, N.M. Dean, and A.A. Levin. **2001**. Pharmacokinetics and pharmacodynamics of an antisense phosphorothioate oligonucleotide targeting Fas mRNA in mice. *J. Pharmacol. Exp. Ther.* 296 : 388–395.

**47** Zhang, H., J. Cook, J. Nickel, R. Yu, K. Stecker, K. Myers, and N.M. Dean. **2000**. Reduction of liver Fas expression by an antisense oligonucleotide protects mice from fulminant hepatitis. *Nature Biotechnol.* 18 : 862–867.

**48** Yu, R.Z., G. Riley, K. Hoc, T.A. Watanabe, J. Matson, S. Murray, S. Booten, S. Bhanot, B. Monia, and R.S. Geary. **2003**. effect of treatment regimen on sub-organ pharmacodynamics of a 2′-MOE modified antisense oligonucleotide, Isis 116847, targeting putative protein tyrosine phosphatase (PTEN) mRNA in rats. In: AAPS Annual Meeting and Exposition. American Association of Pharmaceutical Scientists, Salt Lake City, UT.

**49** Yu, R.Y., E.J. McArdle, G. Riley, K. Hoc, K.C. Nishihara, T.A. Watanabe, S.K. Pandey, S. Bhanot, and R.S. Geary. **2002**. In vivo pharmacodynamics and sub-cellular distribution of a 2-MOE-modified antisense oligonucleotide (ASO), ISIS 116847, targeting putative protein tyrosine phosphatase (PTEN) mRNA in mice and rats. In: Annual Meeting of Society of Toxicology and Chemistry, Baltimore, Maryland.

**50** Wei, N., J. Fiechtner, D. Boyle, A. Kavanaugh, S. Delauter, S. Rosengren, G.S. Firestein, J. Tami, R. Yu, and L. Sewell. **2003**. Synovial biomarker study of ISIS 104838, an antisense oligodeoxynucleotide targeting TNF-alpha, in rheumatoid arthritis. In: 67th Annual Meeting of the American College of Rheumatology (ACR). October 23–28, Orlando, Florida, USA.

**51** Graham, M.J., S.T. Crooke, K.M. Lemonidis, H.J. Gaus, M.V. Templin, and R.M. Crooke. **2001**. Hepatic distribution of a phosphorothioate oligodeoxynucleotide within rodents following intravenous administration. *Biochem. Pharmacol.* 62:297–306.

**52** Morris, M.J., W.P. Tong, C. Cordon-Cardo, M. Drobnjak, W.K. Kelly, S.F. Slovin, K.L. Terry, K. Siedlecki, P. Swanson, M. Rafi, R.S. DiPaola, N. Rosen, and H.I. Scher. **2002**. Phase I trial of BCL-2 antisense oligonucleotide (G3139) administered by continuous intravenous infusion in patients with advanced cancer. *Clin. Cancer Res.* 8:679–683.

**53** Leeds, J.M., M.J. Graham, L. Troung, and L.L. Cummins. **1996**. Quantitation of phosphorothioate oligonucleotides in human plasma. *Anal. Biochem.* 235:36–43.

**54** Bigelow, J.C., L.R. Chirin, L.A. Mathews, and J.J. McCormack. **1990**. High-performance liquid chromatographic analysis of phosphorothioate analogues of oligodeoxynucleotides in biological fluids. *J. Chromatogr. Biomed. Appl.* 533:133–140.

**55** Zinker, B.A., C.M. Rondinone, J.M. Trevillyan, R.J. Gum, J.E. Clampit, J.F. Waring, N. Xie, D. Wilcox, P. Jacobson, L. Frost, P.E. Kroeger, R.M. Reilly, S. Koterski, T.J. Opgenorth, R.G. Ulrich, S. Crosby, M. Butler, S.F. Murray, R.A. McKay, S. Bhanot, B.P. Monia, and M.R. Jirousek. **2002**. PTP1B antisense oligonucleotide lowers PTP1B protein, normalizes blood glucose, and improves insulin sensitivity in diabetic mice. *Proc. Natl. Acad. Sci. USA* 99:11357–11162.

**56** Liang, Y., M.C. Osborne, B.P. Monia, S. Bhanot, W.A. Gaarde, C. Reed, P. She, T.L. Jetton, and K.T. Demarest. **2004**. Reduction in glucagon receptor expression by an antisense oligonucleotide ameliorates diabetic syndrome in db/db mice. *Diabetes* 53:410–417.

**57** Watts, L.M., V.P. Manchem, T.A. Leedom, A.L. Rivard, R.A. McKay, D. Bao, T. Neroladakis, B.P. Monia, D.M. Bodenmiller, J.X. Cao, H.Y. Zhang, A.L. Cox, S.J. Jacobs, M.D. Michael, K.W. Sloop, and S. Bhanot. **2005**. Reduction of hepatic and adipose tissue glucocorticoid receptor expression with antisense oligonucleotides improves hyperglycemia and hyperlipidemia in diabetic rodents without causing systemic glucocorticoid antagonism. *Diabetes* 54:1846–1853.

**58** Crooke, R.M., M.J. Graham, K.M. Lemonidis, C.P. Whipple, S. Koo, and R.J. Perera. **2005**. An apolipoprotein B antisense oligonucleotide lowers LDL cholesterol in hyperlipidemic mice without causing hepatic steatosis. *J. Lipid Res.* 46:872–884.

**59** Bradley, J.D., R. Crooke, L.L. Kjems, M. Graham, R. Leong, R. Yu, D. Paul, and M. Wedel. **2005**. Hypolipidemic effects of a novel inhibitor of human APO-B 100 in humans. In: The American Diabetes Association's 65th Annual Meeting, San Diego, California.

**60** Yacyshyn, B.R., W.Y. Chevy, J. Goff, B. Salzberg, R. Baerg, A.L. Buchman, J. Tami, R. Yu, E. Gibiansky, W.R. Shanahan, and Isis 2302-CS9 Investigators. **2002**. Double blind, placebo controlled trial of the remission inducing and steroid-sparing properties of an ICAM-1 antisense oligodeoxynucleotide alicaforsen (ISIS 2302) in active steroid-dependent Crohn's disease. *Gut* 51(1):30–36.

**61** Yuen, A.R., J. Halsey, G.A. Fisher, J.T. Holmlund, R.S. Geary, T.J. Kwoh, A. Dorr, and B.I. Sikic. **1999**. Phase I study of an antisense oligonucleotide to protein kinase C-$\alpha$ (ISIS 3521/CGP 64128A) in patients with cancer. *Clin. Cancer Res.* 5:3357–3363.

**62** Nemunaitis, J., J.T. Holmlund, M. Kraynak, D. Richards, J. Bruce, N. Ognoskie, T.J. Kwoh, R.S. Geary, F.A. Dorr, D. Von Hoff, and S.G. Eckhardt. **1999**. Phase I evaluation of ISIS 3521, an antisense oligodeoxynucleotide to protein kinase c-alpha, in patients with advanced cancer. *J. Clin. Oncol.* 17:3586–3595.

**63** Kjems, L. **2005**. New targets for glycemic control protein-tyrosine-phosphatase-1B antisense inhibitor. In: American Diabetes Association's 65th Annual Meeting, San Diego, California.

**64** Crooke, R. **2004**. Second-generation antisense drug for cardiovascular disease demonstrates significant and durable reductions in cholesterol. In: 9th Drug Discovery Technology World Congress, Boston, MA.

**65** Yu, R., L. Gibiansky, E. Gibiansky, and R.S. Geary. **2001**. Population pharmacokinetics and pharmacodynamics of ISIS 2302 (Role of population analysis in drug development). In: ASCPT Annual Meeting, Orlando, Florida.

**66** Yu, R.Z., J.Q. Su, J.S. Grundy, R.S. Geary, K.L. Sewell, A. Dorr, and A.A. Levin. **2003**. Prediction of clinical responses in a simulated phase III trial of Crohn's patients administered the antisense phosphorothioate oligonucleotide ISIS 2302: comparison of proposed dosing regimens. *Antisense Nucleic Acid Drug Dev.* 13:57–66.

# 5
# Pharmacokinetics of Viral and Non-Viral Gene Delivery Vectors

*Martin Meyer*, Gururaj Rao*, Ke Ren, and Jeffrey Hughes*

## 5.1
## General Overview of Gene Therapy

Gene therapy refers to the introduction of new genetic material of therapeutic value into somatic cells. This approach is a potentially powerful method for the treatment of diseases for which classical pharmacotherapy is unavailable or not easily applicable. A variety of delivery systems (also known as "vectors"), including viruses and plasmid-based systems, have been evaluated. The function of the vector is to transverse the biological barriers for reaching its attended target, usually the nucleus. The infectivity characteristics that make viruses attractive as a delivery system also pose their greatest drawback; consequently, a significant amount of attention has been directed at plasmid-based approaches. These systems offer advantages over viral methodologies for gene delivery. Most significant is their lack of immunogenicity and the ease of customization of the vector system. In addition, since plasmid-based systems are not infectious agents and are incapable of self-replication, they pose no risk of evolving into new classes of human pathogens.

A current theme in plasmid-based delivery approaches is to mimic Nature's methods for nucleic acid delivery. To date, the best system to emulate Nature has been viral vectors. Briefly, most viral vectors escape immune surveillance, interact with cell membranes (e.g., receptor), internalize (via endocytosis), escape from endosomes, migrate to the nuclear envelope, enter the nucleus, and finally take over cellular functions. Plasmid-based systems (cationic liposomes and cationic polymers) can mimic portions of these events. This chapter will explore the barriers facing gene delivery vectors, with an emphasis of the pharmacokinetic behavior of these systems. In order to understand the *in-vivo* barrier, a brief review of physiology will be provided.

* The first two authors share first authorship on this book chapter.

*Pharmacokinetics and Pharmacodynamics of Biotech Drugs: Principles and Case Studies in Drug Development.* Edited by Bernd Meibohm

ISBN: 3-527-31408-3

## 5.2 Anatomical Considerations

Capillaries are the exchange vessels of the body. They have structural variations to allow different levels of metabolic exchange (of exogenous and endogenous substances) between blood and the surrounding tissues. The structure of the walls varies depending on their resident tissue. There are three major types of blood capillaries: continuous; fenestrated; and sinusoidal (discontinuous) [1]:

- Continuous capillaries are found in the skin, all types of muscle, mesenteries, and the central nervous system (blood–brain barrier). These capillaries are characterized by tight junctions between the endothelial cells and an uninterrupted basement membrane. The restrictive capacity of the capillary walls barely allows extravasation of macromolecules into the parenchyma of these tissues.
- Fenestrated capillaries are present in all exocrine and endocrine glands, gastric and intestinal mucosa, choroid plexuses of brain, renal tubules and glomeruli, and ciliary body of the eye. The passage of macromolecules through these capillary walls is restricted by size (the fenestrae are 40–60 nm in diameter), and by the presence of the basement membrane.
- Discontinuous capillaries are present in the liver, bone marrow, and spleen. These capillaries have relatively large gaps between the endothelial cells (30–500 nm), and have incomplete basement membranes, which make them best suited for extravasation of macromolecules.

Most gene therapy applications require extravasation of the DNA carriers so that only relative small DNA complexes can pass through the blood vessels and interact directly with parenchymal cells after vascular administration [2]. Under pathophysiological conditions, the structure of the vasculature can change. This phenomenon – termed the "enhanced permeation effect" – has been utilized to passively target macromolecules to tumors, since blood vessels in tumors are relatively more leaky.

## 5.3 Naked DNA

The *in-vivo* disposition and tissue distribution of naked plasmid DNA has been studied by several investigators after administration via various routes, including intravenous (IV), intramuscular (IM), intradermal, and intranasal. It is generally difficult to determine pharmacokinetic parameters for naked DNA as it is rapidly and extensively degraded in plasma. Our studies indicated that supercoiled DNA is degraded with a half-life of 1.2 min upon incubation in isolated rat plasma at 37 °C [3]. The open-circular form has a half-life of 21 min, while the linear form degrades with a half-life of 11 min. Other studies show similar results, with most of the supercoiled form of plasmid converting to open-circular and linear forms by 15 min [4]. The clearance from plasma after IV administration is even more rapid. The supercoiled form of plasmid DNA was found to be detectable in plasma only after the administration of a very high dose of plasmid (ca. 7.5 mg/kg) to rats [5]. Similar results were obtained by Kawabata et al. [4], wherein the levels of plasmid DNA dropped to levels lower than the lower limit of detection of the applied assay by about 5 min. This indicated that the plasmid is subjected to distribution mechanisms in addition to degradation that cause its rapid clearance from the blood. Tissue distribution and *in-vivo* tissue localization studies using [$^{32}$P]-labeled DNA at 5 min after injection showed that maximum radioactivity was taken up by non-parenchymal cells in the liver. Co-administration of polyanions such as dextran sulfate inhibited this hepatic uptake of radioactivity, suggesting that uptake into the non-parenchymal cells may involve a scavenger receptor. This result was corroborated by another study using a liver perfusion technique [6]. We investigated the effect of the plasmid topoform on the *in-vivo* pharmacokinetics of naked plasmid DNA, after IV bolus injection in rats [5]. Supercoiled (SC pDNA), open-circular (OC pDNA) and linear (L pDNA) were injected at four different doses, ranging from 250 µg to 2500 µg. Plasmid DNA was extracted from the blood, and the three topoforms were separated using agarose gel electrophoresis. SC pDNA was detectable in the bloodstream only after a 2500 µg dose, and had a (mean ± SD) clearance of 390 ± 50 mL/min and a volume of distribution (Vd) of 81 ± 8 mL. The pharmacokinetics of OC pDNA exhibited non-linear characteristics, with clearance ranging from 8.3 ± 0.8 to 1.3 ± 0.2 mL/min and a Vd of 39 ± 19 mL. L pDNA was cleared at 7.6 ± 2.3 mL/min, and had a Vd of 37 ± 17 mL. AUC analysis revealed that 60 ± 10% of the SC was converted to the OC form, and that OC pDNA was almost completely converted to L pDNA. Clearance of SC pDNA was decreased after liposome complexation to 87 ± 30 mL/min. However, the clearance of OC and L pDNA was increased relative to naked pDNA at an equivalent dose to 37 ± 9 mL/min and 95 ± 37 mL/min, respectively. Plasmid DNA pharmacokinetics have also been studied after alternative modes of administration. Quantitative PCR analysis after intramuscular and intradermal injections of naked plasmid showed that DNA persisted at the injection site and in lymph nodes up to 28 days after injection. After IV injections, plasmid was detected in the blood, lungs, lymph nodes and spleen at 2 days post-injection. However, DNA levels were below detection limits in all tissues by 14 days post-injection. In another study, plasmid DNA biodistribution was studied using quantitative PCR following intranasal administration. At 15 min and 24 h post-administration, maximum levels of DNA were detected in the liver, kidney, and heart. Interestingly, significant levels of DNA were also detected in the brain at both time points. The maximum serum concentration (430 pg/mL) was reached at 90 min post-administration. The AUC after intranasal administration was $1.5 \times 10^3$ pg h/mL compared to $11 \times 10^3$ pg h/mL obtained after IV administration.

Liu et al. [7] found significant differences between the disposition of complexed and uncomplexed [$^{125}$I]-labeled plasmid DNA in mice. An equal amount of radioactivity was found in blood and liver at 15 min after injection of naked plasmid DNA. In contrast, at the same time-point, a major portion of injected DNA was in the lung when DNA–lipid complexes were injected.

Osaka et al. (1996) followed the pharmacokinetics and tissue distribution of [$^{33}$P]-labeled plasmid DNA coding for luciferase. Differences in the distribution of plasmid DNA administered alone or complexed with cationic lipids were studied using whole-body autoradiography and microautoradiography [8]. The results showed that there were distinct differences between the distribution patterns of the two groups at the early time points. The initial plasma half-life for naked DNA (6.6 min) was longer than that for complexed DNA (4.3 min). This was attributed to entrapment of the lipid–DNA complexes in the tissue vasculature. Maximum levels of radioactivity were observed in the liver upon administration of uncomplexed plasmid. In contrast, lung and spleen had the highest levels of radioactivity when a plasmid–lipid complex was administered. In addition, luciferase expression was also markedly different between the [$^{33}$P]-DNA and [$^{33}$P]-DNA/lipid-treated groups. No gene expression was observed upon administration of naked DNA, while expressed protein was observed only in the lung at 1.5 h after administration of the complex. At 24 h after administration, gene expression levels did not correlate well with the levels of radioactivity in all organs examined, except in the lung. This disassociation between the levels of radioactivity (and thus the levels of plasmid DNA-related material) and actual gene expression has been partly elucidated by a few studies using direct methods such as PCR and Southern blotting for detection and quantitation of DNA [9,10]. In these studies, it was observed that the introduced DNA is largely degraded in most tissues by 24 h. This strongly suggests that the radioactivity observed in the tissues at this time point represents degraded, functionally inactive DNA.

## 5.4
## Non-Viral Vectors

Non-viral vectors have attracted considerable interest as gene delivery systems during the past decade [11]. Chemical vectors include polycationic carriers such as liposomes (lipoplexes) and polymers (polyplexes). These carriers avoid the DNA size limitations and immunogenicity associated with viral vectors. Although the transfection efficiency remains lower than viral vectors, this approach seems useful for many *in-vivo* applications. An ideal delivery system has to fulfill a number of functions, such as packaging the DNA, migrating through blood capillary vessels, targeting, cellular uptake, and tracking to the nucleus. After administration, non-viral vectors encounter resistance due to the barriers in gene delivery (Fig. 5.1). Systemic barriers include degradation of DNA by plasma nucleases, opsonization of DNA complexes by negatively charged serum components, uptake by the reticuloendothelial system, and distribution of DNA to non-target tissues [2]. Cellular barriers include internalization at the cell surface, endosomal release, cytoplasmic degradation, and translocation into the nucleus (Fig. 5.2) [12].

Compared to *ex-vivo* gene therapy, the *in-vivo* approach faces further barriers as mentioned above: interactions with blood components, vascular endothelial cells, and uptake by the reticuloendothelial system. Several target cells are not directly

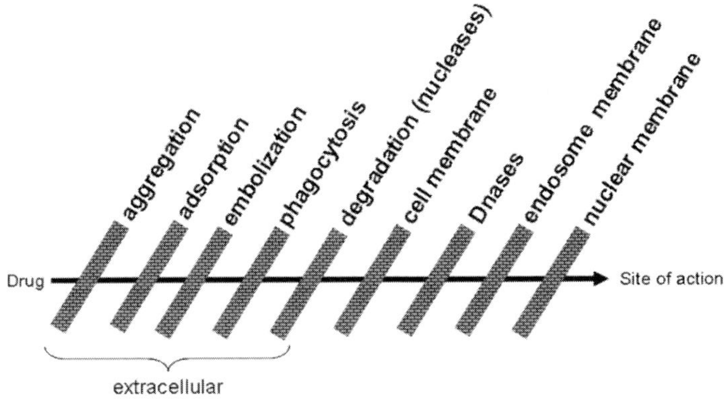


**Fig. 5.1** Barriers in gene delivery.

accessible, and can only be reached via systemic delivery. In this case, the efficient and specific delivery of therapeutic genes to the target site of action is a challenge that will need to be overcome. With respect to the important factors of biodistribution and circulation time, the physico-chemical parameters must be taken in account.

In the case of non-viral gene therapy, efficacy depends on the intrinsic properties of plasmid DNA and the delivery system used. The topology of plasmid DNA is known to affect its *in-vitro* transcriptional efficiency. Higher gene expression

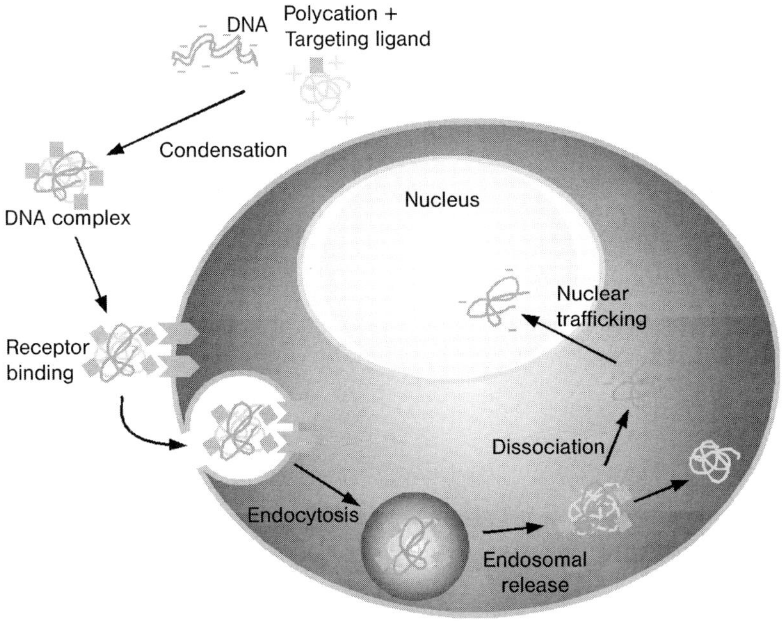


**Fig. 5.2** Steps in gene delivery.

levels are obtained with the supercoiled form of plasmid than with the open-circular or linear forms [13–15]. This does not appear to be related to the efficiency of DNA uptake, because topology does not affect the amount of DNA incorporated into the cell nucleus [16]. One possible reason for the higher efficiencies could be that the supercoiled plasmid interacts more effectively with nuclear transcriptional factors than the other topoforms [14]. Another reason could be the greater intracellular stability of the supercoiled plasmid form under physiological conditions. The covalently closed circular state of the supercoiled plasmid makes it more resistant to nuclease degradation [17–19]. This argument is supported by the observation that transfection efficiencies of covalently closed open-circular plasmid are similar to those obtained with supercoiled DNA [20]. In contrast, nicked open-circular forms of plasmid show lower efficiencies [13]. Transfection efficiencies can also be affected by the regulatory elements (such as promoters and enhancers) present in the plasmid [21]. Transcriptional regulatory elements determine various parameters, such as cell-specificity, efficiency, and duration of transgene expression. For non-viral gene delivery approaches, very strong promoter/enhancers, such as cytomegalovirus (CMV) immediate-early promoter and SV40 early promoter, have been widely used to obtain high levels of transgene expression [22].

Several non-viral carriers have been developed and tested for gene delivery, but none of the currently available vectors contains all the desirable features. Two of the most promising delivery systems include liposomal formulations and cationic polymers [23]. Despite their different chemistry and structure, these vectors show similarities in their biological and pharmacokinetic characteristics. Both charged and neutral liposomal carriers have been used for gene delivery. In the case of neutral and anionic liposomes, DNA is usually encapsulated inside the liposome. Although this is beneficial in that it protects the DNA from degradation, this method has disadvantages such as low encapsulation efficiency and poor transfection rates. From the chemistry point of view, the positively charged functional groups, which aid in the complexation of the negatively charged DNA [24], are important.

Pharmacokinetic aspects on the cellular, organ, and whole-body level should be considered for the development of sophisticated vectors, with physico-chemical properties of the DNA–vector complex (e.g., charge, size) being mainly responsible for the destiny of the gene delivery system in a biological organism [25].

## 5.4.1
## Polymer-Based Vectors

### 5.4.1.1 Introduction

The first reports of using cationic polymers for DNA delivery can be traced back to 1973, before the use of cationic lipids. However, until recently, research in the area of polymer-mediated gene delivery lagged behind that of lipid-based delivery vectors [26]. Recent advances in the development of polymers have regenerated the interest in polymer-mediated gene delivery. Polycationic carriers are either naturally occurring or chemically synthesized compounds. Examples include his-

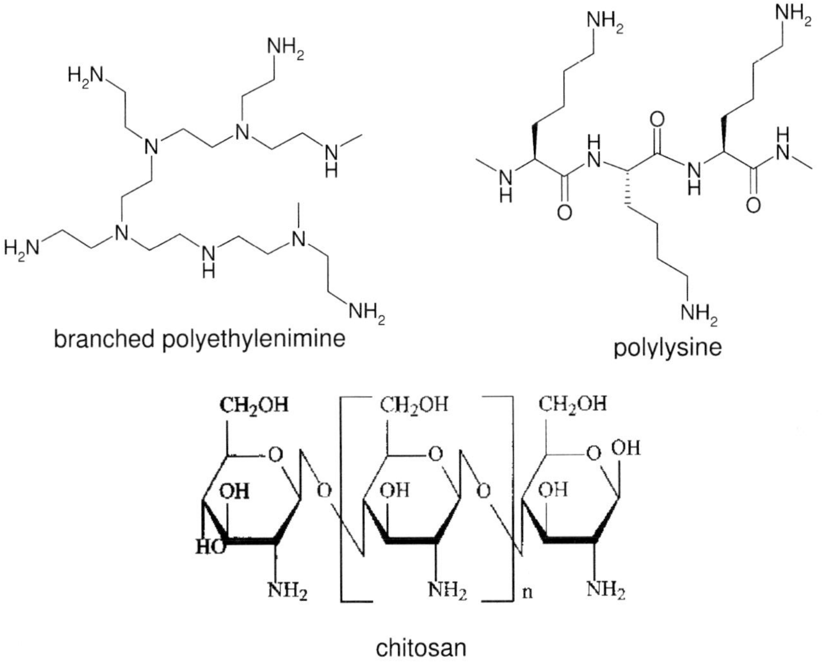


**Fig. 5.3** Structures of cationic polymers used for DNA delivery.

tones, protamines, cationized human serum albumins, chitosans, cationic peptides (e.g., polylysine, polyhistidine and polyarginine), and polyamines such as polyethylenimine (PEI), and polyamidoamide (PAMAM) dendrimers (Fig. 5.3). Among these, PEI is one of the most capable vectors. As a result of its ability to protect DNA and RNA against degradation, as well as the high intrinsic endosomolytic activity as compared to other polycations, PEI has attracted considerable attention and is one of the most extensively studied non-viral vectors [27].

#### 5.4.1.2 **Influence of Charge and Size**

Cationic polymers form complexes with DNA via electrostatic interaction. The concentrations of polycation and DNA in the polyplex are expressed in terms of the N/P ratio; that is, the ratio of positively charged polymer nitrogens to negatively charged DNA phosphates. Depending on the properties of the vector, the N/P ratio and the preparation procedure, a DNA complex has its unique characteristics. The surface charge of the complex is an important factor for *in-vivo* gene delivery and affects the biodistribution. Due to the presence of glycoproteins and glycolipids, the cell surface membrane is negatively charged. On the one hand, this negatively charged membrane is a good target for cationic complexes and promotes cellular uptake. However, on the other hand the ubiquitous negative charge also leads to nonspecific binding of complexes, which presents an obstacle to the cell-specific delivery of genes after *in-vivo* administration. Negatively charged proteins such as albumin are abundant in the blood, and can also bind to cationic non-viral vectors. This leads to neutralization of the cationic charge, the formation of aggregates, and an increase in size of the polyplexes – all of which affect transfection efficiency. Acute toxicity could also result due to aggregation of the positively charged complexes with erythrocytes [28]. In addition, the plasma protein complement C3 has been found to bind to the positively charged DNA complexes and lead to activation of the complement system [29]. Opsonization of foreign particles with plasma proteins represents one of the first steps in the process of removal of foreign particles by the innate immune system, which decreases the circulation time.

Attempts have been made to shield the positive surface charge of polyplexes in order to avoid these nonspecific interactions. One popular method is to attach polyethylene glycol (PEG), a hydrophilic polymer, to the polyplexes. This shielding strategy has been applied to the *in-vivo* gene delivery of polyplexes to tumors after systemic injection [30]. Pegylation results in an almost charge-free surface (N/P range between 2 and 6), stabilizes the particles against aggregation, reduces the surface charge, blocks interactions with plasma and erythrocytes, and increases the circulation time. In addition to hydrophilic polymers, proteins can be used to shield the surface charge. Transferrin, a plasma protein, was able to shield the complex against nonspecific interactions and also target the polyplexes to distant tumors via the transferrin receptor [31].

Size of the gene delivery complexes is also important because it can affect tissue distribution by restricting vascular permeability and organ access (see Section 5.2). The endocytotic entry of complexes into cells is also a size-limited process. Polycations interact with the polyanionic DNA and condense it into compact particles. This ability of vectors to condense DNA into nanoparticles is often crucial for transfection efficiency: The particle size ranges from 50 nm up to 1000 nm, depending upon several factors including the N/P ratio, the polymer used, the total concentration, and the ionic strength of the buffer. Small particles seem to be necessary to allow endocytosis into cells. Condensation of DNA with 25-kDa PEI resulted in particles with diameters of about 80–100 nm. Aggregates up to 2000 nm in size are achieved when using 2-kDa PEI, which reduces the transfection ability [32]. However, some groups report that larger particles increase the *in-vitro* transfection efficiency. Ogris et al. hypothesize that larger complexes show an increased sedimentation rate facilitating cell attachment and uptake *in vitro* [33].

#### 5.4.1.3 **Biodistribution and Gene Expression**

When plasmid DNA is complexed with PEI (resulting in positively charged PEI/DNA-polyplexes) and administered systematically, the lungs and liver are the main organs of biodistribution. Positively charged complexes are cleared from the circulation within minutes after mouse tail vein injection. Shortly after administration, the complexes accumulate mainly in the lungs. The amount of plasmid DNA then gradually decreases in the lung tissue, while it increases in the liver.

The discontinuous vasculature in the liver might play an important role in allowing the macromolecules access to the parenchyma. DNA biodistribution and gene expression can be targeted to different organs. Even though reporter gene expression can be found in a diversity of organs, including the lungs, heart, liver, spleen, and kidneys, the highest level of gene expression is primarily found in the lungs [34]. It seems that the initial "targeting" of the DNA complexes towards the lungs is responsible for high levels of gene expression. Furthermore, the extensive degradation of plasmid DNA in the liver is a major reason for this discrepancy between organ distribution and gene expression. The preferential distribution to the lungs after tail vein injection may be a result of nonspecific interactions of positively charged complexes with blood and cell components, which lead to the formation of aggregates that are trapped in the lung capillaries. This is not surprising, because the lung provides the first capillary bed after IV injection. It should be mentioned that this "passive" targeting is also a reason for acute toxicity: lung embolism was observed after IV administration of positively charged complexes [35].

If organs other than the lungs need to be targeted after systemic administration, strategies must be developed for increasing the circulation time or the addition of targeting ligands. As mentioned above, PEGylation leads to a masking of surface charge and inhibition of nonspecific interactions. The significantly reduced uptake by macrophages of the liver and spleen leads to an increased blood circulation time. To add active targeting to the delivery system, ligands with high specificity and affinity to recognition sites can be linked to the polymers. Antibodies, peptides, proteins and sugar residues have been used for the targeting of polymer-based DNA complexes (Table 5.1).

**Table 5.1** Targeting ligands used with gene delivery systems.

| Ligand | Receptor | Target cells | Reference |
|---|---|---|---|
| Anti-CD3-antibody | CD3 | Lymphocytes | [36] |
| Anti-JL1-antibody | JL1 | Lymphocytes | [37] |
| Anti-platelet endothelial cell adhesion molecule (PECAM) antibody | PECAM | Endothelial cells | [38] |
| Asialoorosomucoid | Asialoglycoprotein-receptor | Hepatocytes | [39] |
| Epidermal growth factor | EGF-receptor | Tumor cells | [40] |
| Mannose | Mannose receptor | Macrophages | [41] |
| RGD-containing peptide | Integrins | Epithelial cells | [42] |
| Transferrin | Transferrin receptor | Tumor cells | [43] |

Several groups have combined PEGylation with the cell-targeting approach (Table 5.2). In this case, a preferential expression in a distant tumor (up to 100-fold higher than in other organs) can be observed if the polyplexes are charge-shielded with PEG and targeted with transferrin [31]. The ligands, as well as the PEG, can be conjugated by different techniques which in turn influences the resulting properties of the complex.

**Table 5.2** Ligands used with the pegylation approach in polymeric gene delivery systems.

| Ligand | Reference |
|---|---|
| Artery wall binding peptide | [44] |
| Epidermal growth factor | [45] |
| Folate | [46] |
| Galactose | [47] |
| RGD peptide | [48] |
| Transferrin | [49] |

Gene delivery, specifically, protein expression using non-viral systems is usually transient. For most cancer gene therapy strategies a short period of gene expression is sufficient, but prolonged and sustained release is necessary for treating diseases such as hypercholesterolemia. Protein expression using PEI as vector peaks at 4 to 24 h after dosing, and is undetectable in the kidney at 14 days after administration [50]. Consequently, a repeated administration of the gene delivery dose is necessary in order to attain a constant level of the expressed protein over a longer time. Clearly, the frequency of dosing depends on the need of the treated disease and also other factors such as the stability of the expressed protein in the target tissue. However, repeated delivery of the vector may cause toxic effects (e. g., induction of cytokines), which is a disadvantage of this approach [51]. In order to avoid this drawback, several sustained-release gene delivery systems such as nanoparticles or polymeric implants are currently under investigation [52–54]. For example, biodegradable nanoparticles formulated using a biocompatible polymer, poly(D,L-lactide-*co*-glycolide), showed sustained gene delivery. Cells transfected with wild-type (wt)-p53 DNA-loaded nanoparticles (antiproliferative therapy for cancer treatment) demonstrated sustained p53 mRNA levels compared to cells transfected with naked wt-p53 DNA or with a common commercially available transfection agent [55].

### 5.4.2 Lipid-Based Vectors

#### 5.4.2.1 Introduction

Since Felgner et al. first reported in 1987 on the *in-vitro* transfection of eukaryotic cells with cationic lipid, cationic liposomes have been used extensively for gene delivery [56]. To date, numerous cationic lipids have been synthesized and tested for gene delivery. In general, cationic lipids are composed of three parts (Fig. 5.4): a positively charged head group; a hydrophobic tail group; and a linker bond between the two.

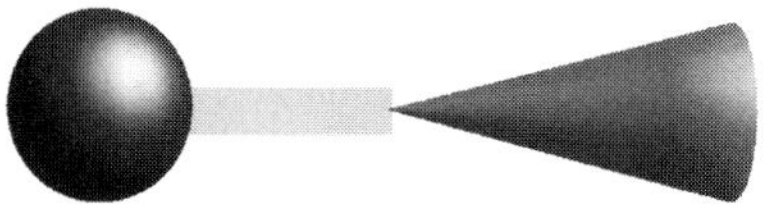


**Fig. 5.4** Basic structure of cationic lipids.

Head groups include quaternary ammonium salt lipids, lipoamines (primary, secondary, tertiary amines), a combination of both, and various other groups. Linkers are usually ethers or esters, but amides, carbamates and disulfides have also been used. Tail groups usually consist of cholesteryl groups or saturated or unsaturated alkyl chains (12 to 18 carbons in length). DC-Chol (Fig. 5.5) was the first cationic lipid to be used for clinical trials. Extensive reviews have covered the various published structures and the different aspects of structure–activity relationships of cationic lipids [57]. Despite these various investigations that spotlight the relationship between cationic lipid formulation and transfection activity, solid results are rarely achieved. However, it has been proven that the addition of neutral co-lipids (helper lipids) increases the transfection efficiency. Cholesterol and dioleylphosphatidylethanolamine (DOPE) are the most common helper lipids [7,58].

It is accepted that cationic liposomes form complexes with the negatively charged DNA molecules through electrostatic charge interaction/hydrophobic-base interactions, thereby preventing degradation. Liposome–DNA complexes are often heterogeneous in size, shape, and composition, and the structure of the lipoplexes may change over time. The structure of the complexes is determined by many factors such as lipid:DNA ratio or the structure of the lipid. In general, three types of structures have been reported:

- Beads on a string structure: positively charged liposomes adhere as an intact bead on a negatively charged DNA strand.
- Spaghetti structure: the liposomes coat the DNA strand with a cylindrically shaped bilayer.
- Multilamellar structure: the DNA strand is intercalated in multilamellar liposome membrane.

In addition, combinations of these structures were found [59].

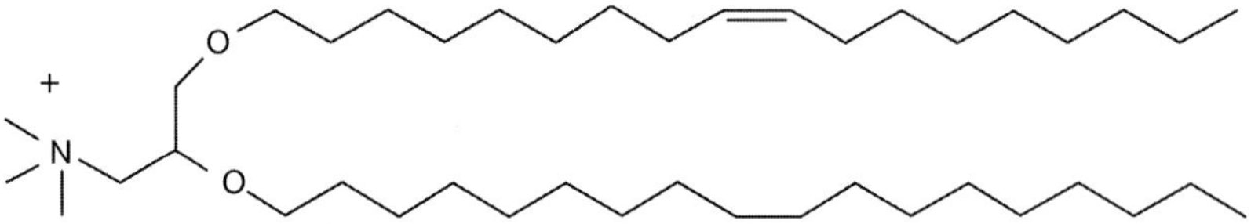

DOTMA: 1,2-Dioleoyloxypropyl-trimethylammonium chloride

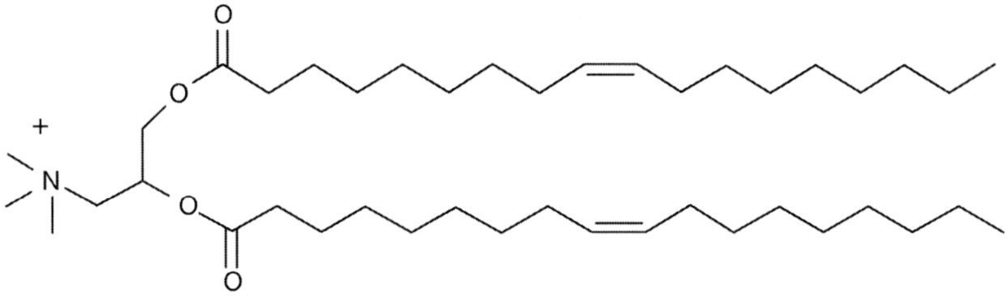

DOTAP: 1,2-Dioleoyloxy-3-(trimethylammonio) propane chloride

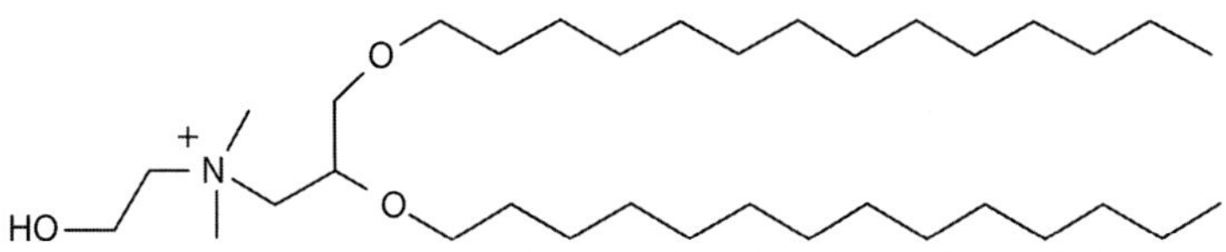

DMRIE: 1,2-Dimyristyloxypropyl-3-dimethyl-hydroxyethyl ammonium bromid

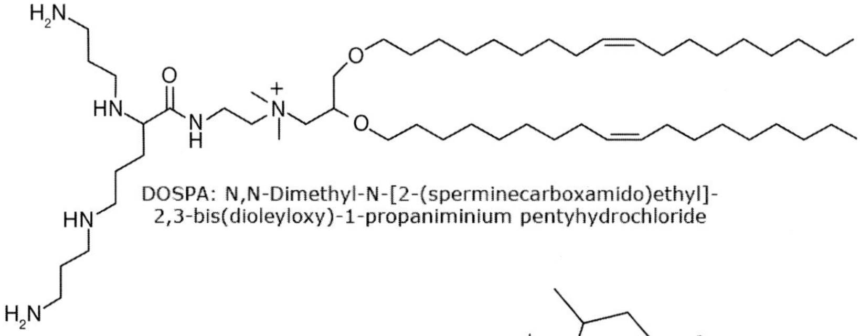

DOSPA: N,N-Dimethyl-N-[2-(sperminecarboxamido)ethyl]-2,3-bis(dioleyloxy)-1-propaniminium pentyhydrochloride

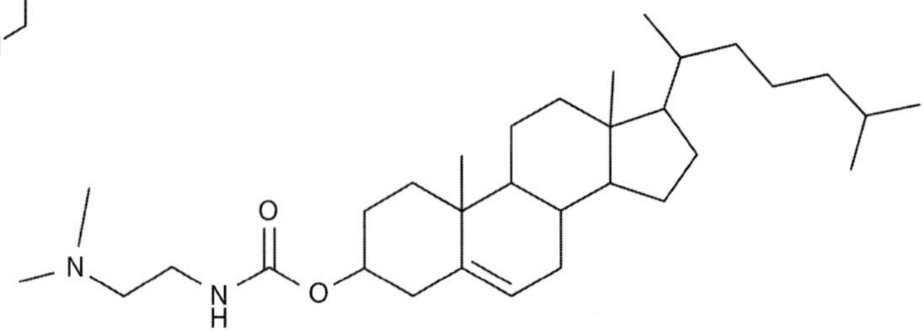

DC-Chol: 3-β-[N-(N',N'-dimethylaminoethanol)carbamoyl] cholesterol

**Fig. 5.5** Cationic lipids developed for gene delivery.

The mechanism of transfection with lipoplexes is similar to that with vectors such as cationic polymers, though some differences exist. The complexes are generally believed to interact with cells through a nonspecific charge interaction. A net positive charge ensures efficient binding to negatively charged cell surfaces. It has been proposed, due to the membranous structure of lipoplexes, that the complexes can directly penetrate the cell membrane through fusion with the membrane or lipid-mediated portion, but it was shown that there is no correlation between the capability for fusion with the plasma membrane and transfection efficiency [60]. The majority of studies suggest that lipoplexes are taken up by cells via endocytosis [61]. Following uptake, the cationic lipids play a role in destabilizing the endosomal membrane [62].

Studies have shown that cationic lipids interact with systems containing anionic phospholipids [63]. Anionic phospholipids can displace cationic lipids from plasmids, which lead to a release of DNA. Additionally, the cationic lipids have the ability to induce non-bilayer hexagonal ($H_{II}$) phase structures. The hexagonal structure has been correlated with high *in-vitro* transfection efficiency, whereas lamellar lipoplexes ($L\alpha$) bind stably to anionic phospholipids vesicles and do not induce the release of DNA [24]. The transition into the hexagonal phase also depends on the presence of helper lipids such as DOPE, as well as the shape and structure of the lipids [64]. Helper lipids promote the hexagonal phase conversion, and structure–activity relationship studies have concluded that lipids with a small head group part relative to the hydrophobic tail prefer the hexagonal phase, leading to a high transfection rate.

#### 5.4.2.2 **Influence of Physico-Chemical Properties**

The *in-vivo* application of lipoplexes has been limited due to serum-associated inhibition of transfection and complement activation. Following systemic administration, positively charged lipid–DNA complexes interact rapidly with plasma proteins that reduce the surface charge of the complexes, rendering them more neutral or negatively charged. They are eliminated due to nonspecific interactions with the reticuloendothelial system, and may prove to be unstable. The introduction of PEG, as in the case of polyplexes, helps to overcome this problem. Other approaches include neutralization of the cationic surface by serum binding protein or glycolipid insertion. Although the addition of PEG does inhibit nonspecific interactions to a certain degree, the cationic charges are not completely shielded and interactions still occur. To solve this problem, negatively charged PEG molecules were integrated in cationic lipoplexes [65]. These lipoplexes had an increased circulation time and a higher tumor-to-lung ratio of gene expression. This is interesting, because the PEG coating of lipoplexes normally decreases cell-surface interactions, which leads to lower transfection efficiencies. It should be mentioned here that PEG lipids stabilize the lamellar phase and can inhibit the transition to the hexagonal phase, thus reducing DNA release following destabilization of the endosomal membrane. In the case of cationic liposome-oligodeoxyribonucleotide (ODN) complexes – which behave differently from DNA complexes – the incorporation of PEG into the liposomes prevents aggregation, increases the particle stability, and improves the biological activity of the ODNs. Meyer et al. [66] also found that the PEG-coated ODN-lipoplexes were stable in plasma and enhanced the cellular uptake of the ODNs in serum-containing cell culture medium. Analogous to polymer-based vectors, many physico-chemical properties such as size or pH influence the transfection efficiency of lipoplexes. In turn, the physico-chemical properties depend on parameters such as lipid:DNA and lipid:co-lipid ratios. For example, it has been shown that highly positively charged complexes exhibit a homogeneous size distribution, whereas complexes prepared from a 1:1 lipid:DNA charge ratio (neutral) are characterized by a heterogeneous size distribution [67].

#### 5.4.2.3 **Biodistribution and Gene Expression**

Different methods (radiolabeled lipoplexes, PCR, Southern blot analysis) have been used to investigate the blood clearance and biodistribution of lipoplex components. The major organs of distribution after IV administration of lipoplexes are the lungs and liver. Mahato et al. [68, 69] reported that at 15 min after the injection of *N*-[1-(2, 3-dioleyloxy)propyl]-*N*,*N*,*N*-trimethylammonium chloride (DOTMA):DOPE complexes in mice, the majority of DNA was found in the lungs and liver (followed by spleen, kidneys, heart, and blood), but the lung accumulation decreased with time. Accumulation in the liver increased significantly with time up to 24 h post-injection, but gene expression in the liver was least 100- to 200-fold lower than that observed in the lungs. The lipoplex particles are removed from the bloodstream by cells of the mononuclear phagocyte system, mainly Kupffer cells in the liver. After phagocytosis, degradation of the DNA leads to a low gene expression in the liver. However, gene expression in the liver has been shown to be significantly increased by using intravenously injected lipoplexes prepared from 1-[2-(oleoyloxy)ethyl]-2-oleyl-3-(2-hydroxyethyl)imidazolinium chloride (DOTIM):cholesterol (1:1 molar ratio) multilamellar vesicles [7]. Mahato et al. [69] investigated the effect of lipid:DNA charge ratio on gene expression at 24 h after tail vein injection in mice by formulating pCMV-hGH (human growth hormone) with oleyloleoyl L-carnitine ester (DOLCE):DOPE or DOTMA:DOPE (2:1, mol:mol; 400 nm extruded) at 0.5:1, 2:1, 3:1, and 5:1 (+:–) charge ratios. The hGH serum levels increased with increasing lipid:DNA charge ratios, with the highest hGH serum levels obtained with a 3:1 (+:–) ratio, but no difference was observed between 3:1 and 5:1 (+:–) ratios. Similar results were obtained in lung and liver when, with a different reporter gene, the results were similar. The same group also determined the influence of cationic lipid:co-lipid ratio, injection volume, plasmid dose, and liposome size. Extrusion of DOTMA:DOPE (2:1, mol:mol) liposomes through 100-, 400-, and 800-nm pore size polycarbonate membranes and complexation of pCMV-hGH with these extruded liposomes at a 3:1 (+:–) charge ratio resulted in mean particle sizes of 231, 256, and 328 nm. The gene expression was five-fold higher with polyplexes prepared using 400-nm extruded liposomes as compared with those prepared with 100-nm extruded liposomes. Clearly, gene expression is influenced by a wide variety of formulation factors.

It has been observed that levels of gene expression and the organs showing maximum levels of gene expression differ, depending upon the lipid used. In one study [70], plasmid DNA complexed with Lipofectin® [DNA:lipid ratio, 1:3 (nmol:μg), dose 4 mg DNA/kg] did not result in detectable levels of gene expression. However, DNA complexed with Lipofectamine® at the same DNA:lipid ratio and same dose resulted in high levels of gene expression in all tissues [70]. Another study [71] using DNA complexed with Lipofectin [DNA:lipid ratio, 1:8 (nmol:μg)] showed maximum levels of gene expression in the lungs, spleen, heart, liver, kidneys, and lymph nodes when a dose of 5 mg DNA/kg was used. However, gene expression levels increased in the lungs and liver but decreased in the spleen, when the dose was increased to 7.5 mg DNA/kg. This is in contrast to the findings of another study [72], where gene expression profiles were followed after administration of DNA complexed with DLS liposomes [dioctadecylamidoglycylspermine (DOGS)/DOPE]. In this study, when the amount of DNA was increased to more than 5 mg/kg, expression levels in spleen and lungs reached a plateau, whereas levels of expression in the liver decreased dramatically. This was attributed to a toxic effect in the liver at higher concentrations. Liu et al. [73] showed that all of the internal organs, including the lungs, liver, spleen, heart and kidneys, expressed the transgene upon systemic administration of 1.25 mg/kg plasmid DNA complexed with a DOTMA-Tween 80 lipid formulation, with the lungs and spleen showing maximum expression. It was determined that gene expression levels increased in all organs upon increasing the dose of plasmid DNA from 1.25 mg/kg to 5 mg/kg when a DNA:lipid ratio of 1:12 (nmol:μg) was used.

Besides the localization of gene expression, duration is an important criterion. Protein expression in the lung can often be detected within the first hours after injection, and decreases after 24 h. Furthermore, for transgene expression after a second administration of lipoplexes, a certain span of time (up to two weeks) is needed to reach the levels of expression similar to the levels after the first injection. The production of anti-inflammatory cytokines after the first administration has been related to the refractory period because these cytokines are able to shut down the viral promoters usually placed before the transgene. Hofland et al. [74] investigated the duration of alkaline phosphatase (AP) expression after IV single administration of a stable lipid–DNA complex. The highest activity was in the lungs and spleen, followed by the heart, muscle, and liver. The AP activity in the lung was apparent only 6 h after injection. In general, however, the AP activity peaked at 24 h after injection, but then fell by approximately 10-fold at day 5. By day 7, residual activity could only be detected in the lungs, heart, and muscle. This lack of sustained transgene expression may be an important limitation to therapy, and must be taken into account when setting up treatment regimens.

## 5.5 Viral Vectors

At present, several types of virus are under investigation for gene delivery vectors, but in this chapter only one of the more promising viral vectors will be described. Recombinant adeno-associated virus (rAAV) has been widely used as a therapeutic gene delivery vector. The main advantages of gene therapy are: (1) stable transgene expression in dividing and post-mitotic cells, without inducing any significant inflammatory toxicity; and (2) a lack of human pathogenicity. rAAV has been selected for use in several clinical trials. The findings of recent studies have suggested that the pharmacokinetic properties and biodistribution of viral vectors require consideration in order to achieve the safest and most effective application of these molecular medicines.

### 5.5.1 rAAV: Properties

Adeno-associated virus (AAV) is a parvovirus with a diameter of ca. 25 nm [75]. It is a single-strand 4.7-kb DNA (ssDNA) genome packaged into three viral capsid proteins: VP1 (87 kDa), VP2 (73 kDa), and VP3 (62 kDa). These proteins form the 60-subunit viral particle in a ratio of 1 : 1 : 20. The linear ssDNA contains two open reading frames (ORFs) flanked by two inverted terminal repeats of 145 nucleotides each. The upstream ORF encodes four overlapping nonstructural replication proteins (Rep), Rep78, Rep68, Rep52, and Rep40 [76]. The downstream ORF codes for the capsid proteins. After injection into host cells, the ssDNA genome of AAV is converted to the double-strand template in cell nuclei, and finally integrated into the host genome at chromosome 19q13.3 [77, 78]. This chromosome-selective integration is lost in rAAV vectors in which the Rep coding sequences are removed. AAV has demonstrated a broad tropism of infection, including lung, neuron, eye, liver, muscle, hematopoietic progenitors, joint synovium, and endothelial cells [79]. rAAV vectors retain much of this tropism, with significant variations seen among serotypes, depending on the tissue. These serotypes differ in the composition of their capsid protein coat. rAAV serotype 2 (rAAV2) has been the most widely studied and best described. It binds to both heparin sulfate proteoglycans and fibroblast growth factor receptors as an essential step for cellular entry [80]. Recently, rAAV5 and rAAV8 have also been investigated and found to bind to different cellular receptors. This probably accounts for their different biodistribution properties when injected into brain and other tissues. The mechanisms whereby other AAV serotypes enter host cells are actively being studied.

The biodistribution of rAAV vectors, primarily for serotype 2, has been studied and was reviewed recently along with the pharmacokinetic properties of other viral vectors [81]. A general aim of distribution studies is to define the target organs to which the vector spreads and cell types within the organ that can be infected. This involves studies of the whole animal, for example after IV administration, and within individual organs, such as after parenchymal injections. The pharmacokinetic properties of a vector at the whole-animal level typically depend on many factors, including the route and duration of administration, the dose, the physical properties of the vector (e.g., size), and cell-tropism. The type of assay used to monitor the distribution of the vector is also important. Quantitative PCR is a standard technique in gene transfer because of its sensitivity and ability to demonstrate the presence of the vector even if the transgene is not expressed. Other assays include radioactive tracers, nonquantitative or semiquantitative PCR, Southern blotting, and immunohistochemistry to study the biodistribution of vectors, as well as functional assays of transgene expression.

IV administration of rAAV2 generally results in the vector accumulating primarily in the liver, although smaller amounts spread to many tissues including the spleen, smooth muscle, striated muscle and kidneys; these were measured using PCR and Southern blotting [82] as well as transgene expression [83]. After one week, most of the remaining viral DNA was found in the liver, with some distributed to muscle. Watson et al. [83] used transgene expression to study rAAV2 biodistribution after IV or intramuscular injection in a mouse model for lysosomal storage disease (MPS VII). The route of vector administration dramatically affected its spread and distribution. Intramuscular injection resulted in high and localized transgene production especially in the liver, while IV injection produced low expression in this tissue. It was interesting to note that the site of IV injection might be important, with portal vein injections leading to more hepatic distribution of AAV2 than tail vein injections. However, the overall pharmacokinetic pattern of transgene expression over eight weeks was similar in animals after the injection of vector particles either into the muscle, tail vein, or portal vein [84].

IV distribution of rAAV2 has been reported in the application of hemophilia B (human factor IX) via portal vein or muscle. PCR was used to detect the gene expression. Stable and persistent human factor IX (hFIX) was observed. The kinetics of expression after injection of vector particles into muscle, tail vein, or portal vein was similar, with hFIX detectable at two weeks and reaching a plateau by eight weeks. Intraportal administration of vector resulted in a higher level of gene expression compared to tail vein or intramuscular injection. PCR results showed a predominant localization of the rAAV FIX genome in the liver and spleen after tail vein injection, with a higher proportion in liver after portal vein injection.

Lai et al. [85] studied rAAV2 biodistribution after intrahepatic *in-utero* injection in rhesus monkey fetuses. The vector genomes were distributed into many tissues, including the brain, astrocytes, and peripheral blood. However, these vectors do not appear to enter the brain in older animals possessing intact blood–brain barriers. Favre et al. [86] also showed that rAAV2 mediated gene expression for at least 18 months in lymph nodes and the liver, but not in the gonads.

Whether rAAV distributes to sperm cells has been of significant concern because of the possibility of introducing mutations in sperm. While AAV DNA can be found in the testes of rodents, rabbits, and dogs following either hepatic arterial or IM injections, these signals appear to be localized to the testis basement membrane and the interstitial space, with no intracellular signal observed. In mice and rats, there was a dose-dependent increase of vector expression in gonadal DNA using PCR. However, in dogs and rabbits, DNA extracted from semen was negative for vector sequences. In clinical studies, human subjects injected IM with an AAV vector at doses up to $2 \times 10^{12}$ vector genomes/kg showed no evidence of vector sequences in semen [87]. These studies suggest that rAAV introduced into skeletal muscle or the hepatic artery does not transduce male germ sperm cells efficiently.

While there has been no report yet of using radiolabeled rAAV to determine biodistribution, this principle has been demonstrated with other viral vectors. Zinn et al. [88] labeled recombinant adenovirus serotype 5 knob with the gamma emitter $^{99m}$Tc. The data showed that the radiolabeling process had no effect on receptor binding. Using IV injections in mice, the experiment confirmed that the liver had a 10-fold higher specific binding than heart, kidney, or lung. The localization of this vector in the liver was also dose-dependent, as seen with rAAV vectors.

### 5.5.2
### rAAV Serotype and Biodistribution

rAAV distribution within tissues appears to be serotype-dependent. In the brain, most of the rAAV serotypes studied to date (rAAV1, 2, 5, and 8) predominantly transduce neurons, though non-neuronal cells can be transduced at lower efficiency either *in vivo* or in primary cultures. In contrast, rAAV4 transduces primarily ependymal cells in the periventricular region, while rAAV5 transduces both neurons and ependymal cells. Pseudotypes of rAAV2/5 and rAAV2/1, in which rAAV2 DNA was encapsulated in the capsids of serotypes 5 and 1, led to greater spread of transgene expression than rAAV2. rAAV2/1 and rAAV2/5 also transduced different populations of neurons in the midbrain and hippocampus [89]. Whether this increased spread is due to differences in the distribution of the vector remains to be determined.

AAV serotype is also important in peripheral tissues. Despite extensive experience with rAAV2 vectors used in the lung, gene expression has been low in cystic fibrosis (CF) gene therapy. Sirninger et al. [90] showed that rAAV5 provided significantly more expression in this tissue. Similarly, intrapleural administration of rAAV5 led to a 10-fold increase in $\alpha_1$-antitrypsin transgene expression compared to rAAV2. In liver, transgene expression was reported to be 10- to 100-fold higher with AAV8 than observed with other serotypes. This improved efficiency correlated with a higher number of transduced hepatocytes, although it is not clear to what extent this reflected greater spread through the organ versus cellular interactions with the vector (e.g., uptake into the cell or intracellular distribution) [91].

## 5.6 Summary

In summary, current gene delivery vectors are extremely diverse. Each vector will have to overcome particular vector-specific barriers to be effective in gene delivery. Several of these barriers arise from the particulate nature of vectors. Depending upon the intended target and administration site of the vector, these particles will need to overcome numerous biological hurdles. Clearly, an understanding of the pharmacokinetic behavior of these vectors will be of utmost importance when designing an effective therapeutic regimen.

## 5.7 References


1 Bennett H.S., Luft J.H., and Hampton J.C. **1959**. Morphological classification of vertebrate blood capillaries. *Am. J. Physiol.* 196(2):381–390.

2 Nishikawa, M., and L. Huang. **2001**. Nonviral vectors in the new millennium: delivery barriers in gene transfer. *Hum. Gene. Ther.* 12:861–870.

3 Houk, B.E., G. Hochhaus, and J.A. Hughes. **1999**. Kinetic modeling of plasmid DNA degradation in rat plasma. *AAPS PharmSci.* 1:E9.

4 Kawabata, K., Y. Takakura, and M. Hashida. **1995**. The fate of plasmid DNA after intravenous injection in mice: involvement of scavenger receptors in its hepatic uptake. *Pharm. Res.* 12:825–830.

5 Houk, B.E., R. Martin, G. Hochhaus, and J.A. Hughes. **2001**. Pharmacokinetics of plasmid DNA in the rat. *Pharm. Res.* 18:67–74.

6 Yoshida, M., R.I. Mahato, K. Kawabata, Y. Takakura, and M. Hashida. **1996**. Disposition characteristics of plasmid DNA in the single-pass rat liver perfusion system. *Pharm. Res.* 13:599–603.

7 Liu, Y., L.C. Mounkes, H.D. Liggitt, C.S. Brown, I. Solodin, T.D. Heath, and R.J. Debs. **1997**. Factors influencing the efficiency of cationic liposome-mediated intravenous gene delivery. *Nat. Biotechnol.* 15:167–173.

8 Osaka, G., K. Carey, A. Cuthbertson, P. Godowski, T. Patapoff, A. Ryan, T. Gadek, and J. Mordenti. **1996**. Pharmacokinetics, tissue distribution, and expression efficiency of plasmid [$^{33}$P]DNA following intravenous administration of DNA/cationic lipid complexes in mice: use of a novel radionuclide approach. *J. Pharm. Sci.* 85:612–618.

9 Lew, D., S.E. Parker, T. Latimer, A.M. Abai, A. Kuwahara-Rundell, S.G. Doh, Z.Y. Yang, D. Laface, S.H. Gromkowski, and G.J. Nabel. **1995**. Cancer gene therapy using plasmid DNA: pharmacokinetic study of DNA following injection in mice. *Hum. Gene Ther.* 6:553–564.

10 McClarrinon, M., L. Gilkey, V. Watral, B. Fox, C. Bullock, L. Fradkin, D. Liggitt, L. Roche, L.B. Bussey, E. Fox, and C. Gorman. **1999**. In vivo studies of gene expression via transient transgenesis using lipid-DNA delivery. *DNA Cell Biol.* 18: 533–547.

11 Anderson, W.F. **1998**. Human gene therapy. *Nature* 392:25–30.

12 Lechardeur, D., and G.L. Lukacs. **2002**. Intracellular barriers to non-viral gene transfer. *Curr. Gene Ther.* 2:183–194.

13 Cherng, J.Y., N.M. Schuurmans-Nieuwenbroek, W. Jiskoot, H. Talsma, N.J. Zuidam, W.E. Hennink, and D.J. Crommelin. **1999**. Effect of DNA topology on the transfection efficiency of poly((2- dimethylamino)ethyl methacrylate)-plasmid complexes. *J. Control. Release* 60:343–353.

14 Kano, Y., T. Miyashita, H. Nakamura, K. Kuroki, A. Nagata, and F. Imamoto. **1981**. In vivo correlation between DNA supercoiling and transcription. *Gene* 13:173–184.

**15** Sekiguchi, J.M., R.A. Swank, and E.B. Kmiec. **1989**. Changes in DNA topology can modulate in vitro transcription of certain RNA polymerase III genes. *Mol. Cell. Biochem.* 85:123–133.

**16** Pina, B., R.J. Hache, J. Arnemann, G. Chalepakis, E.P. Slater, and M. Beato. **1990**. Hormonal induction of transfected genes depends on DNA topology. *Mol. Cell. Biol.* 10:625–633.

**17** Ohse, M., K. Kawade, and H. Kusaoke. **1997**. Effects of DNA topology on transformation efficiency of *Bacillus subtilis* ISW1214 by electroporation. *Biosci. Biotechnol. Biochem.* 61:1019–1021.

**18** Tanswell, A.K., O. Staub, R. Iles, R. Belcastro, J. Cabacungan, L. Sedlackova, B. Steer, Y. Wen, J. Hu, and H. O'Brodovich. **1998**. Liposome-mediated transfection of fetal lung epithelial cells: DNA degradation and enhanced superoxide toxicity. *Am. J. Physiol.* 275:L452–460.

**19** Xie, T.D., L. Sun, H.G. Zhao, J.A. Fuchs, and T.Y. Tsong. **1992**. Study of mechanisms of electric field-induced DNA transfection. IV. Effects of DNA topology on cell uptake and transfection efficiency. *Biophys. J.* 63:1026–1031.

**20** Bergan, D., T. Galbraith, and D.L. Sloane. **2000**. Gene transfer in vitro and in vivo by cationic lipids is not significantly affected by levels of supercoiling of a reporter plasmid. *Pharm. Res.* 17:967–973.

**21** Herweijer, H., G. Zhang, V.M. Subbotin, V. Budker, P. Williams, and J.A. Wolff. **2001**. Time course of gene expression after plasmid DNA gene transfer to the liver. *J. Gene Med.* 3:280–291.

**22** Xu, L., T. Daly, C. Gao, T.R. Flotte, S. Song, B.J. Byrne, M.S. Sands, and K. Parker Ponder. **2001**. CMV-beta-actin promoter directs higher expression from an adeno-associated viral vector in the liver than the cytomegalovirus or elongation factor 1 α promoter and results in therapeutic levels of human factor X in mice. *Hum. Gene Ther.* 12:563–573.

**23** Boussif, O., F. Lezoualc'h, M.A. Zanta, M.D. Mergny, D. Scherman, B. Demeneix, and J.P. Behr. **1995**. A versatile vector for gene and oligonucleotide transfer into cells in culture and in vivo: polyethylenimine. *Proc. Natl. Acad. Sci. USA* 92:7297–7301.

**24** Koltover, I., K. Wagner, and C.R. Safinya. **2000**. DNA condensation in two dimensions. *Proc. Natl. Acad. Sci. USA* 97:14046–14051.

**25** Takakura, Y., M. Nishikawa, F. Yamashita, and M. Hashida. **2002**. Influence of physicochemical properties on pharmacokinetics of non-viral vectors for gene delivery. *J. Drug Target.* 10:99–104.

**26** Henner, W.D., I. Kleber, and R. Benzinger. **1973**. Transfection of *Escherichia coli* spheroplasts. 3. Facilitation of transfection and stabilization of spheroplasts by different basic polymers. *J. Virol.* 12:741–747.

**27** Boussif, O., F. Lezoualc'h, M.A. Zanta, M.D. Mergny, D. Scherman, B. Demeneix, and J.P. Behr. **1995**. A versatile vector for gene and oligonucleotide transfer into cells in culture and in vivo: polyethylenimine. *Proc. Natl. Acad. Sci. USA* 92(16):7297–7301.

**28** Boeckle, S., K. von Gersdorff, S. van der Piepen, C. Culmsee, E. Wagner, and M. Ogris. **2004**. Purification of polyethylenimine polyplexes highlights the role of free polycations in gene transfer. *J. Gene Med.* 6:1102–1111.

**29** Plank, C., K. Mechtler, F.C. Szoka, Jr., and E. Wagner. **1996**. Activation of the complement system by synthetic DNA complexes: a potential barrier for intravenous gene delivery. *Hum. Gene Ther.* 7:1437–1446.

**30** Ogris, M., S. Brunner, S. Schuller, R. Kircheis, and E. Wagner. **1999**. PEGylated DNA/transferrin-PEI complexes: reduced interaction with blood components, extended circulation in blood and potential for systemic gene delivery. *Gene Ther.* 6:595–605.

**31** Kircheis, R., L. Wightman, A. Schreiber, B. Robitza, V. Rossler, M. Kursa, and E. Wagner. **2001**. Polyethylenimine/DNA complexes shielded by transferrin target gene expression to tumors after systemic application. *Gene Ther.* 8:28–40.

**32** Petersen, H., K. Kunath, A.L. Martin, S. Stolnik, C.J. Roberts, M.C. Davies, and T. Kissel. **2002**. Star-shaped poly(ethylene glycol)-block-polyethylenimine copolymers enhance DNA condensation of low molecular weight polyethylenimines. *Biomacromolecules* 3:926–936.

**33** Ogris, M., P. Steinlein, M. Kursa, K. Mechtler, R. Kircheis, and E. Wagner. **1998**. The size of DNA/transferrin-PEI complexes is an important factor for gene expression in cultured cells. *Gene Ther.* 5:1425–1433.

**34** Goula, D., C. Benoist, S. Mantero, G. Merlo, G. Levi, and B.A. Demeneix. **1998**. Polyethylenimine-based intravenous delivery of transgenes to mouse lung. *Gene Ther.* 5:1291–1295.

**35** Kircheis, R., S. Schuller, S. Brunner, M. Ogris, K.H. Heider, W. Zauner, and E. Wagner. **1999**. Polycation-based DNA complexes for tumor-targeted gene delivery in vivo. *J. Gene Med.* 1:111–120.

**36** O'Neill, M.M., C.A. Kennedy, R.W. Barton, and R.J. Tatake. **2001**. Receptor-mediated gene delivery to human peripheral blood mononuclear cells using anti-CD3 antibody coupled to polyethylenimine. *Gene Ther.* 8:362–368.

**37** Suh, W., J.K. Chung, S.H. Park, and S.W. Kim. **2001**. Anti-JL1 antibody-conjugated poly (L-lysine) for targeted gene delivery to leukemia T cells. *J. Control. Release* 72:171–178.

**38** Li, S., Y. Tan, E. Viroonchatapan, B.R. Pitt, and L. Huang. **2000**. Targeted gene delivery to pulmonary endothelium by anti-PECAM antibody. *Am. J. Physiol. Lung Cell. Mol. Physiol.* 278:L504–L511.

**39** Wu, G.Y., and C.H. Wu. **1987**. Receptor-mediated in vitro gene transformation by a soluble DNA carrier system. *J. Biol. Chem.* 262:4429–4432.

**40** Blessing, T., M. Kursa, R. Holzhauser, R. Kircheis, and E. Wagner. **2001**. Different strategies for formation of pegylated EGF-conjugated PEI/DNA complexes for targeted gene delivery. *Bioconjug. Chem.* 12:529–537.

**41** Diebold, S.S., M. Kursa, E. Wagner, M. Cotten, and M. Zenke. **1999**. Mannose polyethylenimine conjugates for targeted DNA delivery into dendritic cells. *J. Biol. Chem.* 274:19087–19094.

**42** Erbacher, P., J.S. Remy, and J.P. Behr. **1999**. Gene transfer with synthetic virus-like particles via the integrin-mediated endocytosis pathway. *Gene Ther.* 6:138–145.

**43** Kircheis, R., A. Kichler, G. Wallner, M. Kursa, M. Ogris, T. Felzmann, M. Buchberger, and E. Wagner. **1997**. Coupling of cell-binding ligands to polyethylenimine for targeted gene delivery. *Gene Ther.* 4:409–418.

**44** Nah, J.W., L. Yu, S.O. Han, C.H. Ahn, and S.W. Kim. **2002**. Artery wall binding peptide-poly(ethylene glycol)-grafted-poly-(L-lysine)-based gene delivery to artery wall cells. *J. Control. Release* 78:273–284.

**45** Lee, H., T.H. Kim, and T.G. Park. **2002**. A receptor-mediated gene delivery system using streptavidin and biotin-derivatized, pegylated epidermal growth factor. *J. Control. Release* 83:109–119.

**46** Benns, J.M., R.I. Mahato, and S.W. Kim. **2002**. Optimization of factors influencing the transfection efficiency of folate-PEG-folate-graft-polyethylenimine. *J. Control. Release* 79:255–269.

**47** Erbacher, P., T. Bettinger, P. Belguise-Valladier, S. Zou, J.L. Coll, J.P. Behr, and J.S. Remy. **1999**. Transfection and physical properties of various saccharide, poly(ethylene glycol), and antibody-derivatized polyethylenimines (PEI). *J. Gene Med.* 1:210–222.

**48** Faraasen, S., J. Voros, G. Csucs, M. Textor, H.P. Merkle, and E. Walter. **2003**. Ligand-specific targeting of microspheres to phagocytes by surface modification with poly(L-lysine)-grafted poly(ethylene glycol) conjugate. *Pharm. Res.* 20:237–246.

**49** Kursa, M., G.F. Walker, V. Roessler, M. Ogris, W. Roedl, R. Kircheis, and E. Wagner. **2003**. Novel shielded transferrin-polyethylene glycol-polyethylenimine/DNA complexes for systemic tumor-targeted gene transfer. *Bioconjug. Chem.* 14:222–231.

**50** Boletta, A., A. Benigni, J. Lutz, G. Remuzzi, M.R. Soria, and L. Monaco. **1997**. Nonviral gene delivery to the rat kidney with polyethylenimine. *Hum. Gene Ther.* 8:1243–1251.

**51** Maheshwari, A., S. Han, R.I. Mahato, and S.W. Kim. **2002**. Biodegradable polymer-based interleukin-12 gene delivery: role of induced cytokines, tumor infiltrating cells and nitric oxide in anti-tumor activity. *Gene Ther.* 9:1075–1084.

**52** Cohen, H., R.J. Levy, J. Gao, I. Fishbein, V. Kousaev, S. Sosnowski, S. Slomkowski, and G. Golomb. **2000**. Sustained delivery and expression of DNA encapsulated in polymeric nanoparticles. *Gene Ther.* 7 : 1896–1905.

**53** Lim, Y.B., S.O. Han, H.U. Kong, Y. Lee, J.S. Park, B. Jeong, and S.W. Kim. **2000**. Biodegradable polyester, poly[α-(4-aminobutyl)-L-glycolic acid], as a non-toxic gene carrier. *Pharm. Res.* 17 : 811–816.

**54** Luo, D., K. Woodrow-Mumford, N. Belcheva, and W.M. Saltzman. **1999**. Controlled DNA delivery systems. *Pharm. Res.* 16 : 1300–1308.

**55** Prabha, S., and Labhasetwar. **2004**. Nanoparticle-mediated wild-type p53 gene delivery results in sustained antiproliferative activity in breast cancer cells. *Molecular Pharmaceutics* 1 : 211–219.

**56** Felgner, P.L., T.R. Gadek, M. Holm, R. Roman, H.W. Chan, M. Wenz, J.P. Northrop, G.M. Ringold, and M. Danielsen. **1987**. Lipofection: a highly efficient, lipid-mediated DNA-transfection procedure. *Proc. Natl. Acad. Sci. USA* 84 : 7413–7417.

**57** Hirko, A., F. Tang, and J.A. Hughes. **2003**. Cationic lipid vectors for plasmid DNA delivery. *Curr. Med. Chem.* 10 : 1185–1193.

**58** Hui, S.W., M. Langner, Y.L. Zhao, P. Ross, E. Hurley, and K. Chan. **1996**. The role of helper lipids in cationic liposome-mediated gene transfer. *Biophys. J.* 71 : 590–599.

**59** Sternberg, B., F.L. Sorgi, and L. Huang. **1994**. New structures in complex formation between DNA and cationic liposomes visualized by freeze-fracture electron microscopy. *FEBS Lett.* 356 : 361–366.

**60** Stegmann, T., and J.Y. Legendre. **1997**. Gene transfer mediated by cationic lipids: lack of a correlation between lipid mixing and transfection. *Biochim. Biophys. Acta* 1325 : 71–79.

**61** Wrobel, I., and D. Collins. **1995**. Fusion of cationic liposomes with mammalian cells occurs after endocytosis. *Biochim. Biophys. Acta* 1235 : 296–304.

**62** Wattiaux, R., M. Jadot, M.T. Warnier-Pirotte, and S. Wattiaux-De Coninck. **1997**. Cationic lipids destabilize lysosomal membrane in vitro. *FEBS Lett.* 417 : 199–202.

**63** Stamatatos, L., R. Leventis, M.J. Zuckermann, and J.R. Silvius. **1988**. Interactions of cationic lipid vesicles with negatively charged phospholipid vesicles and biological membranes. *Biochemistry* 27 : 3917–3925.

**64** Smisterova, J., A. Wagenaar, M.C. Stuart, E. Polushkin, G. ten Brinke, R. Hulst, J.B. Engberts, and D. Hoekstra. **2001**. Molecular shape of the cationic lipid controls the structure of cationic lipid/dioleylphosphatidylethanolamine-DNA complexes and the efficiency of gene delivery. *J. Biol. Chem.* 276 : 47615–47622.

**65** Nicolazzi, C., N. Mignet, N. de la Figuera, M. Cadet, R.T. Ibad, J. Seguin, D. Scherman, and M. Bessodes. **2003**. Anionic polyethyleneglycol lipids added to cationic lipoplexes increase their plasmatic circulation time. *J. Control. Release* 88 : 429–443.

**66** Meyer, O., D. Kirpotin, K. Hong, B. Sternberg, J.W. Park, M.C. Woodle, and D. Papahadjopoulos. **1998**. Cationic liposomes coated with polyethylene glycol as carriers for oligonucleotides. *J. Biol. Chem.* 273 : 15621–15627.

**67** Pires, P., S. Simoes, S. Nir, R. Gaspar, N. Duzgunes, and M.C. Pedroso de Lima. **1999**. Interaction of cationic liposomes and their DNA complexes with monocytic leukemia cells. *Biochim. Biophys. Acta* 1418 : 71–84.

**68** Mahato, R.I. **2005**. Water insoluble and soluble lipids for gene delivery. *Adv. Drug Deliv. Rev.* 57 : 699–712.

**69** Mahato, R.I., K. Anwer, F. Tagliaferri, C. Meaney, P. Leonard, M.S. Wadhwa, M. Logan, M. French, and A. Rolland. **1998**. Biodistribution and gene expression of lipid/plasmid complexes after systemic administration. *Hum. Gene Ther.* 9 : 2083–2099.

**70** Hofland, H.E., D. Nagy, J.J. Liu, K. Spratt, Y.L. Lee, O. Danos, and S.M. Sullivan. **1997**. In vivo gene transfer by intravenous administration of stable cationic lipid/DNA complex. *Pharm. Res.* 14 : 742–749.

**71** Zhu, N., D. Liggitt, Y. Liu, and R. Debs. **1993**. Systemic gene expression after intravenous DNA delivery into adult mice. *Science* 261 : 209–211.

**72** Thierry, A.R., Y. Lunardi-Iskandar, J.L. Bryant, P. Rabinovich, R.C. Gallo, and L.C. Mahan. **1995**. Systemic gene therapy: biodistribution and long-term expression of a transgene in mice. *Proc. Natl. Acad. Sci. USA* 92:9742–9746.

**73** Liu, Y., D. Liggitt, W. Zhong, G. Tu, K. Gaensler, and R. Debs. **1995**. Cationic liposome-mediated intravenous gene delivery. *J. Biol. Chem.* 270:24864–24870.

**74** Hofland, H.E., D. Nagy, J.J. Liu, K. Spratt, Y.L. Lee, O. Danos, and S.M. Sullivan. **1997**. In-vivo gene transfer by intravenous administration of stable cationic lipid/DNA complex. *Pharm. Res.* 14:742–749.

**75** Berns, K.I., and C. Giraud. **1996**. Biology of adeno-associated virus. *Curr. Top. Microbiol. Immunol.* 218:1–23.

**76** McLaughlin, S.K., P. Collis, P.L. Hermonat, and N. Muzyczka. 1988. Adeno-associated virus general transduction vectors: analysis of proviral structures. *J. Virol.* 62:1963–1973.

**77** Leopold, P.L., B. Ferris, I. Grinberg, S. Worgall, N.R. Hackett, and R.G. Crystal. **1998**. Fluorescent virions: dynamic tracking of the pathway of adenoviral gene transfer vectors in living cells. *Hum. Gene Ther.* 9:367–378.

**78** Linden, R.M., P. Ward, C. Giraud, E. Winocour, and K.I. Berns. **1996**. Site-specific integration by adeno-associated virus. *Proc. Natl. Acad. Sci. USA* 93:11288–11294.

**79** Miao, C.H., H. Nakai, A.R. Thompson, T.A. Storm, W. Chiu, R.O. Snyder, and M.A. Kay. **2000**. Nonrandom transduction of recombinant adeno-associated virus vectors in mouse hepatocytes in vivo: cell cycling does not influence hepatocyte transduction. *J. Virol.* 74:3793–3803.

**80** Summerford, C., J.S. Bartlett, and R.J. Samulski. **1999**. AlphaVbeta5 integrin: a co-receptor for adeno-associated virus type 2 infection. *Nat. Med.* 5:78–82.

**81** Gonin, P., and C. Gaillard. **2004**. Gene transfer vector biodistribution: pivotal safety studies in clinical gene therapy development. *Gene Ther.* 11(Suppl 1): S98–S108.

**82** Ponnazhagan, S., M.J. Woody, X.S. Wang, S.Z. Zhou, and A. Srivastava. **1995**. Transcriptional transactivation of parvovirus B19 promoters in nonpermissive human cells by adenovirus type 2. *J. Virol.* 69:8096–8101.

**83** Watson, G.L., J.N. Sayles, C. Chen, S.S. Elliger, C.A. Elliger, N.R. Raju, G.J. Kurtzman, and G.M. Podsakoff. **1998**. Treatment of lysosomal storage disease in MPS VII mice using a recombinant adeno-associated virus. *Gene Ther.* 5:1642–1649.

**84** Nathwani, A.C., A. Davidoff, H. Hanawa, J.F. Zhou, E.F. Vanin, and A.W. Nienhuis. **2001**. Factors influencing in vivo transduction by recombinant adeno-associated viral vectors expressing the human factor IX cDNA. *Blood* 97:1258–1265.

**85** Lai, L., B.B. Davison, R.S. Veazey, K.J. Fisher, G.B. Baskin. **2000**. A preliminary evaluation of recombinant adeno-associated virus biodistribution in rhesus monkeys after intrahepatic inoculation in utero. *Hum. Gen. Ther.* 13: 2027–2039.

**86** Favre, D., N. Provost, V. Blouin, G. Blancho, Y. Cherel, A. Salvetti, and P. Moullier. **2001**. Immediate and long-term safety of recombinant adeno-associated virus injection into the nonhuman primate muscle. *Mol. Ther.* 4:559–566.

**87** Arruda, V.R., P.A. Fields, R. Milner, L. Wainwright, M.P. De Miguel, P.J. Donovan, R.W. Herzog, T.C. Nichols, J.A. Biegel, M. Razavi, M. Dake, D. Huff, A.W. Flake, L. Couto, M.A. Kay, and K.A. High. **2001**. Lack of germline transmission of vector sequences following systemic administration of recombinant AAV-2 vector in males. *Mol. Ther.* 4: 586–592.

**88** Zinn, K.R., J.T. Douglas, C.A. Smyth, H.G. Liu, Q. Wu, V.N. Krasnykh, J.D. Mountz, D.T. Curiel, and J.M. Mountz. **1998**. Imaging and tissue biodistribution of 99mTc-labeled adenovirus knob (serotype 5). *Gene Ther.* 5:798–808.

**89** Burger, C., O.S. Gorbatyuk, M.J. Velardo, C.S. Peden, P. Williams, S. Zolotukhin, P.J. Reier, R.J. Mandel, and N. Muzyczka. **2004**. Recombinant AAV viral vectors pseudotyped with viral capsids from serotypes 1, 2, and 5 display differential efficiency and cell tropism after delivery to different regions of the central nervous system. *Mol. Ther.* 10:302–317.

**90** De, B., A. Heguy, P.L. Leopold, N. Wasif, R.J. Korst, N.R. Hackett, and R.G. Crystal. **2004**. Intrapleural administration of a serotype 5 adeno-associated virus coding for α1-antitrypsin mediates persistent, high lung and serum levels of α1-antitrypsin. *Mol. Ther.* 10:1003–1010.

**91** Gao, G.P., M.R. Alvira, L. Wang, R. Calcedo, J. Johnston, and J.M. Wilson. **2002**. Novel adeno-associated viruses from rhesus monkeys as vectors for human gene therapy. *Proc. Natl. Acad. Sci. USA* 99:11854–11859.

# Part III
# Challenges and Opportunities

*Pharmacokinetics and Pharmacodynamics of Biotech Drugs: Principles and Case Studies in Drug Development.* Edited by Bernd Meibohm

ISBN: 3-527-31408-3

# 6
# Bioanalytical Methods Used for Pharmacokinetic Evaluations of Biotech Macromolecule Drugs: Issues, Assay Approaches, and Limitations

*Jean W. Lee*

## 6.1
## Introduction

Rational drug discovery, combined with knowledge from genomics and proteomics, have expanded drug development from traditional small-molecule drugs to a plethora of macromolecular drugs discovered through biotechnological approaches. These new drugs and approaches include proteins, peptides, monoclonal antibodies (Ab), Ab fragments, as well as antisense oligonucleotides and DNA gene therapy. The processes of bioanalytical method development and validation have been well developed to generate pharmacokinetic (PK) data to provide ADME (adsorption, distribution, metabolism, and elimination) information on small-molecule drugs. The bioanalytical methods for small-molecule drugs, however, are not readily applicable to macromolecules, due to basic differences in their chemistry and biology. Therefore, different bioanalytical approaches must be considered for the method development, validation, and assay implementation of macromolecules.

Sample extraction and chromatography serve as the basic tools of bioanalytical methods for the separation of xenobiotic analytes from matrix components before detection. The procedures are relatively straightforward for small molecules due to their simple and well-defined molecular structures. Sample extraction, followed by liquid chromatography tandem mass spectrometric (LC-MS/MS) detection, when operated in a multiple reaction mode (MRM, also known as Selected Reaction Monitoring), has become the workhorse of bioanalytical methods during the past decade. However, there are limited options available for the sample clean-up of a macromolecule analyte. Unlike small-molecule drugs, most protein therapeutic agents are analyzed without extraction, or with a crude protein precipitation step. Method selectivity often rests upon the ligand avidity or the resolutioning power of the mass spectrometer. Due to a lack of sample clean-up, extensive tests of the matrix effects are required during method development. The ion transmission on a triple quadrupole MS is limited for mass/charge ratios (m/z) of less than a few thousands, and therefore is not the best detection choice for macromolecule drugs. Most biotech macromolecules are determined using a ligand-binding assay (LBA),

*Pharmacokinetics and Pharmacodynamics of Biotech Drugs: Principles and Case Studies in Drug Development.* Edited by Bernd Meibohm

ISBN: 3-527-31408-3

with binding ligands such as antibodies (Ab) for immunoassays (IAs). Innovations in MS instrumentation, with phenomenally high mass accuracy in a time-of-flight (TOF) Fourier transfer ion cyclotron resonance (FT-ICR) and, more recently, in the LTQ Orbitrap$^{TM}$ MS, have extended the mass range for the measurement of macromolecules. Therefore, mass spectrometry also plays a critical role in the development of bioanalytical assays for macromolecule therapeutic agents.

Many biotech therapeutic agents are identical or similar to endogenous molecules, which presents serious analytical challenges such as the possibility of cross-reactivity, the establishment of standard calibrators and quality control (QC) preparations, as well as changes in the PK profile caused by the endogenous component. The catabolic and metabolic pathways of a therapeutic protein may not be known during its drug development process. In addition, biotransformation (e. g., proteolysis, amidation and acetylation) may alter the antigenicity and immunoreactivity of the protein and therefore affect binding assays. The possibility of different metabolic products for the therapeutic agent, the different variant forms of the endogenous protein, and anti-drug Ab can also impact upon the selectivity of the method and data quality, which overall would affect the measured PK profile [1, 2]. Endogenous protein variants and immunogenicity can be both population- and patient-dependent. They may not appear until the late phases of drug development. Thus, in addition to monitoring the assay performance of QC samples, there will be a need to identify any unexpected assay deviation in patient samples, such as the presence of anti-drug Ab. The issues of immunogenicity were recently discussed in several reports and conferences [3]. Overall, the challenges in establishing a solid bioanalytical methods for macromolecule therapeutic agents are numerous and more difficult to control than for small-molecule drugs.

In this chapter a number of approaches will be discussed that have been developed to address the challenges in macromolecule drug assays that impact upon selectivity, accuracy, and precision. The most commonly used methods for protein/peptide drugs – IA and LC-MS/MS – are discussed in detail, with case studies to illustrate the process, issues, assay approaches, and method limitations. Other less common methods and emerging technologies will be briefly mentioned. Biologic therapeutics that are not pure and/or not well characterized pose extra challenges which are beyond the scope of this chapter. Neither will biomarker assays used to support pharmacodynamic (PD) studies be covered in this chapter.

## 6.2
## Bioanalytical Methods for Macromolecule Drug Analysis: Common Considerations

### 6.2.1
### Sample Integrity and Analyte Stability

Macromolecule therapeutic agents tend to be more reactive than small-molecule drugs under various circumstances, from sample collection to analysis. Biological and chemical transformation may occur as a result of enzyme hydrolysis, heat or shearing force (e. g., blood withdrawal using a small-diameter hypodermic needle, or high-speed centrifugation) during blood collection, sample storage, shipping, and through multiple freeze–thaw cycles. It is important to evaluate analyte stability over these processes at the early phase of method development, even with a preliminary technique. Serious risks will be undertaking by starting a clinical study without stability data on sample collection, shipping, at least one freeze–thaw cycle, and short-term storage. Freezing and thawing may cause protein aggregation and the release of proteases or receptors from residual intracellular particles, and this in turn may adversely affect the analytical results. If sample volume permits, multiple aliquots of the same sample should be stored in order to minimize the number of freeze–thaw cycles.

In general, EDTA plasma is preferable to serum or heparinized plasma, or a special collection tube could contain a protease inhibitor cocktail. The samples should be processed rapidly at low temperature to separate the biological fluids of interest from the cell components, and stored at –70 °C. Other endogenous components could be inadvertently converted to a form that is the same or similar to the analyte, resulting in artificially high and variable concentrations. Bioanalytical problems may also arise from the fact that a therapeutic protein can be a modified form (e. g., by pegylation; see Chapter 11) of an endogenous protein [4], and the IA might quantify the common epitope of the therapeutic agent as well as the endogenous form. Therefore, sample integrity should be considered for a macromolecule therapeutic agent if the assay does not distinguish it from the endogenous form.

### 6.2.2
### Surface Adsorption

Peptides and proteins in an aqueous medium tend to adhere to plastic surfaces. For biological fluids containing ample quantities of proteins (e. g., plasma), surface adsorption is not an issue. Experiments should be conducted during method development to investigate any adhesion problems of the analyte in the deproteinized extract, or in a low-protein medium such as urine. Sample collection tubings, test tubes, transferring pipettes, storage containers and chromatographic systems should be evaluated for surface adsorption. For example, the preferences for LC-MS system are: PEEK tubings to stainless steel, ceramic probes to quartz or stainless steel, and polymer- to silica-based packing materials.

Precautions should also be taken with the aqueous solvents used to prepare standard working solutions in order to avoid adsorption problems, especially at low concentrations. To minimize adsorption during standard curve and validation sample (VS)/QC preparations, aliquots of the high-concentration stock solutions should be spiked promptly into the control blank plasma to prepare the standards and VS/QC. Once the compounds are in an environment of protein solutions, the adsorption problem is negligible. If there is a problem, however, adding or pre-rinsing with a solution of protein or a chaotropic agent (e. g., Tween, Triton X-100 or CHAPS) may help to alleviate the problem.

### 6.2.3
### Process of Method Development and Validation of Bioanalytical Methods for Macromolecule Drug Analysis

The process of developing and validating bioanalytical methods for the analysis of macromolecule therapeutic agents is similar to that for small molecules, with the exception that special attention should be paid to the challenging issues discussed below. The general process of method development and validation, which is often dynamic and iterative, is illustrated schematically in Fig. 6.1. Method validation includes all of the procedures required to show that a particular method is "reliable for the intended application" [5]. Pre-analytic plans include an understanding of the intended applications, information on method/reagent availability and sample integrity, and to define the requirements for assay performance on sensitivity, selectivity, linearity, reproducibility, and analyte stability.

Method development includes feasibility and optimization to meet the predefined study requirements. The appropriate standard calibrator range and concentrations of VS and QC (see discussion below) should be established for the dosage form and route of administration. The lessons learned during method development are critical for the development of specific parameters for the performance of the assay. For example, the number of validation batches and acceptance criteria should be described in a validation plan, and the standard operating procedure (SOP) should be written before conducting the pre-study validation. The correct approach should be to "develop a valid (acceptable) method," rather than simply

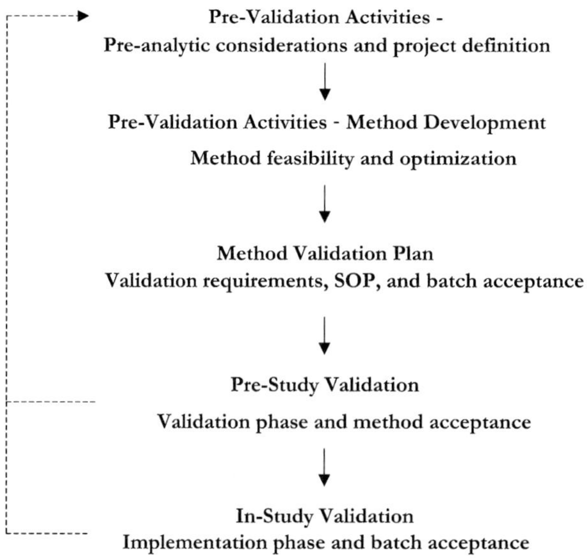


**Fig. 6.1** The processes of method development and validation. Activities from pre-validation planning, method development, pre-study validation to implementation (in-study validation) of bioanalytical methods.

to "validate (accept) a developed method" [6]. The QC data provide the in-study validity for acceptance of the assay batches and monitoring of assay performance.

The industry standard for bioanalytical method validation of small-molecule drugs to provide bioavailability and bioequivalence PK data was presented in a conference report and in a guidance document by the U.S. Food and Drug Administration (FDA) [7, 8]. Method validation and sample analysis in adherence to this guidance are referred to as "GLP-compliant". While the principles of the FDA May 2001 guidance for small-molecule drugs should be applied to macromolecule drugs, due to the heterogeneous nature of macromolecules and the inherent variability of IAs, the guidance cannot be directly applied to macromolecules. Several position papers have been published to discuss the issues in its application, and recommendations were proposed [9,10]. A conference is planned for 2006 by the FDA and the American Association of Pharmaceutical Scientists (AAPS) for consensus discussions.

### 6.2.4 Reference Standards

In general, the reference standard of a macromolecule drug is accompanied by a certificate of analysis, with information on potency (defined by the standardized biological activity per unit weight or volume) or purity in mass with the peptide sequence, salt form and water content. The activity unit may be standardized by the World Health Organization (WHO) or the United States Pharmacopeia (USP). However, not all macromolecule therapeutic agents are characterized by a standardized method with consistency [11–13]. Furthermore, the analytical method used to characterize the reference standard is often different from the analytical method used for biological samples. For example, a cell-based bioassay and an HPLC method are used for analytical characterization, while an IA is used for bioanalytical samples. Due to differences in post-translational modifications, such as the extent of deamidation and glycosylation, proteins from different sources can vary in their potency and immunoreactivity. It is important to note that potency, chromatographic responses, and immunoreactivity may vary for proteins from different sources, upon aggregation, or due to other unforeseen changes in the environment or manufacturing process. It is the responsibility of the bioanalytical laboratory to compare various reference lots using the same validated method. Immunoassay calibrator curves from two lots of reference standard for a protein therapeutic agent are illustrated in Fig. 6.2. It is apparent that the immunoreactivity behaviors of the two reference lots were different, clearly illustrating that assumptions which might have held true for small molecules are not necessarily true for macromolecule therapeutic agents. Therefore, the development of an SOP to include considerations of these possibilities is important. It is also important to realize that many unpredictable factors could affect the therapeutic agent, and a systematic detailed analysis of these factors may be required as part of the overall procedure. For example, if a protein therapeutic agent were suspected to form aggregates under environmental or process changes, a method or methods relevant to its biological activity must be included in the certification analysis plan.

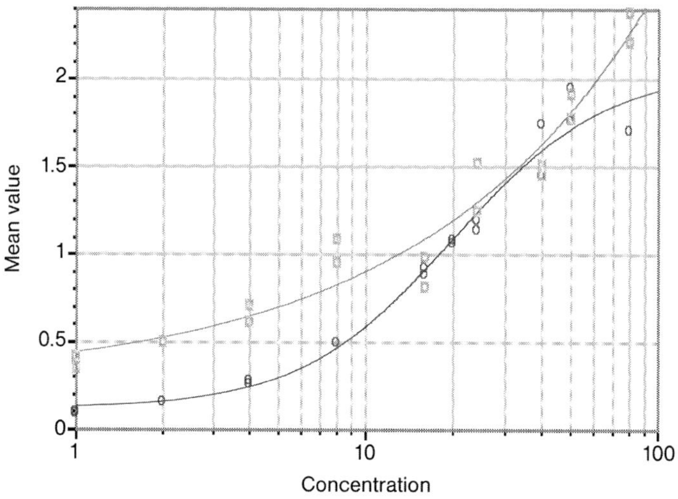


**Fig. 6.2** An example of the impact of reference standard lot inconsistency on assay performance. Y-axis: mean value of the OD reading, X-axis: concentration in pg/mL. Circles: standard curve using calibrators prepared with the original reference standard lot. Squares: standard curve using calibrators prepared with a different lot. The standard curves showed that the new reference standard behaved differently from the original in immunoreactivity with shallower slope, higher background noise and decreased sensitivity. The assay format is noncompetitive.

### 6.2.5 Drug Compounds that Exist Endogenously

Many protein therapeutics are the recombinant or modified products of endogenous components. In some cases, the protein drug may not be distinguishable from the endogenous form by the assay. If the endogenous concentration is much lower than the pharmacological concentrations, the assay accuracy will not be affected by the presence of the endogenous component. However, if the endogenous component concentrations were substantial, it would be difficult to find an analyte-free biological matrix to prepare standard calibrators [1, 5, 7, 8, 14]. In such case, an alternative analyte-free matrix is used, including protein buffers, analyte-depletion by charcoal stripping, high-temperature incubation, acid or alkaline hydrolysis, or affinity chromatography; alternatively, a heterologous matrix lacking the analyte or containing a non-cross-reactive homologue may be used [14–16]. The substitute matrix may not be exactly the same as that of the study samples, presenting the problem of matrix effects. Analytical bias due to matrix effects can be evaluated by the method of standard addition [17].

### 6.2.6
### Validation Samples, Quality Controls, and Assay Range

While a substitute matrix can be used to prepare standard calibrators for a drug compound that exists endogenously, VS/QCs should be prepared in the authentic matrix, regardless. VS data are used during method validation to characterize the intra- and inter-run accuracy/precision and stability. QC data are used for assay performance monitoring and to accept or reject a run during in-study validation. For pre-study validation, no validation batch should be rejected, unless with assignable causes. VS from a pre-study validation can be used later as QC samples.

VS/QCs should be prepared independently from the standard calibrators. Five or more levels of VS, including the low limit of quantitation (LLOQ), low, mid, high, and upper limits of quantitation (ULOQ) concentrations are often prepared. The three concentrations at low, mid, and high within the curve range can be retained for in-study QCs. For analytes that exist endogenously, matrix from multiple individuals should be screened with a preliminary method to identify lots with low or undetectable analyte concentrations. The samples may be pooled and aliquots spiked with varying amounts of reference material to create various concentrations of VS and QC samples.

VS and QC concentrations should reflect the expected study sample. For example, the plasma sample concentrations from intravenous or inhaler dosing would be very different. The concentration of samples around the $C_{max}$ region may exceed the ULOQ. In that case, a VS/QC should be prepared to mimic the high concentration samples and tested for dilution linearity to extend the assay range.

### 6.2.7
### Protein Binding Problems

Protein binding of circulating macromolecule drugs is more complicated and less understood than that of small-molecule drugs. Binding may occur with low and/or high affinity-binding proteins/receptors that are soluble, in the cell membranes, or within cell organelles. Binding proteins are known to cause interference in the IA of protein analytes such as cytokines [18, 19], growth hormone [20], tissue plasminogen activator [13], and insulin-like growth factors-1 and -2 [21, 22]. The binding kinetics and equilibrium are dependent upon the specific configuration of the drug and binding protein, and are affected by factors such as phosphorylation, protein folding, and cofactors. Disagreement in serum concentrations may occur when Ab with different specificities are used for the determination of a therapeutic protein. Extraction procedures used to dissociate small molecules from serum proteins are usually not applicable because they will also denature the protein therapeutic agent. In addition, binding to anti-drug Ab as a result of repeated administration can also cause analytical interference and may alter the PK and PD profiles of the therapeutic protein [17,23,24].

## 6.3 The Bioanalytical Method Workhorses

The most common bioanalytical tools applied for macromolecule drug bioanalysis are IA, HPLC, and LC-MS/MS.

Most macromolecule drugs are antigenic, or they bind to another biomacromolecule ligand such as nucleic acid probes, receptors, and specific binding proteins. Naturally, ligand-binding assays – especially IAs – have been the most commonly used bioanalytical methods for these drugs. The widely used enzyme-linked immunoadsorption assay (ELISA), which exploits the specificity conferred by a capturing Ab and a detecting Ab, is shown schematically in Fig. 6.3. The detecting Ab is conjugated to a reporter enzyme, which converts a substrate into the detection product. The method is flexible, with many enzymes/substrates and assay formats for many method options [1, 14, 25].

HPLC-UV is often used for formulation product analysis, but its sensitivity and selectivity is inadequate for bioanalytical application. A gradient elution of 30 to 60 min is often used to provide column separation from the matrix proteins. HPLC-UV methods are not desirable due to low signal-to-noise (S/N) ratios and long chromatographic run times.

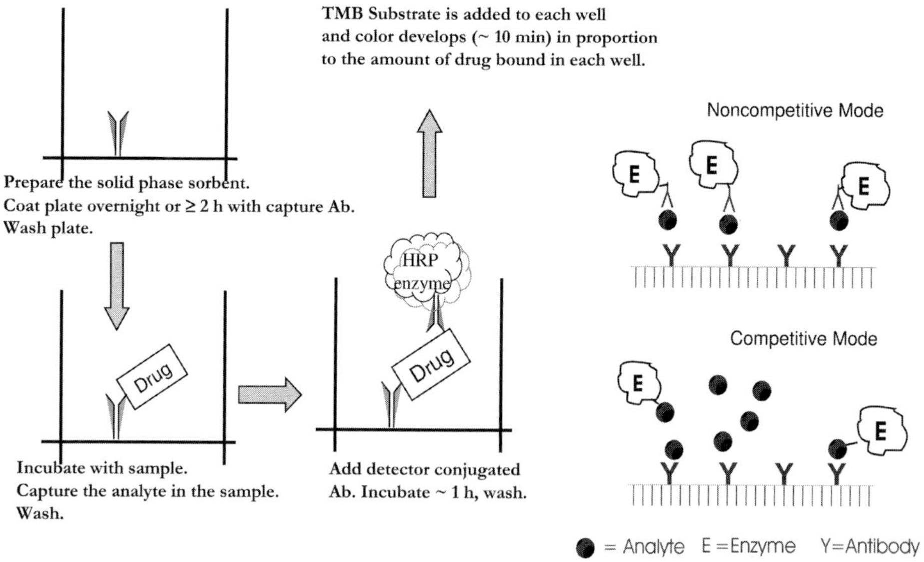


**Fig. 6.3** Scheme of an ELISA sandwich assay. Left panel: A typical assay format using a 96-well microplate, pre-coated with a capture antibody (Ab). The drug analyte in the sample is bound to the immobilized capture Ab. The extraneous matrix components are washed away before addition of the detector second Ab, which is conjugated with horseradish peroxidase (HRP) enzyme. The enzyme catalyzes the release of luminescence from the substrate, such as tetramethylbenzidine (TMB). More details are given in the example in Section 6.3.1. This format represents a non-competitive assay. Right panel: The two formats of noncompetitive and competitive assays.

LC-MS/MS has been the workhorse for the bioanalysis of small drug compounds over the past decade because of its superior sensitivity, selectivity, and almost universal detection of any ionizable molecule. The technology can be applied to peptide drugs of a few thousand Daltons because peptides with multiple charges would have m/z-values well within the limited detection mass range of triple quadrupole MS instruments. The pore size and size distribution of packing materials for chromatographic columns have been optimized for macromolecules to enable rapid solute diffusion for shorter run times and adequate peak shape [26, 27]. However, limited sample clean-up, harsh fragmentation and inconsistent ionization could contribute to quantitative problems. Recently, more sensitive MS instruments with soft ionization [e.g., electrospray (ESI)] and high-mass resolution [e.g., TOF and Fourier transform mass spectrometers (FTMS)] have become available commercially to alleviate these problems. LC-ESI-MS/MS methods have been applied to peptide drugs and have proven to be a reliable workhorse in addition to IAs [28, 29].

Capillary electrophoresis (CE) has high separation efficiency, especially for chromatographically challenging large molecules such as oligonucleotides and glycoproteins. Unfortunately, the advantage of nanoliter sample loading volume becomes its own severe detection limitation. Laser-induced fluorescence detection can be used to provide adequate sensitivity for peptides labeled with a fluorescent probe detected by a monochromatic laser, especially at the near-infrared region. However, this application is limited to peptides with reactive functional groups [30, 31]. CE-MS has been used for the analysis of oligonucleotides and glycoproteins (e.g., human erythropoietin) [32, 33]. This technique has enabled the differentiation of post-translational glycosylation variants between the recombinant and endogenous human erythropoietins [34].

The method of choice is dependent upon the analyte, the assay performance required to meet the intended application, the timeline, and cost-effectiveness. The assay requirements include sensitivity, selectivity, linearity, accuracy, precision, and method robustness. Assay sensitivity in general is in the order of IA > LC-MS/MS > HPLC, while selectivity is IA $\approx$ LC-MS/MS > HPLC. However, IA is an indirect method which measures the binding action instead of relying directly on the physico-chemical properties of the analyte. The IA response versus concentration curve follows a curvilinear relationship, and the results are inherently less precise than for the other two methods with linear concentration–response relationships. The method development time for IA is usually longer than that for LC/MS-MS, mainly because of the time required for the production and characterization of unique antibody reagents. Combinatorial tests to optimize multiple factors in several steps of some IA formats are more complicated, and also result in a longer method refinement time. The nature of IAs versus that of LC-MS/MS methods are compared in Table 6.1. However, once established, IA methods are sensitive, consistent, and very cost-effective for the analysis of large volumes of samples. The more expensive FTMS or TOF-MS methods can be used to complement IA on selectivity confirmation.

**Table 6.1** General differences between chromatographic methods (HPLC and LC-MS/MS) and ligand-binding assays (immunoassays).

| Property | HPLC | LC-MS/MS | Immunoassays |
|---|---|---|---|
| Basis of measure | Physico-chemical (UV, fluorescence) | Physico-chemical (ESI) | Biochemical binding action (many detection modes) |
| Detection method | Direct | Direct | Indirect |
| Analytical reagents | Common and available | Common and available, except IS | Unique, may not be available |
| Analytes | Small and macromolecules | Small peptides and oligonucleotides | Most macromolecules |
| Sample preparation | Extraction for small molecules | Protein precipitation or solid-phase extraction | Usually no extraction |
| Calibration model | Linear | Linear | Nonlinear |
| Assay environment | Aqueous and organic | Usually contains organic | Aqueous (pH 6–8) |
| Development time | Weeks | Weeks | Months (Ab production) |
| Inter-assay variability | Moderate (15–20% CV) | Low (<10% CV) | Higher (>20% CV) |
| Imprecision source | Intra-assay | Intra-assay | Inter-assay |
| Working range | Broadest | Moderately broad | Limited |
| Equipment cost | | Expensive | Inexpensive |
| Analysis mode | | Series, batch | Batch |
| Assay throughput | | Good | Excellent |

During the course of drug development, several bioanalytical methods might have been applied to meet the intended applications. Under certain circumstances, comparison and cross-validation of the methods may be required to provide continuity and proper interpretation of the data. For example, an HPLC method might be used for the high concentration samples from toxicology studies, while a more sensitive LC-MS/MS or ELISA method may be developed and validated for human studies. Alternatively, an LC-MS/MS method might be used for the initial PK profiling of the peptide drug and metabolites, followed by an ELISA method for late phase monitoring of the drug only. Study samples from multiple subjects and QCs should be used for the cross-validation of different methods. Although the results may, or may not, show comparable accuracy, they will provide information for the interpretation and proper use of the data generated from these methods.

The advantages, issues and limitations of these methods are discussed in the following sections.

## 6.3.1 Ligand-Binding Assays: Immunoassays

### 6.3.1.1 Common Method Approach

While many reagents and assay format choices are available for ligand-binding assays, the most commonly used ELISA method is discussed here as an example. Typical ELISA formats of noncompetitive (left panel) and competitive (right panel) sandwich assays are illustrated in Fig. 6.3. The basic reagent components are: (1) capturing Ab; (2) detector Ab; (3) the drug analyte itself; and (4) a solid phase platform (such as microplate wells). An ELISA method for measuring an antibody drug is illustrated in a case study in Section 6.4.1. The ELISA method is a heterogeneous assay in that the analyte is separated from the bulk of the matrix by forming an immunocomplex with the immobilized Ab on the solid phase, followed by several wash cycles. Assays without separation steps are homogeneous assays (e.g., proximal scintillation or fluorescence methods), often used for drug discovery screening. Besides microplates, beads with activated functional groups to covalently bind the capturing agents can be used. Magnetic beads have been used for easy washing and decanting. The capturing ligand can be an Ab or a protein drug, or the ligand protein of a monoclonal Ab drug. The flexible assay formats include homogeneous versus phase-separation assays, competitive versus noncompetitive reactions, binding reaction in solution phase versus solid phase, and the use of general capturing and reporter ligand reagents. In addition, other ligands besides Ab can be used, such as receptors, binding proteins, DNA probes, and aptamers.

There is a plethora of detecting system pairs of various enzyme conjugates and multitudes of corresponding substrates. Some of the common pairs are alkaline phosphatase/*p*-nitrophenylphosphate and horseradish peroxidase (HRP)/tetramethylbenzidine (TMB). To increase sensitivity, the detection signal can be amplified by systems such as biotin-avidin (or streptavidin), acetylcholine esterase, and diaphorase-NADPH-alcohol dehydrogenase-crystal violet. There are many instrumental detection choices, including colorimetric, fluorometric, time-resolved fluorescent, chemiluminescent, electrochemiluminescent, and flow cytometric detection. It is important to consider signal readout and background noise to obtain higher S/N values, as well as a wide dynamic reading range. In order to reduce the need for custom-labeled reagents, a variety of secondary Abs are commercially available as a general reporter to decrease method development time. However, the use of anti-idiotypic antibodies in a sandwich assay would offer better selectivity than a general reporter or capture antibodies. The adoption of diagnostic kits of endogenous proteins for the measurement of therapeutic protein agents in PK studies would also circumvent reagent development. The use of commercial kits for the analysis of a recombinant protein with an endogenous counterpart is illustrated in a case study in Section 6.4.2.

The reader should refer to books and review articles on ligand-binding assays for general IA method development [1, 14, 25, 35, 36]. The approach to method development and validation for macromolecule drugs using ligand-binding assays has been discussed in recent workshops, and a position paper published by the AAPS Ligand Binding Assay Bioanalytical Focus Group [10]. The advantages, issues, and limitations of IA applications to macromolecule drug bioanalysis are detailed in the following section.

#### 6.3.1.2 Advantages of Immunoassays

The most important advantage of IA is the sensitivity for macromolecules (e. g., in the range of attomolar (aM) and yoctomolar (yM), using a small sample volume of ~100 µL per assay) as compared to that of LC-MS/MS methods (e. g., fM and aM using a larger sample volume of ~500 µL per assay). In addition, IA does not require expensive instruments that need to be operated and maintained by specialists, as is necessary for MS. Also, the assay formats are versatile and flexible. The turnaround time for IA is similar to that for LC-MS/MS, or better, as the readouts are parallel (96-well simultaneously) rather than sequential (peak-by-peak and injection-by-injection), and there is minimal signal drift within a batch. IA has the advantage of unlimited molecular size of the macromolecule analytes due to their antigenicity. Once the suitable ligand reagents are produced in sufficient amounts and characterized, a consistent supply can be guaranteed with proper project management. IA is amenable to mass production by automation at relatively low cost, especially for late-phase clinical studies.

#### 6.3.1.3 Issues and Limitations of Immunoassays

##### 6.3.1.3.1 Reagent Characterization, Consistency, and Stability

One of the drawbacks of IA is the considerable time required to produce and characterize the ligand reagents. In general, multiple bleeds of polyclonal Ab or multiple lots of monoclonals are screened for their specific affinity to the desired epitopes. Several lots are then chosen for feasibility tests for an assay platform. For example, Ab specific for the drug molecules, discriminating them from the endogenous forms, post-translation variants, and degradation metabolites, might be selected. An Ab pair would be finalized to be the capture and reporter Ab, and characterized in the method development. It is important to document the quality of the Ab and other critical reagents such as blocking buffer, conjugating reagents, washing buffers, enzyme conjugate, and substrate. Method development includes optimization of the reagent binding stoichiometry (titer Ab and reagents) and factors affecting reaction kinetics and equilibrium (e. g., time, temperature, ionic strength, and pH of buffer). Specificity should be demonstrated against structurally similar endogenous components and known metabolites for the finalized method.

Because the unique antibody reagents are labile and prone to change, reagent stability and lot-consistency should be investigated and documented. The appropriate preparation and storage conditions should be defined to assure consistent supply over the entire course of drug development. In general, reagents should be stored refrigerated, frozen, or lyophilized [35]. Lot-to-lot consistency of the reagents should be established, especially if the reporter labeled Ab are from commercial vendors (e.g., variability of labeling). An inventory of prepackaged reagents (with defined reagent lot number, supplied in vials with expiration dates, analogous to diagnostic supplies) can be planned and reserved upfront for specified studies. In addition, method robustness should be tested in stressed samples that should include patient samples.

#### 6.3.1.3.2 **Selectivity and Matrix Effects**

Selectivity is the ability of an assay to measure the analyte of interest in the presence of other constituents in the sample. Because IAs are often performed without sample extraction, they are more prone to matrix interference than are chromatographic methods with extraction. Matrix interference could come from cross-reactivity with structurally similar components in the sample, or from nonspecific binding to structurally dissimilar components in the matrix. The results are high background noise, loss of sensitivity, and inaccurate and nonreproducible data. Sometimes, the problem may only occur in a few exceptional patient samples that have structurally similar components such as unknown metabolites, or dissimilar components from samples with hyperlipidemia, hemolysis, complement components, rheumatoid factors, binding proteins, autoantibodies, and heterophilic anti-immunoglobulin Ab.

The diligent analyst would develop a robust method with rigorous matrix effect tests on multiple lots, including hemolyzed and lipidemic samples. An initial test would be a spike-recovery evaluation on at least six individual lots. Samples should be spiked at or near the LLOQ, and at a high level near the ULOQ. If matrix interference were indicated by unacceptable relative error (RE) percentage in certain lots, the spiked sample of the unacceptable lots should be diluted with the standard calibrator matrix to estimate the minimum dilution requirement (MDR) at and above which the spike-recovery is acceptable. The spike-recovery test should then be repeated with the test samples diluted at the MDR. Note that this approach will increase the LLOQ for a less sensitive assay. If sensitivity is an issue, then other venues will be required to address the matrix effect problem. For example, the method can be modified to include sample clean-up, antibodies and/or assay conditions may be changed, or the study purpose may be tolerable to acknowledge that the method may not be selective for a few patients (whose data may require special interpretation).

Because the catabolic and metabolic pathways of biotech drugs are often poorly defined and sufficiently sensitive comparator assays are lacking, additional matrix effect tests by parallelism should be conducted with actual study samples. These are often performed on subject samples with aberrant PK profiles. A pool from several time points with sufficient analyte concentration of that subject is serially diluted. The observed concentration times the dilution factors should be within acceptable RE and CV %. Complementary technologies such as high-mass resolution MS methods can be used with IAs to confirm method selectivity.

#### 6.3.1.3.3 Standard Curve Regression Model and Curve Fitting

Physico-chemical measurements using chromatographic methods produce responses that are linear to the concentrations. As IA measures the resulting signals of a reaction, however, the response is a nonlinear function of the analyte concentration. Often, the regression model used to describe this relationship is a four- or five-parameter logistic function, as shown in the sigmoid shape standard curve in Fig. 6.4.

Anchor points (asymptotic low- and high-calibrators outside the working range) are included in the curve fitting for better interpolation near the limits of quantification. The variance of a linear curve function is constant over the concentration range, and weighting is relatively simple (usually by 1/concentration or 1/concentration$^2$). In contrast, the standard deviation of the replicate measurements in a curvilinear model is not a constant function of the mean response. Therefore, special statistical approaches of curve fitting and weighting are required for IA. A weighted, nonlinear, least-squares method was recommended [37, 38]. Individual weights often lead to undesirable statistical properties in the estimates of cali-

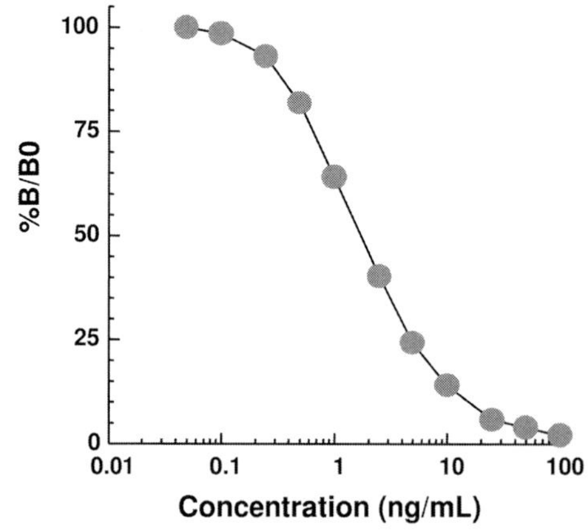


**Fig. 6.4** A four-parameter logistic standard curve depicting a competitive assay. In this format the antibody (Ab) is present in a limited amount. The known amount of labeled drug competes with the drug analyte in the sample for the limited sites of the Ab. The response factor produced by the label antigen–antibody complex (Ag–Ab) is inversely proportional to the concentration of the drug analyte in the sample. In this example the % $B/B_0$ (% label bound at concentration X divided by total bound at concentration zero) is used as the response factor. The parameters of the regression model are: (A) zero concentration response, (B) slope factor (positive), (C) inflection point at middle response ($EC_{50}$) and (D) infinite concentration response. For a non-competitive assay, the signal produced is proportional to the drug analyte concentration in the sample. The capturing Ab is present in an excess amount, and the drug-occupied Ab–Ag–Ab complex is being monitored.

bration curve parameters, particularly in the case of duplicate measurements. Failure to weigh responses properly will result in greater bias and imprecision in analytical results, particularly at analyte concentrations near the limits of quantification [39]. The following items highlight the major points in IA regression models and curve fitting recommended by DeSilva et al. [10]:

- More calibrator points (at least eight) should be used to describe the regression model, and anchor points should be used.
- In general, the functional range of a curvilinear regression model is narrower than that of a linear model; the assay range is extended by the dilution of samples with concentrations higher than the ULOQ. Dilution linearity also demonstrates the lack of a prozone or hook effect of the assay. A hook (prozone) effect is the phenomenon where the responses of higher concentrations of analyte are lower than expected.
- VS data are used to select the regression model and curve fitting. The regression model and weighting that generate the lowest total error (Absolute Bias + Intermediate Precision) of the VS should be chosen.
- IA results are inherently less precise than those of chromatographic assays [40]. Owing to the greater inter-assay imprecision, more validation runs should be used (e. g., at least six batches) to achieve an adequate level of confidence in the estimates of assay performance [41]. Acceptance criteria should be defined *a priori* for the method validation; they should be more lenient than those of chromatographic methods.
- The use of a correlation coefficient is not recommended for model validation [9, 17]. Even for a linear model, unacceptable calibration bias can exist, despite a correlation coefficient $> 0.9999$ [42].

#### 6.3.1.3.4 **Single versus Multiple-Analyte Assays**

Common IAs are geared for one analyte per assay. Multiplex methods such as Luminex LabMAP, ELISPOT, and rolling-circle amplified antibody chips have been used in drug discovery or biomarkers screening [43–45]. However, these technologies have not been used for the determination of drug compound and metabolites for GLP-compliant applications. This could be due to a low demand for metabolite data for biotech macromolecule drugs, as most of these metabolites are pharmacologically inactive. More rigorous validation would be required for multiplex assays to become useful for quantifications of multiple analytes, for example the prodrug, the active drug, variants of the endogenous form, metabolites, and/or isotypes of the anti-drug Ab.

## 6.3.2 HPLC-ESI-MS/MS Methods

### 6.3.2.1 Common Method Approach

ESI and atmospheric pressure chemical ionization (APCI) are the major LC-MS ionization sources. In ESI-MS, the molecules are ionized in the solution phase inside electrically charged droplets before vaporizing into the gas phase; further protonation or deprotonation can then occur in the gas phase [46–48]. In APCI-MS, the molecules in solution are vaporized in a heated gas stream, followed by chemical ionization in the gas phase. Being a "soft" ionization methodology, ESI is preferred for biomacromolecules [48]. The triple quadrupole LC-ESI-MS/MS has been applied for the determination of peptides and oligonucleotides, such as insulin [49], insulinotropins [50], endothelins [51], and multiple oligonucleotides [52]. In general, simple protein precipitation is used for sample clean-up. A gradient elution offers good LC separation of the analyte from other analytes or matrix components. The MRM, using distinctive m/z-values of the selective precursor and product fragment ions, further improves selectivity [53].

### 6.3.2.2 Advantages of HPLC-ESI-MS/MS Methods

In general, it is easier (and faster) to develop an HPLC-ESI-MS/MS method for multiple analytes with the matrix effects identified and under control as compared to IA (see below). Furthermore, selectivity and interference from matrix components and/or metabolites is less of a problem with an LC-MS/MS method compared to IA. With an appropriate internal standard (IS), LC-MS/MS methods are more precise than IA. Moreover, the LC-MS/MS calibration range is broader and can accommodate disproportionate concentration ranges of the drug compound and its metabolites. In contrast, most IA are geared to quantify only one analyte at a time because the method development of multiplex IA is complicated and often cannot be optimized for all analytes of interest.

### 6.3.2.3 Issues and Limitations of LC-ESI-MS/MS Methods

#### 6.3.2.3.1 Matrix Effects

The pressure of a fast turnaround time for the expensive LC-MS instrument and false confidence in MS mass resolution power often leads to compromised methods with shortened chromatographic runs. With limited sample clean-up for macromolecules and inadequate chromatographic separation, matrix components can co-elute with the analyte. They may compete for the limited charge or impede (or promote) movement of the analyte ions to the surface of the droplets, resulting in matrix effects [54]. Matrix effects can impact on selectivity, sensitivity, linearity and reproducibility of the assay. For ESI, competition for ionization can occur both in the mobile phase and the gas phase [55]. The pH, volatility, and surface tension of the mobile phase will affect ionization efficiency. The major suppressing matrix components are salts, which can be easily separated chromatographically from the analyte. The capacity factor (k′) of the analyte should be estimated to assure column separation of the analyte from major hydrophilic matrix components [54–56]. Late-eluting matrix components (hydrophobic components in a reversed-phase chromatography) often result in broad peaks at variable retention times in subsequent chromatograms.

Matrix effects do not need to be completely eliminated as long as they are consistent in the biological matrix among individuals, over the assay concentration range, and allow for sufficient assay sensitivity. Similar to IA, the initial approach to detect matrix effects is to perform spike-recovery experiments at or close to the LLOQ and ULOQ concentrations on at least six individual matrix lots [10]. If the recovery accuracy is inconsistent among the matrix lots, a post-column infusion experiment can be performed to investigate the nature of the interfering component [57, 58]. A diagram of the instrument setup is shown in Fig. 6.5. An analyte solution is infused to the LC effluent of a control matrix extract via a 'Tee' into the MS. The constant signal of the analyte infusion would show perturbation by the matrix effluent. The upper trace in the inset shows a typical matrix suppression peak. The mobile phase and/or water blank served as a control to the matrix extract, as shown in the lower trace. Figure 6.6 shows the matrix effect trace super-

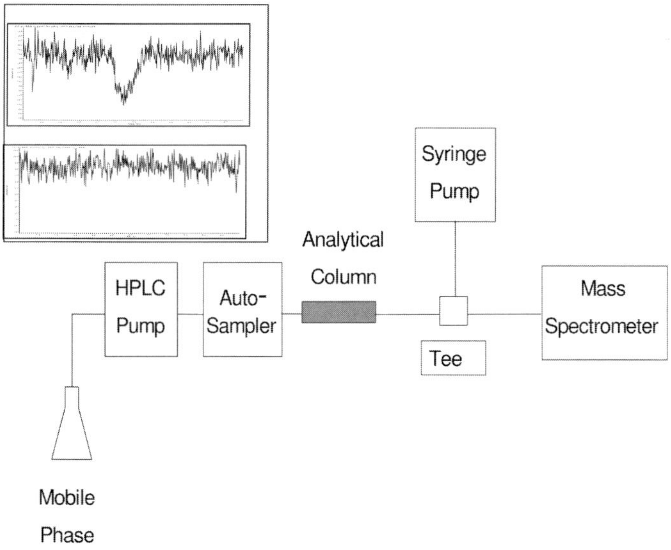


**Fig. 6.5** Diagram of system set-up for post-column infusion test for matrix effect. The analyte in the mobile phase was infused by a syringe pump at about 10 μL/min. The blank matrix extract or the test control (mobile phase or water blank extract) was injected into the analytical column. The effluent from the analytical chromatographic column was mixed with the analyte infused via a "Tee", and the signal response of the analyte was monitored by the mass spectrometer. The lower trace in the inset shows the control blank injection with a constant signal of the infused analyte; the upper trace shows the matrix-suppressing component.

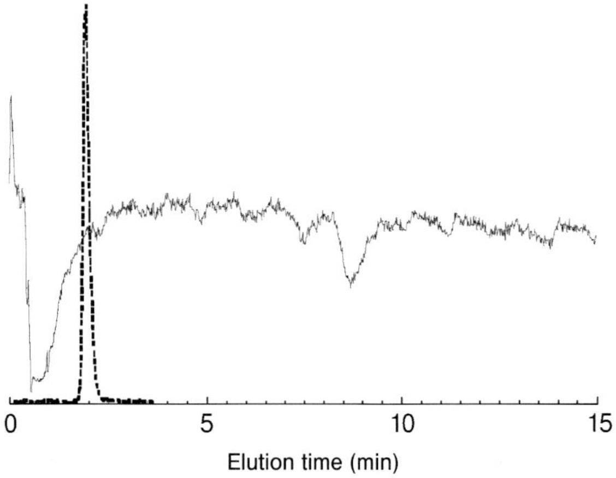


**Fig. 6.6** Detection of matrix-suppression components at the analyte retention time and at late elution. The chromatogram of the drug analyte is superimposed on the post-column infusion trace of the matrix extract. The chromatography was reversed-phase with a short run time of 3 min. The retention time of the analyte (~2.5 min) is at the end of a region where the bulk of hydrophilic suppressing components elutes. Other ion-suppressing components are shown at ~8–9 min. Locating the ion-suppressing components enabled chromatographic resolution of the matrix effect problem.

imposed onto the chromatogram of the analyte, identifying a matrix-suppression component at the analyte retention time, as well as broad late-eluting suppression components. Changing the analytical column to one with higher resolution resolved the analyte peak away from the matrix components. The late-eluting components were diverted to waste with a column-switching valve. The final method showed no significant matrix effect, as confirmed by the spike-recovery test. Besides chromatographic modification, matrix effects can be corrected with the help of a co-eluting IS, preferably a stable isotope-labeled version of the analyte of interest (see Section 6.3.2.3.3).

#### 6.3.2.3.2 Limited Options of Sample Clean-Up

**Protein Precipitation for Peptide Drugs.** The most common sample clean-up for peptide drugs is protein precipitation using organic solvents. Several issues should be considered for this crude method, which are later illustrated in the case study in Section 6.4.3.

- The goal is to achieve good, consistent recovery of the analyte, while the bulk of the matrix protein is optimally removed with minimal volume change. Precipitation is effective by adding a small volume of acetone. However, if the analyte peptide is of moderate size, a strong organic solvent such as acetone could co-precipitate the analyte, and methanol might be more preferable. The supernatant can be concentrated by a dry-down step to improve sensitivity.

- If the peptide drug binds strongly to the matrix protein(s), the peptide would co-precipitate with the proteins. A small volume of a strong acid (such as HCl) can be added to the sample to have the drug–protein complex dissociate prior to precipitation.
- The extraction recovery should not be concentration-dependent.
- The procedure should be amenable to automation. For example, instead of using test tubes and centrifugation, the processes of filtration by a 96-well microplate into a 96-well collection plate, drying, reconstitution and injection onto the LC-MS/MS are more conducive to automation.
- Limited sample clean-up could overload the analytical column, and residual matrix components can accumulate on the column after multiple injections. The residual matrix components can also solidify and deposit over a period of time in the LC-MS ionization source or vacuum interface, resulting in a decrease in ion transfer efficiency. The decrease in instrumentation performance (i.e., signal intensity) can be monitored by the signals of system-suitability samples dispersed within an analytical batch. The practice of replacing the pre-column in every run and "scrubbing" the analytical column periodically with a cleaning mobile phase will help to maintain instrument performance.

**Solid-phase Extraction for Peptides and Proteins.** Specific bonded phases can be designed to capture the protein or peptide analytes onto solid-phase particles. These may include selective molecular recognition of the protein/peptide. For example, cation or anion exchangers have been used in proteomic research to capture basic or acidic tryptic peptides, metal complex beads to capture phosphorylated peptides, ion exchangers for glycopeptides, and affinity chips were used for the known target peptide sequences [59–62]. The solid-phase approach provides means of capturing and concentrating the analyte of interest, and washing away extraneous components. However, the commercial production of reliable, cost-effective packing materials and a good IS to track the processes are required for routine quantitative bioanalytical applications of this approach.

#### 6.3.2.3.3 **Internal Standard**

An IS that co-elutes with the analyte is crucial for normalizing ESI-MS system fluctuations and correcting for matrix effects. An ideal IS should also be chemically similar to cover variations in extraction recovery. Analogue IS are frequently used. For example, an analogue of insulin with a single amino acid residue difference, such as arg-insulin or porcine insulin, was used to increase the assay precision for human insulin [63]. A stable isotope-labeled analyte would be an ideal IS, and this approach has been widely used for small-molecule drugs. The application for proteins/peptides is less straightforward due to the difficulties in producing and purifying the labeled IS. The stable isotope-labeled IS can be prepared by introducing $^{2}H$, $^{13}C$, $^{15}N$, or $^{18}O$ via peptide synthesis, cell culture, or during proteolysis [64]. Deuterated and $^{13}C$-labeled IS are used most commonly. The IS should be purified from the non-labeled form, and should show that there is no interference due to impurity or cross-talk from isotopic abundance. (Less-abundant, heavy isotopes are present in the natural elements, and appear as small peaks in the MS spectra of a drug compound. The amount in a high-concentration sample could be substantial in contributing to the heavy IS peak as "cross-talk" if the IS concentration were set too low.) The IS should also be sufficiently heavier than the analyte in order to make enough difference in the transition m/z-values. In addition, the $^{2}H$ label should be positioned where there is minimal $^{1}H/^{2}H$ exchange.

#### 6.3.2.3.4 **Chromatography: Analytical Column and Mobile Phase Choices**

Reversed-phase analytical columns of $C_{18}$ or $C_8$ bonded silica particles are commonly used for protein and peptide analysis. Unlike small molecules, porous particles with wide bore (300 Å) and large size (5 μm) are required for flow dynamics and mass transfer. Polymeric and silica-based monolithic chromatography columns are preferred choices due to faster mass transfer and higher resolution than packed materials [65,66]. Because of protein/peptide adherence to the silica-based surfaces, peak tailing and broadening often occurs. Trifluroacetic acid (TFA) is usually added as a mobile phase modifier to alleviate this situation. However, whilst TFA enhances peak sharpness it also causes signal suppression in the ESI, reducing sensitivity [67]. A compromise in maintaining peak sharpness without sacrificing sensitivity would be to reduce the amount of TFA in the mobile phase (e.g., $<0.05\%$). Heptafluorobutyric acid may be used as an alternative to TFA. Other alternatives involve the development of better packing materials for the analytical columns, such as $C_{30}$ polymeric-bonded phases or thin porous layers of the Poroshell spheres [26, 27].

Gradient elution is often used to provide high column resolution of the analyte from the large amount of matrix components. For a hydrophilic analyte, an ion-pairing reagent such as TFA is often used to increase column retention in reversed-phase chromatography. However, the low amount of TFA required to avoid ion-suppression may not be sufficient to achieve column retention. In that case, normal-phase or ion-exchange chromatography would be a better choice than reversed-phase chromatography for hydrophilic peptides.

#### 6.3.2.3.5 **MS Parameter Choices**

The quadrupoles in an MS instrument serve as selective mass filters to isolate ions with m/z-values specific for the analytes of interest. The triple quadrupole MS/MS instrument is typically operated by a pneumatically assisted electrospray source with an additional heated auxiliary gas flow for higher flow rates. There is a trade-off between resolution (favored by lower flow) and sensitivity (favored by higher flow) of the quadrupole analyzers. The biological molecules can be protonated or deprotonated at multiple sites to produce ions of 'n' charged states $[M\pm nH]^{n\pm}$. The MRM-MS/MS scan mode has a high duty cycle for the detection of specific ions at high sensitivity. Multiple charged ions can be scanned for the analytes at the appropriate ion energy of the precursor ions at Q1, and several of the abundant product ions can be selected for selectivity and sensitivity tests. The collision energy at Q2 is optimized to produce the specific product ions, which are filtered selectively at Q3. The unique fragment of a peptide can be used to improve resolution. High mass accuracy of the MS instrument offers the differentiating resolution by the filters at Q1 and Q3. Many MRM scans can be grouped together in one method to measure many selected ions, such as for several analytes, and/or several product ions for one analyte. The method development and application of an LC-ESI-MS/MS method of a peptide drug is illustrated in a case study in Section 6.4.

## 6.4 Case Studies

### 6.4.1 Development and Validation of an ELISA Method for an Antibody Drug

Drug "X" is an Ab against a target macromolecule "Y". An ELISA method was developed and validated for the determination of X in human and monkey plasma. Since the purposes of these methods were to support preclinical and clinical PK studies, the method validation and sample assays were conducted under an in-house SOP, which is "GLP-compliant" with IA considerations according to DeSilva et al. [10].

96-well microtiter plates were precoated with a solution of Y. The plates were dried and blocked with 1% casein in phosphate-buffered saline (PBS). Automation with a Tecan Genesis RSP-100® robotic pipettor was used for sample aliquotting and dilution to increase sample throughput, precision, and accuracy. Samples were added to the wells and allowed to bind to Y at room temperature for 2 h. After washing five times with 0.5% Tween in PBS, a secondary Ab against human (or monkey) IgG conjugated to HRP was added, allowed to bind to the immobilized X for 2 h, and washed again. A peroxidase substrate solution was added and incubated at room temperature for 1 h in the dark. The reaction was stopped by adding 5% sodium dodecyl sulfate, and the plate read at 415 nm (reference wavelength 490 nm) within 30 min. A four-parameter logistic regression was used to calculate the content of X in the samples.

In order to assess matrix effects, spike-recovery experiments were performed on samples from 10 individual human and six monkey matrix lots. Human samples were spiked with concentrations at the LLOQ and 10-fold the LLOQ. Monkey samples were spiked with concentrations at the LLOQ, two- and 400-fold the LLOQ. The results of the spike-recovery experiments are shown in Table 6.2. Some unspiked human samples showed a substantial amount of X, and their values were subtracted from the spiked sample results before calculating the spike recovery. For samples from monkey plasma samples, all samples were blank and no correc-

**Table 6.2** Spike-recovery experiments of compound X to human and monkey multiple matrix lots.

| Matrix lot | Human unspiked conc. | Recovery [%] spiked with LLOQ | Recovery [%] spiked with 10 × LLOQ | Monkey unspiked conc. | Recovery [%] spiked with LLOQ | Recovery [%] spiked with 2 × LLOQ | Recovery [%] spiked with 400 × LLOQ |
|---|---|---|---|---|---|---|---|
| 1 | 18.1 | 130 | 83.9 | <LLOQ | 110 | 113 | 92.7 |
| 2 | <LLOQ | 112 | 104 | <LLOQ | 102 | 98.8 | 104 |
| 3 | <LLOQ | 103 | 101 | <LLOQ | 95.5 | 96.6 | 99.1 |
| 4 | 0.7 | 103 | 111 | <LLOQ | 96.3 | 82.1 | 102 |
| 5 | <LLOQ | 97.9 | 101 | <LLOQ | 95.5 | 111 | 97.2 |
| 6 | <LLOQ | 111 | 84.8 | <LLOQ | 100 | 97.4 | 105 |
| 7 | 0.3 | 120 | 96.3 | | | | |
| 8 | 4.9 | 66.2 | 110 | | | | |
| 9 | 6.2 | 87.7 | 106 | | | | |
| 10 | 5.0 | 68.0 | 104 | | | | |

Acceptance criteria: Low limit of quantitation (LLOQ): within ±30% of the corrected mean of the matrix lots.
Higher concentrations: within ±20% of the corrected mean.
Overall: at least 80% of the lots are within acceptance.

tion was necessary. Eight of the 10 lots met the acceptance criteria of the spike-recovery, which satisfied the in-house SOP requirement. The human lots with <LLOQ concentrations were pooled to prepare standards and VS/QC samples.

Method validation was carried out in six (monkey) or seven (human) validation batches. Each batch included eight levels of standard calibrators, four levels of VS at concentrations at the LLOQ, low, mid, and high concentrations within the calibrator range, and two additional concentrations higher than the ULOQ. The >ULOQ VS were run at dilution factors of 20 and 400 to mimic the expected high concentration samples. The accuracy and precision data of the VS in human plasma are listed in Table 6.3.

Method robustness was established to show assay consistency with various supplies of the reference standards and two other critical reagents:

- The target macromolecule Y, which might be subjected to conformational changes: Multiple batches from two suppliers were procured and validated to assure assay consistency and sufficient supply inventory; and
- Preparative batches of the mouse anti-human IgG-HRP conjugate: Titers were determined to optimize every new preparation to maintain the same assay performance.

**Table 6.3** Validation of compound X in human plasma.[a)]

| Batch no. | VS [pg/mL] | 25 | 75 | 225 | 750 | 4000 | 80000 |
|---|---|---|---|---|---|---|---|
| 1 | Mean | 22.9 | 70.2 | 196 | 767 | 4405 | 89669 |
| | CV% | 14.3 | 5.0 | 5.0 | 8.0 | 6.0 | 7.8 |
| | RE% | 8.5 | 6.0 | 13.0 | –2.0 | –10.0 | –12.1 |
| 2 | Mean | 27.2 | 71 | 209 | 703 | 4054 | 89926 |
| | CV% | 6.0 | 11.0 | 11.0 | 2.0 | 6.0 | 7.3 |
| | RE% | –8.8 | 6.0 | 7.0 | 6.0 | –1.0 | –12.4 |
| 3 | Mean | 27.8 | 77 | 186 | 621 | 3924 | 86365 |
| | CV% | 6.0 | 8.5 | 8.3 | 15.1 | 12.0 | 1.9 |
| | RE% | –11.2 | –3.2 | 17.5 | 17.2 | 1.9 | -8.0 |
| 4 | Mean | 21.8 | 74.6 | 193 | 737 | 4042 | 81065 |
| | CV% | 1.9 | 3.8 | 3.4 | 8.4 | 5.6 | 5.4 |
| | RE% | 12.8 | 0.5 | 14.1 | 1.7 | –1.0 | –1.3 |
| 5 | Mean | 25.5 | 77.9 | 182 | 568 | 3824 | 87807 |
| | CV% | 6.0 | 2.6 | 4.0 | 0.6 | 6.3 | 15.8 |
| | RE% | –2.0 | –3.9 | 19.0 | 24.3 | 4.4 | -9.8 |
| 6 | Mean | 27.1 | 75.6 | 197 | 898 | 4254 | 82109 |
| | CV% | 10.2 | 0.6 | 8.6 | 21.3 | 2.3 | 5.5 |
| | RE% | –8.3 | –0.8 | 12.4 | –19.8 | –6.3 | –2.6 |
| 7 | Mean | 28.2 | 73.7 | 223 | 754 | 3651 | 82996 |
| | CV% | 12.5 | 14.1 | 4.6 | 25.6 | 5.8 | 7.2 |
| | RE% | –12.7 | 1.8 | 0.7 | –0.5 | 8.7 | –3.7 |
| Overall | Mean | 24.5 | 72.8 | 198 | 745 | 3798 | 82667 |
| | CV% | 11.4 | 8.7 | 9.3 | 17.1 | 11.6 | 9.4 |
| | RE% | –2.0 | –2.9 | –12.0 | –0.7 | –5.1 | 3.3 |

a) Within- and between-batch accuracy and precision data of VS from seven pre-study validation runs. VS at 4000 and 8000 pg/mL were diluted 20- and 400-fold before assay, respectively. For each batch, mean, CV%, and RE% were calculated from six replicates of VS. The overall statistics were calculated from all data points (n = 42).

### 6.4.2 Development and Validation of a Sandwich Immunoradiometric Method Using Commercial Kits for a Recombinant Peptide Drug

The use of commercial kits eliminates the time required to produce and characterize Ab reagents. However, method validation must be conducted to show that the kit developed for diagnostic use is suitable for the intended PK study application. The kit selection and method validation processes are illustrated here for the bioanalytical application of a recombinant human parathyroid hormone (rhPTH) drug, ALX1–11 (Preos®, NPS Pharmaceuticals).

PTH is a naturally occurring polypeptide with 84 amino acid residues that acts as the major regulator of calcium ion homeostasis [68]. The first two amino acids of PTH from the amino N-terminus are required for biological activity. Proteolytic fragments of PTH are inactive at the PTH-1 receptor, but may cross-react with an IA. The putative peptide PTH (7–84) fragment was reported to exist at 10- to 20-fold higher concentrations than the intact PTH (1–84), depending upon health status [69].

Sandwich immunoradiometric assay kits were developed with Ab against the carboxyl C-terminus immobilized on beads to capture the analyte and radiolabeled Ab against the N-terminus for detection. Although the widely used commercial kit was supposed to be specific for the intact molecule [70, 71], it was found to bind circulatory fragments [72]. Therefore, it is important to validate selectivity of the commercial kits against peptide fragments. Three recently developed commercial kits were assessed on cross-reactivity against N-terminal fragments of (3–84) and (7–84), and the assay performance of standards and VS from four evaluation runs was assessed for their suitability for PK study applications. The results in Table 6.4 show that Kit A was superior to the other two kits in both assessments, and it was selected for subsequent method validation.

Standard calibrators were prepared by spiking the recombinant drug to standard zero from the kit; they were run against the kit standards and found to be comparable. To prepare VS/QCs, PTH was screened in plasma samples from 32

**Table 6.4** Comparison of three parathyroid hormone (PTH) commercial assay kits.

| | Kit A | Kit B | Kit C |
|---|---|---|---|
| Cross-reactivity[a] | | | |
| 1–84 rhPTH | 86% | 95% | 103% |
| 3–84 fragment | 0.6% | 48% | 139% |
| 7–84 fragment | <0.1% | 1.3% | 63% |
| Precision and accuracy from four evaluation batches[b] | | | |
| Validation sample CV% | | | |
| Low | 9.1 | 13.1 | 11.7 |
| Mid | 4.8 | 8.4 | 9.1 |
| High | 3.9 | 9.4 | 11.8 |
| Validation sample RE% | | | |
| Low | 4.2 | –19.1 | –12.9 |
| Mid | 6.4 | 12.3 | 20.4 |
| High | –3.2 | 20.8 | 3.2 |

a) Cross-reactivity test of the kits against the recombinant whole peptide of 1–84, and fragments of 3–84 and 7–84.

b) Precision and accuracy of VS at low, mid, and high concentrations.

rhPTH: Recombinant human parathyroid hormone.

individual lots. The lower-concentration lots were selected, and pooled to form the low VS/QC, which was spiked with rhPTH to prepare the mid and high VS/QCs. The target value of the low VS/QC was determined from the mean of four evaluation batches as 23.4 pg/mL. The theoretical values of the mid and high VS/QCs were calculated from the spiked amount plus the basal value. Six pre-study validation runs were performed. To demonstrate the lack of interference in the matrix, 10 individual lots of control human plasma were tested for spike-recovery of rhPTH at a concentration of 52.5 pg/mL. The spike-recovery was calculated by subtracting the endogenous value of the unspiked samples from that of the spiked sample. All 10 lots were quantitated within 20% of the theoretical spiked value, even from samples with a high basal value of 54.3 pg/mL. The variance of the spike recovery among the 10 lots was only 6% CV.

In order to use commercial reagents in a drug development program, it was important to negotiate and plan with the kit supplier to assure consistency of the Ab reagents, and that sufficient quantities would be reserved. Method robustness included the pre-study validation tests with a second lot of the capture Ab, three analysts, and three batches of radioiodinated detector Ab. Method robustness was further demonstrated by in-study validation, with four additional analysts performing sample analysis using 12 batches of radioiodinated detector Ab over a time span of approximately three years.

### 6.4.3 Development and Validation of LC-MS/MS Method for a Peptide Drug

Enfuvirtide (Fuzeon®, T-20, Ro 29–9800, Hoffmann-La Roche) is a 36-amino acid synthetic peptide with a molecular weight of 4492 Da. It selectively inhibits human immunodeficiency virus (HIV) fusion to the host cell membranes [73]. The N-terminus of the molecule is acetylated and the C-terminus is amidated. A metabolite, M-20, is deamidated at the C-terminus. An ELISA method was initially used during drug development of this compound, but the decision was made to develop and validate an LC-ESI-MS/MS method for the simultaneous determination of enfuvirtide and M-20 for PK studies to support the NDA submission of this product [53]. Some of the issues of LC-ESI-MS/MS application for peptide bioanalysis are highlighted in the following.

Protein precipitation was optimized for peptides of this size to adequately debulk the plasma proteins with sufficient analyte recovery. The commonly used 4 : 1 ratio of acetonitrile to plasma sample was reduced to 2 : 1 to prevent co-precipitation of the peptide. The plasma sample was acidified to dissociate the drug from protein binding before the addition of acetonitrile to the samples in a 96-well microplate. A semi-automated Tomtec Quadra 96® workstation was used to perform all sample processing. After centrifugation at high speed, the supernatant was dried, reconstituted, and re-centrifuged before being injected onto the LC-MS/MS system.

The LC-ESI-MS/MS system comprised an analytical column of $C_{18}$, 50 × 2 mm, 300 Å with 5-μm particle size, a $C_{18}$ guard column, and two pumps. A linear gradient elution was set up with mobile phase A of 0.2% acetic acid in water, and B of 15% methanol in acetonitrile. Both mobile phases contained 0.02% TFA to maintain good column retention and peak shape. The LC effluent was connected to a Micromass Ultima or MDS Sciex-Applied Biosystem API 4000 triple quadrupole mass spectrometer with an ESI source operated in the positive ion MRM mode.

The prominent molecular ion of four positive charges $[M+4H]^{4+}$ had a m/z of 1124.0 for enfuvirtide, and of 1124.2 for M-20. Fragmentation yielded several product ions: an abundant, singly charged fragment at m/z 159 (the immonium ion of tryptophan), and a triply charged ion fragment generated from the molecular ion. There was interference with the transition m/z of 1124 → 159 from the endogenous substances of the control plasma, probably due to the large number of tryptophan-containing peptides in the plasma extract. The triply charged fragment ion m/z 1343 was a large $b_{33}$ fragment (4026 Da) and more specific to enfuvirtide and M-20, providing excellent selectivity and S/N ratio for the analysis with the transition m/z of 1124 → 1343. It was important to have chromatographic resolutions of enfuvirtide and M-20 and their corresponding IS because they have the same MRM transitions. Enfuvirtide and M-20 were well separated with retention times of 2.70 and 3.02 min, respectively.

The wide-pore analytical column provided fast elution times for the large peptides, with good peak shape. TFA in the mobile phase enhanced chromatographic peak sharpness but caused signal suppression in the ESI, reducing MS-MS sensitivity [67]. Mobile phases containing 0.02% TFA and 0.2% acetic acid provided a good compromise of maintaining peak sharpness with adequate sensitivity. The LLOQ was 10 ng/mL, with S/N ratios of ~10. There were no obvious differences in the background noise between the control plasma lots from the $HIV^+$ patients and those of healthy volunteers.

The IS of the analytes were synthesized by labeling leucine (the fourth in the peptide sequence) to produce $d_{10}$ T-20 and $d_{10}$ M-20, with purities of 88.8% and 86.1%, respectively. The hydrogen-containing impurity in the IS contributed to a small signal in the blank control sample. However, this contribution by the $d_{10}$-IS was very small in comparison to the LLOQ signal. It had no effect on the linearity of the standard curve or the accuracy of the LLOQ quantification. In order to minimize the IS contribution to the analyte signal while maintaining adequate IS response, the $d_{10}$-IS was kept at a modest amount, not to exceed the mid-standard concentrations of the analytes.

In order to minimize adsorption during standard curve and QC preparations, aliquots of the stock solutions were spiked promptly into the control blank plasma to prepare the standards and QCs. Once the compounds were in an environment of protein solutions, the adsorption problem was alleviated. The de-proteinized extracted samples did not show an adhesion problem, as reflected by the stability data of the extracts. There was very little tailing of the compound peaks, indicating that they were not adhering to the LC-MS/MS system.

## 6.5 Future Perspectives: Emerging Quantitative Methods

Rapid advances have been made in proteomic research, with growing success in protein quantitation utilizing MS technologies [74–76]. The quantitative tools from proteomics and other analytical innovations have great potential to be translated to macromolecule drug bioanalytical applications, and are discussed in the following sections.

### 6.5.1 Sample Clean-Up

Immunoaffinity debulking reagents are commercially available for the removal of abundant proteins such as serum albumin, immunoglobulins, and transferrin. Depleting the major proteins would reduce the assay chemical noise in a plasma sample. In addition, approaches of using novel solid-phase materials to trap the analytes, selectively wash away extraneous components, and elute with a small volume greatly improve sensitivity and selectivity. Solid-phase extraction materials of controlled sizes such as polymeric ion exchangers, porous carbide particles, and beads with covalently bonded Ab can be good tools. They can be constructed into a 96-well format for automation to increase precision and throughput. The immunoaffinity particles can be used either on- or off-line to MS for sample clean-up and analyte concentration.

### 6.5.2 Innovation in MS Instruments

Innovation in MS instruments for proteins elevated the potential of sensitive assays to pico- and femto-molar concentrations, including nano- or micro-flow HPLC connected to nanospray MS, ion trap hybrid with triple quadrupole MS (QTrap™), LTQ Orbitrap™, and ion mobility (FAIMS) as a second dimension of ion separation. Direct determinations of the intact protein become possible with very high-mass resolution instruments such as TOF-MS and FTMS. The non-scanning types of instruments TOF-MS and FTMS are able to simultaneously monitor multiple m/z values without significant losses of sensitivity and with much higher mass resolution than that of the triple quadrupole instruments. For example, TOF-MS offers the advantage of a wide dynamic range and high resolving power of 10 000 to 15 000 with mass accuracy of 1 to 10 ppm for unlimited analyte mass.

Matrix-assisted laser desorption ionization (MALDI) and surface-enhanced laser desorption ionization (SELDI) have been used online with TOF-MS for protein differential profiles of intact or hydrolyzed biological matrices in proteomics. The potential use of affinity chips, grafted with specific Ab towards the drug compound for MALDI or SELDI, will bring sensitive and selective tools for macromolecules. Specific Ab towards either the intact protein or several signature peptides of the protein can be used for sample clean-up to capture and concentrate the analyte of interest to increase sensitivity and selectivity [59].

Capillary electrophoresis (CE) coupled to MS has the advantage of high resolution and soft ionization for biomolecules, which may be used to differentiate post-translational modifications and variants of intact proteins and oligonucleotides. Different modes of CE (capillary zone electrophoresis, capillary isoelectric focusing, capillary electrochromatography, micellar electrokinetic chromatography, nonaqueous capillary electrophoresis) to MS as well as online preconcentration techniques (transient capillary isotachophoresis, solid-phase extraction, membrane preconcentration) are used to compensate for the restricted detection sensitivity of the CE methodology [77, 78].

Previous applications of ESI-MS could quantify intact proteins at relatively high concentrations, such as that of human transferrin in plasma samples [79] and $\alpha_{2u}$-globulin in urine and kidney samples [80]. With the use of solid-phase extraction combined with a high-sensitivity MS, improvements on sensitivity were achieved, for example for the quantification of rk5 in monkey plasma reaching a LLOQ of 10.29 ng/mL [81].

### 6.5.3
### Quantification using Signature Hydrolytic Peptides

Instead of quantification of the intact protein, measurements can be made by MRM-LC/MS monitoring the representative (signature) peptide(s) from a protein digest with a triple quadrupole MS instrument. Barr et al. first reported the use of a synthetic $^{2}$H-labeled tryptic peptide as reference standard for the quantification of the native apoliprotein A1 by LC-MS [82]. Several applications were found for the determination of HbA$_{1c}$ in human blood [83], rhodopsin in rod outer segment suspensions [84], phosphorylated proteins in cell lysates with the phosphorylated tryptic peptides purified by SDS/PAGE [85], human glutathione S-transferases in human liver cytosol [86], and C-reactive protein in human serum [87]. Although the signal response versus concentration of the MS is specific for individual peptides, an absolute quantitation is possible by utilizing stable isotope-labeled peptides.

Labeling with $^{2}$H, $^{13}$C, $^{15}$N and/or $^{18}$O can be introduced via peptide synthesis, cell culture, or hydrolysis in labeled water [88]. The heavy isotope-labeled peptide can be used as an IS to obtain quantitative measurements of the protein concentration. Typically, the protein sample of interest is digested with trypsin, and the isotope-labeled control peptides are added to the mixture. The signature peptides in the digest can be separated and quantified by HPLC-ESI-MS/MS. Alternatively, MALDI-MS can also be used for tryptic peptide determinations after some separation steps such as gel electrophoresis [89].

The uniqueness of the peptide chosen must be taken into account. Metabolites of the protein drug may have the same sequence, or share a common sequence of the product ions being monitored in the MRM-LC/MS method, possibly resulting in interference. Databases are available to search for tryptic peptide sequences of known proteins. In addition, peptides and their corresponding heavy isotope-labeled IS should be purified and their purity defined in order to be used as reference standard and IS. Alternatively, protein/peptides can be tagged with isotope-coded affinity tags (ICAT™ or iTRAC™ reagents) [74]. Method validation of the MRM-LC/MS should also include reproducibility and efficiency of the enzyme digestion and the peptide extraction recovery from the digestion matrix.

### 6.5.4
### Advances in Ligand Reagents Design and Production

IA remains the major method for bioanalysis of macromolecule drugs. The rate-limiting step of assay development is Ab production. Traditional polyclonal or monoclonal Ab production *in vivo* takes about three to six months. Advancement of *in-vitro* Ab production could reduce the time required for immunization and clone selection. The interference problems of heterophilic Ab (human anti-animal Ab) that are present in a small percentage of normal individuals could also be eliminated.

One *in-vitro* approach is phage display, which offers highly specific recombinant monoclonal Ab production selected from repertoires of libraries created by cloning [90]. Large repertoires of Ab fragments were created from Ab V genes bypassing hybridoma technology and immunization. Molecules of the desired functional properties were rapidly selected and produced. The moderate binding avidity of the Ab fragments ($10^{-5}$ to $10^{-8}$ M) can be further improved by mutating residues of determining regions or by increasing the number of binding sites making dimeric, trimeric or multimeric molecules [91]. A second *in-vitro* approach is the use of aptamers, which are short single-stranded DNA or RNA molecules of selected sequence with affinity for a target molecule. Aptamers offer advantages over traditional Ab in their ease of production, regeneration, and stability due to the chemical properties of nucleic acids versus amino acids [92]. In contrast to Ab, aptamers require the formation of a three-dimensional structure for binding, and are anticipated to have a higher affinity for rapid competitive assays [93]. The selection is based on binding characteristics, including binding constants and kinetics. Surface plasmon resonance instruments such as Biacore can be used to study the Ab binding characteristics. Real-time kinetic studies of biomolecular interactions in multiple channels from microfluidic cells can be performed. The binding constants, $k_{on}$ and $k_{off}$, can be calculated for the custom design of capturing and/or detection ligands.

## 6.6
## Conclusions

Biotech macromolecule drugs are chemically and biologically different from the small-molecule drugs. Special considerations must be given for bioanalytical method development, validation and applications, including the purity of the reference standard, analyte stability and sample integrity, heterogeneity of variant forms existing endogenously, assay method sensitivity, selectivity, accuracy and precision, and robustness. There are general issues of reference standard consistency, sample clean-up methods, matrix effect, surface adsorption, and interferences from unknown metabolites and anti-drug antibodies. Specific method issues, assay approaches, and limitations of the workhorse methods of IA and LC-ESI-MS/MS were further discussed, with case studies as illustrations in this chapter. Innovative technologies from proteomics and Ab productions will provide more effective tools in the near future to remove some of the limitations for faster method development, and more sensitive and selective methods, regardless of the molecular size.

## Acknowledgments

The author wishes to thank the following colleagues who have contributed to the content of this chapter: Rich Sukovaty, Patrick Lin, and George Scott (current and former colleagues at MDS Pharma Services); David Wells (NPS Pharmaceuticals); Surendra Bansal, Stan Kolis and David Chang (Hoffmann-La Roche). She also thanks Daniel Figeys (University of Ottawa) for critical review of the manuscript.

## 6.7 References

1 Findlay JWA, Das I: Validation of immunoassays for macromolecules from biotechnology. *J. Clin. Ligand Assay* (1998) *21*: 249–253.

2 Toon S: The relevance of pharmacokinetics in the development of biotechnology products. *Eur. J. Drug Metab. Pharmacokinet.* (1996) *21*: 93–103.

3 Mire-Sluis AR, Barrett YC, Devanarayan V, et al.: Recommendations for the design and optimization of immunoassays used in the detection of host antibodies against biotechnology products. *J. Immunol. Methods* (2004) *289*: 1–16.

4 Molineux G: Pegylation: Engineering improved biopharmaceuticals for oncology. *Pharmacotherapy.* (2003) *23*(Pt 2): 3S–8S

5 Shah VP, Midha KK, Dighe SV, et al.: Analytical methods validation: bioavailability, bioequivalence, and pharmacokinetic studies. *J. Pharm. Sci.* (1992) *81*: 309–312.

6 Bowsher RR, Smith WC. Personal communication.

7 Shah VP, Midha KK, Findlay JWA, et al.: Bioanalytical method validation – A revisit with a decade of progress. *Pharm. Res.* (2000) *17*: 1551–1557.

8 Guidance for industry on bioanalytical method validation: availability. Federal Reg (2001) *66*: 28526–28527. May guidance.

9 Findlay JWA, Smith WC, LeeJW, et al.: Validation of immunoassays for bioanalysis: a pharmaceutical industry perspective. *J. Pharm. Biomed. Anal.* (2000) *21*: 1249–1273.

10 Desilva B, Smith W, Weiner R, et al.: Recommendations for the bioanalytical method validation of ligand-binding assays to support pharmacokinetic assessments of macromolecules. *Pharm. Res.* (2003) *20*: 1885–1900.

11 Wadhwa M, Thorpe R.: Standardization and calibration of cytokine immunoassays: meeting report and recommendations. *Cytokine* (1997) 9: 791–793.

12 De Kossodo S, Houba V, Grau GE: Assaying tumor necrosis factor concentrations in human serum. A WHO International Collaborative study. *J. Immunol. Methods* (1995) *182*: 107–114.
13 Ledur A, Fitting C, David B, et al.: Variable estimates of cytokine levels produced by commercial ELISA kits: results using international cytokine standards. *J. Immunol. Methods* (1995) *186*: 171–179.
14 Findlay JWA, Das I: Some validation considerations for immunoassays. *J. Clin. Ligand Assay* (1997) *20*: 49–55.
15 Das Sarma, Duttagupta C, Ali E, et al.: Direct microtitre plate enzyme immunoassay of folic acid without heat denaturation of serum. *J. Immunol. Methods* (1995) *184*: 7–14.
16 Carter P: Preparation of ligand-free human serum for radioimmunoassay by adsorption on activated charcoal. *Clin. Chem.* (1978) *24*:362–364.
17 Hartmann C, Smeyers-Verbeke J, Massart DL, et al.: Validation of bioanalytical chromatographic methods. *J. Pharm. Biomed. Anal.* (1998) *17*: 193–218.
18 Mire-Sluis AR, Das RG, Padilla A: WHO cytokine standardization: facilitating the development of cytokines in research, diagnosis and as therapeutic agents. *J. Immunol. Methods* (1998) *216*: 103–116.
19 Piscitelli SC, Reiss WG, Figg WD, et al.: Pharmacokinetic studies with recombinant cytokines. Scientific issues and practical considerations. *Clin. Pharmacokinet.* (1997) *32*: 368–381.
20 Fisker S, Ebdrup L, Orskov H: Influence of growth hormone binding protein on growth hormone estimation in different immunoassays. *Scand. J. Clin. Lab. Invest.* (1998) *58*: 373–382.
21 Daughaday WH, Kapadia M, Mariz I: Serum somatomedin binding proteins: physiologic significance and interference in radioligand assay. *J. Lab. Clin. Med.* (1987) *109*: 355–363.
22 Bank P, Baxter RC, Blum WF, et al.: Validation measurements of total IGF concentrations in biological fluids. Recommendations from the Third International Symposium on Insulin-like Growth Factors. *J. Endocrinol.* (1994) *143*: 1–2.
23 Chapuzet E, Mercier N, Bervos-Martin S, et al.: Validation of chromatographic bioanalytical method. *S.T.P. Pharma. Pratiques* (1997) *7*: 169–194.
24 Rehlaender BN, Cho MJ: Antibodies as carrier proteins. *Pharm. Res.* (1998) *15*: 1652–1656.
25 Lee JW, Colburn WA: Immunoassay techniques. In: Ohannesian L, Streeter AJ (Eds.), *Handbook of Pharmaceutical Analysis.* Marcel Dekker, New York (2001): 225–312.
26 Kirkland JJ, Truszkowski FA, Ricker RD: Atypical silica-based column packings for high-performance liquid chromatography. *J. Chromatogr. A* (2002) *965*: 25–34.
27 Kirkland JJ: Development of some stationary phases for reversed-phase high-performance liquid chromatography. *J. Chromatogr. A* (2004) *1060*: 9–21.
28 Farmen RH, Lee JW. Analysis of drugs in biological fluids during development. *Applied Clinical Trials* (1998), 58–65.
29 Fenn JB, Mann M, Meng CK: Electrospray ionization for mass spectrometry of large biomolecules. *Science* (1989) *246*: 64–71.
30 Patrick JS, Lagu AL: Review applications of capillary electrophoresis to the analysis of biotechnology-derived therapeutic proteins. *Electrophoresis* (2001) *22*: 4179–4196.
31 Sowell J, Salon J, Strekowski L, et al: Covalent and noncovalent labeling schemes for near-infrared dyes in capillary electrophoresis protein applications. *Methods Mol. Biol.* (2004) *276*: 39–75.
32 Moini M: Capillary electrophoresis mass spectrometry and its application to the analysis of biological mixtures. *Anal. Bioanal. Chem.* (2002) *373*: 466–480.
33 Nemunaitis J, Holmlund JT, Kraynak M, et al.: Phase I evaluation of ISIS 3521, an antisense oligodeoxynucleotide to protein kinase C-α, in patients with advanced cancer. *J. Clin. Oncol.* (1999) *17*: 3586–3595.
34 De Frutos M, Cifuentes A, Diez-Masa JC: Differences in capillary electrophoresis profiles of urinary and recombinant erythropoietin. *Electrophoresis* (2003) *24*: 678–680.
35 Chard T: An introduction to radioimmunoassay and related techniques, In: Burdon RH, van Knippenberg PH (Eds.), *Laboratory Techniques in Biochemistry and*

*Molecular Biology*, 3rd edn, (1987) Vol. 6, Part. 2: 38–39. Elsevier, New York.
**36** Crowther JR: *The ELISA Guidebook* (2000) Humana Press, Totowa, NJ.
**37** Rodbard D, Frazier GR: Statistical analysis of radioligand assay data. *Methods Enzymol.* (1975) *37*: 3–22.
**38** Dudley RA, Edwards P, Ekins RP, et al.: Guidelines for immunoassay data processing. *Clin. Chem.* (1985) *31*: 1264–1271.
**39** Carroll RJ, Cline DBH: An asymptotic theory for weighted least squares with weights estimated by replication. *Biometrika* (1988) *75*: 35–43.
**40** Braggio S, Barnaby RJ, Grossi P, et al.: A strategy for validation of bioanalytical methods. *J. Pharm. Biomed. Anal.* (1996) *14*: 375–388.
**41** Kringle RO: An assessment of the 4-6-20 rule for acceptance of analytical runs in bioavailability, bioequivalence, and pharmacokinetic studies. *Pharm. Res.* (1994) *11*: 556–560.
**42** Smith WC, Sittampalam GS: Conceptual and statistical issues in the validation of analytic dilution assays for pharmaceutical applications. *J. Biopharm. Statistics* (1998) *8*: 509–532.
**43** Ray CA, Bowsher RR, Smith WC, et al. Development, validation and implementation of a multiplex immunoassay for the simultaneous determination of five cytokines in human serum. *J. Pharm. Biomed. Anal.* (2005) *36*: 1037–1044.
**44** VaqeranoJE, Peng M, Chang JW, et al.: Digital quantification of the enzyme-linked immunospot (ELISPOT). *Biotechniques* (1998) *25*: 830–836.
**45** Kingsmore SF, Patel DD: Multiplexed protein profiling on antibody-based microarrays by rolling circle amplification. *Curr. Opin. Biotechnol.* (2003) *14*: 74–81.
**46** Bruins AP: Liquid chromatography-mass spectrometry with ionspray and electrospray interfaces in pharmaceutical and biomedical research. *J. Chromatogr.* (1991) *554*: 39–46.
**47** Amad H, Cech NB, Jackson GS, et al.: Importance of gas-phase proton affinities in determining the electrospray ionization response for analytes and solvents. *J. Mass Spectrom.* (2000) *35*: 784–789.
**48** Cole RB: Some tenets pertaining to electrospray ionization mass spectrometry. *J. Mass Spectrom.* (2000) *35*: 763–772.
**49** Darby SM, Miller ML, Allen RO, et al.: A mass spectrometric method for quantitation of intact insulin in blood samples. *J. Anal. Toxicol.* (2001) *25*: 8–14.
**50** Wolf R, Rosche F, Hoffmann T, et al.: Immunoprecipitation and liquid chromatographic-mass spectrometric determination of the peptide glucose-dependent insulinotropic polypeptides GIP1–42 and GIP3–42 from human plasma samples. New sensitive method to analyze physiological concentrations of peptide hormones. *J. Chromatogr. A* (2001) *926*: 21–27.
**51** Carrascal M, Schneider K, Calaf RE, et al.: Quantitative electrospray LC-MS and LC-MS/MS in biomedicine. *J. Pharm. Biomed. Anal.* (1998) *17*: 1129–1138.
**52** Fountain KJ, Gilar M, Gebler JC: Analysis of native and chemically modified oligonucleotides by tandem ion-pair reversed-phase high-performance liquid chromatography/electrospray ionization mass spectrometry. *Rapid Commun. Mass Spectrom.* (2003) *17*: 646–653.
**53** Chang D, Kolis SJ, Linderholm KH, et al.: Bioanalytical method development and validation for a large peptide HIV fusion inhibitor (Enfuvirtide, T-20) and its metabolite in human plasma using LC-MS/MS. *J. Pharm. Biomed. Anal.* (2005) *38*: 487–496.
**54** Matuszewski BK, Constanzer ML, Chaver-Eng CM: Strategies for the assessment of matrix effect in quantitative bioanalytical methods based on HPLC-MS/MS. *Anal Chem.* (2003) *75*: 3019–3030.
**55** Wang G, Cole RB: Solutions, gas-phase, and instrumental parameter influence on charge-state distributions in electrospray ionization mass spectrometry. In: Cole RD (Ed.), *Electrospray Ionization Mass Spectrometry.* John Wiley & Sons, New York (1997): 137–174.
**56** Matuszewski BK, Constanzer ML, Chaver-Eng CM: Matrix effect in quantitative LC/MS/MS analyses of biological fluids: a method for determination of finasteride in human plasma at picogram per milliliter concentrations. *Anal. Chem.* (1998) *70*: 882–889.

**57** Bonfiglio R, King RC, Olah TV, et al.: The effects of sample preparation methods on the variability of the electrospray ionization response for model drug compounds. *Rapid Commun. Mass Spectrom.* (1999) *13*: 1175–1185

**58** Rollag JG, Nachi R, Schlesiger L, et al.: The resolution of ion suppression caused by drug-interaction compounds. (2001) ASMS Conference Proceedings.

**59** Anderson NL, Anderson NG, Haines LR, et al.: Mass spectrometric quantitation of peptides and proteins using stable isotope standards and capture by anti-peptide antibodies (SISCAPA). *J. Proteome Res.* (2004) *3*: 235–244.

**60** Martin R. Larse N, Tine E, et al.: Highly selective enrichment of phosphorylated peptides from peptide mixtures using titanium dioxide microcolumns. *Molec. Cell. Proteomics* (2005) *4*: 873–886.

**61** Dayal B, Ertel NH: ProteinChip technology: a new and facile method for the identification and measurement of high-density lipoproteins apoA-I and apoA-II and their glycosylated products in patients with diabetes and cardiovascular disease. *J. Proteome Res.* (2002) *1*: 375–380.

**62** Warren EN, Elms PJ, Parker CE, et al.: Development of a protein chip: a MS-based method for quantitation of protein expression and modification levels using an immunoaffinity approach. *Anal. Chem.* (2004) *76*: 4082–4092.

**63** Bilati U, Pasquarello C, Corthals GL, et al.: Matrix-assisted laser desorption/ionization time-of-flight mass spectrometry for quantitation and molecular stability assessment of insulin entrapped within PLGA nanoparticles. *J. Pharm. Sci.* (2005) *94*: 688–694.

**64** Bucknall M, Fung KYC, Duncan MW: Practical quantitative biomedical applications of MALDI-TOF mass spectrometry *J. Am. Soc. Mass Spectrom.* (2002) *13*: 1015–1027.

**65** Walcher W, Toll H, Ingendoh A, et al.: Operational variables in high-performance liquid chromatography-electrospray ionization mass spectrometry of peptides and proteins using poly(styrene-divinylbenzene) monoliths. *J. Chromatogr. A.* (2004) *1053*: 107–117.

**66** Xiong L, Zhang R, Regnier FE: Potential of silica monolithic columns in peptide separations. *J. Chromatogr. A.* (2004) *1030*: 187–194.

**67** Apffel A, Fischer SM, Goldberg G, et al.: Enhanced sensitivity for peptide mapping with electrospray liquid chromatography-mass spectrometry in the presence of signal suppression due to trifluoroacetic acid-containing mobile phases. *J. Chromatogr. A* (1995) *712*: 177–190.

**68** Fitzpatrick LA, Bilezikian JP: Actions of parathyroid hormone. In: Belezikian JP, Raisz LG, Rodan GA (Eds.), *Principles of Bone Biology.* Academic Press, New York (1996), Chapter 25.

**69** Potts JT, Jr., Bringhurst FR, Gardella TJ, et al.: Parathyroid hormone: Physiology, chemistry, biosynthesis, secretion, metabolism, and mode of action. In: Degroot LJ (Ed.), *Endocrinology.* W.B. Saunders, Philadelphia, PA (1996): 920–965.

**70** Nussbaum SR, Zahradnik RJ, Lavigne JR, et al.: Highly sensitive two-site immunoradiometric assay of parathyrin and its clinical utility in evaluation patients with hypercalcemia. *Clin. Chem.* (1987) *33*: 1364–1367.

**71** Brown RC, Aston JP, Weeks I, et al.: Circulation intact parathyroid hormone measured by a two-site immunochemiluminotetric assay. *J. Clin. Endocrinol. Metab.* (1987) *65*: 407–414.

**72** Lepage R, Roy L, Brossard JH, et al.: A non-(1–84) circulating parathyroid hormone (PTH) fragment interferes significantly with intact PTH commercial assay measurements in uremic samples. *Clin. Chem.* (1998) *44*: 805–809.

**73** Kilby JM, Hopkins S, Venetta TM, et al.: Potent suppression of HIV-1 replication in humans by T-20, a peptide inhibitor of gp41-mediated virus entry. *Nature Medicine* (1998) *4*: 1302–1307.

**74** Tao WA, Aebersold R: Advances in quantitative proteomics via stable isotope tagging and mass spectrometry. *Curr. Opin. Biotechnol.* (2003) *14*: 110–118.

**75** Goshe MB, Smith RD: Stable isotope-coded proteomic mass spectrometry. *Curr. Opin. Biotechnol.* (2003) *14*: 101–109.

**76** Lill J: Proteomic tools for quantitation by mass spectrometry. *Mass Spectrom. Rev.* (2003) *22*: 182–194.

**77** Stutz H: Advances in the analysis of proteins and peptides by capillary electrophoresis with matrix-assisted laser desorption/ionization and electrospray-mass spectrometry detection. *Electrophoresis* (2005) *26*: 1254–1290.

**78** Hernandez-Borges J, Neususs C, Cifuentes A, Pelzing M. On-line capillary electrophoresis-mass spectrometry for the analysis of biomolecules. *Electrophoresis* (2004) *25*: 2257–2281.

**79** Bergen HR, Lacey JM, O'Brien JF, et al.: Online single-step analysis of blood proteins: The transferrin story. *Anal. Biochem.* (2001) *296*: 122–129.

**80** Mao Y, Moore RJ, Wagnon KB, et al. Analysis of 2u-globulin in rat urine and kidneys by liquid chromatography-electrospray ionization mass spectrometry. *Chem. Res. Toxicol.* (1998) *11*: 953–961.

**81** Ji QC, Gage EM, Rodila R, et al.: Method development for the concentration determination of a protein in human plasma utilizing 96-well solid-phase extraction and liquid chromatography/tandem mass spectrometric detection. *Rapid Commun. Mass Spectrom.* (2003) *17*: 794–799.

**82** Barr JR, Maggio VL, Patterson DG, Jr., et al.: Isotope dilution-mass spectrometric quantification of specific proteins: model application with apolipoprotein A-I. *Clin. Chem.* (1996) *42*: 1676–1682.

**83** Jeppsson J, Kobold U, Barr J, et al.: Approved IFCC reference method for the measurement of HbA1c in human blood. *Clin. Chem. Lab. Med.* (2002) *40*: 78–89.

**84** Barnidge DR, Dratz EA, Martin T, et al.: Absolute quantification of the G protein-coupled receptor rhodopsin by LC/MS/MS using proteolysis product peptides and synthetic peptide standards. *Anal. Chem.* (2003) *75*: 445–451.

**85** Gerber SA, Rush J, Stemman O, et al.: Absolute quantification of proteins and phosphoproteins from cell lysates by tandem MS. *Proc. Natl. Acad. Sci. USA* (2003) *100*: 6941–6945.

**86** Zhang F, Bartels MJ, Stott WT: Quantitation of human glutathione S-transferases in complex matrices by liquid chromatography/tandem mass spectrometry with signature peptides. *Rapid Commun. Mass Spectrom.* (2004) *18*: 491–498.

**87** Kuhn E, Wu J, Karl J, et al.: Quantification of C-reactive protein in the serum of patients with rheumatoid arthritis using multiple reaction monitoring mass spectrometry and 13C-labeled peptide standards. *Proteomics* (2004) *4*: 1–12.

**88** Sechi S, Oda Y: Quantitative proteomics using mass spectrometry. *Curr. Opin. Chem. Biol.* (2003) *7*: 70–77.

**89** Oda Y, Huang K, Cross FR, et al.: Accurate quantitation of protein expression and site-specific phosphorylation. *Proc. Natl. Acad. Sci. USA* (1999) *96*: 6591–6596.

**90** McCafferty J, Griffiths AD, Winter G, et al.: Phage antibodies: filamentous phage displaying antibody variable domains. *Nature* (1990) *348*: 552–554.

**91** Lee CV, Sidhu SS, Fuh G: Bivalent antibody phage display mimics natural immunoglobulin. *J. Immunol. Methods* (2004) *284*: 119–132.

**92** Murphy MB, Fuller ST, Richardson PM, et al.: An improved method for the in vitro evolution of aptamers and applications in protein detection and purification. *Nucleic Acids Res.* (2003) *31*: e110

**93** Baldrich E, Acero JL, Reekmans G, et al.: Displacement enzyme linked aptamer assay. *Anal. Chem.* (2005) *77*: 4774–4784.

# 7
# Limitations of Noncompartmental Pharmacokinetic Analysis of Biotech Drugs

*Arthur B. Straughn*

## 7.1
## Introduction

Traditionally, linear pharmacokinetic analysis has used the n-compartment mammillary model to define drug disposition as a sum of exponentials, with the number of compartments being elucidated by the number of exponential terms. More recently, noncompartmental analysis has eliminated the need for defining the rate constants for these exponential terms (except for the terminal rate constant, $\lambda_z$, in instances when extrapolation is necessary), allowing the determination of clearance (CL) and volume of distribution at steady-state ($V_{ss}$) based on geometrically estimated Area Under the Curves (AUCs) and Area Under the Moment Curves (AUMCs). Numerous papers and texts have discussed the values and limitations of each method of analysis, with most concluding the choice of method resides in the richness of the data set.

A basic assumption related to both methods of analysis is that the elimination of drug from the body is exclusively from the sampling compartment (i. e., blood/plasma), and that rate constants are first order. However, when some or all of the elimination occurs outside the sampling compartment – that is, in the peripheral or tissue compartment(s) – these types of analysis are prone to error in the estimation of $V_{ss}$, but not CL. In compartmental modeling, the error is related to the fact that no longer do the exponents accurately reflect the inter-compartmental and elimination (micro) rate constants. This "model misspecification" will result in an error that is related to the relative magnitudes of the distribution rate constants and the peripheral elimination rate constant. However, less widely understood is the fact that this model misspecification will also result in errors in noncompartmental pharmacokinetic analysis.

This is particularly important when considering the disposition of "biotech" drugs such as peptides, proteins or nucleic acids, which have in many instances been shown to distribute to specific tissue sites, redistribute slowly back into the systemic circulation compartment, and are prone to a much greater proportion of elimination from the tissue sites rather than from the systemic circulation (i. e.,

*Pharmacokinetics and Pharmacodynamics of Biotech Drugs: Principles and Case Studies in Drug Development.* Edited by Bernd Meibohm

ISBN: 3-527-31408-3

the sampling compartment). Besides problems in defining $V_{ss}$ when elimination from peripheral compartments is present, additional complications arise when nonlinear distribution, binding, and/or elimination exist, or the administration of drug is not directly into the sampling compartment. It is the purpose of this chapter to use simulations to illustrate – but not solve – the problem of determining $V_{ss}$ from plasma concentration data when elimination from tissue sites represented by peripheral compartments is present.

## 7.2
## The Concept of Volume of Distribution

It is fairly simple to define volume of distribution ($V$) as the proportionality constant relating the total amount of drug in the body at a specified time ($A_t$) to the concentration in the plasma at that same time ($C_t$):

$$V = \frac{A_t}{C_t} \tag{1}$$

This mathematical definition allows one to visualize a space into which a drug distributes but represents a theoretical or apparent, and not a real, space. This is because the calculation is based on the concentration found in the real volume of the sampling compartment (i. e., plasma or blood), which in most cases by itself is different from concentrations in other body spaces. This will result in many drugs exhibiting a volume of distribution that is far larger than the plasma or blood volume, total body water, or even body volume. However, this way of expressing the volume of distribution is most useful in characterizing a drug's relative affinity for space outside the blood, and may be of value in deciding if the drug has a potential for site-specific activity in the tissue. It may also be of value in determining a loading appropriate dose (not dosage regimen) to achieve a desired target concentration.

Although Eq. (1) is a simple mathematical relationship, there are numerous limitations to its appropriate application, based on the assumptions one makes about the pharmacokinetic model to which it is applied. In the case where one assumes instantaneous equilibrium of drug between the tissue and the plasma or blood (i. e., a one-compartment model), the concentration in the sampling compartment is, by definition, proportional to the tissue concentration at all times after dosing, and $V$ determined for any $A_t$ and $C_t$ pair will be constant. Since $A_t$ at time 0 is the dose ($D$), it is common to express volume of distribution in a one-compartment model as:

$$V = \frac{D}{C_0} \tag{2}$$

However, if the distribution of drug requires a finite amount of time to reach equilibrium between the sampling compartment and peripheral or tissue compartments (i.e., a two- or more compartment model), the application of Eq. (2) will only account for the volume of the sampling compartment [i.e., plasma ($V_p$) or central compartment ($V_c$)]. When equilibrium of drug among the compartments is reached – that is, when steady-state is reached and the ratio of concentrations in the plasma and tissue is constant – the volume of distribution can be expressed as:

$$V_{ss} = V_p + \frac{f_{up}}{f_{ut}} \cdot V_t \tag{3}$$

where $f_{up}$ and $f_{ut}$ are unbound fractions of drug in the plasma and tissue, respectively, $V_p$ is the volume of the plasma water (ca. 3 L), and $V_t$ is the volume of body water outside the blood (ca. 35 L).

This physiological approach to defining volume is useful to help explain which drug might be more likely to reside outside the blood based on tissue affinity ($f_{up}/f_{ut}$), but does not readily lend itself to a calculation because of the difficulty in determining either free or total concentrations in the tissue. A simpler mathematical expression of $V_{ss}$ would be one related to equilibrium conditions achieved with a continuous intravenous (IV) infusion ($R_0$) at steady-state where:

$$V_{ss} = \frac{A_{ss}}{C_{ss}} \tag{4}$$

Although it is a relatively simple matter to assess $C_{ss}$ in a clinical setting, the evaluation of $A_{ss}$ would generally require sacrifice of the organism. As outlined in the following section, the mathematical expression of $V_{ss}$ may be accurately obtained after IV bolus dosing, with or without regard to the number of compartments, as long as elimination is exclusively from the central compartment and all rate processes are first order.

## 7.3 Calculation of $V_{ss}$

Wagner [1] has shown, with IV bolus dosing, that the $V_{ss}$ for a n-compartment open mammillary model with first-order elimination from the central compartment is:

$$V_{ss} = D \cdot \left( \frac{\sum \frac{A_i}{\lambda_i^2}}{\sum \frac{A_i}{\lambda_i}} \right)^2 \tag{5}$$

where $A_i$ and $\lambda_i$ represent the respective coefficients and exponents of the polyexponential equation describing the concentration versus time curve.

In the case where only two exponential terms exist (i. e., a two-compartment model), Eq. (5) reduces to:

$$V_{ss} = \frac{D}{C_0} \cdot \left(1 + \frac{k_{12}}{k_{21}}\right) \tag{6}$$

where $D/C_0$ is $V_c$ and $k_{12}$ and $k_{21}$ are the first-order distribution rate constants from the central to tissue and tissue to central compartments, respectively.

More recently, Benet and Galeazzi [2] have described a noncompartmental volume of distribution as:

$$V_{ss} = \frac{D \cdot \text{AUMC}}{\text{AUC}^2} \tag{7}$$

where AUMC and AUC are the areas under the $C_t \times t$ versus $t$ curve and $C_t$ versus $t$ curve from 0 to infinity, respectively.

It should be noted that both the AUMC and AUC may be determined geometrically using the trapezoidal rule with extrapolation from the last $C_t$ to infinity where:

$$\text{AUMC}_{t-\infty} = \frac{t \cdot C_t}{\lambda_z} + \frac{C_t}{\lambda_z^2} \tag{8}$$

$$\text{AUC}_{t-\infty} = \frac{C_t}{\lambda_z} \tag{9}$$

Note that $\lambda_z$ is the terminal log-linear exponent from Eq. (5) and $C_t$ is the last concentration from the data set.

Equation (7) is more commonly expressed as:

$$V_{ss} = \frac{D}{\text{AUC}} \cdot \text{MRT} \tag{10}$$

where MRT (Mean Residence Time) is AUMC/AUC.

The equations above apply strictly to drugs administered as a single IV bolus dose, but for drug administered as an infusion or via oral route, or after multiple dosing, the calculation of AUMC must be adjusted to account for drug input [i. e., infusion time ($T$) or absorption rate constant $K_a$ and extent of bioavailability $F$], as shown by Straughn [3]. Although, in theory, AUC will not be affected by the route, the AUMC will be overestimated, and this will result in an overestimation of $V_{SS}$.

It can be shown, of course, that Eq. (10) is identical to Eq. (5) in the n-compartment model, but the use of Eq. (10) requires fewer assumptions to be made in determining $V_{SS}$. That is, Eq. (10) only requires that disposition rates are first order, while Eq. (5) requires additional assumptions related to the underlying multi-compartment pharmacokinetic model.

## 7.4 Pitfalls in Calculating $V_{ss}$

As stated above, the $V_{SS}$ calculation using Eqs. (5) or (10) is valid only when elimination exclusively occurs from the sampling (plasma/blood) compartment. When some or all elimination occurs from the tissue compartment (Fig. 7.1), the concentration versus time profile will still be characterized by a bi-exponential equation; however, the ability of modeling systems to quantify the micro rate constants is lost. That is to say, essentially identical bi-exponential concentration time profiles are possible with and without elimination from the tissue compartment. Therefore, when modeling from a plasma profile only, there is no way of determining if the exit of drug from the body is exclusive to the central compartment.

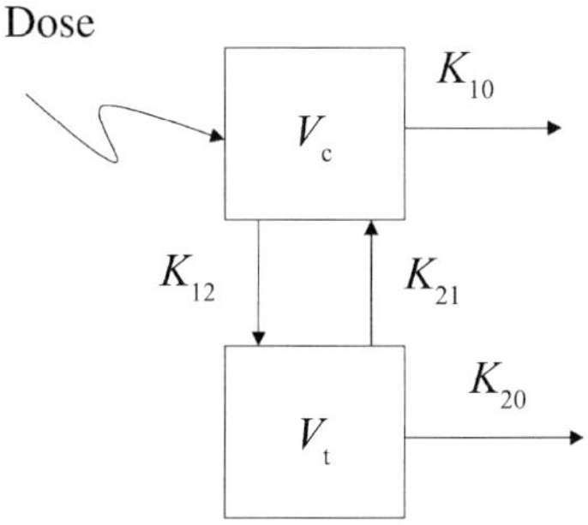


**Fig. 7.1** Two-compartment open model with plasma and tissue elimination.

From an examination of Eq. (6) for a two-compartment model it is evident that $V_{SS}$ is dependent on the quantification of $K_{12}$ and $K_{21}$. For this model $K_{12}$ and $K_{21}$ can be determined by nonlinear regression analysis of plasma concentration–time data, either by deriving them from the fitted values of the coefficients and exponentials of the bi-exponential expression describing the concentration–time data, or by coding them directly into the modeling program. For the case where tissue elimination exists, it is possible to code into the model the existence of a $K_{20}$, but the convergence process will not be able to resolve the appropriate micro rate constant.

To evaluate the error associated with model misspecification related to the tissue elimination dilemma, a set IV of bolus plasma concentration profiles was generated by numerical integration (Stella V 7.0.3, High Performance Systems Inc., Lebanon, NH, USA) based on the model in Fig. 7.1, and using various values for the micro rate constants ($k_{10}$, $k_{20}$, $k_{12}$, and $k_{21}$). $V_{SS}$ was then determined from the AUCs and AUMCs using the noncompartmental approach (Eq. 10). The "true" $V_{SS}$ [Eq. (4)] was calculated from the $C_{SS}$ and $A_{SS}$ generated by simulating a constant IV infusion using the respective micro constants used to generate the IV bolus plasma concentration profiles.

As a side note, it can be shown that a slight modification of Eq. (6) can be used to calculate the "true" $V_{SS}$ by incorporating $K_{20}$ as follows:

$$V_{ss} = V_c \cdot \left(1 + \frac{k_{12}}{k_{21} + k_{20}}\right) \tag{9}$$

All simulations were based on an IV bolus dose of 1000 mg, an infusion dose rate of 10 mg/h, and a $V_c$ of 1 L. The micro rate constants used for the simulations were chosen to generate a family of bi-exponential curves with terminal half lives ranging from approximately 6 h to 3 days and exhibiting a fraction of drug eliminated by the tissue ($f_{e,T}$) from approximately 30% to 99%. The parameters for 10 simulations are listed in Table 7.1.

**Table 7.1** Simulation results and parameters used.

| Cond. | $K_{10}$ [h] | $K_{12}$ [h] | $K_{21}$ [h] | $K_{20}$ [h] | $f_{e,T}$ [a] | $V_{ss}$ NC [b] [L] | $V_{ss}$ True [L] | Ratio True/NC | ss TB ratio [c] | Terminal $t_{1/2}$ [h] |
|---|---|---|---|---|---|---|---|---|---|---|
| 9 | 0.01 | 0.1 | 0.10 | 0.01 | 0.476 | 1.83 | 1.91 | 1.04 | 0.91 | 69 |
| 8 | 0.10 | 1.0 | 0.10 | 0.01 | 0.476 | 9.26 | 10.09 | 1.09 | 9.09 | 39 |
| 2 | 0.10 | 0.1 | 0.10 | 0.10 | 0.333 | 1.25 | 1.50 | 1.20 | 0.50 | 6.9 |
| 3 | 0.10 | 1.0 | 0.10 | 0.10 | 0.833 | 3.50 | 6.00 | 1.71 | 5.00 | 6.9 |
| 6 | 0.01 | 0.1 | 0.01 | 0.01 | 0.833 | 3.50 | 6.00 | 1.71 | 5.00 | 69 |
| 1 | 0.01 | 0.1 | 0.01 | 0.10 | 0.901 | 1.08 | 1.91 | 1.77 | 0.91 | 8.4 |
| 7 | 0.01 | 1.0 | 0.01 | 0.01 | 0.980 | 26.0 | 51.00 | 1.96 | 50.00 | 69 |
| 10 | 0.10 | 1.0 | 0.01 | 0.01 | 0.833 | 26.0 | 51.00 | 1.96 | 50.00 | 64 |
| 4 | 0.01 | 1.0 | 0.01 | 0.10 | 0.989 | 1.83 | 10.09 | 5.51 | 9.09 | 7.0 |
| 5 | 0.10 | 1.0 | 0.01 | 0.10 | 0.901 | 1.83 | 10.09 | 5.51 | 9.09 | 6.9 |

a) $f_{e,T}$ = fraction eliminated by tissue.
b) $V_{ss}$ NC = volume of distribution at steady state determined by noncompartmental calculation.
c) ss TB ratio = tissue-to-blood ratio of drug at steady state after continuous IV infusion.

Note that in a two-compartment model with tissue elimination the $f_{e,T}$ is:

$$f_{e,T} = 1 - \left(\frac{k_{10}}{k_{10} + k_{20} \cdot \left(\frac{k_{12}}{k_{21} + k_{20}}\right)}\right) \tag{10}$$

The tissue-to-blood ratio at steady-state (ss TB ratio) may be determined as:

$$\text{ss TB ratio} = \frac{k_{12}}{k_{21} + k_{20}} \tag{11}$$

## 7.5 Results and Discussion

Figures 7.2 and 7.3 illustrate the profiles generated from the parameters given in Table 7.1, with Fig. 7.2 (conditions 1–5) showing the profiles with the shorter terminal half-lives and Fig. 7.3 (conditions 6–10) showing the longer terminal half-lives. The data in Table 7.1 indicate that the noncompartmental method for determining $V_{ss}$ will always underestimate the "true" $V_{ss}$ if elimination from the tissue is present. The magnitude of this error (shown in Table 7.1 as the ratio of the True to Noncompartmental $V_{ss}$) is generally less for drugs with less extensive tissue elimination ($f_{e,T}$) and larger for smaller tissue to blood transfer rate constants ($k_{21}$). However, this error appears to be unrelated to any of the other parameters. Although a more extensive evaluation of parameters effects may reveal

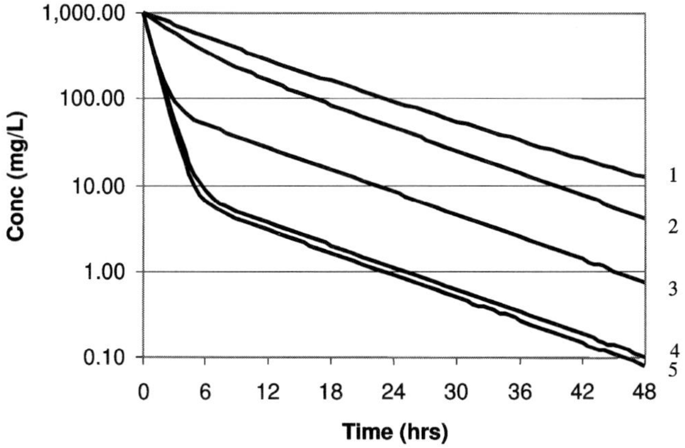


**Fig. 7.2** Simulations for conditions 1–5 (see Table 7.1 for conditions).

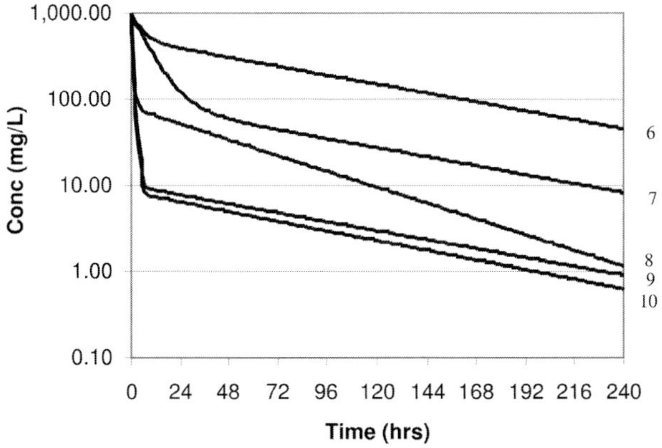


**Fig. 7.3** Simulations for conditions 6–10 (see Table 7.1 for conditions).

other more precise trends in this error, it is clear from just these few simulations that the overall shape of the concentration versus time profile will do little to reveal the presence or absence of such an error. Therefore, even with knowledge of the extent of tissue distribution and elimination, the calculation of $V_{SS}$ from plasma concentration data using the noncompartmental approach will result in errors of unknown magnitude.

## 7.6 Conclusions

From these simulations based on a two-compartment model with both plasma and tissue elimination, a $V_{SS}$ determined by utilizing noncompartmental methods will have a value less than the "true" $V_{SS}$. These simulations also show that the shape of the plasma concentration–time curve and the relative magnitudes of the plasma and tissue elimination rate constants do not correlate with the error. However, this error does tend to be greater for hypothetical drugs that are more extensively eliminated by tissue routes.

## 7.7 References


1 Wagner, J.G. **1976**. Linear pharmacokinetic equations allowing direct calculation of many needed pharmacokinetic parameters from the coefficients and exponents of polyexponential equations which have been fitted to the data. *J. Pharmacokinet. Biopharm.* 4: 443–467.

2 Benet, L.Z., and R.L. Galeazzi. **1979**. Noncompartmental determination of the steady-state volume of distribution. *J. Pharm. Sci.* 68: 1071–1074.

3 Straughn, A.B. **1982**. Model-independent steady-state volume of distribution. *J. Pharm. Sci.* 71: 597–598.

# 8
# Bioequivalence of Biologics

*Jeffrey S. Barrett*

## 8.1
## Introduction

The term "bioequivalence" has many definitions depending on the setting in which it is referred. It can imply a concept, a metric, a criterion, a study design, and/or a regulatory decision among possible intended uses. Within the context of a small-molecule drug product, the various meanings related to bioequivalence are reasonably well understood, and in some way relate to the *in-vivo* assessment of active moieties (parent compound and/or metabolites) that have pharmacologic activity. Biologic drug products are typically large molecules or complex proteins that are synthetic or recombinant versions of natural biological substances. Common biologics include insulin, human growth hormone, vaccines, erythropoietin, blood coagulation factors, and cell/tissue-based therapies [1], as discussed elsewhere in this book. Unlike their small-molecule chemical drug counterparts, biologics are difficult to produce with consistency, even by the same manufacturer. It is estimated that by 2010, approximately half of all newly approved drug products will be of biological origin [2].

Biopharmaceuticals represent an \$ 18 billion industry in the United States, and approximately \$ 30 billion worldwide [3, 4]. The market is growing at about 10% annually [3, 4]. During the 2004 fiscal year, the United States Food and Drug Administration (FDA) approved 23 biologics and is likely to maintain or increase this pace in the future. Over the next few years, patents are due to expire on about 18 biologics having total annual sales of between \$ 10 billion and \$ 15 billion (see Table 8.1) [5–6]. Amgen's Epogen (epoetin-α) treatment for anemia, with more than \$ 2 billion in annual sales [7], goes off-patent next year, as does Johnson & Johnson's erythropoietin version, Procrit, with \$ 1.7 billion in annual sales. Biogen's Avonex (interferon β-1 a) for multiple sclerosis, with \$ 761 million in annual sales, lost patent exclusivity in 2005. Schering-Plough's Intron A (interferon α-2 b) for leukemia and hepatitis B and C, with \$ 1.4 billion in annual sales, lost patent protection in 2004.

Likewise, generic drug makers are anxious to enter the lucrative arena of biologics. Yet, there is currently no means of easy access by the established regulatory

*Pharmacokinetics and Pharmacodynamics of Biotech Drugs: Principles and Case Studies in Drug Development.* Edited by Bernd Meibohm

ISBN: 3-527-31408-3

**Table 8.1** Biologic product expirations (US and EU Patents; as of 2005) [5].

| Manufacturer | Product (brand name/generic name) | Indication | Market Exclusivity Expiration US | EU |
|---|---|---|---|---|
| Genentech | Nutropin (somatropin) | Growth disorders | Expired | Expired |
| Abbott | Abbokinase (urokinase | Ischemic events | Expired | Expired |
| Eli Lilly | Humulin (humin insulin) | Diabetes | Expired | Expired |
| Genzyme | Ceredase (alglucerase); Cerezyme (imiglucerase) | Gaucher disease | Expired | Expired |
| AstraZeneca | Streptase (streptokinase) | Ischemic events | Expired | Expired |
| Biogen/Roche | Intron A (interferon α-2 b) | Hepatitis B and C | Expired | Expired (France); 2007 (Italy) |
| Serono | Serostim (somatropin) | AIDS wasting | Expired | NA |
| Eli Lilly | Humatrope (somatropin) | Growth disorders | Expired | NA |
| Amgen | Epogen, Procrit, Eprex (erythropoietin) | Anemia | 2013 | Expired, 2004 |
| Roche | NeoRecormon (erythropoietin) | Anemia | NA | 2005 |
| Genentech | TNKase (tenecteplase) | Acute myocardial infarction | 2005 | 2005 |
| InterMune | Actimmune (interferon γ-1 b) | Chronic granulomatous disease (CGD), malignant osteopetrosis | 2005, 2006, 2012 | Expired, 2004 |
| Genentech | Activase (alteplase) | Acute myocardial infarction | 2005, 2010 | 2005 |
| Chiron | Proleukin (interleukin-2) | HIV | 2006, 2012 | 2005 |
| Amgen | Neupogen (filgrastim) | Anemia, leukemia, neutropenia | 2015 | 2006 |

mechanism used to register generic small-molecule drugs. Manufacturers of small-molecule generic products may cite data from the original product to gain drug approval when the generic product is only slightly different from the original drug. Because of differences inherent to biologic drug products, guidelines used to approve low-molecular-weight generics cannot be legitimately applied to generic versions of biologics (biogenerics). No legal pathway exists in the United States for the approval of biogenerics; however, the FDA is expected to draft guidelines for so-called "follow-on biologics" in 2006. These recommendations are expected to be similar to the guidelines issued by the FDA's European counterpart, the European Medicines Agency (EMEA), requiring more stringent approval for biogeneric drugs than for typical small-molecule generic products [8].

At the center of the conflict over potential FDA regulatory pathways for approval of biogenerics are conflicting views regarding the feasibility of determining bioequivalence without requiring extensive clinical testing. The use of chemical specifications as criteria for determining bioequivalence has worked well for traditional small-molecule drugs. Most small-molecule drug products have molecular weights less than 1.5 kDa and can easily be evaluated solely on a specified and controlled chemical structure. Structural classification is generally determined readily by nuclear magnetic resonance, mass spectrometry, infrared spectrometry, X-ray crystallography, or other well-known physical methods. Biologics, in contrast, typically contain active macromolecules and therefore exhibit far more complexity and potential variability. Furthermore, such larger molecules may have more than one active epitope. Additionally, for many biologics the causes and effects of variations due to post-translational modifications are not fully characterized.

Opponents of generic biologics therefore contend that current analytical characterization is not robust enough to ensure the safety and potency of biogenerics. Generic manufacturers would not use the same production and purification schemes and would not have access to the proprietary good manufacturing practice (GMP) and good laboratory practice (GLP) protocols of the manufacturer of the innovator product. Opponents argue further that biologics are too complex for generics manufacturers to successfully reverse-engineer the process.

The term *generic* will likely not be applied to biologics as the word carries a specific legal and scientific meaning as discussed in this chapter. Some researchers and manufacturers have proposed the term *therapeutically equivalent biologics* (TEB). The FDA other groups seem to prefer *follow-on protein products* (FOPPs) or *follow-on biologics* (FOBs). Another proposed term is SEPP or *subsequent entry protein pharmaceutical.* Two other suggestions are *second-generation biologics* and *biosimilar drugs* (term used currently by European Union regulators). In any event, it is clear that the application of the concept of bioequivalence for biologics will be different from what has been and continues to be applied for small-molecule drug products. This chapter reviews the nature of biologics from the standpoint of characteristics typically used to judge whether a new drug product (generic product) can be substituted for an existing agent (innovator or reference product). The standard bioequivalence criterion is reviewed with an example of a classical bioequivalence approach applied to a biologic product as a case study illustrating these concerns.

## 8.2 Prevailing Opinion: Science, Economics, and Politics

Based on the Food, Drug and Cosmetic Act, bioavailability has been defined as "the rate and extent to which the active ingredient or active moiety is absorbed from a drug product and becomes available at the site of action. For drug products that are not intended to be absorbed into the bloodstream, bioavailability may be assessed by measurements intended to reflect the rate and extent to which the active ingredient or active moiety becomes available at the site of action." The estimation of bioavailability is based upon the assessment of drug exposure via measured blood or plasma concentrations following a given dose from a test formulation relative to the reference formulation. Bioequivalence trials are used primarily to infer therapeutic equivalence of different formulations based on the similarity of the pharmacokinetic characteristics of the formulations tested.

Bioequivalence, as a regulatory requirement, has evolved over the past 35 years and continues to be the topic of much discussion. Evidence of differences in bioavailability from various oral formulations of the same therapeutic agents had become apparent by the early 1960s. Implicit in the evolution of bioequivalence as a science and a regulatory mechanism to ensure product quality, and likewise instill consumer confidence with generic drug products, is the reality that the paths to the current regulatory environment were most assuredly designed with homogeneous, small-molecule drugs in mind. Biologic drugs challenge some of the fundamental assumptions regarding the translation of safety and efficacy to proposed substitutable drug products.

Currently, generic biologics cannot be marketed in the United States. The Drug, Price Competition and Patent Restoration Act of 1984, commonly known as the "Hatch-Waxman Act", and its corresponding regulations govern the approval and market-entry of generic drugs. The Hatch-Waxman Act's Abbreviated New Drug Application (ANDA) provisions specifically exclude biologics, however. The exclusion of biologics within the scope of the Hatch-Waxman Act's ANDA provisions is most likely because only a few biotechnology-derived drugs existed when the act was enacted in 1984. Recent socioeconomic and political changes, along with continuous technological improvements in the ability to produce and test biologically derived drugs, have significantly increased the likelihood that a regulatory framework permitting generic biologics in the US marketplace will eventually emerge.

The FDA's Center for Drug Evaluation and Research (CDER) has indicated that the ANDA regulatory scheme does not allow the FDA to obtain enough evidence to approve a generic biopharmaceutical, due in part to CDER's inability to request additional preclinical or clinical testing under an ANDA. Nevertheless, proponents of generic biopharmaceuticals have asserted that section 505(b)(2) of the Hatch-Waxman Act provides a new drug application pathway that may serve as an ANDA-like approach to approval of generic versions of complex biologics originally approved under biologic license applications (BLAs). In October 1999, the FDA announced that the section 505(b)(2) pathway may serve as a way to obtain regulatory approval of a generic version of certain biological products originally approved under a new drug application (NDA). A section 505(b)(2) application would permit an applicant to obtain approval of an NDA based on the FDA's earlier findings of safety and efficacy from a previously approved NDA and thus would not require an applicant to obtain a right of reference from the original applicants. The section 505(b)(2) applicant must also provide any additional clinical data needed to demonstrate that differences between the original drug and its copycat have not changed its safety and effectiveness.

As with an ANDA application, a section 505(b)(2) applicant must include appropriate patent certifications and explain the basis upon which it believes that it does not infringe any valid claim of a so-called "Orange Book" listed patent. Under 35 U.S.C. § 271(e)(2), the holder of the original NDA could then bring suit against the later applicant and obtain a 30-month stay of the section 505(b)(2) approval. Products marketed under approved section 505(b)(2) applications, like ANDA-based products, may receive an "AB" substitutability rating in the Orange Book if the product has the same active ingredient(s), dosage form, strength, and bioequivalence.

## 8.3 Biologics: Time Course of Immunogenicity

One of the major concerns with the desire to evaluate follow-on biologics for an *in-vivo* quality control assessment is that the design commonly employed for small-molecule drugs assumes that an acute exposure of the test agent will be an adequate setting to confer chronic exposure and, more importantly, safety. As mentioned above, biologic drugs present a challenge to this small-molecule assumption as they typically exhibit some form of immunogenicity over an often unpredictable time event scale. Immunogenicity to protein-based therapeutics may manifest as: (1) anaphylactic reaction; (2) reduced efficacy of the therapeutic protein; or (3) as production of antibodies (Abs) that bind to and reduce the production of the native endogenous protein [9,10]. As an example we can examine the third type of immunogenicity listed above resultant from erythropoietin administration [11–13]. Other examples are compiled in Table 8.2.

Recombinant human erythropoietin (EPO) has been in use since 1988 in the treatment of anemia, without any reports of significant immunogenicity. There were a few cases (three reported prior to 1998) of pure red cell aplasia (PRCA) in patients with chronic kidney disease. However, the number rose considerably in 1998 and reached a peak in 2002 (about 21 cases in two years). Patients with EPO-associated PRCA develop antibodies to EPO that neutralize both the administered EPO and also the endogenous EPO. This results in a depletion of the erythroid precursor cells in the bone marrow, making these patients transfusion-dependent. Almost all of the observed cases of PRCA occurred in patients with chronic kidney disease receiving Eprex (epoetin-$\alpha$; OrthoBiotech, a division of Janssen-Cialeg, marketed outside the US) subcutaneously. The duration between the initiation of therapy with Eprex and development of PRAC ranged from 0.3 to 82 months (median 9 months). EPO-associated PRCA was not observed in patients receiving Eprex intravenously. Following subsequent investigations, the presence of leachates in the formulation, arising from an interaction between polysorbate 80 (a component in the formulation) and the uncoated rubber stopper was found to be responsible for the observed antigenicity.

This highlights how the combination of formulation constituents can result in immunogenicity to the protein, and how this manifests a major concern in the

**Table 8.2** Immunogenicity outcomes from select biologic agents: Time course of observation and method of investigation.

| Example | Reference(s) | Objective | Study design/Duration | Assay | Outcome |
|---|---|---|---|---|---|
| Insulin | 15 | Immunogenicity of rh-insulin relative to porcine insulin | 100 subjects receiving rh-insulin vs. 121 receiving porcine insulin over 12 months | Species-specific binding assay | ~44% rh-insulin patients developed antibodies (Abs) reaching a plateau at 6 months.<br>60% patients receiving porcine insulin developed Abs in 12 months |
| Growth hormone | 16 | Evaluate Abs in children treated with rh-GH and met-GH | 46 GH-deficient children treated for at least 12 months (20 naïve and 26 previous treatment with pituitary extracted GH) | RIA | Abs generated in 75% of treatment naïve group within 1 year; 12% of the pretreated group (only in the second year).<br>Abs remained through duration of treatment with met-GH, but decreased and eventually become undetectable in subjects receiving rh-GH. Abs did not affect clinical efficacy |
| | 17 | Immunogenicity over long-term treatment | 304 GH deficiency and 91 with Turner's syndrome ~54 months | ELISA | 3% GH-deficient patients generated Abs within 3–12 months (subsequently declined).<br>No patient with Turner's syndrome generated detectable Abs |
| | 18 | Immunogenicity of six different GH formulations | 260 children with GH deficiency | RIA | Formulations differ by immunogenicity response<br>16% of the patients treated with Lilly product generate Abs<br>Abs increased up to 6–9 months followed by a decrease and are undetectable at 30 months (Lilly product) |
| Anti-TNF | 19 | Immunogenicity | Review | | Abs generated and detected in 3–16% of patients receiving Etanercept (infliximab) within 12 months of treatment |
| | 20 | Safety and schedule of infliximab | 442 patients with Crohn's disease treated Q8weeks for 46 weeks (observed 54 weeks) | ELISA | 14% (64/442) of the patients had detectable levels of Abs and 46% had inconclusive results (in 54 weeks)<br>ELISA confounded by drug |

**Table 8.2** (continued)

| Example | Reference(s) | Objective | Study design/Duration | Assay | Outcome |
|---|---|---|---|---|---|
| Erythropoietin (Eprex) | 13 | Change in formulation | Review of case studies | | EPO used since 1988 without reports of significant immunogenicity<br>Few cases (3 prior to 1998) of PRCA in patients with chronic kidney disease; rise in 1998 peaking in 2002 (~21 cases in 2 years)<br>Most cases of PRCA from Eprex given by SC route (OrthoBiotech)<br>Duration between therapy/development of PRAC ranged from 0.3 to 82 months (median 9). PRCA not observed in patients given Eprex by IV route<br>Presence of leachates in the formulation found to be responsible |
| Interferon-α | 21 | Immunogenicity over long term treatment | Review of case studies | | 20–30% of patients develop Abs to interferon<br>Patients with renal cell carcinoma, non-neutralizing Abs generated in 8 weeks (median time); neutralizing Abs in about 14 weeks<br>Patients with hairy cell leukemia, generated Abs starting 7 months |

Abbreviations: EPO: erythropoietin; GH: growth hormone; PRCA: pure red cell aplasia; RIA: radioimmunoassay; SC: subcutaneous; TNF: tumor necrosis factor.

manufacture and processing of biosimilar protein drugs. Also, the duration over which such immunogenic reactions can occur makes long-term immunogenicity evaluation of the biosimilars, for both innovator and follow-on sponsors, an important aspect in determining the safety of the product. Apart from the multitude of formulation considerations, the disease being treated adds an additional level of uncertainty to the onset of generation of antibodies to protein therapeutics. For example, in patients with renal cell carcinoma receiving interferon-α, non-neutralizing antibodies were generated in about 8 weeks (median time) of treatment, and neutralizing Abs developed in about 14 weeks. In contrast, patients with hairy cell leukemia receiving interferon-α generated antibodies starting at 7 months, further supporting the need for immunogenicity evaluations in the target population.

It is unlikely that immunogenic responses can be generalized based on our current limited understanding of the involved processes. An increased risk of immunogenicity is perceived with xenopeptides, however, as human-derived therapies also induce antibody formation that in some cases has been associated with severe clinical consequences. In reality, similar to clinical experiences with peptide-based agents in general, antibody responses against xenopeptide hormone therapies in the majority of cases have been benign in nature, with minimal clinical impact [14]. This does not diminish the need to understand the time course of these events, but it does play a role in the evaluation of consumer risk – a central tenet in the regulation of follow-on biologic drug products.

## 8.4 Pharmaceutical Equivalence

Generic drug products are typically evaluated with respect to both pharmaceutical equivalence and bioequivalence. While "bio"equivalence is generally the focus of debates regarding biologics, pharmaceutical equivalence is not without issue and is, in part, also a motivation for changing the nomenclature for biologics. Specifically, pharmaceutical equivalence ensures that two formulations have the same active ingredients, same strength, same dosage form and route of administration, comparable labeling, and meet compendial or other standards of identity, strength, quality, purity, and potency. For small-molecule drug products, pharmaceutical equivalence reflects the first – and occasionally the minimal – hurdle over which a sponsor company must evaluate its product against an approved reference product. In some cases, the characterization of pharmaceutical equivalence is sufficient to evaluate manufacturing or process changes without the necessity of an *in-vivo* bioequivalence trial [22, 23].

In contrast, for biologic drug products manufacturing changes cannot be accurately classified as major or minor because any change can have a significant effect on quality, safety, and/or efficacy. In the case of biosimilar epoetins, such significant variability exists among individual products that testing of each individual compounds rather than establishment of class standards is recommended. Defining analyses to ensure product similarity is necessary to permit consistency of testing. The EMEA has recommended specific testing that varies based on drug characteristics [24, 25]. Determinants of product comparability testing based on EMEA guidelines are as follows:

- Quality findings
- Nature of the product
- Dosage regimen
- Route of administration
- Therapeutic window identified in dose-ranging studies
- Short- versus long-term use
- Extent of knowledge of structure–activity relationships
- Previous experience with immunogenic activity
- Mechanism of action
- Patient population(s)
- Availability of preclinical and clinical data

### 8.4.1
### How Changes in Quality Might Affect Safety and Efficacy

The type and extent of preclinical and clinical studies should be determined on a case-by-base basis. Establishing comparability among biologic products generally requires a full characterization of physico-chemical properties, identification of impurities, and quantification of biologic activity with both *in-vitro* and *in-vivo* testing. Because of the inherent variability of biologic processes, batch-to-batch consistency must also be ensured. If quality attributes (e.g., purity, potency, identity, and stability) cannot be adequately assessed with analytic studies, then preclinical and/or clinical studies will likely be needed.

Most of the emphasis on pharmaceutical product quality has been on the need to improve analytical testing and to evaluate the correlation of such testing with clinical findings. Combe et al. [26] recently reviewed literature reports of both analytic and clinical studies conducted with biosimilar epoetin products currently marketed outside the United States and Europe in light of the recently implemented EMEA guidelines. The analytic studies reported that products differed widely in composition, did not always meet self-declared specifications, and exhibited batch-to-batch variation. Although several clinical studies demonstrated correction of anemia with biosimilar epoetins by using an open-label or placebo-controlled study design, only four of 22 studies were competitor-controlled. Most of the studies were small (median 41 patients; range 18 to 1079 patients) and of short duration (median 12 weeks; range 6 weeks to 1 year). Hence, while efforts to characterize biologic products have improved greatly [27] and, in some cases, may be adequate to characterize a biologic drug product and its manufacturing process, these are not portable to across or within class.

## 8.5 Bioequivalence: Metrics and Methods for Biologics?

As discussed above, the metrics for assessing small-molecule drug products are well established and, despite some limitation, provide a reasonable assessment of *in-vivo* drug product performance, ultimately ensuring a degree of confidence in the generic formulation with minimal consumer risk. The focal point of the criteria for small molecules has been the pharmacokinetic behavior of the active molecular entities as measured by drug bioavailability determination. Measures of bioavailability for single-dose studies include the area under the blood, serum or plasma drug concentration–time curve (AUC) and the peak blood, serum or plasma drug concentration ($C_{max}$). Other metrics are used for steady-state bioequivalence evaluations and, more recently, the FDA has advocated early-exposure metrics in bioequivalence assessments. In addition, urinary recovery has been used as a surrogate for systemic drug exposure. A list of the many metrics used to assess bioequivalence, as well as the studies under which they are commonly generated and their basic operating characteristics [28], is provided in Table 8.3.

**Table 8.3** Metrics for bioequivalence evaluation [28].

| Metric | Study designs | Operating characteristics |
|---|---|---|
| AUC: Area under the curve for drug[a)] concentration[b)] against time – may be qualified by a specific time (e.g., from 0 to 12 h, $AUC_{12}$ – partial areas)[c)] | Single-dose crossover (fasted or fed)<br>Parallel group<br>Steady-state (single dose vs. steady-state comparison may be required) | Generally viewed as extent metric; partial areas utilized as early exposure metric based on AUC partition<br>Partial area may permit discrete characterization of complicated absorption processes.<br>Systematic errors possible when inappropriate data density utilized to generate AUC |
| AUC or $AUC_{0-\infty}$: AUC from zero to infinity, obtained by extrapolation | Single-dose crossover<br>Single-dose parallel | Extent of exposure metric<br>Less sensitive metric when terminal phase accounts for a large portion of AUC |
| $AUC_{0-\tau}$: AUC during dosing interval ($\tau$) at steady state | Steady-state | Steady-state exposure metric; dependent on reasonable demonstration of steady-state attainment |
| $C_{max}$: Observed maximum or peak concentration | Single-dose crossover<br>Parallel group<br>Steady-state (optional) | Rate metric (historical)<br>Subject to inaccuracy due to discrete sampling<br>Insensitive to absorption time course differences |

**Table 8.3** (continued)

| Metric | Study designs | Operating characteristics |
|---|---|---|
| $C_{max}$/AUC: Ratio of $C_{max}$ to AUC | | Scaled parameter<br>Variance estimate complicated due to transformation (ratio) of correlated parameters |
| $C_{min}$: Minimum plasma concentration | Steady-state | Sampling and assay dependencies possible |
| $C_{ave}$: average plasma concentration | Steady-state | Extent metric; trade-off between sensitivity and sampling bias |
| Fluctuation: $(C_{max} - C_{min})/C_{ave}$ | Steady-state | Peak–trough exposure normalized to average concentration<br>Dependent on attainment of steady-state |
| Swing: $(C_{max} - C_{min})/C_{min}$ | Steady-state | Peak–trough exposure normalized to trough<br>Failure may identify absorption (rate) differences<br>Dependent on attainment of steady-state |
| MAT: Mean absorption time | Single-dose crossover | Less sensitive to discrete sampling issues<br>May mask absorption profile differences |
| MRT: Mean residence time | Single-dose crossover | Extent metric; ratio of moments |
| *Ae*: Cumulative urinary recovery of drug – may be defined for a specific period, e.g., *Ae* from 0 to 12 h, $Ae_{12}$) | Single-dose crossover | Urine collection often more problematic (variable) than plasma, serum, blood sampling<br>Useful when systemic profile cannot be captured or influenced by non-absorption dependencies (i.e., binding to plasma ACE) |
| $T_{max}$: Time after administration of the drug at which $C_{max}$ is observed | Single-dose crossover<br>Parallel group<br>Steady-state | Distribution defined by median and range; non-normal distribution given discrete sampling scheme |

**a)** Drug refers to parent drug and/or metabolite.
**b)** Measured concentration in plasma, serum, whole blood or other relevant fluid matrix.
**c)** Current FDA guidance recommends partial area truncated at population median of $T_{max}$ for the reference formulation (at least two quantifiable samples collected before expected peak for accurate estimate).

Using log-transformed data, bioequivalence is established by showing that the 90% confidence interval of the ratio of geometric mean responses (usually AUC and $C_{max}$) of the two formulations is contained within the limits of 0.8 to 1.25 [22]. Equivalently, it could be said that bioequivalence is established if the hypothesis that the ratio of geometric means is less than or equal to 0.8 is rejected with

$\alpha = 0.05$, and the hypothesis that the ratio of geometric means is greater than or equal to 1.25 is rejected with $\alpha = 0.05$. Thus, this criterion has been termed a "two one-sided test procedure". Although either presentation is correct, the confidence interval appears to be the preferred option, presumably because of the ease of interpretation. The confidence interval criterion provides a reliable indicator of the likelihood that the true average responses of two formulations are within 20% of each other. It places no restrictions on the trial design.

The duration of studies is also a key design feature, ensuring that testing is sufficiently long enough to allow detection of differences from the reference product and to account for the intended duration of drug use in standard clinical practice. Thus, a short-duration bioequivalence trial reliant on decision rules constructed on pharmacokinetic metrics will not provide the same surrogacy of safety and efficacy for biologics as it will for small-molecule drug products. More realistic study designs given the objective to assess time-to-event (immunogenicity) can be proposed for biologics, but will require agent-specific criteria in order to be relevant [29].

## 8.6 Case Study: Low-Molecular-Weight Heparins

To illustrate the application of a conventional bioequivalence approach to a biologic product, we refer here to a two-way crossover study with the low-molecular-weight heparin (LMWH), tinzaparin [30]. The objective of the study was to examine the *in-vivo* response of two formulations of tinzaparin, both of which had been evaluated in patients. It is common practice within the pharmaceutical industry to modify formulations throughout the drug development process en route to the final market image. While it is desirable to conclude such efforts prior to initiation of clinical trials that will represent the pivotal clinical data supporting the efficacy of a drug candidate, this is often difficult to achieve in practice. Likewise, there exists a multitude of reasons for which a drug sponsor may have to bridge clinical trial results via comparability of bioavailability of two or more formulations in a sample population – essentially a bioequivalence approach. This was the setting for the tinzaparin study discussed herein. The results of this study were submitted with the NDA and deemed acceptable in the evaluation of clinical data supporting the safety and efficacy of the drug product. The "test formulation" in this study became the eventual, commercial Innohep (tinzaparin sodium) drug product.

LMWHs are heterogeneous mixtures of glycosaminoglycans derived typically from porcine mucosal spleen. LMWHs are depolymerized heparin preparations, produced by either chemical or enzymatic methods. The resulting molecules have a mean molecular mass of 4–8 kDa. More than 60% of the polysaccharides have molecular masses between 2 and 8 kDa, resulting in a reduction in thrombin-neutralizing capacity (anti-factor IIa activity). The anticoagulant properties of heparin depend primarily on the presence of specific pentasaccharide sequences with a high affinity to antithrombin. The depolymerization methods used to prepare LMWH result in destruction of some antithrombin-binding sites, thereby redu-

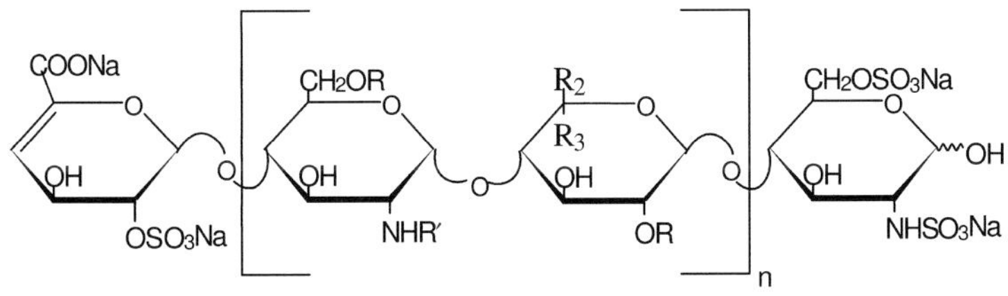


**Fig. 8.1** The chemical structure of a representative chain of tinzaparin.
n = 1 to 25, R = H or $SO_3Na$, R′ = H or $SO_3Na$ or $COCH_3$
$R_2$ = H and $R_3$ = COONa or $R_2$ = COONa and $R_3$ = H

cing the anticoagulant activity of LMWHs. Chemical modifications of the end groups and internal structure, degree of sulfation, and charge density vary from product to product and affect the characteristics of the resulting LMWH. Depending on the method of preparation, the LMWHs are different mixtures of various polysaccharides with different anti-factor Xa, anti-factor IIa activities, endothelial tissue factor pathway inhibitor (TFPI) release, differences in other cellular effects, and hence different biological actions [31, 32]. Their principal indication is centered around the treatment and prophylaxis of thrombosis, although they have numerous pharmacologic actions with great potential in varied indications, including cancer.

LMWHs contain both biologically active and biologically inactive species. Thus, pharmacodynamics (PD) determination, using anti-factor Xa activity, even in conjunction with a battery of additional biological assays, does not reflect the actual plasma levels of these multicomponent drug products. These various components also exhibit multiple biological actions, each with a distinct time course, further confounding the correlation of PD to pharmacokinetics (PK). Hence, assays developed for a single pharmacological activity more appropriately describe anticoagulant PD, but not the PK. There are also analytical problems associated with the direct measurement of a LMWH. First, the quantitative recovery of low concentrations (1–10 μg/mL) of LMWH from plasma, containing many heparin-binding proteins [33], is quite difficult. Second, analytical techniques capable of quantitatively determining each species in a LMWH have not been available. Third, detection sensitivities of the currently available analytical methods have been insufficient for PK determinations of LMWHs.

Plasma anti-factor Xa activity is an accepted surrogate for the concentration of molecules that contain the high-affinity binding site for antithrombin, while anti-factor IIa activity is correlated with the fraction of molecules that potentiate the inhibition of thrombin. US Pharmacopeia anticoagulant potency tests and amidolytic measurements of the various LWMHs have demonstrated differences in anti-Xa activities for each LMWH. However, since anti-IIa and anti-Xa activities of each LMWH do not represent the total antithrombotic and antihemorrhagic effects of the respective agents, the International Society on Thrombosis and Haemostasis recommends that vial labeling be based on weight. Labeling should also include specific anti-IIa and anti-Xa activity as assessed against the International

Standards and the recommended therapeutic dose. The antithrombotic potency and potential bleeding effects of one product cannot be extrapolated to another on the basis of weights in milligrams of anti-Xa or anti-IIa activities [34].

Tinzaparin, a sodium salt of a LMWH produced via heparinase digestion has an average molecular weight between 5500 and 7500 Da; the proportion of chains with molecular weight lower than 2000 Da is not more than 10% in the marketed tinzaparin formulation (Fig. 8.1). While this fraction is generally considered to be pharmacologically inactive, this has never been evaluated *in vivo*. The importance of the <2000 Da fraction on the anti-coagulant pharmacodynamics of tinzaparin assessed by anti-Xa and anti-IIa activity was studied in a two-way crossover trial. This comparison also reflects the desire to assess bioequivalence between formulations representing the diversity of drug product characteristics evaluated clinically during the development of tinzaparin.

In this trial, 30 healthy volunteers received a single 175 IU/kg subcutaneous dose of tinzaparin containing approximately 3.5% of the <2000 Da fraction and a tinzaparin-like LMWH containing 18.3% of the <2000 Da fraction [30]. The anti-Xa:anti-IIa ratios of the drug substances were comparable at 1.5 and 1.7 for tinzaparin and the tinzaparin-like LMWH, respectively (Table 8.4). Both formulations were safe and well tolerated, and there were no significant adverse events in subjects receiving either LMWH (Table 8.5). An important distinction between LMWHs and protein-based biologic drug products is the immunogenic potential, which is quite low for LMWHs and likewise does not present the same safety concerns or risk as for protein therapeutics.

The mean plasma anti-factor Xa and anti-factor IIa activities following single 175 IU/kg subcutaneous administration of test and reference formulations is shown graphically in Fig. 8.2 A and B, respectively. The mean maximum plasma anti-Xa activity ($A_{max}$) was approximately 0.818 IU/mL at 4 h after tinzaparin injection. Anti-Xa activity fell to undetectable levels by 24–30 h in all subjects. The mean maximum plasma anti-IIa activity was 0.308 IU/mL at 5 h post-dose, and anti-IIa activity fell to undetectable levels by 24 h in all subjects. Inter-subject var-

**Table 8.4** Formulation characteristics of test and reference LMWH (tinzaparin) formulations evaluated in a single-dose crossover (bioequivalence) design.

| **Treatment** | **Drug substance potency and MW distribution** | | | | |
|---|---|---|---|---|---|
| | **Ratio Xa/IIa** | **Average MW [Da]** | **% Da <2000** | **2000–8000** | **>8000** |
| Reference product: | | | | | |
| Tinzaparin (tinzaparin sodium) | 1.5 | 7250 | 3.5 | 65.4 | 31.1 |
| Test product: | | | | | |
| Tinzaparin-like LMWH | 1.7 | 5650 | 18.3 | 59.7 | 22.0 |

MW: molecular weight.

**Table 8.5** Observed safety profile following single 175 IU/kg subcutaneous administration of tinzaparin or a tinzaparin-like LMWH test product to healthy volunteers.

| **Preferred term** | **Tinzaparin** | | **Tinzaparin-like LMWH** | |
|---|---|---|---|---|
| **Total subjects** | **30** | | **30** | |
| | **N** | **%** | **N** | **%** |
| Subjects with one or more AEs | 16 | 53.3 | 16 | 53.3 |
| Injection-site hematoma | 10 | 33.3 | 6 | 20.0 |
| Headache | 4 | 13.3 | 3 | 10.0 |
| Dizziness | 1 | 3.3 | 4 | 13.3 |
| Injection-site pain | 2 | 6.7 | 3 | 10.0 |
| Somnolence | 0 | 0 | 3 | 10.0 |
| Diarrhea | 0 | 0 | 2 | 6.7 |
| Hematoma | 0 | 0 | 2 | 6.7 |
| Phlebitis | 2 | 6.7 | 0 | 0 |
| Abdominal pain | 0 | 0 | 1 | 3.3 |
| Asthenia | 1 | 3.3 | 0 | 0 |
| Injection-site inflammation | 1 | 3.3 | 0 | 0 |
| Nausea | 1 | 3.3 | 0 | 0 |
| Paresthesia | 1 | 3.3 | 0 | 0 |
| Pharyngitis | 1 | 3.3 | 0 | 0 |
| Rhinitis | 1 | 3.3 | 0 | 0 |
| Taste perversion | 0 | 0 | 1 | 3.3 |

AE: adverse effect.

iation was lower (<18% for both anti-Xa and anti-IIa metrics) than in previous fixed-dose administration studies. Based on average equivalence criteria, the two LMWH preparations were determined to be bioequivalent using either anti-Xa or anti-IIa activity as biomarkers (Table 8.6). Within-subject comparisons of $A_{max}$ and AUC identified no discernable pattern in either the rate ($A_{max}$) or exposure (AUC) of anti-Xa or anti-IIa activity. The calculated intra-subject variabilities were low (<14% for anti-Xa activity; <18% for anti-IIa activity), yielding little evidence for a significant subject-by-formulation interaction. Differences in the percentage of molecules in the <2000 Da molecular weight fraction of tinzaparin did not translate into differences in anti-Xa and anti-IIa activity *in vivo*.

Anti-Xa activity has been used extensively for monitoring heparin administration to patients, and remains a necessary feature of the current heparin dose-adjustment paradigm [35]. Moreover, anti-Xa and anti-IIa activity have been used successfully in the exploration of patient covariates which may suggest regimen modifications for certain disease states [35, 36] and subpopulations [37, 38]. These parameters also appear to be sufficient to judge the *in-vivo* performance of such agents. One reason for the lack of predictability of anti-Xa and anti-IIa activity

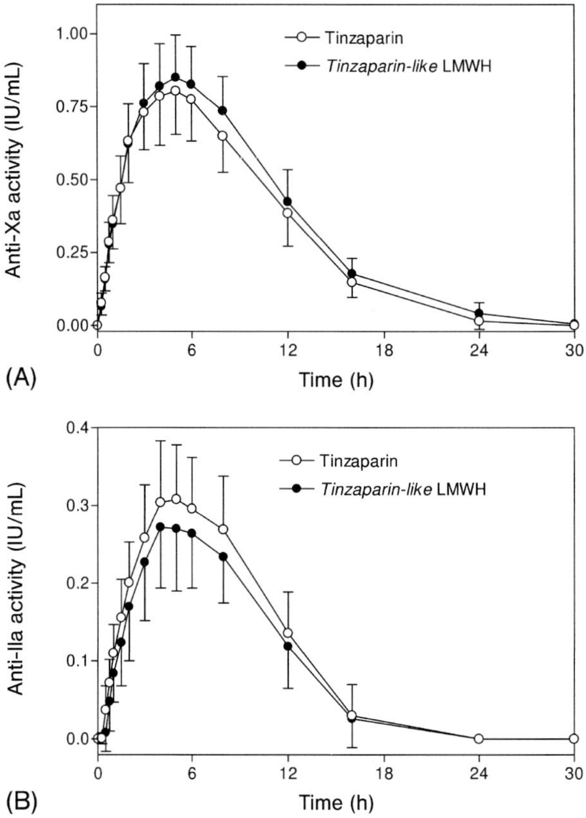


**Fig. 8.2** Mean (± SD) anti-factor Xa (A) and anti-factor IIa (B) activity in healthy volunteers (n = 30) administered 175 IU/kg tinzaparin and a tinzaparin-like LMWH by subcutaneous injection.

with clinical outcomes might be the numerous methodologies available and the inadequate sampling schemes used in some trial designs. In the absence of new mechanistic predictors, these markers have served as the best means of quantifying relevant drug product and manufacturing changes. Comparisons across agents are less straightforward, and the biomarker response from one such agent may not be portable to another with respect to outcomes that these markers may contribute to but certainly do not predict. Recent clinical data support this suggestion [39]. Indeed, it may very well be that a composite of several markers is required to make such comparisons.

**Table 8.6** Pharmacokinetic and bioequivalence metrics from a single-dose, two-way crossover study of LMWH (tinzaparin) formulations administered to healthy volunteers.

| | | Anti-Xa metrics | | Anti-IIa metrics | |
|---|---|---|---|---|---|
| | | Tinzaparin | Tinzaparin-like LMWH | Tinzaparin | Tinzaparin-like LMWH |
| $T_{max}$ (h) | Median | 5.0 | 5.0 | 5.0 | 5.0 |
| | Range | 3.0–8.0 | 3.0–8.0 | 3.0–8.0 | 3.0–8.0 |
| | %CV | 23.50 | 23.93 | 28.75 | 28.96 |
| $A_{max}$ (IU/mL) | Mean (SD) | 0.869 (0.236) | 0.887 (0.141) | 0.330 (0.073) | 0.301 (0.074) |
| | Range | 0.631–1.917 | 0.625–1.200 | 0.212–0.456 | 0.200–0.467 |
| | %CV | 27.15 | 15.84 | 22.08 | 24.51 |
| | Mean ratio[a] | 103.5 | | 90.8 | |
| | 90% CI | (97.4, 110.0) | | (84.1, 98.1) | |
| AUC(0–last) (IU h/mL) | Mean (SD) | 8.641 (1.535) | 9.768 (1.626) | 2.901 (0.581) | 2.507 (0.538) |
| | Range | 6.618–12.174 | 6.918–12.987 | 1.894–4.255 | 1.604–3.505 |
| | %CV | 17.76 | 16.65 | 20.04 | 21.47 |
| | Mean ratio[a] | 113.2 | | 86.2 | |
| | 90% CI | (108.5, 118.1) | | (81.3, 91.3) | |
| AUC (0-∞) (IU h/mL) | Mean (SD) | 9.696 (1.741) | 10.317 (1.557) | 3.787 (1.509) | 3.115 (0.652) |
| | Range | 6.892–13.764 | 7.664–13.486 | 2.138–10.720 | 2.121–4.491 |
| | %CV | 17.96 | 15.09 | 39.84 | 20.95 |
| | Mean ratio[a] | 108.6 | | 87.3 | |
| | 90% CI | (104.8, 112.5) | | (82.2, 92.7) | |
| $t_{1/2}$ (h) | Mean (SD) | 4.4 (2.8) | 4.2 (0.9) | 5.7 (7.4) | 4.3 (1.5) |
| | Range | 2.4–18.4 | 2.5–6.9 | 2.1–42.5 | 2.0–8.4 |
| | %CV | 63.65 | 22.37 | 131.29 | 35.57 |

a) The ratio is designated as test/reference with the tinzaparin formulation as the reference and the tinzaparin-like formulation as the test, respectively.

## 8.7 Conclusions

That documentation of pharmaceutical- and bio-equivalence should be provided to regulatory authorities is not at issue. However, the means by which these data can and should be demonstrated remain the subject of discussion. Political, economic and scientific hurdles pervade, and this issue remains unresolved. The imposition of existing small-molecule equivalence criteria on the registration of generic biologic drug products is unlikely to provide an acceptable degree of consumer protection. Likewise, the conventional bioequivalence trial used to infer therapeutic equivalence of different formulations based on the similarity of the pharmacokinetic characteristics of the formulations tested is likely ineffective without additional requirements. At issue is both the complexity of the drug substance characterization and the potential for immune-mediated toxicity. The characterization of the active pharmaceutical ingredient would seem to be solvable by the agreement on metrics that define the *in-vitro* performance and *in-vivo* response regardless of the complexity of analytical testing. This would seem to suggest a return to agent-specific guidance provided by the FDA, consistent with expectations for the sponsor with respect to any post-marketing formulation changes. The safety issue is more problematic as it seeks to assess the similarity (or not) of a hopefully, low-frequency adverse event. Metrics for comparison are less defined and more likely to require larger trials of longer duration than the standard single-dose, fasted bioequivalence trial. Hence, while the desire to promote generic or follow-on biologics is good and in the best interest of the patient consumer, this desire cannot prevail against assumptions that cannot be made or otherwise tested following small-molecule guidelines.

## 8.8 References


1 Nihalani P, Agrawal M, Anvekar A: Multisource biopharmaceuticals: a new perspective. *Clin. Res. Reg. Affairs* (2002) *19*: 367–380.

2 IMSHealth.com. Biogenerics: a difficult birth? http://open.imshealth.com/webshop2/IMSinclude?I_article_20040518a.asp.

3 *The Law of Biologic Medicine: Hearing Before the Subcommittee. on the Judiciary*, 108th Cong. 1 (2004) (statement of Sen. Orrin Hatch, Chairman, Subcommittee. on the Judiciary).

4 Wasson A: Taking biologics for granted? Takings, trade secrets, and off-patent biological products. Duke Law & Technology Review (2005) Rev 0004 http://www.law.duke.edu/journals/dltr/articles/2005dltr0004.html.

5 Schellekens H, Ryff JC: 'Biogenerics': the off-patent biotech products. *Trends Pharmacol. Sci.* (2002) *23*: 119–121.

6 Schellekens H: When biotech proteins go off-patent. *Trends Biotechnol.* (2004) *22*: 406–410.

7 Biopharma: Biopharmaceutical products in the U.S. market. EPO/rDNA, Amgen. Biopharma.com. Available at: http://www.biopharma.com/sample_entries/118.html.

8 Committee for Proprietary Medicinal Products: Guideline on Comparability of Medicinal Products Containing Biotechnology-derived Proteins as Active Substance. Non-clinical and Clinical Issues. Evaluation of Medicines for Human Use, The European Agency for the Evaluation of Medicinal Products, London (2003).

9 Schellekens H, Casadevall N: Immunogenicity of recombinant human proteins: causes and consequences. *J. Neurol.* (2004) *251* (Suppl. 2): ii/4–ii/9.

10 Schellekens H: Immunogenicity of therapeutic proteins: clinical implications and future prospects. *Clin. Ther.* (2002) *24* (11): 457–462.

11 Schellekens H: Biosimilar epoetins: how similar are they? *Eur. J. Hosp. Pharm.* (2004) *3*: 43–47.

12 Haselbeck A: Epoetins: differences and their relevance to immunogenicity. *Curr. Med. Res. Opin.* (2003) *19*: 430–432.

13 Boven K, Knight J, Bader F, Rossert J, Eckardt KU, Casadevall N: Epoetin associated pure red cell aplasia in patients with chronic kidney disease: solving the mystery. *Nephrol. Dial. Transplant.* (2005) *20* (Suppl 3): iii33–iii40.

**14** Schnabel CA, Fineberg SE, Kim DD: Immunogenicity of xenopeptide hormone therapies. *Peptides* (2006) *27* (7): 1902–1910.

**15** Fineberg SE, Galloway JA, Fineberg NS, Rathbun MJ, Hufferd S: Immunogenicity of recombinant human insulin. *Diabetologia* (1983) *25*: 465–469.

**16** Massa G, Vanderschueren-Lodeweyckx M, Bouillon R: Five year follow up of growth hormone antibodies in growth hormone deficient children treated with recombinant human growth hormone. *Clin. Endocrinol.* (1993) *38*(2): 137–142.

**17** Zeisel HJ, Lutz A, von Petrykowski W: Immunogenicity to mammalian cell derived recombinant growth hormone preparation during long term treatment. *Hormone Res.* (1992) *37* (Suppl 2): 47–55.

**18** Rougeot C, Marchand P, Dray F, Girard F, Job JC, Pierson M, et al.: Comparative study of biosynthetic growth hormone immunogenicity in growth hormone deficient children. *Hormone Res.* (1991) *35*(2): 78–81.

**19** Anderson PJ: Tumor necrosis factor inhibitors: clinical implications of their different immunogenicity profiles. *Semin. Arthritis Rheum.* (2005) *34* (5 Suppl 11): 19–22.

**20** Hanauer SB, Feagan BG, Lichtenstein GR, et al.: Maintenance infliximab for Crohn's disease: the ACCENT I randomized trial. *Lancet* (2002) *359*: 1541–1549.

**21** Figlin RA, Itri LM: Anti-Interferon antibodies: a perspective. *Semin. Hematol.* (1988) *25*(3 Suppl 3): 9–15.

**22** Guidance for Industry: Bioavailability and Bioequivalence Studies for Orally Administered Drug Products – General Considerations, U.S. Department of Health and Human Services, FDA (2002).

**23** Guidance for Industry: Waiver of In Vivo Bioavailability and Bioequivalence Studies for Immediate-Release Solid Oral Dosage Forms Based on a Biopharmaceutics Classification System, U.S. Department of Health and Human Services, FDA (2000).

**24** Committee for Proprietary Medicinal Products: Guideline on Comparability of Medicinal Products Containing Biotechnology-derived Proteins as Active Substance. Quality Issues. Evaluation of Medicines for Human Use, The European Agency for the Evaluation of Medicinal Products, London (2003).

**25** Committee for Proprietary Medicinal Products: Note for Guidance on Preclinical Safety Evaluation of Biotechnology-derived Pharmaceuticals. Human Medicines Evaluation Unit, The European Agency for the Evaluation of Medicinal Products, London (1997).

**26** Combe C, Tredree RL, Schellekens H: Biosimilar epoetins: an analysis based on recently implemented European medicines evaluation agency guidelines on comparability of biopharmaceutical proteins. *Pharmacotherapy* (2005) *25*(7): 954–962.

**27** Maiorella BL, Ferris R, Thomson J, et al.: Evaluation of product equivalence during process optimization for manufacture of a human IgM monoclonal antibody. *Biologicals* (1993) *21*: 197–205.

**28** Barrett JS: Bioavailability and bioequivalence studies. In: Bonate PL, Howard DR (Eds.), *Pharmacokinetics in Drug Development: Clinical Study Design and Analysis, Volume 1*. AAPS Press, Arlington, VA (2004).

**29** Patterson SD, Zariffa NM, Montague TH, Howland K: Nontraditional study designs to demonstrate average bioequivalence for highly variable drug products. *Eur. J. Clin. Pharmacol.* (2001) *57*: 663–670.

**30** Barrett JS, Kornhauser DH, Hainer JW, Gaskill J, Hua TA, Strogel P, Johansen K, van Lier JJ, Knebel W, Pieniaszek HJ: Anticoagulant pharmacodynamics of tinzaparin following 175 IU/kg subcutaneous administration to healthy volunteers. *Thromb. Res.* (2001) *101*: 243–254.

**31** Abbate R, Gori AM, Farsi A, Attanasio M, Pepe G: Monitoring of low-molecular-weight heparins in cardiovascular disease. *Am. J. Cardiol.* (1998) *82*: 33L–36L.

**32** Fareed J, Walenga JM, Hoppensteadt D, Huan X, Racanelli A: Comparative study on the in vitro and in vivo activities of seven low-molecular-weight heparins. *Haemostasis* (1988) *18* (Suppl 3): 3–15.

**33** Capila I, Linhardt RJ: Heparin-protein interactions. *Angew Chemie Int. Ed.* (2002) *41*: 390–412.

**34** Verstraete M: Pharmacotherapeutic aspects of unfractionated and low molecular weight heparins. *Drugs* (1990) *40*: 498–530.

**35** Simonneau G, Sors H, Charbonnier B, Page Y, Laaban J-P, Bosson J-L, Mottier D, Beau B: A comparison of low-molecular-weight heparin with unfractionated heparin for acute pulmonary embolism. *N. Engl. J. Med.* (1997) *337*: 663–669.

**36** Hirsch J, Warkentin TE, Raschke R, Granger C, Ohman EM, Dalen JE: Heparin and low-molecular-weight heparin: mechanisms of action, pharmacokinetics, dosing considerations, monitoring, efficacy, and safety. *Chest* (1998) *114*: 489S–510 S.

**37** Barrett JS, Gibiansky E, Hull RD, Planès A, Pentikis H, Hainer JW, Hua TA, Gastonguay M: Population pharmacodynamics in patients receiving tinzaparin for the prevention and treatment of deep vein thrombosis. *Int. J. Clin. Pharmacol. Ther.* (2001) *39*(10): 431–446.

**38** Laporte S, Mismetti P, Piquet P, Doubine S, Touchot A, Decousus H: Population pharmacokinetic of nadroparin calcium (Fraxiparine®) in children hospitalised for open heart surgery. *Eur. J. Pharm. Sci.* (1999) *8*: 119–125.

**39** Bara L, Planes A, Samama M-M: Occurrence of thrombosis and haemorrhage, relationship with anti-Xa, anti-IIa activities, and D-dimer plasma levels in patients receiving low molecular weight heparin, enoxaparin or tinzaparin, to prevent deep vein thrombosis after hip surgery. *Br. J. Haematol.* (1999) *104*: 230–240.

# 9
# Biopharmaceutical Challenges: Pulmonary Delivery of Proteins and Peptides

*Kun Cheng and Ram I. Mahato*

## 9.1
## Introduction

Over the past few decades, recombinant DNA technology has produced a large number of proteins and peptides, which can be used for the treatment of various genetic and acquired diseases [1]. Because of their large size and susceptibility to proteolytic degradation in the gastrointestinal tract, most pharmaceutical formulations of protein and peptide drugs in the market are injectable [2]. However, administration by injection is not preferable and has poor patient compliance, especially for chronic diseases requiring frequent and long-term treatment. Numerous efforts have been made to identify alternative noninvasive administrations of proteins and peptides. Among these, the pulmonary route has received a great deal of attention.

The lung as delivery site for peptide and protein drugs possesses several favorable characteristics, including a highly vascularized tissue, low intrinsic enzymatic activity, large absorptive surface (>100 $m^2$ in humans), and a greater tolerance to foreign substances and higher permeability of the alveolar epithelium compared to other administration sites. In addition, molecules absorbed in the lung bypass the portal circulation and thus avoid first-pass metabolism in the liver [3–5]. Consequently, many activities have focused on the pulmonary delivery of peptides and proteins. Recombinant human deoxyribonuclease (Pulmozyme®) was approved in 1993 by the Food and Drug Administration (FDA) as the first protein drug to be delivered via the pulmonary route for the treatment of cystic fibrosis. Formulations for the pulmonary delivery of several proteins and peptides including insulin, calcitonin, interferons, parathyroid hormone, and leuprolide are currently in clinical studies for the treatment of pulmonary and systemic diseases (Table 9.1).

Several pharmaceutical and physiological barriers must be overcome for the successful pulmonary delivery of peptide and protein drugs [3]. For example, many of these macromolecular drugs have relatively low permeability when they are administered without any absorption enhancers [4]. Furthermore, the clinical toxicology of peptides/proteins in the lung, especially for chronic disease, should be of some concern [6]. Therefore, cost-benefit ratios should be evaluated in the

*Pharmacokinetics and Pharmacodynamics of Biotech Drugs: Principles and Case Studies in Drug Development.* Edited by Bernd Meibohm

ISBN: 3-527-31408-3

**Table 9.1** Summary of peptide and proteins used for pulmonary delivery.

| Peptide/ Protein | Indication | Species | Methods | Results | Reference |
|---|---|---|---|---|---|
| Insulin | Diabetes | Human (416 type-1 diabetic patients) | Inhalation (Exubera®) | Inhaled insulin effective, well-tolerated and well-accepted | [60] |
| Insulin | Diabetes | Human (18 type-I diabetic patients) | Inhalation (AERx®) | Pharmacodynamic system efficiency 12.7%. Dose–response relationship close to linear | [64] |
| Insulin | Diabetes | Human (Healthy volunteers; type-II diabetic patients | Inhalation (Technosphere) | Relative bioavailability: 50% for the first 3 h and 30% for the entire 6-h period | [121] |
| Insulin | Diabetes | Human (26 type-II diabetic patients) | Inhalation | 3-month treatment significantly improved glycemic control compared with baseline | [122] |
| Leuprolide acetate | Treatment of prostate cancer | Human male volunteers | Aerosols | Absolute bioavailability ranging from 4% to 18% | [71] |
| Leuprolide acetate | Treatment of prostate cancer | Rat; Human male volunteer | Inhalation, intranasal administration | Successful systemic delivery via lung<br>Bioavailability of suspension aerosols four-fold greater than that of solution aerosol | [72] |
| Detirelix | Potent LHRH antagonist | Sheep | Intratracheal; inhalation | Successful systemic absorption via intratracheal administration or inhalation | [73] |
| | | Dog | Intratracheal; inhalation | Absorption from lung slow ($T_{max}$ 6.5 h; relative bioavailability 29 ± 10%) | [74] |
| Cetrorelix | Potent LHRH antagonist | Rat | Aerosol | Bioavailability ranging from 48.4 ± 27.0% to 77.4±44.0% | [77] |
| 1-Deamino-cysteine-8-D-arginine vasopressin (dDAVP) | Treatment of diabetes insipidus, Alzheimer's disease, modulation of blood pressure | Rat | Aerosol | dDAVP transport over respiratory tract via passive transepithelial transport process | [83] |
| Salmon calcitonin (SCT) | Reduce bone resorption | Rat | Intratracheal; dry powder and solution | Absorption enhancers in dry powder more efficient than in solution | [68] |

**Table 9.1** (continued)

| Peptide/ Protein | Indication | Species | Methods | Results | Reference |
|---|---|---|---|---|---|
| Alpha 1-antitrypsin ($\alpha$1A) | Treatment of $\alpha$1A deficiency | Sheep | Aerosol | Aerosolized $\alpha$1A able to pass through alveolar epithelium and gain access to interstitial compartment of lung | [107] |
| Interferon-$\alpha$ (INF-$\alpha$) | Treatment of bronchiolo-alveolar carcinoma | Human (10 patients) | Aerosol | No tumor responses detected according to standard criteria | [90] |
| Interferon--gamma (INF-$\gamma$) | Antitumor | Human (Phase I trial) | Inhalation | Inhalation increases alveolar concentration of INF-$\gamma$ without major side effects | [99] |
| Consensus interferon (Con-IFN) | Treatment of viral infections | Rat | Aerosol | $T_{max}$ 25–30 min; estimated bioavailability ca. 70% | [101] |

early stages of the development process, particularly if the drug of interest is already available in injectable form. Nevertheless, the high permeability of many peptides and proteins to the lung has opened up new avenues for the development of their pulmonary delivery systems.

In this chapter, we will discuss the biophysical basis and recent advances in pulmonary delivery of peptides and proteins.

## 9.2
## Structure and Physiology of the Pulmonary System

The pulmonary system is responsible for oxygenation of the blood and the removal of carbon dioxide from the body. The respiratory system is divided into the proximal conducting (upper) airways, and the distal respiratory (lower) tracts (Fig. 9.1). The conducting airways (generation 0~16) includes respiratory airways from the pharynx down to the bronchioles, while the distal respiratory part (generation 17~23) contains terminal bronchioles, respiratory bronchioles and alveoli (Table 9.2) [3]. There is extensive branching from a single trachea to the final 300~486 millions of alveoli sacs, as shown in Fig. 9.2 [7], which provides a surface area of over 100 $m^2$ for gas exchange between the air and blood [8].

The lung contains more than 40 different types of cells, amongst which epithelial cells are vital for maintenance of the pulmonary blood–gas barrier. The epithelium also provides absorptive and secretive functions. The diversity of epithelial cell types is summarized as airway epithelium and alveolar epithelium cells.

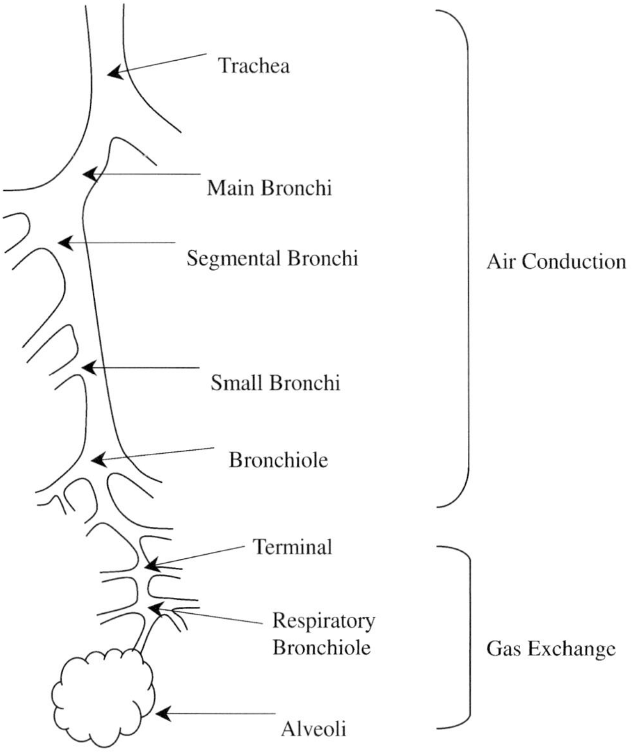


**Fig. 9.1** The "tree structure" of the lung (modified from [119] and [29]).

### 9.2.1 Airway Epithelium

The pharynx, larynx, trachea and bronchi are lined with pseudostratified, ciliated columnar epithelium that contain at least eight cell types, including mucous secretory goblet and Clara cells, which produce a protective mucus layer of 5–10 μm thickness (see Table 9.2). Subepithelial secretory glands, present in the bronchial submucosa, also contribute to the mucus blanket [9]. Through coordinated ciliary movement a propulsive wave is created, which continuously moves the mucus layer up towards the larynx. Consequently, the mucosal surface of trachea and bronchi is constantly swept to remove inhaled materials. As the bronchi divide into bronchioli, the ciliated columnar respiratory epithelium is much thinner and changes to a simpler non-ciliated cuboidal epithelium. The epithelium in the terminal and respiratory bronchioles consists of ciliated, cuboidal cells and a small number of Clara cells. However, Clara cells become the most predominant type in the most distal part of the respiratory bronchioles [10].

**Table 9.2** Airway branching and epithelium cell types in the lung (modified from [24], [119]).

| Site | | Generation | Diameter [mm] | Epithelial cell type | Thickness [μm] |
|---|---|---|---|---|---|
| Conducting airways | Trachea | 0 | 20–18 | Pseudostratified cell<br>Ciliated cell<br>Columnar cell<br>Goblet cell<br>Serous cell<br>Mucous cell | 20–40 |
| | Main bronchi | 1 | 13 | | |
| | Lobar bronchi | 2–3 | 7–5 | | |
| | Segmental bronchi | 4 | 4 | | |
| | Small bronchi | 5–11 | 3–1 | | |
| | Bronchioles, terminal bronchioles | 12–16 | 1.0–0.4 | Ciliated cuboidal cell<br>Clara cell<br>Serous cell | 10 |
| Distal respiratory airways | Respiratory bronchioles | 17–19 | | | |
| | Alveolar ducts | 20–22 | 0.3 | Alveolar type I<br>Alveolar type II | 0.1–0.5 |
| | Alveolar sacs | 23 | 0.3–0.225 | | |

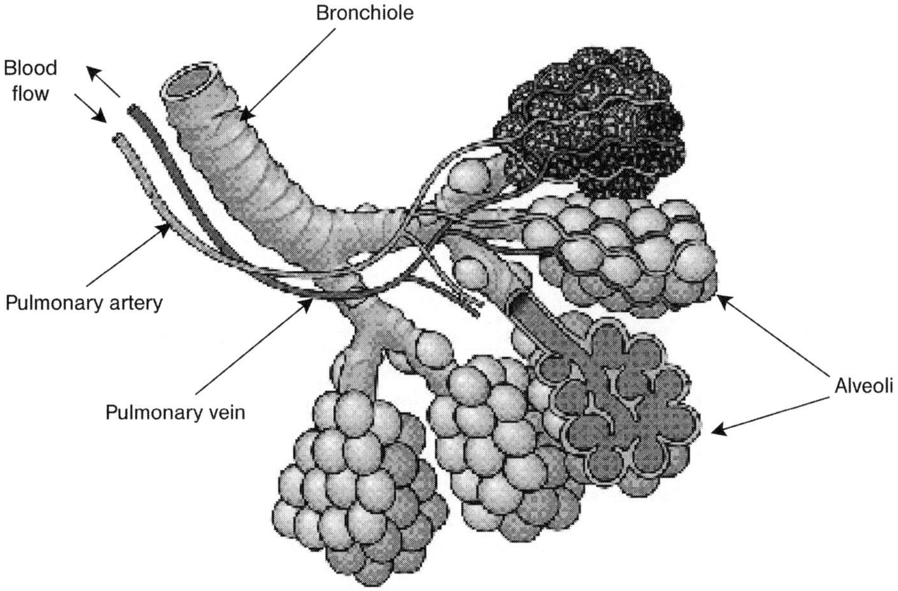


**Fig. 9.2** General structure and cross-section of alveoli.

### 9.2.2
### Alveolar Epithelium

At the distal respiratory site, the alveolar epithelial cell layer is much flatter (0.1 ~ 0.5 μm) and composed of two major cell types, squamous type I and agranular type II pneumocytes. Type I pneumocytes are non-phagocytic and highly flattened cells with broad and thin extensions. They occupy ~95% of the alveolar luminal surface, although they are less numerous than type II cells. The remaining surface is occupied by type II pneumocytes, which have blunt microvilli and contain multivesicular bodies [3, 11].

Type I pneumocytes, joined with endothelial cells by fused basement membranes, offer a very short airways–blood pathway for the diffusion of gases and drug molecules. They are known to contain numerous endocytotic vesicles which play an important role in the absorption process of proteins and transcellular movement of transporters [12, 13]. The functions of type II pneumocytes are well studied and include

- the production, secretion and reuptake of pulmonary surfactant, a mixture of phospholipids (90%) and proteins (10%) to reduce the surface tension at the air–fluid interface;
- differentiation into type I pneumocytes after epithelial barrier injuries; and
- regulation of immune responses in the lung [13–15].

Hydrophilic surfactant proteins A (SP-A) and D (SP-D), secreted by type II pneumocytes, interact specifically with a wide range of microorganisms and play important roles in the innate, natural defense system of the lung [16]. Both mRNA and protein levels of SP-A and SP-D increase dramatically in response to lung infection, injury and endotoxin challenge [17]. Type II pneumocytes also express class II major histocompatibility complex (MHC) antigens and intracellular adhesion molecule (ICAM-1), which may facilitate pulmonary immune responses [15].

## 9.3
## Barriers to Pulmonary Absorption of Peptides and Proteins

Barriers to pulmonary absorption of proteins and peptides include respiratory mucus, mucociliary clearance, pulmonary enzymes/proteases, alveolar lining layer, alveolar epithelium, basement membrane, macrophages and other cells [3, 18]. The molecular weight cutoff of tight junctions for alveolar type I cells is 0.6 nm, while endothelial junctions allow the passage of larger molecules (4–6 nm). In order to reach the bloodstream in the endothelial vasculature, proteins and peptides must cross this alveolar epithelium, the capillary endothelium, and the intervening extracellular matrix.

Alveolar epithelium and local proteases are believed to be the major barriers for the efficient absorption of inhaled proteins and peptides. Many novel and potent absorption enhancers have been investigated for the peptide/protein absorption from the lung [4]. On the other hand, Vanbever et al. [19] showed that alveolar macrophages serve as another major barrier for the systemic transport of proteins from the lung, especially for moderate to large-sized proteins. Consequently, physico-chemical means of minimizing the uptake of proteins/peptides by alveolar macrophages are proposed as novel strategies for enhancing the pulmonary absorption of these compounds [19].

## 9.4 Strategies for Pulmonary Delivery

Peptides and proteins can be delivered to the lung for localized or systemic effects by either intratracheal instillation or aerosol inhalation.

### 9.4.1 Intratracheal Instillation

Intratracheal instillation is a simple method for the direct administration of a small amount of solution into the lung by cannula. Simplicity and accuracy of the administered drug dose are the major advantages. The major disadvantages of intratracheal instillation include localized and uneven drug distribution in the lung, and more deposition occurring in the conducting airways with relatively less drug reaching the distal regions of the lung [18]. Only a small absorptive area is used for the absorption from deposition. Accordingly, bioavailability after intratracheal instillation is lower than that achieved with aerosol delivery [20, 21]. Taken together, studies using intratracheal instillation can provide information about protein stability, systemic absorption and toxicity in initial studies.

### 9.4.2 Aerosol Inhalation

Therapeutic aerosol preparations are two-phase colloidal systems consisting of very fine liquid droplets or solid particles dispersed in a gaseous medium [22]. The site of deposition of the inhaled aerosol depends on its physico-chemical characteristics such as particle size, shape, charge, density, and hygroscopicity. Among these factors, aerosol particle size is one of the most important in determining drug deposition and distribution in the lung. Aerosol particle deposition in the respiratory tract also depends upon biological factors such as lung morphology, breath-holding, inspiratory flow rate, tidal volume, and disease state [3]. The particle size distribution is usually characterized by the mass median aerodynamic diameter (MMAD) and geometric standard deviation (GSD). MMAD is a function of particle size, shape and density. Strict control of the MMAD ensures reproducibility of drug deposition and retention within desired regions of the lung [1]. A GSD >1.2 indicates a heterodisperse aerosol, while a GSD of 1 indicates a monodisperse aerosol [23]. Most therapeutic aerosols are heterodisperse.

#### 9.4.2.1 Aerosol Deposition Mechanisms

Aerosol particles deposit in the lung by three principal mechanisms: inertial impaction; gravitational sedimentation; and Brownian diffusion. Particles with a larger MMAD are deposited by the first two mechanisms, while smaller particles access the peripheral region of the lung by diffusion.

##### 9.4.2.1.1 Inertial impaction

Inertial impaction usually occurs in the first 10 generations of the lung, and is the dominant deposition mechanism for particles greater than a few microns in size [3, 23]. An aerosol particle with a large momentum will not be able to follow changes in direction of the inspired air upon inhalation into bifurcations, and therefore will be impacted onto the mucosal surfaces. Inertial impaction is the main deposition mechanism in the upper airways and near-bronchial branching points. The impaction increases with increasing breathing frequency, air flow velocity and aerosol particle size [24]. The contribution of inertial impaction is nil in the alveolar region because of negligible air velocity.

##### 9.4.2.1.2 Gravitational Sedimentation

Gravitational sedimentation is the predominant deposition mechanism in the last five to six generations of airways (e.g., the bronchioles and alveolar region), where the air velocity is very low [23]. It allows deposition of particles into the more peripheral parts of the respiratory tract. Inhaled particles of less than ~3 μm MMAD are too small to be deposited by inertial impaction on their way through the airways; instead, they will be subjected to sedimentation under gravitational forces, which overcomes the total force of the air resistance. An increase in particle size and a decrease in breathing rate will increase gravitational sedimentation [24].

##### 9.4.2.1.3 Brownian Diffusion

Deposition by diffusion is the main mechanism for particles smaller than ~0.5 μm, and is important in bronchioles, alveoli, and bronchial bifurcations. Aerosol particles are displaced by a random collision of gas molecules; this results in particle collision with the airway walls [24]. Deposition by diffusion increases with the decrease in particle size, and breath-holding following inhalation was also found to increase this deposition [25].

#### 9.4.2.2 Devices for Pulmonary Drug Delivery

Delivery devices play a major role in the efficiency of pulmonary delivery, and major advances have been made in the development of new devices in recent years. The most commonly used devices for pulmonary drug delivery include nebulizers, metered-dose inhalers (MDIs) and dry-powder inhalers (DPIs). These devices vary as much in their sophistication as they do in their effectiveness. Each type of device has its own advantages and disadvantages. The selection of a suitable aerosol device for pulmonary delivery was reviewed by Dolovich et al. [26].

#### 9.4.2.2.1 **Nebulizers**

Nebulizers convert aqueous solutions or micronized suspensions of drug into an aerosol for inhalation. They require minimal patient coordination, but are cumbersome, non-portable, and time-consuming to use [26]. There are two principal types of nebulizers, air-jet (high-velocity air stream dispersion) and ultrasonic (ultrasonic energy dispersion) nebulizers. Both air-jet and ultrasonic nebulizers produce aerosol at a constant rate regardless of the respiration cycle, which leads to loss of approximately two-thirds of aerosol during the expiration and breath-holding phases. Two improved nebulizers, such as breath-enhanced nebulizers and dosimetric nebulizers, overcome this limitation, as they direct the patient's inhaled air within the nebulizer to enhance aerosol volume during inhalation phase and release aerosol exclusively during the inhalation phase, respectively.

Other types of nebulizers rely upon compressed gas to vaporize a solution that is then available for inhalation by the patient (Fig. 9.3). The stability of proteins and peptides is a potential limitation in this case. Nebulization exerts high shear stress on these macromolecules, which can lead to their denaturation. This is a particular problem because 99% of the droplets generated are recycled back into the reservoir to be nebulized during the next dosing. Furthermore, the physical properties of drug solutions (e. g., ionic strength, viscosity, osmolarity, pH and surface tension) may affect the nebulization efficiency [26]. The droplets produced by nebulizers are rather heterogeneous, which results in very poor drug delivery to the lower respiratory tract. They often require several minutes of use to administer the desired dose of medicine. These drawbacks have led to the development of newer devices such as the AERx (Aradigm, Hayward, CA) and the Respimat

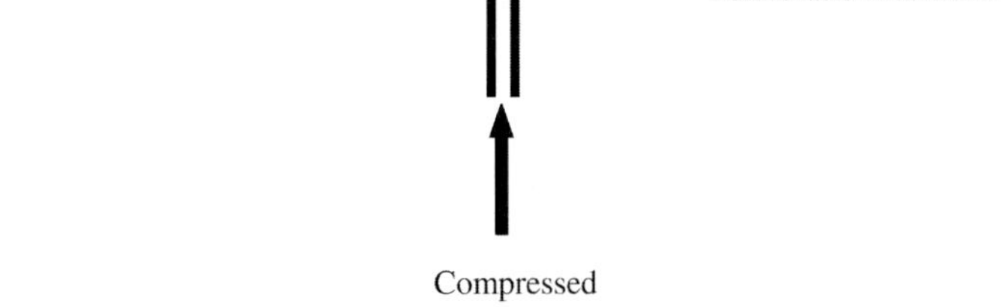


**Fig. 9.3** Schematic representation of a typical air-driven nebulizer (modified from [120]).

(Boehringer, Germany), which generate an aerosol mechanically, while vibrating mesh technologies such as AeroDose (Aerogen Inc, Mountain View, CA) have been used successfully to deliver proteins to the lungs. The recent introduction of recombinant human DNase α (rhDNase Pulmozyme) by Genentech (San Francisco, CA, USA) exemplifies the application of nebulizers for peptide and protein delivery to the respiratory tract [27]. rhDNase reduces the viscosity of airway secretions by cleaving the extracellular fibrillar aggregates of DNA from autolyzing neutrophils in cystic fibrosis cases.

#### 9.4.2.2.2 Metered Dose Inhalers

Metered dose inhalers (MDIs) generate aerosol for inhalation by expelling a metered dose of pressurized liquid propellant containing drug via an orifice in the proper particle size distribution. They are the most commonly used inhalation aerosol devices today (Fig. 9.4) [22, 28], and are portable and easy to use [26]. A typical MDI comprises a canister, a metering valve, actuator, spacer and a holding chamber. In addition, dose counters and content indicators are required in recently issued FDA guidelines. During MDI manufacture, more aerosol formulation than claimed is commonly added which is sufficient for additional 20 to 30 sprays. However, these additional doses are inconsistent and unpredictable. Therefore, the new requirement by FDA will allow patients to track the number of actuations used and avoid using the product beyond the recommended number of doses [28].

MDIs utilize propellants, such as chlorofluorocarbons (CFC) and hydrofluoroalkanes (HFAs) to emit the drug solution through a nozzle [29]. High velocity of the generated aerosol spray results in substantial oropharyngeal deposition by impaction, and therefore loss of drug. This can be avoided by adding a spacer device to reduce the aerosol velocity. Spacers can also overcome difficulties in the coordination of inhalation and actuation, leading to improved dosing reproducibility. Individual doses are measured volumetrically by a metering chamber within the valve.

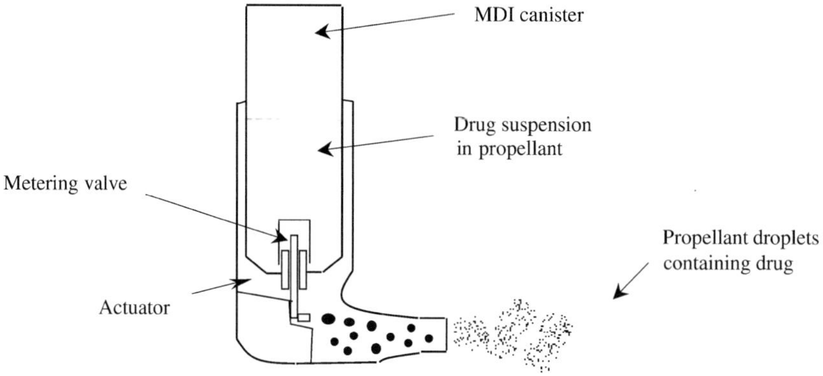


**Fig. 9.4** Schematic representation of a typical metered-dose inhaler (modified from [29]).

MDI delivery efficiency depends on the patient's inspiratory flow rate, breathing pattern and hand-mouth coordination. Increases in tidal volume and decreases in respiratory frequency enhance the peripheral deposition in the lung. Most patients need to be trained to use the MDI correctly, as up to 70% of patients fail to do so [26, 30].

#### 9.4.2.2.3 **Powder Inhalers**

Dry powder inhalers (DPIs) are one of the most popular methods of protein delivery to the lungs. DPIs generate aerosols by drawing air through loose dry powder drugs. Compared to MDIs, they are easier to use, but they require a rapid rate of inhalation to provide the necessary energy for aerosolization, which may be difficult for pediatric or distressed patients [26, 30]. DPIs range from unit dose systems employing only the patient's breath to generate the aerosol, to multiple-dosing reservoir devices that actively impart energy to the powder bed to introduce aerosol particles into the patient's respiratory airflow. For stability reasons, unit-dose devices are most suitable for protein delivery. The schematic design of a new powder inhaler (Novolizer™) is shown in Figure 9.5.

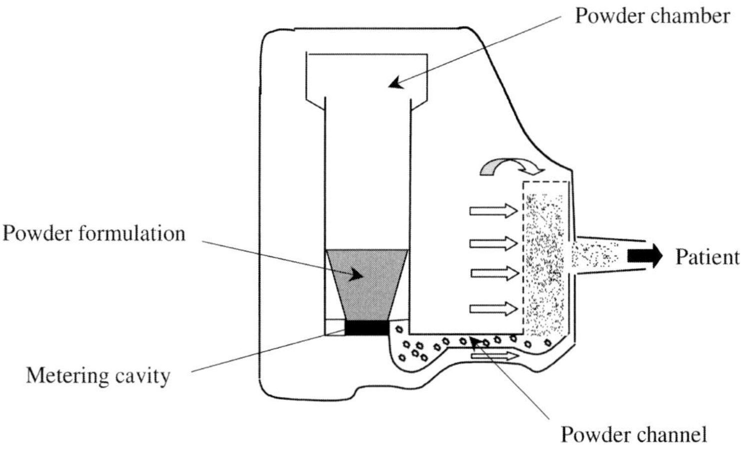


**Fig. 9.5** Schematic representation of a typical dry powder inhaler (modified from the market product Novolizer™).

Lung deposition of drug particles varies among different DPIs. DPIs are complex systems, and their performance depends on the powder de-agglomeration principle, dry powder formulation and the airflow generated by the patient [31]. Both improvements in the device and formulations have been attempted to achieve superior lung deposition. Carrier particles, such as lactose, are commonly added to reduce cohesive forces in the micronized drug powder. When air is directed through the powder, turbulent airflow detaches small drug particles from the carrier particles [3]. Most of the therapeutic dry powders for DPIs are currently made with particles of small geometric diameter (<5 μm) and mass density of 1 ± 0.5 g/cm$^3$. However, a new type of large porous particle, with a geometric diameter >5 μm and a low mass density (<0.4 g/cm$^3$) has been developed for DPIs and showed highly efficient drug delivery into deep lung [32]. Large geometric diameters of the particles reduce inter-particle interactive forces and ease their dispersion.

## 9.5 Experimental Models

An understanding of the absorption and metabolism of peptide and protein drugs in the lung is crucial for the development of an effective formulation for pulmonary delivery. Basically, the absorption and metabolism of drugs in the lung can be investigated with three models: isolated perfused lung models; cell culture models; and *in-vivo* animal models. Isolated perfused lung and cell culture models are simple to establish, but it is difficult to extrapolate results directly to *in-vivo* physiology [33]. Intact lung models most closely resemble the *in-vivo* situation, but do not allow distinction between permeation barriers presented by the alveolar epithelium and other tissues of the lung as a result of the models' complexity. Furthermore, the characterization of alveolar epithelial transport mechanisms in the intact lung is rendered difficult by inaccessibility of the distal airways for precise dosing and sampling, along with the unknown experimental surface area.

### 9.5.1 Isolated Perfused Lung Model

Isolated perfused lung models have been widely used to study the absorption, metabolism, and clearance of drugs administered to the lung [33–35]. The most commonly used species are rat, guinea pig, and rabbit. Briefly, the pulmonary artery and vein are cannulated and the lung is removed from the chest cavity, and maintained in an artificial thorax. This allows for ventilation of the lung as well as for pulmonary administration of drug preparations. The lung is perfused throughout the experiment, and the perfusate can be sampled at predetermined time points. The main advantage of the isolated perfused lung model is that pulmonary absorption and metabolism of a drug can be investigated without the influence of its systemic elimination and metabolism. Furthermore, the experimental conditions are easily controlled and the structural and cellular integrity of the lung is maintained [34].

### 9.5.2 Cell Culture Models

The lung comprises about 40 different cell types, amongst which type I and type II alveolar epithelial cells are the major types targeted by pulmonary drug delivery systems. Type I cells play an important role in the absorption process of proteins, while type II cells produce surfactant, regulate the immune response, and serve as a progenitor of type I cells [12, 13]. Many cell culture systems of alveolar epithelial cells have been developed, providing a simple way in which to study the transport and metabolism mechanisms of a drug at cellular levels. For a review on respiratory epithelial cell culture models, see Mathias et al. [36].

Cell lines of both airway and alveolar epithelial origin are available (immortal cells, transformed or carcinoma-derived), the most commonly used being the CALU-3 cell line derived from airway epithelium [37], and the A549 cell line derived from a bronchioalveolar carcinoma [38]. However, these cultured cell lines share only limited similarity in morphology, biochemical characteristics, and barrier properties (low epithelial resistance due to the lack of tight junctions) with the epithelia *in vivo*. Consequently, the results of absorption and metabolism studies based on cultured cell lines should be interpreted with caution.

The primary cell culture model is a more valid model for the study of absorption and transport processes of a drug via the pulmonary route. It provides a tight epithelial barrier with morphological and functional properties resembling those of the *in-vivo* condition. Primary alveolar epithelial cells from rats [39], rabbits [40] and humans [41] which display morphological and biochemical characteristics similar to the native epithelium have been isolated and can be used for drug transport studies.

## 9.6 Pulmonary Delivery of Peptides and Proteins

### 9.6.1 Mechanism of Peptide Absorption after Pulmonary Delivery

Several peptides of 0.5–10 kDa have been administered to the lungs, including a number of hormones, particularly metabolic hormones such as glucagon, insulin, and calcitonin. To improve absorption by minimizing metabolic breakdown, in some cases analogues of the native peptides have been designed. The co-administration of proteolytic enzyme inhibitors can also be utilized to reduce enzymatic degradation of peptides upon their exposure to the lung.

There is growing interest in studies of the transport of small peptides across lung epithelia, and critical insights regarding pulmonary peptide delivery were obtained by the administration of model peptides to the lung. The transport of dipeptides and tripeptides in the lung has been studied using *in-situ* perfusion preparations of rat lung, brush-border membrane vesicles prepared from alveolar type II cells [42, 43], as well as in monolayers of primary cultured rat pneumocytes [44]. In all cases, evidence for active transport of intact peptide through the pulmonary epithelium was found, although the reported contribution of passive diffusion to overall transport, as well as metabolic cleavage of the compounds, was found to vary with the compounds used and the models used to study them.

The co-transport of di- and tripeptides with protons has been well studied in the small intestine and kidneys, and showed enhanced absorption of peptides by the imposition of an inwardly directed proton gradient. The phenomenon was also observed in the delivery of a model peptide (D-phe-L-Ala) in the lung and absorption of peptides into lung brush-border membrane vesicles was found increased by an inwardly directed proton gradient. This suggests that the transporter expressed in the lung is similar to that of the kidneys and small intestine [43]. When the transport of two model peptides, Gly-D-Phe and Gly-L-Phe, across primary cultured monolayers of rat alveolar epithelial cells was investigated, both peptides were transported in cell culture through paracellular pathways, while Gly-L-Phe also showed a small component of transcellular pathway. The extent of degradation of peptides during transport appears to be peptide-dependent: peptides containing amino acids in L-conformation are less resistant to peptidases present at the cell membrane than their D-conformers [44]. The substitution of D- for L-amino acids in peptides of interest may increase stability, but may also decrease affinity of the peptide towards its putative epithelial transporter.

To investigate the effect of molecular weight, concentration and dose of peptides on their pulmonary absorption, Byron et al. [45, 46] have studied a series of synthetic peptidase-resistant poly-αβ-DL-aspartamides (PHEA), ranging from 4 to 43 kDa molecular weight, in the isolated rat lung model. Aqueous solutions of these polymers were administered to the isolated perfused rat lung and transfer to the perfusate was measured. Polymer transfer rates were dependent on the starting molecular weight distribution, with larger molecules (up to 11.65 kDa) being absorbed more slowly. Dose-ranging studies to investigate concentration dependency indicated the presence of a saturable, carrier-mediated transport process for PHEA with a $V_{max}$ of approximately 180 μg/h. For the lower molecular-weight polypeptides (MW 3.98 kDa), the molecular size of the administered material was the same before and after absorption into the perfusate. However, for F-PHEA with a MW of 7.2 kDa the species transferred to the perfusate appeared to be smaller molecules than those originally administered, indicating a molecular weight limit of rapid transport of about 7 kDa for this hydrophilic polymer.

Hoover et al. [47] assessed the absorption of a series of model D-phenylalanine analogues that were resistant towards enzymatic hydrolysis, after intrapulmonary administration of aerosolized peptide solution into the rat lung. Compared to oral administration, all peptides showed better absorption from the lung than from the gut, with pulmonary absorption ranging from 55.1 to 68.5% for the methylated series of these model peptides. However, some members of the non-methylated peptide series were metabolized during the absorption process, which was not observed in the intestinal absorption studies. Therefore, the advantage of pulmonary over oral administration must be examined on a case-by-case basis.

### 9.6.2
### Mechanisms of Protein Absorption after Pulmonary Delivery

It is generally agreed that proteins can cross the alveolar–capillary barrier of the lung, but the quantitative importance and mechanisms of uptake via this route are not clear. Several mechanisms have been proposed for protein transport in the lung, including paracellular diffusion and transcytosis as the two most likely mechanisms. Several investigators have addressed these issues by studying the transport characteristics of albumin, IgG and horseradish peroxidase (HRP).

Albumin is widely used for the investigation of protein transport mechanisms in the lung. Absorption studies of intact albumin with rat primary epithelial cell cultures showed that transport across the cell monolayer is asymmetric (net absorption), saturable, and highly sensitive to temperature. Transport of albumin occurs by two mechanisms, depending upon concentration. At low concentration, the absorption of albumin across the epithelial barrier appears to be mediated by a ~60 kDa albumin glycoprotein binding protein (gp60) which is expressed in the alveolar epithelial cell membranes [48, 49]. At high concentration, however, albumin absorption is proportional to its concentration without saturation and is insensitive to endocytosis inhibition, thus indicating a passive paracellular mechanism [50].

As a nonspecific fluid-phase endocytosis marker, HRP was used to evaluate the transport characteristics across rat alveolar epithelia cell monolayers. HRP was transported relatively intact (about 50%) across the alveolar barrier via nonspecific fluid-phase endocytosis. The permeability coefficient of HRP was decreased upon lowering the temperature [51], but after conjugation with transferrin via a disulfide linkage, HRP uptake by alveolar cell monolayers was significantly increased. Receptor-mediated internalization of the conjugated HRP was verified by competition for the transferrin receptor [52].

### 9.6.3 Pulmonary Delivery of Peptides and Proteins

#### 9.6.3.1 Insulin

Since the first introduction of insulin to treat diabetic patients in 1923, much effort has been made to seek alternative convenient and painless routes for insulin administration instead of daily injections. In this respect the pulmonary route has received the most attention, and substantial evidence has shown inhaled insulin to be an effective, well-tolerated, noninvasive alternative route [53–56]. Insulin therapy is required for patients with type 1 diabetes. Although some patients with type 2 diabetes can control their disease with oral antidiabetics, many will eventually also require insulin. Thus, inhaled insulin shows promise for type 2 diabetic patients [54, 56]. There are two principal inhalation systems for insulin, namely aqueous solution and dry powder. The dry powder form (Exubera®) has been approved by FDA and the European Medicines Agency (EMEA) in January 2006.

##### 9.6.3.1.1 Animal Studies

Compared with insulin aqueous solution, low-viscosity insulin containing hyaluronate (0.1–0.2%) greatly enhanced the pharmacological availability of insulin via pulmonary delivery routes to rats [57]. Morimoto et al. [57] subsequently examined the effects of intratracheal administration of different concentrations and pH values of low-viscosity solutions of hyaluronate on the pulmonary absorption of rh-insulin in rats. Hyaluronate (2140 kDa) solutions (0.1% and 0.2%, w/v) at pH 7.0 significantly enhanced the bioavailability of insulin compared to the aqueous solution of insulin at pH 7.0. It was concluded that a hyaluronate preparation of low viscosity could serve as a useful vehicle for the pulmonary delivery of peptide drugs.

Chen et al. [58] studied the hypoglycemic efficacy of pulmonary delivery of insulin in dry powder aerosol form. Insulin dry powder was insufflated in rat lung through an incision in the trachea in test animals, while subcutaneous (SC) insulin injections were administered to controls. The areas above the curve of the blood glucose concentrations at 7 h after insulin administration were used to evaluate insulin efficacy. The percentage minimum blood glucose levels, compared to glucose levels before insulin administration for pulmonary delivery of insulin at doses of 20, 10, 5, and 2.5 U/kg, were 6.5, 16.6, 24.6 and 57.0%, respectively. The area under the curve (AUC) of insulin 5 U/kg by the pulmonary delivery was very close to that for SC administration at the same dose level. A linear relationship between the AUC and the logarithmic dose of pulmonary-delivered insulin was observed, and pulmonary delivery of insulin was concluded to be effective.

In order to overcome the inherent problem associated with pulmonary aqueous solution and dry powder aerosols, Choi et al. [2] developed an ethanol suspension of insulin for inhalation, in which the solid insulin is suspended in ethanol and aerosolized with a commercial compressor nebulizer. The aerosol insulin particles were found to be 1.5 μm, with a geometric standard deviation of 1.3 μm. Exposure of rats to 10 mg/mL insulin aerosol resulted in a drastic fall in blood glucose and a marked rise in serum insulin level. The bioavailability of insulin/ethanol aerosol was 33% relative to SC injection, and comparable to that of insulin aerosols in aqueous solution and dry powder form. No acute toxic effects were detected in the rat lungs or airways [2].

#### 9.6.3.1.2 **Human Studies**

Laube et al. conducted a single-blind, non-randomized, placebo-controlled pilot study consisting of seven type 2 diabetic patients using aerosol administration. The results showed average plasma glucose and insulin levels separately for study patients after inhalation of placebo and insulin aerosols during the fasting state. Both glucose and insulin levels measured after insulin inhalation were significantly different from levels measured after placebo inhalation ($p < 0.05$). These findings suggest that postprandial glucose levels can be maintained below diabetic levels by delivering 1.5 U/kg insulin through the lungs 5 min before the ingestion of a meal [53].

New formulations for inhaled insulin are currently under development by several companies, and are at various stages of clinical trials. Among these, AERx® is in advanced Phase III trials [59], while Exubera® has been approved in the USA and Europe. The clinical pharmacokinetics and pharmacodynamics of inhaled insulin were recently reviewed by Eldon et al. [59].

Nektar Therapeutics, in collaboration with Pfizer Inc. and Aventis SA, developed Exubera, which is an inhaled, rapid-acting, dry powder insulin for type 1 and type 2 diabetes. In a large-scale Phase III study with Exubera [60], 416 subjects with type 1 diabetes were screened at 41 centers across the USA and Canada. This was an open-label, 24-week, parallel group and multicenter study. Inhaled insulin was administered as a dry powder formulation before meals, in combination with a single bedtime injection of long-acting insulin. As shown in Figure 9.6, inhaled insulin provided glycemic control comparable with that obtained with subcutaneously injected insulin, which is in accordance with previous Phase II studies [61]. Treatment satisfaction for the insulin inhalation group was much higher than for the control group. Mild to moderate cough was observed with high-frequency inhaled insulin, but the incidence of this decreased as treatment progressed [60].

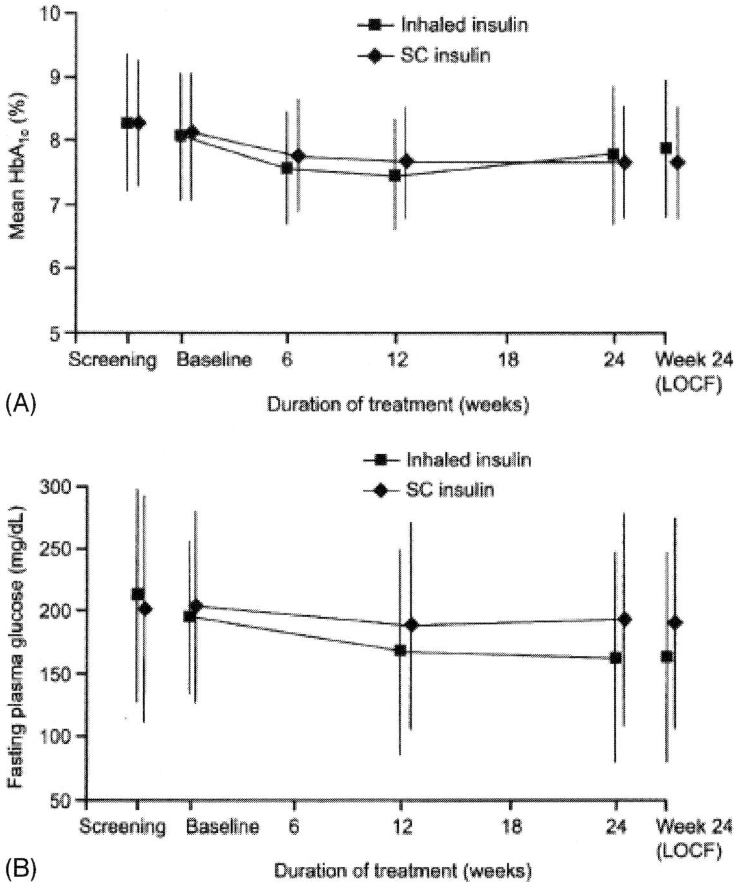


**Fig. 9.6** (A) Hemoglobin A1 c ($HbA_{1c}$) levels in the inhaled and subcutaneous (SC) insulin groups. (B) Mean change in fasting plasma glucose concentration in the inhaled and SC insulin groups. (Reproduced with permission from [60]).

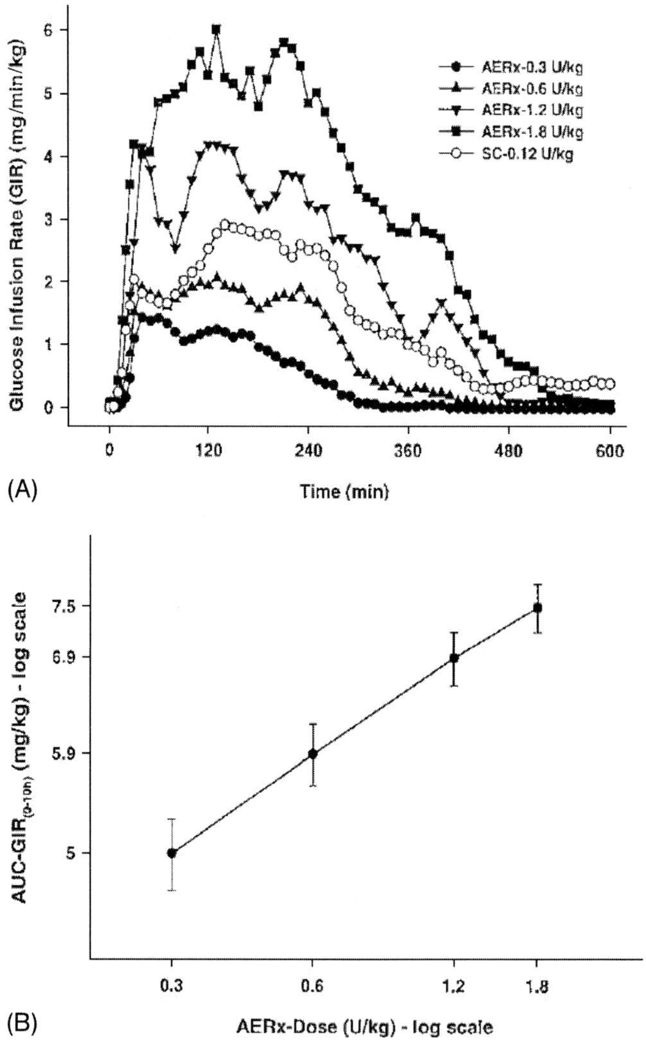


**Fig. 9.7** Dose–response of AERx iDMS inhaled and injected insulin. (A) Glucose infusion rates (GIR) over 10 h after administration of the different treatments. (B) Dose–response relationship of the AUC-GIR$_{0-10\,h}$ for inhaled insulin; vertical bars indicate 95% confidence intervals. (Reproduced with permission from [64]).

AERx was developed by Aradigm Corporation in collaboration with Novo Nordisk A/S. It is a liquid insulin formulation which can create aerosols of 1–3 μm particle diameter [62]. In a study involving 23 healthy volunteers, AERx was used to deliver aqueous insulin aerosols to the lungs at two concentrations (250 U/mL and 500 U/mL), and was compared with a SC injection of insulin solution. The results showed the absorption of insulin to be more rapid after pulmonary dosing than after SC injection [63]. Furthermore, AERx was studied in a randomized, open-label, five-period crossover trial which included 18 type I diabetic patients. Insulin was inhaled via an AERx insulin Diabetes Management System (AERx iDMS) at different doses (0.3, 0.6, 1.2, and 1.8 U/kg body weight). Inhaled insulin provided an almost dose–linear relationship for pharmacokinetic and pharmacodynamic parameters (Fig. 9.7), and the pharmacodynamic system efficiency of inhaled insulin was 12.7% [64].

The effect of pulmonary disease on the absorption of inhaled insulin (AERx iDMS) was also investigated with a two-part, open-label trial including 28 healthy and 17 asthmatic subjects. The AUC of inhaled insulin was significantly higher for healthy subjects than for asthmatics, though there was no difference in maximum serum insulin levels. The asthmatic group also had higher intrasubject variations in pharmacokinetic parameters than the healthy group. The results suggested that diabetic patients with asthma may need to inhale higher doses than patients with normal respiratory function [65]. A Phase III study of AERx inhaled insulin was initiated in 2002 with 300 patients. However, in 2003 the FDA adopted new GMP guidelines for the sterile production of inhalation products, and consequently the Aradigm Corporation needed to optimize their devices and repeat the Phase III study [62].

In conclusion, the pulmonary delivery of insulin offers an efficient and convenient therapy for diabetic patients. The feasibility of inhaled insulin is based mainly on the lungs' large absorption area of alveoli and their extremely thin walls full of intercellular spaces that make them more permeable than other mucosal sites to large proteins. Generally, inhaled insulin showed a more rapid absorption than insulin administered by SC injection [59]. One major concern for pulmonary insulin delivery is the unknown long-term effects of inhaled insulin within the respiratory tract. Thus, possible long-term problems should be considered when insulin is administered in this manner [66].

#### 9.6.3.2 **Salmon Calcitonin**

Salmon calcitonin (SCT) is a synthetic 32-amino acid peptide used for the treatment of osteoporosis and hypercalcemia, generally in the form of a SC injection. Metabolic degradation and low permeability are thought to be possible barriers to the absorption of SCT. In fact, SCT-degrading enzyme activity was found to be higher on the membrane surfaces of the lungs than in the cytosol fraction; the respective enzymes are predicted to be serine proteases and metalloenzymes based on the *in-vitro* action profile of protease inhibitors. The inactivation of SCT mediated by membrane enzymes is suggested from a correlation between the *in-vitro* activity of protease inhibitors and the *in-vivo* effect of SCT [67]. Absorption enhancement was observed following intratracheal co-administration of SCT with protease inhibitors and absorption enhancers in rats using plasma calcium levels as an efficacy index. Absorption enhancers such as unsaturated fatty acids (e.g., oleic acid and polyoxyethylene oleyl ether) and protease inhibitors, namely chymostatin, bacitracin, potato carboxypeptidase inhibitors and phosphoramidon, resulted in considerable enhancements of absorption [67]. These results are in good agreement with the findings of Niven and Byron [46], that the pulmonary absorption enhancement was in the order oleic acid > sorbitan trioleate = oleyl alcohol in the isolated perfused lung of rats.

Consequently, absorption enhancers were used in dry powder and liquid formulations to enhance the pulmonary absorption of SCT. Without absorption enhancers, SCT absorption from dry powder or solution was similar to that observed after intratracheal administration. However, the absorption was more improved from dry powder than from solution when absorption enhancers (oleic acid, lecithin, citric acid, taurocholic acid, dimethyl-β-cyclodextrin, octyl-β-D-glucoside) were co-administered intratracheally. Such improved absorption could be due to the fact that enhancers added to the dry powder dissolved at high concentration because only a trace volume of fluid lining the alveolar epithelium was available for their dissolution. However, the potential implications of such a mechanism on lung toxicity, especially in lung edema, is yet to be investigated in detail [68].

Deftos et al. [69] evaluated the pulmonary route for the delivery of SCT in 10 normal male volunteers with dry powder inhaler and compared it with intramuscular (IM) administration of SCT. The bioavailability of SCT administered via the pulmonary route was found to be only 28% of that administered by the IM route (Fig. 9.8). However, the bioactivity (monitored by the changes in serum calcium level) of SCT administered by the pulmonary route was 66% of that given by the IM route. In this case, bioactivity was deemed to be more clinically relevant than bioavailability, the discrepancy most likely being due to immunoassay-dependent deviations in the measurement of SCT concentrations.

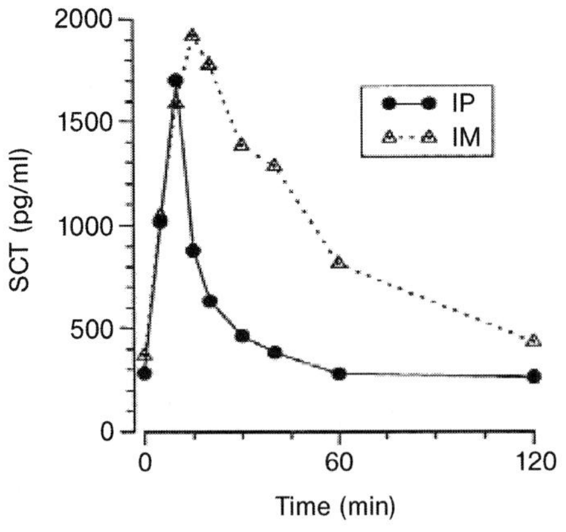


**Fig. 9.8** Serum salmon calcitonin (SCT) concentrations in 10 normal males following intrapulmonary (IP) and intramuscular (IM) administration of SCT. (Reproduced with permission from [69]).

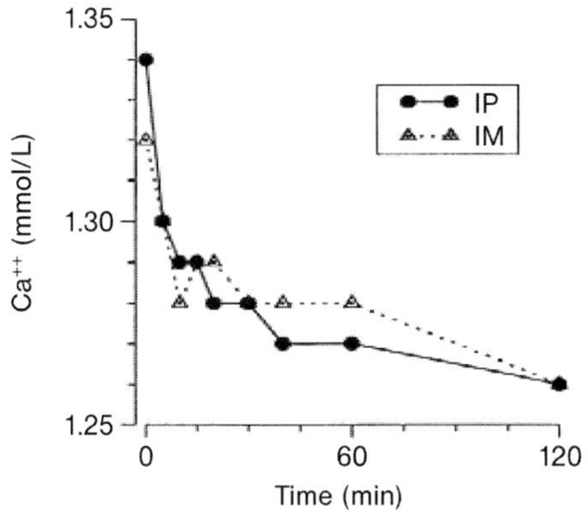


**Fig. 9.9** Ionized calcium ($Ca^{2+}$) concentrations in 10 normal males following intrapulmonary (IP) and intramuscular (IM) administration of SCT. (Reproduced with permission from [69]).

#### 9.6.3.3 Luteinizing Hormone-Releasing Hormone (LHRH) Agonists/Antagonists

LHRH is a decapeptide hormone (pGlu-His-Trp-Ser-Tyr-Gly-Leu-Arg-Pro-Gly-$NH_2$) which is synthesized in the hypothalamus and induces the release of gonadotropins, luteinizing hormone (LH) and follicle-stimulating hormone (FSH) into the systemic circulation. LHRH agonists are commonly used to treat prostate cancer, while antagonists of LHRH are useful for many indications in men and women, such as breast and prostate cancer.

##### 9.6.3.3.1 Leuprolide (LHRH Agonist)

Leuprolide acetate, a highly potent synthetic analogue of LHRH, is a nonapeptide (5-oxo-Pro-His-Trp-Ser-Tyr-D-Leu-Leu-Arg-Pro-ethylamide acetate) with a molecular weight of ~1200 Da. Leuprolide acetate has shown promise for the treatment of infertility, postmenopausal breast cancer, and prostate cancer. Very low oral bioavailability of leuprolide acetate has led to an interest in using the lung as a site for the systemic delivery of leuprolide. Okada et al. [70] have shown that a mixed micellar solution of leuprolide acetate has only 0.05% oral bioavailability compared to IV leuprolide acetate. This low oral bioavailability may be attributed to poor membrane permeability as well as to significant enzymatic deactivation in the intestine.

Adjei and Garren [71] studied the pulmonary bioavailability of aerosol-administered leuprolide acetate in dogs, and found significant plasma levels and a concentration-corresponding decrease in plasma gonadotropins. Studies in human males [72] showed the mean bioavailability of a leuprolide suspension aerosol (~26% compared to IV administration) to be four-fold greater than that of a solution aerosol (6.6%). The absolute bioavailability of leuprolide in healthy human male volunteers ranged from 4% to 18%, while bioavailability corrected for respirable fraction ranged from 35% to 55%; these data showed that the pulmonary route might have high potential for the systemic delivery of leuprolide [71].

#### 9.6.3.3.2 Detirelix (LHRH Antagonist)

Detirelix, a highly potent LHRH antagonist, is an amphipathic decapeptide with a molecular weight of 1540 Da. Detirelix was absorbed systemically when administered by intratracheal instillation or as aerosol in un-anesthetized awake sheep, with the average bioavailability being about 10% [73]. The administration of detirelix-containing aerosols resulted in similar pharmacokinetic profiles as observed for intratracheal instillation in anesthetized dogs [74], which indicated that the absorption was a rate-limiting process. The absorption of detirelix from the lung after instillation was slow ($T_{max}$ 6 ± 3.6 h), with a relative bioavailability of 29 ± 10%. Furthermore, histopathologic examination showed the lungs to be normal during a five-month period of repeated pulmonary administration.

#### 9.6.3.3.3 Cetrorelix (LHRH Antagonist)

Cetrorelix is a synthetic hormone that blocks the effects of gonadotropin-releasing hormone (GnRH). Cetrorelix was found to be effective in inhibiting LH release *in vivo*; it blocks ovulation in cycling rats, and suppresses LH levels in ovariectomized rats [75]. Thus, it is highly efficacious for the treatment of gynecological disorders and LH-sensitive tumors [76]. Lizio et al. [77] studied the pulmonary absorption of cetrorelix acetate in rats using a non-surgical intratracheal instillation method. After administration, the drug was absorbed rapidly, with testosterone plasma concentrations decreasing to subnormal levels at 24, 34, and 72 h, respectively. The pharmacokinetic data of cetrorelix were also investigated at two different doses (0.5 and 1.0 mg/kg body weight). Compared to IV administration, the bioavailability of intratracheal instillation was 75.8 ± 45.4% and 59.0 ± 18.3% at doses of 0.5 and 1.0 mg/kg, respectively. These data proved that cetrorelix can be absorbed from the lungs, and that its subsequent pharmacological activity was maintained.

Later, Lizio et al. [78] used a new aerosol delivery system (ASTA-ADS) to investigate the pulmonary absorption and tolerability of four different cetrorelix formulations delivered as nebulized aerosols to orotracheally cannulated rats. After only 5 min exposure to the cetrorelix aerosol, serum testosterone concentrations were reduced to subnormal levels over a 24-h period. After dose adjustment (dose delivered minus exhaled amount), the bioavailabilities for pulmonary delivery ranged from 48.4 ± 27.0% to 77.4 ± 44.0% compared to IV administration. In addition, the lung function parameters did not reveal any formulation-related changes. Overall, the results of cetrorelix aerosol administration compared well with those obtained with intratracheal instillation of cetrorelix solution [77].

#### 9.6.3.4 Vasopressin

Vasopressin is a nonapeptide used in the treatment of diabetes insipidus [79], of Alzheimer's disease [80], and in the modulation of blood pressure [81]. Vasopressin is known to be absorbed very poorly after oral administration due to its degradation by proteolytic enzymes. Yamahara et al. [82] have studied the transepithelial transport of arginine vasopressin (AVP) across cultured rat alveolar epithelial cell monolayers. The results showed that AVP is transported via a passive diffusional pathway, and also undergoes proteolytic degradation on the apical surface of the alveolar epithelium, most likely by aminopeptidases [82]. Thus, the co-administration of appropriate inhibitors of aminopeptidases might increase the overall transport of intact AVP across the alveolar epithelial barrier.

Folkesson et al. [83] studied the passage of a nonapeptide, 1-deaminocysteine-8-D-arginine vasopressin (dDAVP, MW 1067 Da) after inhalation in both healthy and lung-injured rats. The dDAVP serum levels peaked at 0.5–1 h, with a total recovery (compared to the dose administered) of 84.3 ± 12.9% that increased linearly with time, suggesting a transepithelial transport process for dDAVP. The high bioavailability of dDAVP is likely due to passive diffusion utilizing the large surface area available in the distal respiratory tract of the mammalian lung. The passage of dDAVP through the lung appeared to increase with the maturity of the rats, and was largely unaffected by inflammatory changes in the lung [83].

#### 9.6.3.5 Granulocyte Colony-Stimulating Factor (G-CSF)

G-CSF is a glycoprotein which stimulates the survival, proliferation, differentiation, and function of neutrophil granulocyte progenitor cells and mature neutrophils. It plays an important physiological regulatory role in the host response to infection. Recombinant human G-CSF (rhG-CSF) is derived from *Escherichia coli* (MW 18.8 kDa). G-CSF can be delivered in therapeutically or prophylactically effective quantities by pulmonary administration using a variety of delivery devices, including nebulizers, MDIs and powder inhalers [21]. Since the aerosol administration of G-CSF results in a significant elevation of neutrophil levels, this route can be utilized to treat neutropenia, as well as to combat or prevent infections.

Following intratracheal instillation in hamsters, G-CSF was absorbed rapidly and a significant increase in white blood cell count observed. The bioavailability was estimated to be 62% of the dose reaching the lung lobes [84]. Niven et al. [85] investigated the efficacy and pharmacokinetics of G-CSF powder formulations and solution administered via the pulmonary route in rabbits. Both powder and solution formulations of G-CSF delivered in this way exhibited an earlier peak time than G-CSF given subcutaneously, though the powder formulation was less efficient in reaching the lung lobes than the instilled solution.

Techniques used to administer G-CSF have an important impact on its absorption profile. The plasma G-CSF concentration–time profiles were strikingly different for aerosol and intratracheal deliveries. The peak plasma level was much higher and achieved earlier, and the estimated bioavailability was also significantly greater, after aerosol than after intratracheal administration [21]. It is believed that permeation and enzymatic degradation are rate-limiting steps in the absorption of G-CSF after intratracheal administration. The co-administration of surfactants or protease inhibitors may increase the absorption of G-CSF [86].

#### 9.6.3.6 Interferons

Interferon (IFN) is a cytokine that is used as a drug for a variety of purposes, including the treatment of kidney cancer and tuberculosis. Following challenge or infection by viruses, IFNs are rapidly induced in cells. Conversely, viral replication can be inhibited by the addition of exogenous natural and recombinant IFNs in cell culture.

**IFN-α** is a pleiotrophic cytokine with infection control activities. It can be delivered to the deep lung, without systemic side effects, via aerosol. Twenty human immunodeficiency virus (HIV)-negative subjects with active pulmonary tuberculosis were included in an open, parallel study using aerosolized IFN-α or conventional chemotherapy plus aerosolized IFN-α [87]. Patients receiving chemotherapy plus IFN-α showed more favorable effects than those receiving chemotherapy alone. Similar results were also observed after aerosolized IFN-α treatment of patients with multi-drug-resistant pulmonary tuberculosis [88]. Aerosolized IFN-α has also been used to treat locally advanced bronchioloalveolar cancer. Although results indicated that the aerosolized IFN-α form was feasible, the anti-tumor activity was limited [89,90]. Several other studies concluded that there was no therapeutic advantage of such a delivery system for the treatment of advanced lung cancer. In recent years, interest has also been expressed in the potential treatment of HIV-infected patients with inhaled IFN-α [91]. The lack of *in-vivo* efficacy in many cases may relate to the limitations of delivery, clearance, stability and access to infected cells, which in turn results in subtherapeutic concentrations at the desired site, or systemic toxicity. More effective pharmacotherapy with IFN-α may be possible by modification of formulations and/or delivery systems.

**IFN-β** is a glycosylated protein of 18 kDa and is currently approved in the USA for the treatment of multiple sclerosis (MS). Clinically, it is administered as a once-weekly IM injection [92]. The bioavailability and safety of IFN-β delivered by pulmonary administration was evaluated in nonhuman primates using intratracheal instillation. Bioavailability was approximately 10%, and no significant lung toxicity was found at dosing frequencies of one to three times per week. It follows that pulmonary delivery of IFN-β might be a promising alternative to IM administration [92]. Inhaled IFN-β (using a jet nebulizer) was also studied for the local treatment of thoracic malignancies in eight patients; a high tolerance and non-detectable IFN-β in the serum suggested that this route might be promising for the local treatment of lung diseases [93].

**IFN-γ**, a 20 to 25 kDa lymphokine, is synthesized naturally by activated T cells, and is critical in the immune response against *Mycobacterium tuberculosis*. Beck et al. [94] have demonstrated the efficacy of aerosol-administered murine IFN-γ in pneumocystis-infected mice, while the results of studies in rodents have indicated an antitumor effect [95] and anti-infective potential of IFN-γ [96]. Deposition studies indicated that aerosolized IFN-γ can be effectively delivered to the lower respiratory tract, and that IFN-γ given by this route does not reach the systemic circulation or induce any local or systemic side effects. Thus, pulmonary IFN-$\gamma$ delivery is promising in improving the defenses of the lungs at risk or with infection [97, 98]. This is in accordance with the findings of a Phase I trial of the inhalation of aerosolized IFN-$\gamma$ conducted by Halme et al. [99]; these authors suggested that aerosolized IFN-$\gamma$ can be used to treat respiratory tract tumors or infections. Inhalation of IFN-$\gamma$ was also investigated as a treatment for pulmonary metastatic tumors or renal cell carcinoma [100]. However, since the long-term effects of recombinant human IFN-$\gamma$ (rhIFN-$\gamma$) are not known, caution must be exercised in the development of an aerosol delivery system for this agent.

**Consensus interferon (CIFN)** is a novel, recombinant type I interferon containing 166 amino acids. It may be used in the treatment of viral infections [101] and hepatitis C [102]. Significant protection was achieved against viral infections via aerosol administration of CIFN with an ultrasonic nebulizer. The bioavailability is about 70% with a $T_{max}$ of 25–30 min, demonstrating the feasibility of using an aerosolized IFN to treat systemic viral infections [101]. Although both air-jet and ultrasonic nebulization can lead to noncovalent aggregation of CIFN, the extent of aggregation can be controlled by optimizing the formulations (addition of surfactant) or by improving the nebulization procedure (cooling of the solution during nebulization) [103].

#### 9.6.3.7 TSH, FSH, and HCG

The high molecular weight and low lipophilicity of hormones such as thyrotropin-stimulating hormone (TSH, MW 24–30 kDa), FSH (MW 36 kDa) and human chorionic gonadotropin (HCG, MW 45 kDa) necessitate their administration by injection. Komada et al. [104] proposed the lung as an alternative site for their absorption into the systemic circulation. An aqueous solution and/or a dry powder of these hormones were delivered into the exposed trachea of anesthetized rats. TSH, FSH, and HCG, when delivered in solution of neutral pH, resulted in a bioavailability of 2.5, 2.3, and 0.2%, respectively. When administered with a surfactant or under acidic conditions, transpulmonary absorption of these agents from solution was two- to 30-fold times greater than the above-mentioned values. Following intratracheal administration as a dry powder, the bioavailability of TSH, FSH, and HCG were 1.6, 0.6, and 0.1%, respectively. The bioavailability increased with a decrease in their molecular weight. Hence, the molecular weight of these agents may be more influential than their surface charge for absorption via alveolar epithelia.

#### 9.6.3.8 Elastase Inhibitors

##### 9.6.3.8.1 Alpha-1 Antitrypsin ($\alpha$1A)

$\alpha$1A deficiency is a chronic hereditary disease characterized by progressive emphysema due to an imbalance in the protease/antiprotease levels of the epithelial lining layers of the lungs [105,106]. Specifically, the inhibition of elastase in the lower respiratory tract is compromised by deficient levels of α1A. The recombinant form of the molecule consists of a 45 kDa single polypeptide that is identical to the human molecule (52 kDa). The natural and recombinant α1A is not susceptible to breakdown during nebulization and is readily absorbed into the systemic circulation. This was not unexpected, since IV therapy resulted in demonstrable levels of the protease inhibitor in bronchoalveolar lavage fluid. Following the administration of aerosolized rα1A in sheep (100 mg), peak plasma levels occurred at 12 h [107]. The attraction of pulmonary delivery of rα1A lies in the fact that the dose regimen may be reduced, and targeting to the lungs can be achieved. Dissolution-controlled kinetics may add to the retention of protein within the lungs if appropriately formulated. Smith and coworkers [108] have predicted by modeling delivery and using results obtained from dogs and sheep that the administration of aerosolized α1A may provide an efficient method of augmenting alveolar antiprotease levels.

#### 9.6.3.8.2 Secretory Leukoprotease Inhibitor (SLPI)

SLPI or antileukoprotease is a 12 kDa natural protein with the ability to inhibit various serine proteases. It is an especially potent inhibitor of neutrophil elastase [109], which results in protection of the airway epithelium as well as the alveolar region of the lung [110, 111]. Recombinant SLPI (rSLPI) is identical to the native protein with a pI of 9.1, indicating that this protein carries a positive charge at physiological pH. This increases the probability of SLPI binding to the predominantly anionic surface of the lung, and also helps to explain the slow clearance of SLPI from the airways ($t^1/_2$ ~12 h in sheep [112], 4–5 h in rats [113]). Aerosol therapy of SLPI was shown significantly to suppress the neutrophil elastase burden of the respiratory tract, as well as to reduce the level of interleukin (IL)-8, an important inflammatory mediator in chronic obstructive pulmonary disease (COPD) and cystic fibrosis (CF) patients [114]. However, studies with sheep showed that only one-third of SLPI in the respiratory tract epithelial lining fluid is functional, and that it does not protect the respiratory epithelium effectively [115]. Further studies in CF patients have shown that, in order to be effective, the SLPI must be administered in an impractically high amount (100 mg) and on a regular basis [116].

## 9.7 Limitations of Aerosol Delivery

In general, the creation of aerosols is technically difficult, expensive, and time-consuming. Moreover, patients need to learn specific inhalation techniques for the correct use of inhaler devices [26], and many have difficulty in using MDIs properly. Aerosol preparations are associated with significant losses of drug. Furthermore, due to the inertial impaction of the administered aerosol particles, only small fractions (10~15%) of the dose leaving the inhalation device will reach the small airways and alveoli. Aerosols are, by their nature as disperse systems, unstable; they tend to settle and to show adhesion to themselves or to other surfaces. In addition, the relatively high humidity in the respiratory tract (approaching 100% in the alveoli) can affect aerosol particle size as water condenses on the particles, or by the hygroscopic growth of solid soluble particles and hypertonic droplets [117, 118]. Safety issues are another concern for the therapeutic application of peptides/proteins via the pulmonary route, especially for those formulations that use absorption enhancers, enzyme inhibitors or microparticles [1].

## 9.8 Summary

The pulmonary delivery of proteins and peptides, as a promising alternative to parenteral administration, has attracted much attention in recent years. However, the formulation and aerosolization of these maocromolecules remains a major challenge as many such molecules are susceptible to chemical and/or physical degradation. Therefore, during investigations into the formulation of aerosols containing proteins and peptides, attention is focused on maintaining their biological activities and improving pulmonary deposition and absorption. Complex physicochemical properties, such as flow dynamics within the human airways, in addition to the chemistry and physics of the particles being absorbed, will continue to be a major challenge for delivering proteins and peptides to desired areas within the lungs. These challenges must be successfully addressed in order to realize the full potential of the pulmonary delivery of peptides and proteins, and to bring this now scientifically well-established concept to fruition. Traditionally, inhalation has been used as a means of delivering drugs to treat asthma, but a new era of inhalation is emerging with the delivery of proteins and peptides to treat not only pulmonary diseases but also systemic diseases, including diabetes.

## 9.9 References


1 Agu, R.U., M.I. Ugwoke, M. Armand, R. Kinget, and N. Verbeke. **2001**. The lung as a route for systemic delivery of therapeutic proteins and peptides. *Respir. Res.* 2 : 198–209.

2 Choi, W.S., G.G. Murthy, D.A. Edwards, R. Langer, and A.M. Klibanov. **2001**. Inhalation delivery of proteins from ethanol suspensions. *Proc. Natl. Acad. Sci. USA* 98 : 11103–11107.

3 Taylor, G., and L. Kellaway. **2001**. Pulmonary Drug Delivery. In: A. Hillery, A. Lloyd, and J. Swarbrick (Eds), *Drug Delivery and Targeting: For Pharmacists and Pharmaceutical Scientists.* Taylor & Francis, London, pp. 270–300.

4 Hussain, A., J.J. Arnold, M.A. Khan, and F. Ahsan. **2004**. Absorption enhancers in pulmonary protein delivery. *J. Control. Release* 94 : 15–24.

**5** Gonda, I. **2000**. The ascent of pulmonary drug delivery. *J. Pharm. Sci.* 89:940–945.

**6** Cleland, J.L., A. Daugherty, and R. Mrsny. **2001**. Emerging protein delivery methods. *Curr. Opin. Biotechnol.* 12:212–219.

**7** Mercer, R.R., M.L. Russell, and J.D. Crapo. **1994**. Alveolar septal structure in different species. *J. Appl. Physiol.* 77:1060–1066.

**8** Gehr, P., M. Bachofen, and E.R. Weibel. **1978**. The normal human lung: ultrastructure and morphometric estimation of diffusion capacity. *Respir. Physiol.* 32:121–140.

**9** Breeze, R.G., and E.B. Wheeldon. **1977**. The cells of the pulmonary airways. *Am. Rev. Respir. Dis.* 116:705–777.

**10** Mercer, R.R., M.L. Russell,V.L. Roggli, and J.D. Crapo. **1994**. Cell number and distribution in human and rat airways. *Am. J. Respir. Cell. Mol. Biol.* 10: 613–624.

**11** Clarke, S., and D. Pavia. **1984**. *Aerosols and the Lung: Clinical and Experimental Aspects.* Butterworths, London.

**12** Kim, K.J., and A.B. Malik. **2003**. Protein transport across the lung epithelial barrier. *Am. J. Physiol. Lung Cell. Mol. Physiol.* 284:L247–L259.

**13** McElroy, M.C., and M. Kasper. **2004**. The use of alveolar epithelial type I cell-selective markers to investigate lung injury and repair. *Eur. Respir. J.* 24:664–673.

**14** Groneberg, D.A., C. Witt, U. Wagner, K.F. Chung, and A. Fischer. **2003**. Fundamentals of pulmonary drug delivery. *Respir. Med.* 97:382–387.

**15** Cunningham, A.C., D.S. Milne, J. Wilkes, J.H. Dark, T.D. Tetley, and J.A. Kirby. **1994**. Constitutive expression of MHC and adhesion molecules by alveolar epithelial cells (type II pneumocytes) isolated from human lung and comparison with immunocytochemical findings. *J. Cell Sci.* 107 (Pt 2):443–449.

**16** Crouch, E., and J.R. Wright. **2001**. Surfactant proteins a and d and pulmonary host defense. *Annu. Rev. Physiol* 63:521–554.

**17** McCormack, F.X., and J.A. Whitsett. **2002**. The pulmonary collectins, SP-A and SP-D, orchestrate innate immunity in the lung. *J. Clin. Invest.* 109:707–712.

**18** Sayani, A.P., and Y.W. Chien. **1996**. Systemic delivery of peptides and proteins across absorptive mucosae. *Crit. Rev. Ther. Drug Carrier Syst.* 13:85–184.

**19** Lombry, C., D.A. Edwards,V. Preat, and R. Vanbever. **2004**. Alveolar macrophages are a primary barrier to pulmonary absorption of macromolecules. *Am. J. Physiol. Lung Cell. Mol. Physiol.* 286:L1002–L1008.

**20** Colthorpe, P., S.J. Farr, I.J. Smith, D. Wyatt, and G. Taylor. **1995**. The influence of regional deposition on the pharmacokinetics of pulmonary-delivered human growth hormone in rabbits. *Pharm. Res.* 12:356–359.

**21** Niven, R.W., K.L. Whitcomb, L. Shaner, A.Y. Ip, and O.B. Kinstler. **1995**. The pulmonary absorption of aerosolized and intratracheally instilled rhG-CSF and monoPEGylated rhG-CSF. *Pharm. Res.* 12:1343–1349.

**22** Sciarra, J.J., and C.J. Sciarra. **2000**. Aerosols. In: Gennaro, A.R. (Ed.), *Remington: The Science and Practice of Pharmacy.* Lippincott Williams & Wilkins, Baltimore, pp. 963–979.

**23** Labiris, N.R., and M.B. Dolovich. **2003**. Pulmonary drug delivery. Part I: physiological factors affecting therapeutic effectiveness of aerosolized medications. *Br. J. Clin. Pharmacol.* 56:588–599.

**24** Suarez, S., and A.J. Hickey. **2000**. Drug properties affecting aerosol behavior. *Respir. Care* 45:652–666.

**25** Newman, S.P., J.E. Agnew, D. Pavia, and S.W. Clarke. **1982**. Inhaled aerosols: lung deposition and clinical applications. *Clin. Phys. Physiol. Meas.* 3:1–20.

**26** Dolovich, M.B., R.C. Ahrens, D.R. Hess, P. Anderson, R. Dhand, J.L. Rau, G.C. Smaldone, and G. Guyatt. **2005**. Device selection and outcomes of aerosol therapy: Evidence-based guidelines: American College of Chest Physicians/American College of Asthma, Allergy, and Immunology. *Chest* 127:335–371.

**27** Green, J.D. **1994**. Pharmaco-toxicological expert report Pulmozyme rhDNase Genentech, Inc. *Hum. Exp. Toxicol.* 13 Suppl 1:S1–S42.

**28** Smyth, H. **2005**. Propellant-driven metered-dose inhalers for pulmonary drug delivery. *Expert Opin. Drug Deliv.* 2:53–71.

**29** Geller, D.E. **2002**. New liquid aerosol generation devices: systems that force pressurized liquids through nozzles. *Respir. Care* 47:1392–1404; discussion 1404–1395.

**30** Labiris, N.R., and M.B. Dolovich. **2003**. Pulmonary drug delivery. Part II: the role of inhalant delivery devices and drug formulations in therapeutic effectiveness of aerosolized medications. *Br. J. Clin. Pharmacol.* 56:600–612.

**31** Frijlink, H., and A.D. Boer. **2004**. Dry powder inhalers for pulmonary drug delivery. *Expert Opin. Drug Deliv.* 1:67–85.

**32** Edwards, D.A., J. Hanes, G. Caponetti, J. Hrkach, A. Ben-Jebria, M.L. Eskew, J. Mintzes, D. Deaver, N. Lotan, and R. Langer. **1997**. Large porous particles for pulmonary drug delivery. *Science* 276:1868–1871.

**33** Folkesson, H.G., M.A. Matthay, B.R. Westrom, K.J. Kim, B.W. Karlsson, and R.H. Hastings. **1996**. Alveolar epithelial clearance of protein. *J. Appl. Physiol.* 80:1431–1445.

**34** Tronde, A., E. Krondahl, H. von Euler-Chelpin, P. Brunmark, U.H. Bengtsson, G. Ekstrom, and H. Lennernas. **2002**. High airway-to-blood transport of an opioid tetrapeptide in the isolated rat lung after aerosol delivery. *Peptides* 23:469–478.

**35** Gerde, P., P. Ewing, L. Lastbom, A. Ryrfeldt, J. Waher, and G. Liden. **2004**. A novel method to aerosolize powder for short inhalation exposures at high concentrations: isolated rat lungs exposed to respirable diesel soot. *Inhal. Toxicol.* 16:45–52.

**36** Mathias, N.R., K.J. Kim, and V.H. Lee. **1996**. Targeted drug delivery to the respiratory tract: solute permeability of air-interface cultured rabbit tracheal epithelial cell monolayers. *J. Drug Target.* 4:79–86.

**37** Shen, B.Q., W.E. Finkbeiner, J.J. Wine, R.J. Mrsny, and J.H. Widdicombe. **1994**. Calu-3: a human airway epithelial cell line that shows cAMP-dependent Cl-secretion. *Am. J. Physiol.* 266: L493–L501.

**38** Mason, R.J., and M.C. Williams. **1980**. Phospholipid composition and ultrastructure of A549 cells and other cultured pulmonary epithelial cells of presumed type II cell origin. *Biochim. Biophys. Acta* 617:36–50.

**39** Matsukawa, Y., V.H. Lee, E.D. Crandall, and K.J. Kim. **1997**. Size-dependent dextran transport across rat alveolar epithelial cell monolayers. *J. Pharm. Sci.* 86:305–309.

**40** Shen, J., K.J. Elbert, F. Yamashita, C.M. Lehr, K.J. Kim, and V.H. Lee. **1999**. Organic cation transport in rabbit alveolar epithelial cell monolayers. *Pharm. Res.* 16:1280–1287.

**41** Elbert, K.J., U.F. Schafer, H.J. Schafers, K.J. Kim, V.H. Lee, and C.M. Lehr. **1999**. Monolayers of human alveolar epithelial cells in primary culture for pulmonary absorption and transport studies. *Pharm. Res.* 16:601–608.

**42** Helliwell, P.A., D. Meredith, C.A. Boyd, J.R. Bronk, N. Lister, and P.D. Bailey. **1994**. Tripeptide transport in rat lung. *Biochim. Biophys. Acta* 1190:430–434.

**43** Meredith, D., and C.A. Boyd. **1995**. Dipeptide transport characteristics of the apical membrane of rat lung type II pneumocytes. *Am. J. Physiol.* 269:L137–L143.

**44** Morimoto, K., H. Yamahara, V.H. Lee, and K.J. Kim. **1993**. Dipeptide transport across rat alveolar epithelial cell monolayers. *Pharm. Res.* 10:1668–1674.

**45** Byron, P.R., Z. Sun, H. Katayama, and F. Rypacek. **1994**. Solute absorption from the airways of the isolated rat lung. IV. Mechanisms of absorption of fluorophore-labeled poly-α,beta-[N(2-hydroxyethyl)-DL-aspartamide]. *Pharm. Res.* 11:221–225.

**46** Niven, R.W., F. Rypacek, and P.R. Byron. **1990**. Solute absorption from the airways of the isolated rat lung. III. Absorption of several peptidase-resistant, synthetic polypeptides: poly-(2-hydroxyethyl)-aspartamides. *Pharm. Res.* 7:990–994.

**47** Hoover, J.L., B.D. Rush, K.F. Wilkinson, J.S. Day, P.S. Burton, T.J. Vidmar, and M.J. Ruwart. **1992**. Peptides are better absorbed from the lung than the gut in the rat. *Pharm. Res.* 9:1103–1106.

48 Kim, K.J., Y. Matsukawa, H. Yamahara, V.K. Kalra, V.H. Lee, and E.D. Crandall. **2003**. Absorption of intact albumin across rat alveolar epithelial cell monolayers. *Am. J. Physiol. Lung Cell Mol. Physiol.* 284:L458–L465.

49 John, T.A., S.M. Vogel, R.D. Minshall, K. Ridge, C. Tiruppathi, and A.B. Malik. **2001**. Evidence for the role of alveolar epithelial gp60 in active transalveolar albumin transport in the rat lung. *J. Physiol.* 533:547–559.

50 Hastings, R.H., H.G. Folkesson, and M.A. Matthay. **2004**. Mechanisms of alveolar protein clearance in the intact lung. *Am. J. Physiol. Lung Cell Mol. Physiol.* 286:L679–L689.

51 Matsukawa, Y., H. Yamahara, V.H. Lee, E.D. Crandall, and K.J. Kim. **1996**. Horseradish peroxidase transport across rat alveolar epithelial cell monolayers. *Pharm. Res.* 13:1331–1335.

52 Deshpande, D., D. Toledo-Velasquez, L.Y. Wang, C.J. Malanga, J.K. Ma, and Y. Rojanasakul. **1994**. Receptor-mediated peptide delivery in pulmonary epithelial monolayers. *Pharm. Res.* 11: 1121–1126.

53 Laube, B.L., G.W. Benedict, and A.S. Dobs. **1998**. The lung as an alternative route of delivery for insulin in controlling postprandial glucose levels in patients with diabetes. *Chest* 114:1734–1739.

54 Laube, B.L. **2001**. Treating diabetes with aerosolized insulin. *Chest* 120:99S–106 S.

55 Cefalu, W.T. **2001**. Novel routes of insulin delivery for patients with type 1 or type 2 diabetes. *Ann. Med.* 33:579–586.

56 Cefalu, W.T. **2001**. Inhaled insulin: a proof-of-concept study. *Ann. Intern. Med.* 134:795.

57 Morimoto, K., K. Metsugi, H. Katsumata, K. Iwanaga, and M. Kakemi. **2001**. Effects of low-viscosity sodium hyaluronate preparation on the pulmonary absorption of rh-insulin in rats. *Drug Dev. Ind. Pharm.* 27:365–371.

58 Chen, X.J., J.B. Zhu, G.J. Wang, M.X. Zhou, F.M. Xin, N. Zhang, C.X. Wang, and Y.N. Xu. **2002**. Hypoglycemic efficacy of pulmonary delivered insulin dry powder aerosol in rats. *Acta Pharmacol. Sin.* 23:467–470.

59 Patton, J.S., J.G. Bukar, and M.A. Eldon. **2004**. Clinical pharmacokinetics and pharmacodynamics of inhaled insulin. *Clin. Pharmacokinet.* 43:781–801.

60 Quattrin, T., A. Belanger, N.J. Bohannon, and S.L. Schwartz. **2004**. Efficacy and safety of inhaled insulin (Exubera) compared with subcutaneous insulin therapy in patients with type 1 diabetes: results of a 6-month, randomized, comparative trial. *Diabetes Care* 27:2622–2627.

61 Skyler, J.S., W.T. Cefalu, I.A. Kourides, W.H. Landschulz, C.C. Balagtas, S.L. Cheng, and R.A. Gelfand. **2001**. Efficacy of inhaled human insulin in type 1 diabetes mellitus: a randomised proof-of-concept study. *Lancet* 357:331–335.

62 2004. Insulin inhalation: NN **1998**. *Drugs R D* 5:46–49.

63 Farr, S.J., A. McElduff, L.E. Mather, J. Okikawa, M.E. Ward, I. Gonda, V. Licko, and R.M. Rubsamen. **2000**. Pulmonary insulin administration using the AERx system: physiological and physicochemical factors influencing insulin effectiveness in healthy fasting subjects. *Diabetes Technol. Ther.* 2:185–197.

64 Brunner, G.A., B. Balent, M. Ellmerer, L. Schaupp, A. Siebenhofer, J.H. Jendle, J. Okikawa, and T.R. Pieber. **2001**. Dose-response relation of liquid aerosol inhaled insulin in type I diabetic patients. *Diabetologia* 44:305–308.

65 Henry, R.R., S.R. Mudaliar, W.C. Howland, 3rd, N. Chu, D. Kim, B. An, and R.R. Reinhardt. **2003**. Inhaled insulin using the AERx Insulin Diabetes Management System in healthy and asthmatic subjects. *Diabetes Care* 26:764–769.

66 Harsch, I.A., E.G. Hahn, and P.C. Konturek. **2001**. Syringe, pen, inhaler – the evolution of insulin therapy. *Med. Sci. Monit.* 7:833–836.

67 Kobayashi, S., S. Kondo, and K. Juni. **1994**. Study on pulmonary delivery of salmon calcitonin in rats: effects of protease inhibitors and absorption enhancers. *Pharm. Res.* 11:1239–1243.

68 Kobayashi, S., S. Kondo, and K. Juni. **1996**. Pulmonary delivery of salmon cal-

citonin dry powders containing absorption enhancers in rats. *Pharm. Res.* 13:80–83.

**69** Deftos, L.J., J.J. Nolan, B.L. Seely, P.L. Clopton, G.J. Cote, C.L. Whitham, L.J. Florek, T.A. Christensen, and M.R. Hill. **1997**. Intrapulmonary drug delivery of salmon calcitonin. *Calcif. Tissue Int.* 61:345–347.

**70** Okada, H., I. Yamazaki, Y. Ogawa, S. Hirai, T. Yashiki, and H. Mima. **1982**. Vaginal absorption of a potent luteinizing hormone-releasing hormone analog (leuprolide) in rats I: absorption by various routes and absorption enhancement. *J. Pharm. Sci.* 71:1367–1371.

**71** Adjei, A., and J. Garren. **1990**. Pulmonary delivery of peptide drugs: effect of particle size on bioavailability of leuprolide acetate in healthy male volunteers. *Pharm. Res.* 7:565–569.

**72** Adjei, A., D. Sundberg, J. Miller, and A. Chun. **1992**. Bioavailability of leuprolide acetate following nasal and inhalation delivery to rats and healthy humans. *Pharm. Res.* 9:244–249.

**73** Schreier, H., K.J. McNicol, D.B. Bennett, Z. Teitelbaum, and H. Derendorf. **1994**. Pharmacokinetics of detirelix following intratracheal instillation and aerosol inhalation in the unanesthetized awake sheep. *Pharm. Res.* 11:1056–1059.

**74** Bennett, D.B., E. Tyson, C.A. Nerenberg, S. Mah, J.S. de Groot, and Z. Teitelbaum. **1994**. Pulmonary delivery of detirelix by intratracheal instillation and aerosol inhalation in the briefly anesthetized dog. *Pharm. Res.* 11:1048–1055.

**75** Bajusz, S., V.J. Csernus, T. Janaky, L. Bokser, M. Fekete, and A.V. Schally. **1988**. New antagonists of LHRH. II. Inhibition and potentiation of LHRH by closely related analogues. *Int. J. Pept. Protein Res.* 32:425–435.

**76** Szende, B., K. Lapis, T.W. Redding, G. Srkalovic, and A.V. Schally. **1989**. Growth inhibition of MXT mammary carcinoma by enhancing programmed cell death (apoptosis) with analogs of LH-RH and somatostatin. *Breast Cancer Res. Treat.* 14:307–314.

**77** Lizio, R., T. Klenner, G. Borchard, P. Romeis, A.W. Sarlikiotis, T. Reissmann, and C.M. Lehr. **2000**. Systemic delivery of the GnRH antagonist cetrorelix by intratracheal instillation in anesthetized rats. *Eur. J. Pharm. Sci.* 9:253–258.

**78** Lizio, R., T. Klenner, A.W. Sarlikiotis, P. Romeis, D. Marx, T. Nolte, W. Jahn, G. Borchard, and C.M. Lehr. **2001**. Systemic delivery of cetrorelix to rats by a new aerosol delivery system. *Pharm. Res.* 18:771–779.

**79** Harper, M., C.G. Hatjis, R.G. Appel, and W.E. Austin. **1987**. Vasopressin-resistant diabetes insipidus, liver dysfunction, hyperuricemia and decreased renal function. A case report. *J. Reprod. Med.* 32:862–865.

**80** van Bree, J.B., S. Tio, A.G. de Boer, M. Danhof, J.C. Verhoef, and D.D. Breimer. **1990**. Transport of desglycinamide-arginine vasopressin across the blood-brain barrier in rats as evaluated by the unit impulse response methodology. *Pharm. Res.* 7:293–298.

**81** Kelly, J.M., J.M. Abrahams, P.A. Phillips, F.A. Mendelsohn, Z. Grzonka, and C.I. Johnston. **1989**. [125I]-[d(CH2)5, Sar7]-AVP: a selective radioligand for V1 vasopressin receptors. *J. Recept. Res.* 9:27–41.

**82** Yamahara, H., K. Morimoto, V.H. Lee, and K.J. Kim. **1994**. Effects of protease inhibitors on vasopressin transport across rat alveolar epithelial cell monolayers. *Pharm. Res.* 11:1617–1622.

**83** Folkesson, H.G., B.R. Westrom, M. Dahlback, S. Lundin, and B.W. Karlsson. **1992**. Passage of aerosolized BSA and the nona-peptide dDAVP via the respiratory tract in young and adult rats. *Exp. Lung Res.* 18:595–614.

**84** Niven, R.W., F.D. Lott, and J.M. Cribbs. **1993**. Pulmonary absorption of recombinant methionyl human granulocyte colony stimulating factor (r-huG-CSF) after intratracheal instillation to the hamster. *Pharm. Res.* 10:1604–1610.

**85** Niven, R.W., F.D. Lott, A.Y. Ip, and J.M. Cribbs. **1994**. Pulmonary delivery of powders and solutions containing recombinant human granulocyte colony-stimulating factor (rhG-CSF) to the rabbit. *Pharm. Res.* 11:1101–1109.

**86** Machida, M., M. Hayashi, and S. Awazu. **2000**. The effects of absorption enhancers on the pulmonary absorption of recombinant human granulocyte colony-stimulating factor (rhG-CSF) in rats. *Biol. Pharm. Bull.* 23:84–86.

**87** Giosue, S., M. Casarini, L. Alemanno, G. Galluccio, P. Mattia, G. Pedicelli, L. Rebek, A. Bisetti, and F. Ameglio. **1998**. Effects of aerosolized interferon-α in patients with pulmonary tuberculosis. *Am. J. Respir. Crit. Care Med.* 158:1156–1162.

**88** Giosue, S., M. Casarini, F. Ameglio, P. Zangrilli, M. Palla, A.M. Altieri, and A. Bisetti. **2000**. Aerosolized interferon-α treatment in patients with multi-drug-resistant pulmonary tuberculosis. *Eur. Cytokine Netw.* 11:99–104.

**89** Kinnula,V., K. Cantell, and K. Mattson. **1990**. Effect of inhaled natural interferon-α on diffuse bronchioalveolar carcinoma. *Eur. J. Cancer* 26:740–741.

**90** van Zandwijk, N., E. Jassem, R. Dubbelmann, M.C. Braat, and P. Rumke. **1990**. Aerosol application of interferon-α in the treatment of bronchioloalveolar carcinoma. *Eur. J. Cancer* 26:738–740.

**91** Barnes, E., G. Webster, S. Whalley, and G. Dusheiko. **1999**. Predictors of a favorable response to α interferon therapy for hepatitis C. *Clin. Liver Dis.* 3: 775–791.

**92** Martin, P.L., S. Vaidyanathan, J. Lane, M. Rogge, N. Gillette, B. Niggemann, and J. Green. **2002**. Safety and systemic absorption of pulmonary delivered human IFN-beta1a in the nonhuman primate: comparison with subcutaneous dosing. *J. Interferon Cytokine Res.* 22:709–717.

**93** Halme, M., P. Maasilta, K. Mattson, and K. Cantell. **1994**. Pharmacokinetics and toxicity of inhaled human natural interferon-beta in patients with lung cancer. *Respiration* 61:105–107.

**94** Beck, J.M., H.D. Liggitt, E.N. Brunette, H.J. Fuchs, J.E. Shellito, and R.J. Debs. **1991**. Reduction in intensity of *Pneumocystis carinii* pneumonia in mice by aerosol administration of gamma interferon. *Infect. Immun.* 59:3859–3862.

**95** An, Z., X. Wang, P. Astoul, T. Danays, A.R. Moossa, and R.M. Hoffman. **1996**. Interferon gamma is highly effective against orthotopically-implanted human pleural adenocarcinoma in nude mice. *Anticancer Res.* 16:2545–2551.

**96** Jaffe, H.A., R. Buhl, A. Mastrangeli, K.J. Holroyd, C. Saltini, D. Czerski, H.S. Jaffe, S. Kramer, S. Sherwin, and R.G. Crystal. **1991**. Organ specific cytokine therapy. Local activation of mononuclear phagocytes by delivery of an aerosol of recombinant interferon-gamma to the human lung. *J. Clin. Invest.* 88: 297–302.

**97** Condos, R., F.P. Hull, N.W. Schluger, W.N. Rom, and G.C. Smaldone. **2004**. Regional deposition of aerosolized interferon-gamma in pulmonary tuberculosis. *Chest* 125:2146–2155.

**98** Raju, B., Y. Hoshino, K. Kuwabara, I. Belitskaya, S. Prabhakar, A. Canova, J.A. Gold, R. Condos, R.I. Pine, S. Brown, W.N. Rom, and M.D. Weiden. **2004**. Aerosolized gamma interferon (IFN-gamma) induces expression of the genes encoding the IFN-gamma-inducible 10-kilodalton protein but not inducible nitric oxide synthase in the lung during tuberculosis. *Infect. Immun.* 72:1275–1283.

**99** Halme, M., P. Maasilta, H. Repo, M. Ristola, E. Taskinen, K. Mattson, and K. Cantell. **1995**. Inhaled recombinant interferon gamma in patients with lung cancer: pharmacokinetics and effects on chemiluminescence responses of alveolar macrophages and peripheral blood neutrophils and monocytes. *Int. J. Radiat. Oncol. Biol. Phys.* 31: 93–101.

**100** Kawata, N., Y. Takimoto, H. Hirakata, T. Fuse, D. Hirano, and Y. Yamanaka. **1994**. [Clinical trial of inhalant recombinant interferon-gamma in patients with pulmonary metastasis from renal tumor (preliminary report)]. *Hinyokika Kiyo* 40:773–776.

**101** Niven, R.W., K.L. Whitcomb, M. Woodward, J. Liu, and C. Jornacion. **1995**. Systemic absorption and activity of recombinant consensus interferons after intratracheal instillation and aerosol administration. *Pharm. Res.* 12:1889–1895.

**102** Kao, J.H., P.J. Chen, M.Y. Lai, and D.S. Chen. **2000**. Efficacy of consensus interferon in the treatment of chronic hepatitis C. *J. Gastroenterol. Hepatol.* 15:1418–1423.

**103** Ip, A.Y., T. Arakawa, H. Silvers, C.M. Ransone, and R.W. Niven. **1995**. Stability of recombinant consensus interferon to air-jet and ultrasonic nebulization. *J. Pharm. Sci.* 84:1210–1214.

**104** Komada, F., S. Iwakawa, N. Yamamoto, H. Sakakibara, and K. Okumura. **1994**. Intratracheal delivery of peptide and protein agents: absorption from solution and dry powder by rat lung. *J. Pharm. Sci.* 83:863–867.

**105** Knight, K.R., J.G. Burdon, L. Cook, S. Brenton, M. Ayad, and E.D. Janus. **1997**. The proteinase-antiproteinase theory of emphysema: a speculative analysis of recent advances into the pathogenesis of emphysema. *Respirology* 2:91–95.

**106** Coakley, R.J., C. Taggart, S. O'Neill, and N.G. McElvaney. **2001**. Alpha1-antitrypsin deficiency: biological answers to clinical questions. *Am. J. Med. Sci.* 321:33–41.

**107** Hubbard, R.C., and R.G. Crystal. **1990**. Strategies for aerosol therapy of α 1-antitrypsin deficiency by the aerosol route. *Lung* 168 Suppl:565–578.

**108** Smith, R.M., L.D. Traber, D.L. Traber, and R.G. Spragg. 1989. Pulmonary deposition and clearance of aerosolized α-1-proteinase inhibitor administered to dogs and to sheep. *J. Clin. Invest.* 84:1145–1154.

**109** Wright, C.D., J.A. Kennedy, R.J. Zitnik, and M.A. Kashem. **1999**. Inhibition of murine neutrophil serine proteinases by human and murine secretory leukocyte protease inhibitor. *Biochem. Biophys. Res. Commun.* 254:614–617.

**110** Agerberth, B., J. Grunewald, E. Castanos-Velez, B. Olsson, H. Jornvall, H. Wigzell, A. Eklund, and G.H. Gudmundsson. **1999**. Antibacterial components in bronchoalveolar lavage fluid from healthy individuals and sarcoidosis patients. *Am. J. Respir. Crit. Care Med.* 160:283–290.

**111** Gossage, J.R., E.A. Perkett, J.M. Davidson, B.C. Starcher, D. Carmichael, K.L. Brigham, and B. Meyrick. **1995**. Secretory leukoprotease inhibitor attenuates lung injury induced by continuous air embolization into sheep. *J. Appl. Physiol.* 79:1163–1172.

**112** Vogelmeier, C., R. Buhl, R.F. Hoyt, E. Wilson, G.A. Fells, R.C. Hubbard, H.P. Schnebli, R.C. Thompson, and R.G. Crystal. **1990**. Aerosolization of recombinant SLPI to augment antineutrophil elastase protection of pulmonary epithelium. *J. Appl. Physiol.* 69:1843–1848.

**113** Gast, A., W. Anderson, A. Probst, H. Nick, R.C. Thompson, S.P. Eisenberg, and H. Schnebli. **1990**. Pharmacokinetics and distribution of recombinant secretory leukocyte proteinase inhibitor in rats. *Am. Rev. Respir. Dis.* 141:889–894.

**114** McElvaney, N.G., H. Nakamura, P. Birrer, C.A. Hebert, W.L. Wong, M. Alphonso, J.B. Baker, M.A. Catalano, and R.G. Crystal. **1992**. Modulation of airway inflammation in cystic fibrosis. In vivo suppression of interleukin-8 levels on the respiratory epithelial surface by aerosolization of recombinant secretory leukoprotease inhibitor. *J. Clin. Invest.* 90:1296–1301.

**115** Vogelmeier, C., R.C. Hubbard, G.A. Fells, H.P. Schnebli, R.C. Thompson, H. Fritz, and R.G. Crystal. **1991**. Anti-neutrophil elastase defense of the normal human respiratory epithelial surface provided by the secretory leukoprotease inhibitor. *J. Clin. Invest.* 87:482–488.

**116** Doring, G. **1999**. Serine proteinase inhibitor therapy in α(1)-antitrypsin inhibitor deficiency and cystic fibrosis. *Pediatr. Pulmonol.* 28:363–375.

**117** Morrow, P.E. **1986**. Factors determining hygroscopic aerosol deposition in airways. *Physiol. Rev.* 66:330–376.

**118** Ferron, G.A. **1994**. Aerosol properties and lung deposition. *Eur. Respir. J.* 7:1392–1394.

**119** Hillery, A., A. Lloyd, and J. Swarbrick (Eds.) **2001**. *Drug Delivery and Targeting for Pharmacists and Pharmaceutical Scientists*. Taylor and Francis, London.

**120** Hess, D.R. **2000**. Nebulizers: principles and performance. *Respir. Care* 45:609–622.

**121** Pfutzner, A., A.E. Mann, and S.S. Steiner. **2002**. Technosphere/Insulin – a new approach for effective delivery of human insulin via the pulmonary route. *Diabetes Technol. Ther.* 4:589–594.

**122** Cefalu, W.T., J.S. Skyler, I.A. Kourides, W.H. Landschulz, C.C. Balagtas, S. Cheng, and R.A. Gelfand. **2001**. Inhaled human insulin treatment in patients with type 2 diabetes mellitus. *Ann. Intern. Med.* 134:203–207.

# 10
# Biopharmaceutical Challenges: Delivery of Oligonucleotides

*Lloyd G. Tillman and Gregory E. Hardee*

## 10.1
## Introduction

An understanding of oligonucleotide pharmacokinetic and biopharmaceutic principles is critical in the effective design and development of dosage forms for antisense oligonucleotide (ASO) drugs. In the development of a biotech therapeutic, particularly a macromolecule such as ASO, one must consider the indication and associated target tissue(s), which the drug must reach. In order to provide clinical efficacy, adequate concentrations must be achieved in these target tissues, over a defined duration, but these concentrations must remain below the minimum toxic concentration to avoid the potential of adverse effects. To achieve this balance and remain within the therapeutic index requires that drug therapy addresses a number of dynamic factors – ultimately affecting the rate and amount of drug reaching the systemic circulation and the local site of action. The pharmacokinetic principles discussed in Chapter 4 mainly relate to the kinetics of drug disposition (i. e., distribution and elimination) after the drug has been presented to the systemic circulation. The parenteral administration of simple solution formulations – either by intravenous or subcutaneous routes – results in a broad distribution across systemic tissues, with associated activities demonstrated for example in the liver, synovium, and tumor. It has also been noted that the systemic biodistribution outcome is largely independent of either formulation variables or the route of administration once absorption of the native oligonucleotide into the central compartment is completed. However, many sites remain poorly served by systemic delivery routes, and therefore a local administration may be required for those therapeutic indications resident in tissues such as the lung, eye, brain, and colon.

For certain indications the target tissues are directly accessible for local administration techniques facilitating high local concentrations. This resultant concentration gradient of drug – across the target tissue surface – constitutes a driving force for the oligonucleotide diffusivity to better traverse tissue and cell membranes in line with Fick's First Law of diffusion. The appropriateness of the basic

*Pharmacokinetics and Pharmacodynamics of Biotech Drugs: Principles and Case Studies in Drug Development.* Edited by Bernd Meibohm

ISBN: 3-527-31408-3

diffusion principles inherent with Fick's laws of diffusion is justified on the basis of the demonstrated dose (concentration)–response relationships seen with ASOs. This is seen more directly in the examples discussed in Chapter 4, which showed a rank-order relationship between the pharmacokinetic and pharmacodynamic metrics. The typically seen temporal hysteresis of this relationship accounts for both the slow rate of flux of ASO into the tissues and, of course, the delayed manner of oligonucleotide action in the down-regulation of mRNA and protein synthesis. Oligonucleotide diffusivity is a function of molecular structure, charge, hydrodynamic radius, protein binding equilibria, as well as the biophysical aspects of the administration site such as surface area, temperature, and the characteristics of the barrier membrane (i. e., passive diffusion versus saturable transfer). In light of these variables involved with local therapies, it is evident that there are both formulations and delivery techniques to more efficiently deliver increased quantities of ASO into tissues for local effect. Simply put – "concentration matters".

Until recently, nonparenteral routes have failed to deliver sufficient quantities of ASO to be systemically therapeutic. The recent advent of novel oral delivery technologies, coupled with the increased tissue residence time for second-generation ASOs, allows oral delivery to achieve therapeutic levels for select systemic indications. This chapter will initially outline certain more conventional aspects of parenteral dosage forms, and then focus on formulation technologies that more specifically address local treatment. For the oral route, we will pass to the biopharmaceutic considerations for both local delivery to the gut and systemic delivery via absorption from solid dosage forms. Incumbent with the discussion on formulations is the need initially to overview the physico-chemical properties of ASOs, which in large part determine their biopharmaceutic characteristics.

## 10.2
## ASOs: The Physico-Chemical Properties

In the development of any drug product it is essential that the fundamental physical and chemical properties be determined of both the pure active ingredient and of the various possible dosage formulations. Such data often dictates the release characteristics and underscores the choices considered for dosage form design. ASOs represent a new category of compounds where these properties have yet to be fully characterized, but it is evident that the physico-chemical properties appear quite uniform within a chemical class of ASOs. That is, with respect to the inherent properties of the nascent or unformulated oligonucleotide, there is sufficient evidence that we may assign common properties for the ASOs of a given class of oligonucleotide chemistry. Data consistently demonstrate that all "first-generation" or phosphorothioate oligo*deoxy*nucleotides exhibit similar physico-chemical properties. Likewise, newer chemistries appear to have conformity within their defined chemistry, such as sugar-modified chemistries (e. g., 2′-*O*-alkoxyalkanes) or backbone-modified chemistries (e.g., methylphosphonates versus phosphorothioates). This important feature of class conformity allows for the consistent application of biopharmaceutic and pharmacokinetic principles across the innumerable base sequences possible within any single oligonucleotide chemical class. For example, rates of drug release from a dosage form can be estimated from precedent data as well as distribution and elimination estimates – which many times are dependent upon biophysical interactions such as protein binding. The exception to this axiom may be in the instances where intra- or intermolecular Watson–Crick base-pairing occurs. Such base-pairing, if significant, may result in a stable secondary structure (if self-complementary) or in a noncovalent polymeric structure consistent with "nearest neighbor" thermodynamic data [1]. Such altered states need to be reckoned with if ASO sequences permit such associations beyond four or five base-pairs per 20-mer ASO.

ASOs are synthesized as complex mixtures of diastereomers. In the solid state, they are amorphous, electrostatic, hygroscopic solids with low-bulk densities, possessing very high surface areas, and poorly defined or no melting points. Their good chemical stability allows them to be stored as lyophilized or spray-dried powders, or as concentrated, sterile solutions. For example, the 21-mer ASO Vitravene$^{TM}$ (fomivirsen sodium intravitreal injectable) is approved in the USA for a storage condition from 2 to 25 °C [2].

Due to their polyanionic nature, ASOs are extremely water-soluble under neutral and basic conditions. Consequently, drug-product concentrations are often only limited by an increase in solution viscosity at very high concentrations (e.g. >300 mg/mL) [3]. Not surprisingly, extremes of pH and ionic strength influence the apparent solubility. In acidic environments such as the stomach, inter-nucleotide linkages are partially neutralized by protonation. With this consequent decrease in their polyanionic status there is a marked decrease in oligonucleotide solubility, an event that can be easily reversed by raising the pH.

Phosphorothioate oligonucleotide degradation has been primarily attributed to two mechanisms: desulfurization, and acid-catalyzed hydrolysis. Desulfurization of the backbone, observed at elevated temperatures and under intense ultraviolet (UV) light, leads to (pharmacologically active) ASOs that primarily contain one or more phosphate diester linkage where oxygen has replaced sulfur [4]. It must be noted that, while such ASOs are pharmacologically active, the accumulation of phosphodiester linkages – particularly if on adjacent or terminal nucleotides – renders the compound more susceptible to metabolism by nucleases. Thus, the presence of numerous phosphodiester linkages may impact and shorten the compound's *in-vivo* half-life, particularly if the phosphodiesters are located in the central gap of the ASO. This gap is between the nuclease resistant wings of the molecule where 2′-*O*-alkoxyalkane ribonucleosides afford protection from enzymatic cleavage. Indeed, there is evidence that chimeric ASOs containing phosphodiester linkages along with 2′-*O*-alkoxyalkane derivatives may improve biological potency [5]. Under acidic conditions, a phosphodiester ASO may be more prone to depurination, resulting in ASOs containing one or more abasic sites. The loss of bases decreases the base-pairing binding affinity to the target mRNA, thereby rendering the ASO less effective. Abasic sites are also more susceptible to base-mediated cleavage of the adjacent phosphorothioate linkages. Formulation development activities must therefore ensure adequate stability and therefore need to include analytical techniques with monitoring and control of these attributes. Both, the desulfurization and hydrolytic pathways can be monitored by a variety of analytical techniques, including capillary gel electrophoresis, anion-exchange and ion-pair liquid chromatography and liquid chromatography-mass spectroscopy.

All of the physical-chemical characteristics discussed here are important when designing viable, efficacious formulations with acceptable storage and *in-vivo* stability.

## 10.3 Local Administration

### 10.3.1 Ocular Delivery

A key milestone in the development of antisense therapeutics was the 1998 FDA approval of the first antisense drug, Vitravene (fomivirsen), which was used for the local treatment of eye infections caused by the HIV-induced cytomegalovirus (CMV) [6]. Local treatment for this product is accomplished by intravitreal injection of 50 µL of a sterile, aqueous solution with a 30-gauge needle. Fomivirsen is a 21-mer phosphorothioate oligodeoxynucleotide and, as such, is in the first-generation ASO chemistry class. As noted in Chapter 4, this chemistry has a shorter half-life than the second-generation ASOs. Generally, first-generation chemistries clear from the plasma in hours and have tissue half-lives on the order of one to two days, but in the case of fomivirsen, its clearance from the vitreous of rabbits took seven to ten days [7]. This vitreal clearance was mediated by a combination of tissue distribution and metabolism. At equilibrium in the eye, fomivirsen concentrations were greatest in the retina and iris. Fomivirsen was detectable in retina within hours after intravitreal injection, with retinal concentrations increasing over three to five days. Metabolism is the primary route of elimination from the eye. Metabolites of fomivirsen are detected in the retina and vitreous in animals. Systemic exposure to fomivirsen following single or repeated intravitreal injections in monkeys was below the limits of quantitation. In monkeys treated every other week for up to three months with fomivirsen there were isolated instances when fomivirsen's metabolites were observed in liver, kidney, and plasma at a concentration near the level of detection [8].

An alternative to direct intravitreal injection is to use a noninvasive method of ASO delivery employing iontophoresis. The advantage of using low current density iontophoresis is that the ASO intraocular distribution will allow for therapeutic concentrations to be achieved across a large number of tissues, including the anterior uvea and the cornea as well as neural retinal areas. Voigt et al. [9] evaluated this intraocular ASO delivery approach in the rat model of endotoxin-induced uveitis which is characterized by NOSII (nitrous oxide synthase II) up-regulation. Iontophoresis facilitated the penetration of both anti-NOSII ASO and a control ASO with scrambled nucleotide sequence across the intraocular tissues of the eyes. Both active and control ASO were 21-base oligodeoxynucleotides. This biodistribution was demonstrated by histological examination of the fluorescently labeled ASOs and by the down-regulation of NOSII in the iris/ciliary body of the active treated eyes compared to the saline or scrambled ASO-treated eyes. The application of iontophoresis demonstrates a unique and noninvasive approach for the efficient delivery of ASO into the eye – increasing the likelihood of successful therapy for debilitating eye diseases manifesting posterior segments of the eye.

### 10.3.2 Local Gastrointestinal Delivery

Direct delivery of ASO to the gastrointestinal (GI) tract, by way of oral or rectal dosage forms, may lead to significantly higher local intestinal lumen ASO concentrations than when dosed by parenteral routes. These dosage forms may therefore be effective approaches to treat a number of local GI diseases. Such pathologies include certain malabsorption syndromes, inflammatory diseases, and carcinomas. A number of dosage forms exist that may be used to more conveniently target ASOs for a variety of local GI indications.

#### 10.3.2.1 Rectal Dosing

For the rectal route, dosage forms include suppositories, liquid enemas, or gels and foams. While varying in patient convenience and "feel", these also differ in their biopharmaceutic release profile and regional mucosal surface area coverage. Generally, *suppositories* cover less surface area but may be ideal for specific indications such as proctitis and perhaps proctosigmoiditis [10]. *Enemas* and *gels* cover all of the sigmoid colon and typically reach to the splenic flexure, proximal to the descending colon. *Foams* are known to consistently reach the descending colon but depending on formulation have been known to spread proximal to the splenic flexure as determined by gamma scintigraphy [I. Willding, personal communications, August 2002]. Foams are also generally better tolerated by the patient than enemas owing to the ease of use and a less "full" feeling post administration.

Besides biopharmaceutic distribution, the decision for a dosage form also depends on other factors such as mitigating pathophysiologies and patient compliance. For example, with distally presenting ulcerative colitis it is not uncommon for patients to have proximal bowel constipation complicated with distal diarrhea. This can impact the choice of therapeutic as well as involve the use of other aids such as a vigorous laxative in the event of fecal loading in the descending colon. Compliance also plays a large role in therapeutic outcome and, not surprisingly, patients prefer suppositories over enemas and prefer single daily treatments over twice-daily treatments [11].

#### 10.3.2.2 Oral Dosing

A number of solid dosage formulation technologies, including *capsules* and *tablets*, are available for the oral route to target local GI indications. Generally, these employ controlled-release attributes, thereby allowing for a regional release of dosage unit contents. The accuracy of such oral formulations generally varies in line with the triggering mechanism of the dosage form technology. Typically, this catalyst is pH, but such dosage forms have also been engineered to release on the basis of other variables such as bacterial or enzyme digestion or by way of pH-independent dissolution. Such dosage forms can release ASOs distally in the colon for the treatment of distal pathologies, or they can release more proximally in the small intestine to treat diseases such as Crohn's disease or ileus. In the development of "oralocal" ASO dosage forms – oral route for local or luminal gut therapy – it is also important to consider the ASO chemistry since its metabolic half-life in the gut milieu is critical to clinical outcome. One experiment in a dog colitis model demonstrated that an orally administered second-generation 2′-*O*-methoxyethyl (MOE)-gapmer ASO chemistry could achieve high concentrations in colon tissue after repeat dosing, producing tissue concentrations almost equivalent to those with an enema control [12]. Thus, in the development of an oralocal ASO formulation, the foremost considerations are of the target tissue location (i.e., proximal versus distal) and the type of ASO chemistry to be delivered.

A *retention enema* formulation to treat inflammation locally within the colon is currently under investigation at Isis Pharmaceuticals, Inc. Absorption into local tissues was achieved in animal enema models using ASO in an aqueous buffered, viscous formulation. At 2 h after a 1-h retention, levels of intact oligonucleotide in healthy tissue were about 100-fold higher than those achieved with an intravenous (IV) injection of an equivalent or higher dose (Table 10.1) [12]. Furthermore, the presence of ASO was still evident the following day in this single-dose study. In further support of this nonsystemic therapeutic approach are observations in mice [colitis induced with dextran sulfate solution (DSS)] and rats [colitis induced with trinitrobenzene sulfonic acid (TNBS)] that locally administered solutions of ASO accumulate to a greater extent in inflamed GI tissue than in normal tissue [12]. In the rat colitis model, ISIS-2302 solutions were administered by IV injection or intracolonic instillation. Oligonucleotide concentrations were determined in the liver and colon (Fig. 10.1). While there was no significant difference in tissue or plasma concentrations between normal and colitic rats when given oligonucleotide by the IV route, large differences were found between normal and colitic rats for intracolonic administration. Concentrations in the colon, in particular, were considerably higher in the diseased tissue. The data are consistent with the previous comment that diseased tissue might be targeted by the oral route using oralocal dosage forms.

Based upon these observations and following Phase I safety studies, a Phase II randomized, double-blind, controlled clinical trial was undertaken to treat active ulcerative colitis with escalating doses of ISIS 2302. Based on an intent-to-treat analysis, 59% (13/22) of patients treated daily with 240 mg ISIS 2302 enema over six weeks achieved a positive response, as measured by Disease Activity Index (DAI)

**Table 10.1** Dog colon tissue biopsy concentrations of intact ISIS 2302 oligonucleotide at 3 and 24 h after single-dose retention enema (10 mg/kg) or intravenous (dosage indicated) administrations.

| Formulation | Tissue concentrations [μg/g][a) 3 h] | 24 h |
|---|---|---|
| Enema (5 mg/mL oligonucleotide) | | |
| 1.5% Hydroxypropylmethylcellulose | 660 | 7 |
| 1.0% Carrageenan | 558 | 3 |
| Emulsion with Captex, Labrasol, and Crill | 224 | 1 |
| 0.5% Tween 80, 0.75% HPMC | 621 | 6 |
| 5% Sorbitol, 0.75% HPMC | 417 | 1 |
| IV dosing | | |
| 2 mg/kg | 2 | n.a. |
| 10 mg/kg | 11 | 0.6 |

a) Measurements by capillary gel electrophoresis.
n.a. = not available.

scores. Six months later at the end of the study, 77% (10/13) of these patients continued to demonstrate an improved DAI. Patients treated with ISIS 2302 enema who achieved a response maintained their response, on average, for longer than six months compared to a duration of response of less than three months for placebo-treated patients. It is notable that patients treated with 240 mg ISIS 2302 enema experienced minimal to no systemic absorption of drug (<1% of dose), showing that the drug is acting locally to treat this local disease [13].

### 10.3.3 Pulmonary Delivery

Efficient delivery to lung tissues is best realized upon delivering by way of the pulmonary route. The parenteral delivery of ASO does not lead to substantial lung uptake, and is therefore not an ideal route for treatment of disease in lung tissue. Following a 0.8 mg/kg IV bolus injection, the concentration of ASO in the lung tissue has been shown to be less than 5 μg/g [14], a concentration approaching pharmacologically active levels. Direct administration to the pulmonary airways through inhalation would potentially increase local concentrations, and also serve as a less invasive route of administration. Delivery options include continuous and single-dose nebulizers, metered-dose inhalers, and dry powder inhalers. Such devices are far less invasive than parenteral administration and, in addition to potentially increasing efficacy, are likely to increase patient compliance.

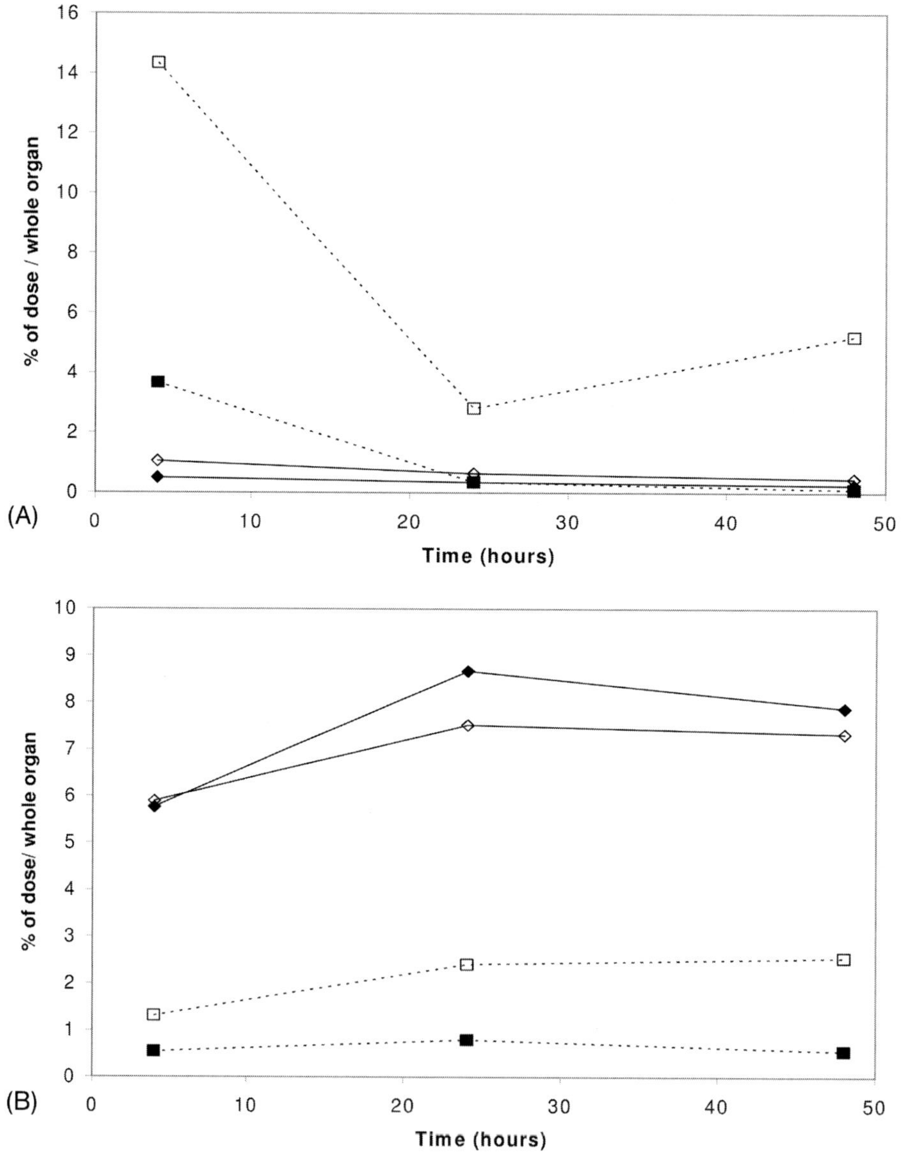


**Fig. 10.1** Rat colon tissue (A) or liver tissue (B) oligonucleotide concentrations as a percentage of administered radioactive dose for normal rats (filled symbols) or treated with trinitrobenzene sulfonic acid (open symbols) as a model for colitis. Solid lines represent intravenous route; broken lines designate rectal dosing (both 100 mg/kg ISIS 2302 with nonexchangeable $^{3}H$ label).

#### 10.3.3.1 Formulation Considerations

Wu-Pong and Byron [15] have contributed a review of the issues associated with pulmonary delivery of ASOs. Since ASOs are freely soluble and highly hygroscopic, it would be reasonable to assume that initial dosage forms will rely upon the aerosolization of simple aqueous solutions. Our data indicate that commercially available nebulization devices will generate suitable aerosolizations of ASO solutions at concentrations up to 180 mg/mL [3]. Ultrasonic and jet nebulizations were found to have essentially no effect on the phosphorothioate stability of the ASO over 40 min, which is longer than typical treatment times.

Depending on particle size, delivery may be primarily to the upper airways or to the alveoli [15, 16]. Particles with diameters of approximately 5–10 μm will generally deposit in the upper airways, while those in the range of approximately 0.5–5 μm approach the alveoli. Particles smaller than 0.5 μm are often exhaled, while particles larger than 10 μm are generally deposited in the oral capacity and subsequently swallowed.

#### 10.3.3.2 Deposition and Uptake

ASO was successfully delivered to cells of the mouse lung using simple aqueous formulations aerosolized with commercial nebulization equipment [17]. Single or multiple nose-only administration led to significant oligonucleotide concentrations in lung tissues (Table 10.2) and bronchiolar lavage fluid with low or mild toxicity, depending on dose. Using immunohistochemistry, ASO could be visualized in pneumocytes, vascular endothelium, and alveolar macrophages. Single

**Table 10.2** Oligonucleotide concentrations in murine lung tissue (n = 3–5 mice) after single, 30-min, nose-only inhalation exposure to ISIS-2105.

| Dose [a)] [mg/kg] | Time [h] | Oligonucleotide total concentration [μ/g tissue] | % Intact [b)] |
|---|---|---|---|
| 1.2 | EOE | 55.7 | 79 |
| | 2 | 52.4 | 63 |
| | 8 | 34.3 | 48 |
| | 24 | 21.8 | 25 |
| 12 | EOE | 153.6 | 86 |
| | 2 | 86.0 | 72 |
| | 8 | 68.7 | 49 |
| | 24 | 78.3 | 33 |

**a)** Dose level expressed as the pulmonary dose, i.e. amount estimated to be deposited in the lung. Necropsies were conducted at end of exposure (EOE) or at selected time intervals after exposure.

**b)** Relative amount of parent, 20-mer total oligonucleotide (ISIS 2105 plus metabolites).

and multiple exposures differed only in the amount of oligonucleotide present, and not the distribution pattern. The pulmonary dose was determined to be 10% of the respirable or nasally inhaled dose. This is a reasonable deposited fraction for this aerosol presentation owing to its mass median aerodynamic diameter (MMAD), which ranged from 0.3 to above 5 μm. Experiments with pulmonary doses as low as 1–3 mg/kg showed that significant concentrations of ASO can be deposited (>50 μg oligonucleotide per g tissue) and maintained (20 h or greater half-life) in lung, with little if any systemic exposure. ASO was localized to the bronchiolar epithelium and alveolar epithelium and endothelium. At doses of 12 mg/kg, systemic distribution was found, with plasma levels as high as 4.2 μg/mL and liver concentrations as high as 30 μg/g. Toxicity was mild at the 12 mg/kg level, and minimal to absent at doses of 3 mg/kg or below. Taken together, the results of these studies indicate that the pulmonary delivery of ASO directly to the lungs for local activity can be accomplished safely and with doses lower than those required with IV administration. It appears that pulmonary dosing of ASOs for local therapeutic benefit is within reach of currently available technologies. Furthermore, the use of newer oligonucleotide chemistries, such as the sugar-modified (2′-*O*-alkoxyalkane) derivatives that significantly enhance tissue accumulation will allow higher steady-state concentrations or less frequent dosing to be achieved, due to their longer tissue elimination half-lives.

In addition to these studies that characterize deposition and uptake, there are several reports in the literature that demonstrate antisense pharmacology using the pulmonary route. The administration of an aerosolized phosphorothioate oligonucleotide (antisense to adenosine $A_1$ receptor mRNA) desensitized rabbits to subsequent challenge with either adenosine or dust mite allergen. Rabbits treated with the sequence-specific ASO demonstrated a dose-dependent desensitization, while animals treated with a mismatched sequence control showed no desensitization [18, 19]. Specifically, these authors showed that aerosols created with ordinary nebulizers (which generate 80% of droplets <5 μm in diameter) effectively reach the bronchial smooth muscle in sufficient quantities to specifically attenuate the target up to 75% [19]. Similar results were found using bradykinin $B_2$ receptor-specific ASO and a mismatched control [19]. In another approach to ASO delivery to the lungs, pharmacologic activity was demonstrated by inserting an antisense mRNA complementary to the ras proto-oncogene into adenovirus vector particles. These vectors were delivered intratracheally to nude mice that had been previously inoculated with a human lung carcinoma. While 90% of the mice treated with a virus expressing a control sequence grew tumors, 87% of those treated with the virus expressing the antisense sequence were tumor-free [20].

The literature reviewed here show that both formulated and unformulated (simple aqueous solution) ASO can be delivered to the lungs, where they subsequently suppress local gene expression, opening a wide variety of diseases to antisense therapy.

### 10.3.4
### Delivery to the Brain

In general, the blood–brain barrier (BBB) is only permeable to lipophilic molecules of molecular weight <600 Da, while even small water-soluble molecules generally cannot be transported [21]. This would preclude the direct delivery of ASO to the brain from the systemic circulation without the assistance of specialized drug delivery systems.

Unmodified phosphodiester oligodeoxynucleotides have been directly injected intracerebrally to down-regulate the expression of neurotransmitters in a sequence-specific manner [22]. While phosphodiester oligodeoxynucleotides are rapidly degraded in the brain, phosphorothioate oligodeoxynucleotides are more metabolically stable, but are nevertheless cleared via the cerebrospinal fluid bulk flow. Whitesell et al. [23] demonstrated extensive brain penetration and marked cellular uptake after continuous intracerebral infusions with a mini-osmotic pump. Direct intracerebral administration has been utilized by Sommer et al. [24] to study ASO directed to c-fos, and by Mustafa and Dar [25] to down-regulate the adenosine A1 receptor expression. One report in the literature details the accumulation of systemically administered ASO in implanted subcutaneous and intracranial glioblastoma tumors in mice [26]. A similar ASO was shown not to cross the BBB in normal animals [27]; thus, it was hypothesized that the presence of a glioma may sufficiently disrupt the BBB to allow pharmacological concentrations of oligonucleotide to accumulate within the tumor. Accumulation was sufficient in this model to demonstrate antitumor activity.

Boado et al. [28] devised delivery systems based on conjugates of streptavidin and the OX26 monoclonal antibody directed to the transferrin receptor as a carrier for the transport of ASO. These delivery systems were found to transport peptide nucleic acid antisense molecules, but not ASO, across the BBB. These authors attributed this difference to preferential binding of phosphorothioate oligonucleotide to plasma protein instead of the antibody complex, which reduced their transport.

### 10.3.5
### Topical Delivery

The barrier properties of human skin have long been an area of multidisciplinary research. Skin is one of the most difficult biological barriers to penetrate and traverse, primarily due to the presence of the stratum corneum. The stratum corneum is composed of corneocytes laid in a brick-and-mortar arrangement with layers of lipid. The corneocytes are partially dehydrated, anuclear, metabolically active cells completely filled with bundles of keratin with a thick and insoluble envelope replacing the cell membrane [29]. The primary lipids in the stratum corneum are ceramides, free sterols, free fatty acids and triglycerides [30], which form lamellar lipid sheets between the corneocytes. These unique structural features of the stratum corneum provide an excellent barrier to the penetration of most molecules, particularly large, hydrophilic molecules such as ASOs.

A number of approaches can be pursued in order to penetrate the stratum corneum barrier. These approaches either alter the ASO's physico-chemical properties directly, the biophysical properties of the barrier, or provide some other driving force or influence onto both the drug and the stratum corneum barrier. Altering the thermodynamic properties of oligonucleotide molecules is possible by either utilizing a hydrophobic counter cation (e.g., benzalkonium) or by chemically modifying the oligonucleotide chemistry to eliminate the anionic backbone charges (e.g., morpholino or PNA oligonucleotide chemistries). These modifications have resulted in increased penetration across isolated hairless mouse skin. This was explained on the basis of greater partitioning into the lipid phase [31]. Alternatively, physically changing the stratum corneum to impair its integrity can result in improved skin penetration. There have been a number of creative approaches to impair the integrity of the skin without overtly causing long-term damage. Some examples of these breaching methods to induce transient openings in the stratum corneum include both ultrasound-induced sonoporation [32] and electroporation [33]. Iontophoresis, which involves the application of an electric field across the skin to induce electrochemical transport of charged molecules, has been studied extensively for the transdermal delivery of ASOs [34]. Lastly, the use of topical formulations containing permeation enhancers (PEs) offers several advantages in regard to the tailoring of formulation compositions and on the practical ease of application – particularly being able to avoid the use of application devices.

Chemical PEs have recently been studied for increasing transdermal delivery of ASOs or other polar macromolecules [35]. Chemically induced transdermal penetration results from a transient reduction in the barrier properties of the stratum corneum. The reduction may be attributed to a variety of factors such as the opening of intercellular junctions due to hydration [36], solubilization of the stratum corneum [37, 38], or increased lipid bilayer fluidization [39, 40]. Combining various surfactants and co-solvents can be used to achieve skin penetration, purportedly resulting in therapeutically relevant concentrations of ASO in the viable epidermis and dermis [41]. In summary, it appears feasible to deliver ASO to the skin using a number of different delivery techniques and formulations.

### 10.3.6 Other Local Delivery Approaches

When considering efficient dose utilization, the use of local delivery should be superior to systemic administration for targeting ASOs to accessible tissues. There are numerous possibilities available for nonsystemic therapies wherein a large number of pathologies may be treatable with ASOs. A partial listing of possible local treatments is provided in Table 10.3.

ASOs have been administered by many of these routes – generally with marked success even with the shorter half-life first-generation oligonucleotide chemistries. Besides Vitravene for CMV retinitis, first-generation ASOs have also been administered locally for ulcerative colitis (rectal), human papillomavirus (HPV) infec-

**Table 10.3** Administration routes and dosage forms allowing for local disease treatments.

| Route | Target organ | Formulation options |
|---|---|---|
| Ocular, intravitreal | Eye tissues | Eye drops; injectable |
| Pulmonary | Lungs | Various aerosol devices |
| Oral | Gastrointestinal tract | Oral dosage forms; controlled release |
| Rectal | Distal colon | Enema; suppository |
| Topical | Skin | Creams; device-mediated |
| Vaginal | Uterine, vaginal mucosa | Suppository; tablets |
| Buccal, sublingual, oral cavity | Buccal mucosa; gingival sulcus | Bioadhesive patch; biodegradable solution matrix, tablet |
| Tympanostomy tube | Inner ear | Sterile drops |
| Intrathecal or intracerebral ventricular catheter | Brain, spinal cord | Sterile solution |
| Intraurethral | Bladder, urethra | Solution; suspensions |
| Intravenous | Lungs | Sterile suspension for injection |
| Intra-arterial | Afferential organs | Sterile suspension for injection |
| Nasal | Nasal tissues | Nasal spray; nebulizer |
| Parenteral depot or placement (e.g., coronary) | Various: subungual, surgical fields, vessels | Sustained-release injectable; biodegradable matrices or splints |
| Intraperitoneal | Abdominal organs, peritoneum | Sterile solution |
| Ex-vivo tissue loading | Homologous transplants | Bathing solution; pressure-mediated tissue loading |

tions or genital warts (intradermal), psoriasis (topical), and for the treatment of coronary vessel decay (using *ex-vivo* tissue loading during coronary surgery).

## 10.4 Systemic Delivery

### 10.4.1 Parenteral Routes

In animal studies, there is overwhelming evidence of biological activity of ASOs upon IV injection of simple solutions. Given the excellent solution stability and solubility possessed by phosphorothioate oligodeoxynucleotides, it has been relatively straightforward to formulate these first-generation drug products in support of early clinical trials. Simple, buffered solutions have been successfully used in clinical studies by IV, intravitreal, and subcutaneous (SC) injections. Simple aqueous formulations of ASO are generally administered by slow infusion, rather than IV bolus, to control peak plasma levels below those associated with acute toxicity. Sterile solutions for injection are prepared by aseptic filtration and are also suitable for lyophilization or freeze-drying for subsequent reconstitution prior to use. These formulations are stored in glass vials but it is reasonable to consider and qualify alternate primary packaging, such as blow-fill-seal plastics, prefilled syringes, or other packaging configurations.

Besides the above-described conventional and relatively simple solution formulations, there is a huge effort underway, using novel formulations, to promote more efficient cellular uptake and delivery of biotechnology compounds, including ASOs. For pharmacologic activity, cellular uptake of ASO is necessary which has generally been perceived as a challenge for ASOs. *In-vitro* observations of phosphorothioate oligodeoxynucleotides led many working in the area to conclude that, because of metabolic instability, size, and charge, ASOs would not be adequately taken up into the targeted tissues or cells necessary to exert pharmacologic activity. These points notwithstanding, an ever-expanding body of *in-vivo* animal and clinical data is demonstrating that even simple saline solutions, parenterally administered, can lead to effective ASO uptake and pharmacologic activity. Yet the efficiency of this cellular uptake process is limited. Thus, the development of novel formulations and delivery routes is opening a wider variety of target tissues and cells to treatment with ASOs through increased efficiency and stability. Such formulations include those specifically addressing cellular uptake and trafficking on a subcellular level. They include a variety of novel formulations, such as functional liposomes (long-circulating or Stealth™ liposomes, charge-based liposomes, fusogenic and cell-surface targeted liposomes), protein-coated microbubbles, and cationic complexes and microspheres. The discussion of such novel formulations affecting cellular-based drug targeting, uptake and release is beyond the scope of this section, and the reader is referred to other sources for details on the subject [42–46].

#### 10.4.1.1 **Sustained-Release Subcutaneous Formulations**

Subcutaneous administration is preferable to continuous IV infusion due to the convenience and lower costs of administration. For SC administration, a higher potency is required than that generally demonstrated by first-generation oligonucleotides in order to reduce the depot mass to provide a reasonable volume for injection. Interesting formulation progress has been reported in the literature. A SC injection of the phosphorothioate *c-myc* ASO, complexed with zinc and microencapsulated with the biodegradable polymer poly(D,L-lactide-*co*-glycolic acid), was found to be significantly more efficacious than an IV administration of unencapsulated drug in human melanoma and leukemia xenografts in immunocompromised mice [47]. The microencapsulated ASO provided a sustained-release formulation that was effective at lower doses and with less frequent dosing than the unformulated (solution) ASO.

### 10.4.2 Oral Delivery

Oral dosage forms are ideal for patient convenience and compliance. For traditional small molecular-weight molecules, this route has been used for both systemic delivery and delivery to local GI tissues, as mentioned earlier. For large, polyanionic, hydrophilic oligonucleotide molecules, a variety of issues must be considered when designing such dosage forms, particularly to enable systemic absorption by the oral route. Among these are chemical instability and precipitation at gastric pH, metabolic instability within the intestinal environment, low intestinal permeability, high protein binding, and first-pass hepatic clearance. These challenges have led many to conclude that effective oral systemic administration of most biotechnology therapeutics with characteristics of ASOs is not feasible [48]. Indeed, the early reports of oral bioavailability approaching 35% from gavage solutions containing radioactive phosphorothioate oligodeoxynucleotides in rats were consequently perceived to be the uptake of degraded forms [49]. The above-mentioned barriers to oral bioavailability of ASO were delineated by Nicklin et al. [50]. Of these, two stand out as critical: instability in the GI tract; and low permeability across the intestinal mucosa. However, progress is being made to address and/or understand each of these barriers by way of changes to oligonucleotide chemistry and use of appropriate formulations.

ASOs are rapidly digested by the ubiquitous nucleases found within the gut. As a consequence, these first-generation ASOs require stabilization in order to achieve a reasonable GI residence time to allow absorption to occur. To impart enzyme resistance, modifications can be made to the phosphate backbone, the nucleotide sugar, or both. It has been demonstrated that such modifications need to be incorporated at only a subset of the sites available on the oligonucleotide molecule. Such hybrid structures have demonstrated significantly improved nuclease stability. This was shown by Zhang et al. [51, 52] for both backbone modifications (methylphosphonates) and for sugar-modified (2′-*O*-methyl) ASOs. However, care must be taken to ensure that mRNA binding affinity is not impaired by such modifications – particularly modifications involving an increase in the size of the 2′-*O*-substituent. It was determined that 2′-*O*-alkoxyalkyl substituents retain or improve upon the RNA binding affinity and have greatly improved nuclease resistance. The MOE derivative was proposed as a promising candidate from this class of derivatives [5], and has since been evaluated for presystemic enzyme stability within the digestive system of the rat. This stability is proportional to the degree of sugar substitution (Fig. 10.2).

Relative to the unmodified phosphorothioate oligodeoxynucleotides, these data demonstrate the increasing benefit of MOE substituents present as either a hemimer, with MOEs at the 3′-termini only; or a gap-mer, with MOEs present at both the 3′- and 5′-termini; and lastly, a full MOE substitution at every sugar location. These MOE gap-mer chemistries are demonstrating clinical efficacy by the subcutaneous

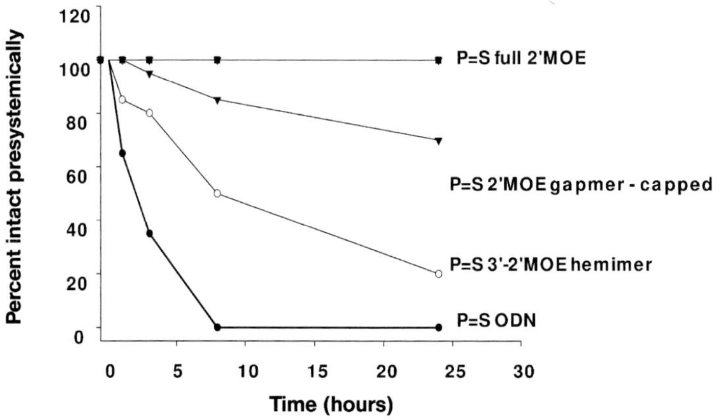


**Fig. 10.2** Stability of various oligonucleotide chemistries in the rat intestine prior to absorption. P=S designates phosphorothioate ASO backbone chemistry.

route and, in the case of ISIS 301012, which targets apoB-100 mRNA, the potency is such that oral dosing for systemic therapy of hyperlipidemia is feasible [53].

An alternative approach to improving enzyme stability of ASO is by way of formulations that impose a physical barrier, preventing or delaying the nuclease enzymes from reaching the oligonucleotide cleavage site. Examples of such formulations include: (1) conventional modified-release dosage forms, which incorporate the oligonucleotide within the solid matrix of the dosage unit, and (2) complexed ASOs with cationic species formulated through a process referred to as complex coacervation. In addition to enzyme stability, such formulations have been purported to impart additional characteristics favorable to oligonucleotide absorption. These include: modified delivery (e.g., site-specific or increased retention by bioadhesion); a change in oligonucleotide physico-chemical properties (e.g., charge neutralization or lipophilicity); and a change or impact on the absorbing membrane itself (e.g., opening of tight junctions or fluidization of mucosal membrane). These concepts are discussed more fully below.

#### 10.4.2.1 **Permeability**

The physico-chemical properties of ASOs present a significant barrier to their GI absorption into the systemic circulation or the lymphatics. These properties include their large size and molecular weight (i.e., ca. 7 kDa for 20-mers), hydrophilic nature (log $D_{o/w}$ approximating –3.5) and multiple ionization p$K_a$-values. The titration of a 20-mer phosphorothioate oligonucleotide on a Sirius Glp$K_a$ instrument noted over 17 p$K_a$s for the unmodified oligodeoxynucleotide sequence and over 32 p$K_a$s for the MOE hemi-mer form of the same sequence (unpublished data). While the baseline permeability values for 20-mer oligodeoxynucleotides approximate $2 \times 10^{-6}$ cm/s in the rat small intestine (mucosal-to-serosal), and can be

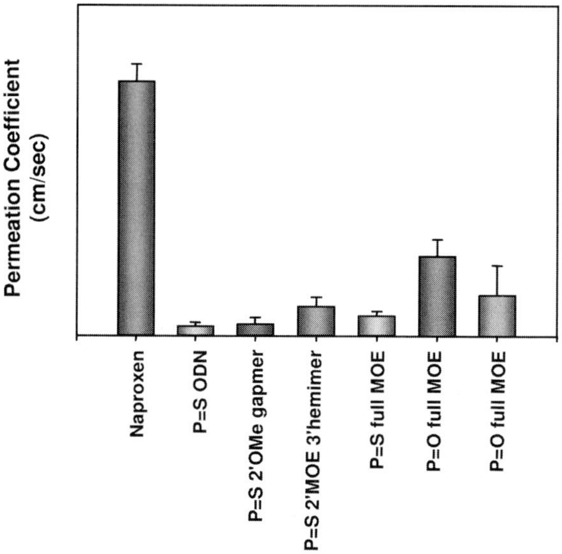


**Fig. 10.3** Effect of oligonucleotide chemistry on permeability in the rat intestine *in situ* [54]. P=O designates phosphodiester ASO backbone chemistry; P=S designates phosphorothioate ASO backbone chemistry.

improved by chemistry modifications (Fig. 10.3), these values are still far below those of compounds exhibiting high permeability, such as naproxen [54].

The use of formulations can further improve upon this permeability. When formulating ASO drugs to improve oral bioavailability, the mechanism of oligonucleotide absorption, either paracellular via the epithelial tight junctions or transcellular, by direct passage through the lipid membrane bilayer, must be considered. By using paracellular and transcellular models appropriate for water-soluble, hydrophilic macromolecules, it was suggested that ASOs predominantly traverse the GI epithelium via the paracellular route [55]. In this regard, formulation design considerations involve the selection of those PEs which facilitate paracellular transport and meet other formulation criteria including: suitable biopharmaceutics, safety considerations, manufacturability, physical and chemical stability, and, practicality of the dosage form configuration (i.e., regarding production costs, dosing regimen, patient compliance, etc.).

Initial attempts at selecting PEs have identified certain surfactants, such as bile salts and fatty acids, which appear to facilitate oligonucleotide absorption. The advantages of these components are many, in that they are endogenous to foods and body constituents, plus the literature is rich with information about the use and exposure of these two classes of compounds [56]. The precise mechanism of action for these PEs is unknown, but is believed to involve a disruption of the mucus layer barrier, an increase in the fluidity of the mucosal membrane, and potentially an opening of the paracellular tight junctions. The mucolytic effect coupled with the increased membrane fluidity imparted by these excipients appears to allow increased concentrations of ASO to enter the villi crypt wells. Experimental data suggest that tight junctions at the villi crypts are larger and therefore more permeable than their counterparts at the villi tips [57]. When PE-formulated drug is administered orally to mice, the ASO can be visualized by immunohistochemistry throughout the brush-border areas of the murine ileum [12]. This enhanced intimacy of the ASO to the absorbing surface increases its GI residence time and therefore its potential absorption via entry into the paracellular tight junctions.

#### 10.4.2.2 **Systemic Bioavailability**

Of the various methods that may be used to determine bioavailability for ASOs, the best refer to tissue levels as the most relevant metric for calculating an estimate of absolute bioavailability. As mentioned in Chapter 4, ASOs distribute rapidly to the tissues, with an extremely slow transfer rate back into the central circulation. In addition, the elimination of ASOs occurs predominantly by nucleases in the tissue compartment. Thus, bioavailability based on plasma concentrations does not provide an accurate estimate of absolute bioavailability for ASOs if plasma concentrations cannot be quantified at extremely low concentrations for a prolonged period of time to adequately assess systemic exposure. The direct use of tissue levels in combination with physiologic pharmacokinetic modeling, however, may allow the accurate determination of bioavailability for ASOs.

Preclinical studies across species noted that bioavailabilities calculated from tissue-derived data were consistently higher than plasma-derived bioavailabilities. This difference varied across tissues, but in general amounted to about a two-fold difference between plasma and tissue bioavailability (Fig. 10.4). While the tissue

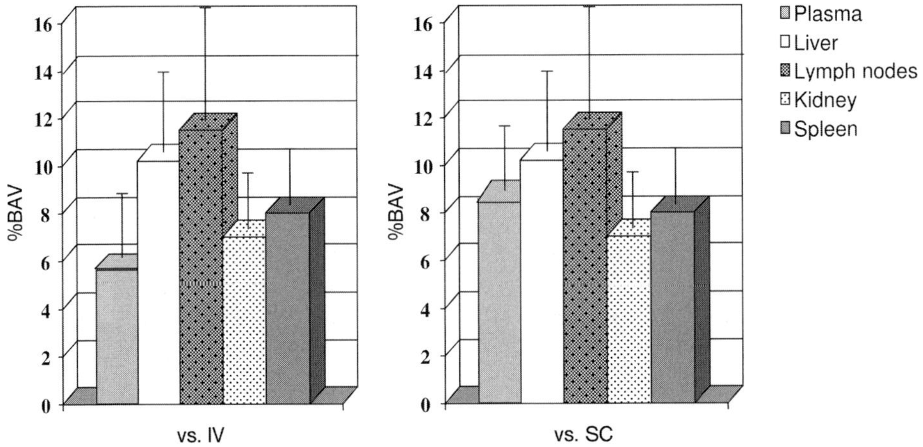


**Fig. 10.4** Monkey plasma and tissue bioavailabilities (BAV) determined by either an intravenous (IV) or subcutaneous (SC) AUC baseline after seven daily intrajejunal administrations of ISIS 104838. Note that the plasma BAV was more in line with tissue(s) for the SC-based data.

bioavailability can certainly be defined as the correct metric for ASO bioavailability, practicalities dictate that plasma bioavailability be used as a surrogate to measure progress on oral formulation performance in human clinical studies.

Further to this, experiments strongly suggest that the more appropriate measure of true/tissue bioavailability is from the use of a SC AUC reference rather than from the use of an IV reference. The rationale for this change is based upon the nonlinear distribution kinetics of ASOs, and a further explanation is provided here.

Assuming that a drug can be characterized by linear pharmacokinetics, the standard approach to determining absolute bioavailability F is based on plasma AUCs of oral data relative to IV data, dose normalized [58]. Accordingly, F would be calculated using the following equation:

$$F = (AUC_{po}/Dose_{po})/(AUC_{iv}/Dose_{iv}) \times 100\%$$

While many aspects of nonlinear pharmacokinetic behavior may impact on the above equation, the more relevant pharmacokinetic processes for ASOs are absorption and distribution at or below therapeutic or nontoxic plasma concentrations. Nonlinear absorption or distribution processes can affect AUC terms in a nonproportional manner when different doses are compared, thereby resulting in an inaccurate determination of bioavailability. This has been shown to occur on numerous occasions for compounds such as ascorbic acid or naproxen [59–62]. Such cases require an understanding of the capacity-limiting cause of the nonlinearity and the pharmacokinetic processes upon which this impacts, in case of ASOs absorption and intercompartmental distribution processes from the central compartment into the peripheral tissues. With this understanding, various methods may then be applied to best approximate the rate of change of the plasma concentrations from one sampling time to the next allowing for an estimate of the absolute BAV.

As mentioned in Chapter 4, experiments have determined that the distribution of ASOs into tissues is nonlinear. This revelation invalidates the above BAV equation in that it is dependent on linear pharmacokinetics and the principle of superposition. A way to circumvent this problem is to decrease the drug input function (i. e., systemic presentation of the ASO) such that ASO plasma concentrations are maintained below the level at which saturation, and thus nonlinearity of the distribution processes, occurs. Drug administration by SC rather than IV administration has a reduced drug input rate and can produce such a scenario. The corresponding plasma-derived data are then suitable for the determination of absolute bioavailability – consistent with linear pharmacokinetic principles and the following equation:

$$F = (AUC_{po}/Dose_{po})/(AUC_{sc}/Dose_{sc}) \times 100\%$$

Therefore, for ASO and other nonlinear distribution drugs it may be more appropriate to base oral bioavailabilities on a SC- rather than an IV-derived reference.

As such, oral bioavailabilities noted in this chapter are determined accordingly unless otherwise noted.

Recent efforts have demonstrated that significant oral bioavailability can be achieved in humans. These as yet unpublished results have shown that the performance of uniquely designed formulations containing PEs perform in a manner similar to that seen in numerous animal model systems.

One useful animal model in this context is the intrajejunally (IJ) ported canine. While not strictly oral administration, the IJ-dosed dog is considered important in that the dog model is a large species and therefore more representative than the rat model of what might be expected in humans. In this model, significant oligonucleotide absorption has been observed after IJ administration of a solution containing both PEs and ISIS 2302, a 20-mer first-generation ASO. Plasma concentrations of this intact ASO and of another ASO, ISIS 15839, which has the same sequence but with a second-generation ASO chemistry are presented in Fig. 10.5. The second-generation chemistry for ISIS 15839 is such that this ASO possesses eight consecutive 2′-MOE modified sugars on the 3′ end of the oligonucleotide. The pharmacokinetic profiles indicate that the so-called "hemi-mer MOE" oligonucleotide has a plasma half-life which is longer than that of the unmodified first-generation ISIS 2302 ASO.

When this formulation was assembled as a conventional tablet it took over 1 h to dissolve, and the resultant plasma concentrations after oral dosing were negligible (data not shown). Upon reformulation to promote immediate and complete dissolution in less than 15 min, the absolute bioavailability approached 3%, as indicated by the plasma concentration plots in Fig. 10.5. These results, obtained

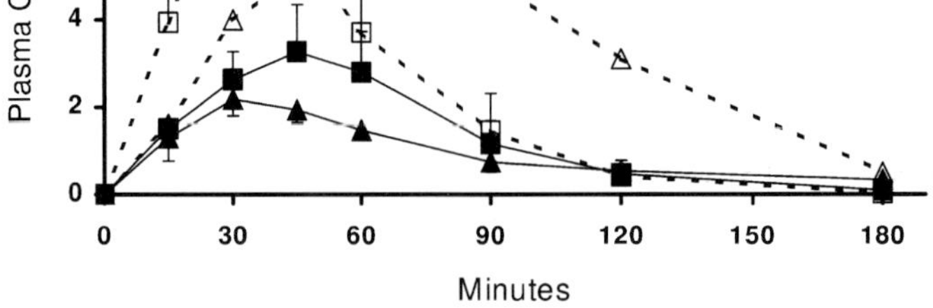


**Fig. 10.5** Intact oligonucleotide in dogs after either intrajejunal (IJ) administration of a solution formulation (open symbols, 10 mg/kg) or oral administration of a tablet formulation (filled symbols, ~15 mg/kg) of two oligonucleotide chemistries. ISIS-2302 (squares) is a phosphorothioate oligodeoxynucleotide; ISIS-15839 (triangles) is a hemi-mer MOE derivative with the same sequence. Error bars indicate standard error.

with a rudimentary solid dosage form, yielded less than half of the bioavailability obtained with the solution IJ dosing. From these results, it may be concluded that the following factors are important for oligonucleotide absorption in the above PE system: coincident presentation of oligonucleotide with the PE components; high concentrations along the absorbing gut mucosa; and enteric protection to prevent component dilution and oligonucleotide exposure to low pH which has been associated with degradation and precipitation of both active ASO and PE. It should be noted again that plasma concentrations represent the absorption and distribution of ASOs, but not the true tissue or body elimination kinetics. Therefore, a more relevant metric for the pharmacokinetics to be based upon is that of tissue levels. This is particularly true for oligonucleotide chemistries that are rapidly cleared from the bloodstream and accumulate in target tissues or organs.

In line with this hypothesis, a study was performed to evaluate both plasma pharmacokinetics and tissue distribution of the above-mentioned first- and second-generation ASO chemistries, ISIS 2302 and ISIS 15839. Pharmacokinetic parameters for each chemistry were assessed in beagle dogs during and following 14-day oral dosing of tablets containing oligonucleotide and PEs in enteric-coated tablets [63]. The *in-vitro* performance of these tablets was such that they remained intact in acid medium, yet rapidly dissolved (~15 min) at pH 6.5. In plasma, ISIS 15839 gave approximately 150% the bioavailability found for ISIS 2302 (Table 10.4). However, at the end of the study, ASO tissue concentrations for both liver and kidney cortex were approximately 10-fold higher for ISIS 15839 than for ISIS 2302 (Table 10.5).

In this multiple dosing study, the distribution of ASO out of the plasma circulation was very rapid, with a distribution half-life of 30–60 min. As mentioned earlier, the bioavailability assessments based upon plasma data may be inaccurate and/or misleading for this class of compound. The bioavailability determined from sentinel tissues, such as the kidney and liver, may be a more credible representation of the true systemic exposure as these tissues reflect true elimination kinetics as well as accumulation to steady-state concentrations during chronic dosing. On this basis, our data indicate that bioavailabilities based upon organ concentrations (after oral and IV dosing) would be 50% higher for the kidney and 100% higher for the liver than those calculated by plasma AUCs.

**Table 10.4** Summary of plasma oligonucleotide pharmacokinetics after the first single 200-mg oral dose in a multiple dose study.

| Oligonucleotide | AUC [μg min/mL] | $C_{max}$ [μg/mL] | Plasma BAV [%][a)] |
|---|---|---|---|
| ISIS 15839 | 91 | 1.2 | 1–4 |
| ISIS 2302 | 62 | 1.1 | 0.1–3 |

a) Bioavailability (BAV) as ratio of plasma $AUC_{oral}$ relative to plasma $AUC_{IV}$ (dose-adjusted).

**Table 10.5** Tissue oligonucleotide concentrations and approximate bioavailabilities (BAV) based on data obtained at the end of two-week daily dosing study in dogs.

| Oligonucleotide chemistry (20-base phosphorothioates) | Tissue concentration [μg/g] Liver | Kidney cortex | %BAV [a)] |
|---|---|---|---|
| ISIS 2302: Antisense to ICAM-1 (deoxy) | 4.9 | 12 | 1.3 |
| ISIS 15839: Antisense to ICAM-1 (hemi-mer MOE of ISIS 2302) | 33 | 109 | 5.5 |

**a)** Average BAV as ratio of final tissue concentrations (oral to IV), dose adjusted.

In consideration of the practical difficulties involved in the experimental determinations of organ bioavailability, we have performed simulations of organ accumulation after oral dosing of first-generation phosphorothioate oligodeoxynucleotide versus a MOE gapmer (Fig. 10.6). Each simulation was generated assuming a random input metric of 2 to 15% oral bioavailability (geometric mean 7%) with all other pharmacokinetic parameters for these two chemistries remaining fixed. The simulations suggest that, even with variable absorption, the long tissues' half-life of second-generation ASOs, which results in enhanced drug accumulation in tissue, makes it feasible to administer the newer chemistry MOE gapmer compounds by the oral route.

To place a context to this model we have recently dosed humans with first-generation ASO solutions containing PEs. This clinical study involved the administration of test solutions directly to the human jejunum by way of intubation. Preliminary analysis of the resultant ASO plasma concentrations after dosing via the IJ route versus the IV route indicate bioavailabilities that are significantly higher than those obtained by the comparable dog IJ model [12]. Furthermore, it has been demonstrated in intubated humans that the presence of an intestinal tube will affect GI motility. Read et al. [64] determined that, in the presence of an intestinal tube, peristalsis is heightened in the small intestine, thereby enhancing the transit of food to the colon. The impact of such enhanced mid-gut motility on the absorption of formulated ASO solutions is unknown, but is believed to diminish absorption due to the concomitant dilutional effect that such motility would have on the formulation's PEs. For these reasons it is possible that these human data, as well as the dog model, are both underpredicting the bioavailabilities of orally dosed formulation in humans. The current data suggest that oral dosing using new chemistries of ASOs will ultimately be feasible for a variety of indications, and that oral dosing using first-generation ASOs may be possible for selected indications, such as for certain sentinel organs or for local GI indications.

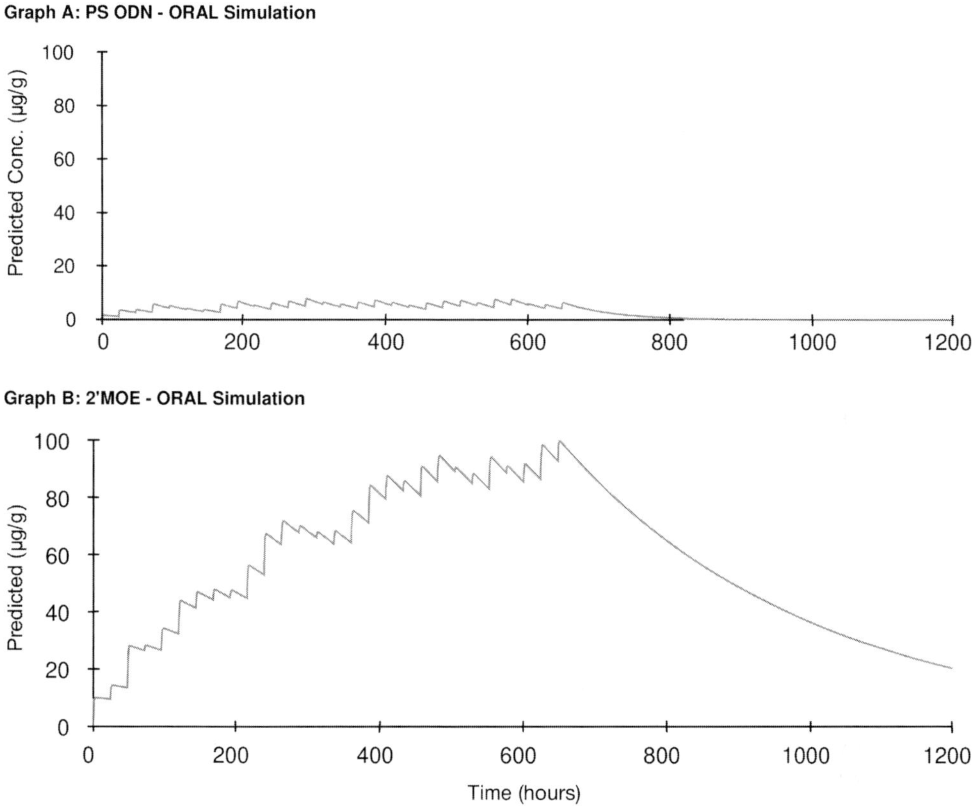


**Fig. 10.6** Simulation of tissue oligonucleotide concentrations for a daily oral dosing regimen of one-month duration for a oligodeoxynucleotide (A) or a MOE gapmer chemistry (B) at 2 mg/kg using a variable bioavailability input metric randomly generated from 2 to 15% (geometric mean 7%).

## 10.5 Conclusions

To date, the clinical administration of ASOs has focused on local or parenteral routes using simple saline solutions. Intravenous dosing in animals has been shown to lead to an accumulation of oligonucleotide in specific cells, such as endothelial and phagocytic cells, and organs such as the kidneys and liver. These observations suggest optimal targets for clinical applications, for example liver diseases such as hepatitis C or ophthalmic indications such as the intraocular treatment of CMV retinitis.

Animal models have shown that formulations and delivery routes can be used to alter the tissue distribution of ASOs. By making use of the knowledge gained through these studies, ASOs may be delivered selectively to a variety of target tissues including some that would otherwise not be easily accessible, such as in the lungs, skin, and colon. The development of such local therapies has the potential for increased patient convenience and compliance, as well as reduced treatment costs due to the more efficient delivery of the ASO therapeutic.

## 10.6 References


1 SantaLucia, J., Jr. **1998**. A unified view of polymer, dumbbell, and oligonucleotide DNA nearest-neighbor thermodynamics. *Proc. Natl. Acad. Sci. USA* 95 : 1460–1465.
2 Vitravene$^{TM}$ **1998**. Formivirsen sodium intravitreal injectable. Monograph. Ciba Vision, Novartis.
3 Hardee, G.E. Unpublished results.
4 Krotz, A.H., R.C. Mehta, and G.E. Hardee. **2005**. Peroxide-mediated desulfurization of phosphorothioate oligonucleotides and its prevention. *J. Pharm. Sci.* 94 : 341–352.
5 Altmann, K.-H., N.M. Dean, D. Fabbro, S.M. Freier, T. Geiger, R. Haner, D. Husken, P. Martin, B.P. Monia, M. Muller, F. Natt, P. Nicklin, J. Phillips, U. Pieles, H. Sasmor, and H. Moser, E. **1996**. Second generation of antisense oligonucleotides: From nuclease resistance to biological efficacy in animals. *Chimia* 50 : 168–176.
6 de Smet, M.D., C.J. Meenken, and G.J. van den Horn. **1999**. Fomivirsen – a phosphorothioate oligonucleotide for the treatment of CMV retinitis. *Ocular Immunology and Inflammation* 7 : 189–198.
7 Leeds, J.M., S.P. Henry, L. Truong, A. Zutsi, A.A. Levin, and D.J. Kornbrust. **1997**. Pharmacokinetics of a potential human cytomegalovirus therapeutic, a phosphorothioate oligonucleotide, after intravitreal injections in the rabbit. *Drug Metab. Dispos.* 25 : 921–926.
8 Levin, A.A. **1999**. A review of issues in the pharmacokinetics and toxicology of phosphorothioate antisense oligonucleotides. *Biochim. Biophys. Acta* 1489 : 69–84.
9 Voigt, M., Y. de Kozak, M. Halhal, Y. Courtois, and F. Behar-Cohen. **2002**. Down-•regulation of NOSII gene expression by iontophoresis of anti-sense oligonucleotide in endotoxin-induced uveitis. *Biochem. Biophys. Res. Commun.* 295 : 336–341.
10 Brown, J., S. Haines, and I.R. Wilding. **1997**. Colonic spread of three rectally administered mesalazine (Pentasa) dosage forms in healthy volunteers as assessed by gamma scintigraphy. *Aliment. Pharmacol. Ther.* 11 : 685–691.
11 Travis, S.P.L. **2001**. The management of refractory distal colitis. *Proc. R. Coll. Physicians Edinb.* 31 : 222–228.
12 Tillman, L.G. Unpublished results.
13 van Deventer, S.J., J.A. Tami, and M.K. Wedel. **2004**. A randomised, controlled, double blind, escalating dose study of alicaforsen enema in active ulcerative colitis. *Gut* 53 : 1646–1651.
14 Geary, R.S., J.M. Leeds, J. Fitchett, T. Burckin, L. Troung, C. Spainhour, M. Creek, and A.A. Levin. **1997**. Pharmacokinetics and metabolism in mice of a phosphorothioate oligonucleotide antisense inhibitor of C-raf-1 kinase expression. *Drug Metab. Disp.* 25 : 1272–1281.
15 Wu-Pong, S., and P.R. Byron. **1996**. Airway-to-biophase transfer of inhaled oligonucleotides. *Adv. Drug Deliv. Rev.* 19 : 47–71.
16 Yu, J., and Y.W. Chien. **1997**. Pulmonary drug delivery: physiologic and mechanistic aspects. *Crit. Rev. Ther. Drug Carrier Syst.* 14 : 395–453.
17 Templin, M.V., A.A. Levin, M.J. Graham, P.M. Aberg, B.I. Axelsson, M. Butler, R.S. Geary, and C.F. Bennett. **2000**. Pharmacokinetic and toxicity profile of a phosphorothioate oligonucleotide following inhalation delivery to lung in mice. *Antisense Nucleic Acid Drug Dev.* 10 : 359–368.
18 Nyce, J.W. **1997**. Respirable antisense oligonucleotides as novel therapeutic agents for asthma and other pulmonary diseases. *Expert Opin. Investig. Drugs* 6 : 1149–1156.
19 Nyce, J.W., and W.J. Metzger. **1997**. DNA antisense therapy for asthma in an animal model. *Nature* 385 : 721–725.
20 Georges, R.N., T. Mukhopadhyay, Y. Zhang, N. Yen, and J.A. Roth. **1993**.

Prevention of orthotopic human lung cancer growth by intratracheal instillation of a retroviral antisense K-ras construct. *Cancer Res.* 53:1743–1746.

**21** Pardridge, W.M. **1995**. Transport of small molecules through the blood-brain barrier: biology and methodology. *Adv. Drug Delivery Rev.* 15:5–36.

**22** Le Corre, S.M., P.W. Burnet, R. Meller, T. Sharp, and P.J. Harrison. **1997**. Critical issues in the antisense inhibition of brain gene expression in vivo: experiences targetting the 5-HT1A receptor. *Neurochem. Int.* 31:349–362.

**23** Whitesell, L., D. Geselowitz, C. Chavany, B. Fahmy, S. Walbridge, J.R. Alger, and L.M. Neckers. **1993**. Stability, clearance, and disposition of intraventricularly administered oligodeoxynucleotides: Implications for therapeutic application within the central nervous system. *Proc. Natl. Acad. Sci. USA* 90:4665–4669.

**24** Sommer, W., X. Cui, B. Erdmann, L. Wiklund, G. Bricca, M. Heilig, and K. Fuxe. **1998**. The spread and uptake pattern of intracerebrally administered oligonucleotides in nerve and glial cell populations of the rat brain. *Antisense Nucleic Acid Drug Dev.* 8:75–85.

**25** Mustafa, S.J., and D. M.S. **1999**. Adenosine A1 receptor antisense oligonucleotide treatment of alcohol and marijuana-induced psychomotor impairments. In US Patent: 5932557.

**26** Yazaki, T., S. Ahmad, A. Chahlavi, E. Zylber-Katz, N.M. Dean, S.D. Rabkin, R.L. Martuza, and R.I. Glazer. **1996**. Treatment of glioblastoma U-87 by systemic administration of an antisense protein kinase C-$\alpha$ phosphorothioate oligodeoxynucleotide. *Mol. Pharmacol.* 50: 236–242.

**27** Crooke, S.T., M.J. Graham, J.E. Zuckerman, D. Brooks, B.S. Conklin, L.L. Cummins, M.J. Greig, C.J. Guinosso, D. Kornbrust, M. Manoharan, H.M. Sasmor, T. Schleich, K.L. Tivel, and R.H. Griffey. **1996**. Pharmacokinetic properties of several novel oligonucleotide analogs in mice. *J. Pharmacol. Exp. Ther.* 277:923–937.

**28** Boado, R.J., H. Tsukamoto, and W.M. Pardridge. **1998**. Drug delivery of antisense molecules to the brain for treatment of Alzheimer's disease and cerebral AIDS. *J. Pharm. Sci.* 87:1308–1315.

**29** Holbrook, K.A., and K. Wolf. **1993**. The structure and development of skin. In: Fitzpatrick, T.B., et al. (Eds.), *Dermatology in General Medicine*. McGraw-Hill, New York, pp. 97–145.

**30** Lampe, M.A., A.L. Burlingame, J. Whitney, M.L. Williams, B.E. Brown, E. Roitman, and P.M. Elias. **1983**. Human stratum corneum lipids: characterization and regional variations. *J. Lipid Res.* 24:120–130.

**31** Lee, Y.M., S.H. Lee, G.I. Ko, J.B. Kim, and D.H. Sohn. **1996**. Effect of benzalkonium chloride on percutaneous absorption of antisense phosphorothioate oligonucleotides. *Arch. Pharmacol. Res.* 19:435–440.

**32** Mitragotri, S., D. Blankschtein, and R. Langer. **1995**. Ultrasound-mediated transdermal protein delivery. *Science* 269: 850–853.

**33** Regnier, V., T. Le Doan, and V. Preat. **1998**. Parameters controlling topical delivery of oligonucleotides by electroporation. *J. Drug Target.* 5:275–289.

**34** Banga, A.K., and M.R. Prausnitz. **1998**. Assessing the potential of skin electroporation for the delivery of protein- and gene-based drugs. *Trends Biotechnol.* 16:408–412.

**35** Furhman, L., et al. **1995**. Evaluation of several liposomal formulations and preparation techniques for the dermal delivery of phosphorothioate antisense oligonucleotides in hairless mouse skin in vitro. In AAPS Annual Meeting, Miami Beach, FL, USA.

**36** Roberts, M.S., M. Walker, and Water. **1993**. The most natural penetration enhancer. In: Hadgraft, J. (Ed.), *Pharmaceutical Skin Penetration Enhancers*. Marcel Dekker, New York, pp. 1–30.

**37** Catz, P., and D.R. Friend. **1988**. Mechanism of skin penetration enhancer: Ethyl acetate. *Pharm. Res.* 6:108.

**38** Millns, J.L., and H.I. Maibach. **1982**. Mechanisms of sebum production and delivery in man. *Arch. Dermatol. Res.* 272:351–362.

**39** Beastall, J.C., J. Hadgraft, and C. Washington. **1988**. Mechanism of action of Azone as a percutaneous penetration enhancer: lipid bilayer fluidity and transition temperature effects. *Int. J. Pharm.* 43:207–213.

**40** Michniak, B.B., et al. **1993**. In-vitro evaluation of a series of azone analogs as dermal penetration enhancers. *Int. J. Pharmaceutics* 91:85–93.

**41** Mehta, R.C., K.K. Stecker, S.R. Cooper, M.V. Templin, Y.J. Tsai, T.P. Condon, C.F. Bennett, and G.E. Hardee. **2000**. Intercellular adhesion molecule-1 suppression in skin by topical delivery of antisense oligonucleotides. *J. Invest. Dermatol.* 115:805–812.

**42** Slepushkin, V.A., S. Simoes, P. Dazin, M.S. Newman, L.S. Guo, M.C. Pedroso de Lima, and N. Duzgunes. **1997**. Sterically stabilized pH-sensitive liposomes. Intracellular delivery of aqueous contents and prolonged circulation in vivo. *J. Biol. Chem.* 272:2382–2388.

**43** Bennett, C.F., M.-Y. Chiang, H. Chan, J.E.E. Shoemaker, and C.K. Mirabelli. **1992**. Cationic lipids enhance cellular uptake and activity of phosphorothioate antisense oligonucleotides. *Mol. Pharmacol.* 41:1023–1033.

**44** Zelphati, O., and F.C. Szoka, Jr. **1996**. Intracellular distribution and mechanism of delivery of oligonucleotides mediated by cationic lipids. *Pharm. Res.* 13:1367–1372.

**45** Lysik, M.A., and S. Wu-Pong. **2003**. Innovations in oligonucleotide drug delivery. *J. Pharm. Sci.* 92:1559–1573.

**46** Brown, L.R., Gillis, K.A., et al. **2005**. PROMAXX microsphere nucleic acid delivery to prevent type 1 diabetes. In: Controlled Release Society Annual Meeting, Miami Beach, FL, USA.

**47** Putney, S.D., J. Brown, C. Cucco, R. Lee, T. Skorski, C. Leonetti, T. Geiser, B. Calabretta, G. Zupi, and G. Zon. **1999**. Enhanced anti-tumor effects with microencapsulated c-myc antisense oligonucleotide. *Antisense Nucleic Acid Drug Dev.* 9:451–458.

**48** Humphrey, M.J. 1986. The oral bioavailability of peptides and related drugs. In: Tomlinson, E. (Ed.), *Delivery Systems for Peptide Drugs*. Plenum Press, New York, pp. 139–151.

**49** Agrawal, S., and R. Zhang. **1998**. Pharmacokinetics and bioavailability of antisense oligonucleotides following oral and colorectal administrations in experimental animals. In: Crooke, S.T. (Ed.), *Antisense Research and Applications*. Springer-Verlag, Berlin, Heidelberg, pp. 525–543.

**50** Nicklin, P.L., S.J. Craig, and J.A. Phillips. **1998**. Pharmacokinetic properties of phosphorothioates in animals – absorption, distribution, metabolism and elimination. In: Crooke, S.T. (Ed.), *Antisense Research and Applications*. Springer-Verlag, Berlin, pp. 141–168.

**51** Zhang, R., Z. Lu, H. Zhao, X. Zhang, R.B. Diasio, I. Habus, Z. Jiang, R.P. Iyer, D. Yu, and S. Agrawal. **1995**. In vivo stability, disposition and metabolism of a 'hybrid' oligonucleotide phosphorothioate in rats. *Biochem. Pharmacol.* 50:545–556.

**52** Zhang, R.W., R.P. Iyer, D. Yu, W.T. Tan, X.S. Zhang, Z.H. Lu, H. Zhao, and S. Agrawal. **1996**. Pharmacokinetics and tissue disposition of a chimeric oligodeoxynucleoside phosphorothioate in rats after intravenous administration. *J. Pharmacol. Exp. Ther.* 278:971–979.

**53** Bradley, J.D., Crooke, R., Kjems, L.L., Graham, M., Leong, R., Yu, R., Paul, D., Wedel, M. **2005**. Hypolipidemic effects of a novel inhibitor of human ApoB-100 in humans. *Diabetes* 54:977.

**54** Khatsenko, O., R. Morgan, L. Troung, C. York-Defalco, H. Sasmor, B. Conklin, and R.S. Geary. **2000**. Absorption of antisense oligonucleotides in rat intestine: effect of chemistry and length. *Antisense Nucleic Acid Drug Dev.* 10:35–44.

**55** Anderberg, E.K., T. Lindmark, P. Artursson. **1993**. Sodium caprate elicits dilatations in human intestinal tight junctions and enhances drug absorption by the paracellular route. *Pharm. Res.* 10:857–864.

**56** Lee, V.H., A. Yamamoto, and U.B. Kompella. **1991**. Mucosal penetration enhancers for facilitation of peptide and protein drug absorption. *Crit. Rev. Ther. Drug Carrier Syst.* 8:91–192.

**57** Hollander, D. **1992**. The intestinal permeability barrier. A hypothesis as to its regulation and involvement in Crohn's disease. *Scand. J. Gastroenterol.* 27:721–726.

**58** Gibaldi, M., and D. Perrier. **1982**. *Pharmacokinetics*. Dekker, New York.

**59** Mayersohn, M. **1972**. Ascorbic acid absorption in man – pharmacokinetic implications. *Eur. J. Pharmacol.* 19 : 140–142.

**60** Runkel, R., E. Forchielli, H. Sevelius, M. Chaplin, and E. Segre. **1974**. Nonlinear plasma level response to high doses of naproxen. *Clin. Pharmacol. Ther.* 15 : 261–266.

**61** Brass, E.P. **2000**. Supplemental carnitine and exercise. *Am. J. Clin. Nutr.* 72: s618–s623.

**62** Traber, M.G. **1997**. Regulation of human plasma vitamin E. *Adv. Pharmacol.* 38 : 49–63.

**63** Geary, R.S. Unpublished results.

**64** Read, N.W., M.N. Al Janabi, T.E. Bates, and D.C. Barber. **1983**. Effect of gastrointestinal intubation on the passage of a solid meal through the stomach and small intestine in humans. *Gastroenterology* 84 : 1568–1572.

# 11
# Custom-Tailored Pharmacokinetics and Pharmacodynamics via Chemical Modifications of Biotech Drugs

*Francesco M. Veronese and Paolo Caliceti*

## 11.1
## Introduction

Biotechnological products, peptide, proteins and oligonucleotides, are of great therapeutic value since the specificity of their biological activity and, on the basis of advances in genetic engineering, their ease of production. Unfortunately, their use suffers from several limitations, including degradation by proteolytic enzymes and rapid renal excretion or removal by the reticuloendothelial system. In an attempt to overcome these limitations, several alternative approaches have been made:

- Replacement of those amino acid residues that are sites of proteolytic degradation by genetic engineering.
- Synthesis of truncated and more stable protein sequences.
- Entrapment into particles such as liposomes or insoluble polymeric microspheres.
- Covalent linking of the surface to polymers.

The first and the second approaches have provided some positive results but, unfortunately, they depend upon the structure of the protein. Typical examples of sequence modifications to improve stability and pharmacokinetics are the preparation of humanized antibodies, where part of the mouse sequence is substituted by the human form, and the granulocyte colony-stimulating factor muteins, where up to seven amino acids are substituted. Examples of truncated sequence proteins with improved characteristic are the 7–36 analogues of glucagon-like peptides or the 1–29 sequence growth hormone-releasing factor [1, 2].

The third approach has provided several successful results, though its application is usually limited to low molecular-weight peptides or non-peptide drugs. The liposome formulations, however, suffer from difficulties of storage or reconstruction and unpredictable distribution. Immobilization into insoluble polymeric particles is difficult in the case of structured proteins, as it normally involves the use of organic solvents. Moreover, protein denaturation and degradation may occur

*Pharmacokinetics and Pharmacodynamics of Biotech Drugs: Principles and Case Studies in Drug Development.* Edited by Bernd Meibohm

ISBN: 3-527-31408-3

during preparation or storage. The polymer may also react with the protein or peptide, as demonstrated in a transesterification reaction that takes place in the case of somatostatin and lactide polymers.

The final method, the covalent linkage of polymer to biotechnological drugs, can be considered of general applicability for overcoming pharmacokinetic problems with biotechnological drugs. The success of this approach has been demonstrated by the many hundreds of publications and patents filed in this field, the number of products presently available on the market, and the many other drug products currently in advanced states of clinical experimentation. The method of polymer coupling has also been found useful for improving the pharmacokinetics of low molecular-weight drugs of synthetic or of biotech origin.

Drug–polymer conjugation will form the basis of discussions in the present chapter. First, the chemical basis underlining the procedures of coupling, which are useful for understanding the potentials and limitations of the method, will be described. Second, selected examples of proteins, peptides, oligonucleotides and low molecular-weight drugs will be discussed where polymer conjugation was successfully exploited to enhance the pharmacokinetic and pharmacodynamic properties of these drug compounds.

## 11.2
## Polymers Used in Biotechnological Drug PEGylation

The conjugation of high polymeric mass to protein drugs is generally aimed at preventing the protein being recognized by proteolytic enzymes and antibodies, and also at reducing glomerular filtration of the drug. Both, natural and synthetic macromolecules have been used, the most frequently studied being dextran [3], human albumin [4], succinic acid-*co*-maleic anhydride [5], polyvinyl pyrrolidone [6], and polyethylene glycol [7] (Fig. 11.1).

Most of these polymers have multi-functional character, which results in cross-linked heterogeneous products. In contrast, monomethoxy polyethylene glycol (PEG) presents only one reactive terminal group per polymer chain. Once PEGylated with these compounds, the protein acquires a brush-like shape, with the hydrophilic PEG chains extended from the protein to the solvent.

For this reason, although the first polymer–protein conjugates used clinically were streptokinase conjugated to dextran [8] and neocarzinostatins linked to succinic acid-*co*-maleic anhydride [5], from the 1980s onwards PEG became the polymer of choice, and all products that subsequently appeared on the market were based on this reagent.

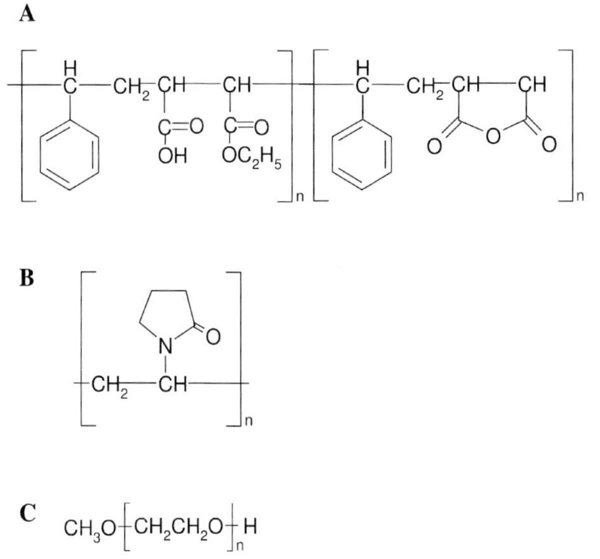

**Fig. 11.1** Structures of polysuccinic acid-maleic acid anhydride (A), polyvinyl pyrrolidone (B), and polyethylene glycol (C).

## 11.3 Advantages of PEG as Drug Carrier

There are several different advantages to linking high-mass PEG to drugs, and these are much dependent upon the type of product under investigation. The first and expected effect of polymer conjugation is an increased retention time in blood as a consequence of the size enlargement, which reduces glomerular filtration. This effect, which is particularly evident for low molecular-weight compounds such as peptides or non-peptide small molecules and oligonucleotides, is also common to proteins. A second effect, which is much desired for peptides and proteins, is surface masking of the drug, providing increased stability towards proteolysis, binding to antibodies and to the immune system. A third effect, which potentially is important for all types of drugs, is the possibility of linking the drug to one end of the PEG chain and a targeting agent at the opposite end, thereby directing the conjugate to target sites in the body while maintaining the advantages of the polymeric prodrug. Passive targeting to a tumor or to inflamed tissues may also be achieved by the mechanism known as the EPR effect (enhanced permeability and retention), on the basis of the polymer high mass that confers to the conjugate pharmacokinetic properties more favorable for tissue distribution.

## 11.4
## Chemical Aspects Critical for the Pharmacokinetics of Drug Conjugates

The chemical structure of the bond linking the drug to the polymer is of critical importance in the *in-vivo* behavior of conjugates, as it dictates the release rate from the polymer, or may even prevent it. Thus, it seems useful briefly to discuss the most important strategies of drug conjugation to PEG, although they can also be used for other polymers.

Activated PEGs are available commercially from many sources, in particular Nektar Therapeutics, which now supplies the starting material for most such drug products that are FDA approved or in advanced clinical development.

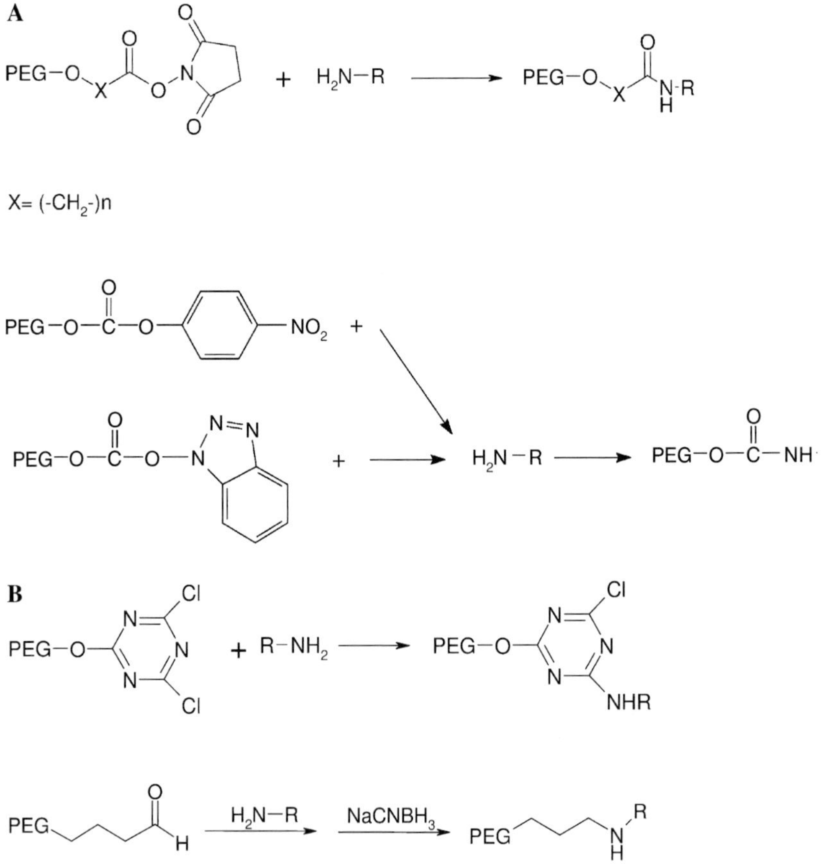


**Fig. 11.2** (A) Acylating reagents: PEG binding yields stable amide bonds but with loss of positive amino charge. The reagents may be on an activated carboxylate or a carbonate. (B) Alkylating reagents: PEG binding yields stable secondary amines while maintaining the positive amino charge. Although trichloro-triazine derivatives where initially used, nowadays other derivatives are used, mainly PEG aldehyde, also in the forms of acetals.

The most common sites of covalent binding of PEG to peptides and proteins are the epsilon-amino groups of lysine, the α-amino groups, and the thiol groups of cysteine. Acylating or alkylating PEGs will react with amines to yield neutral amides (Fig. 11.2 a) or positively charged secondary amines (Fig. 11.2 b), respectively. Both linkages are very stable *in vivo*. While the different charge of the conjugate is not critical in the case of proteins, it may play a role in the pharmacokinetics and tissue distribution in the case of small peptides or nonpeptide drugs. In these conjugates, a charge modification at one site may greatly affect the molecule's hydrophilicity.

An interesting site of conjugation, though seldom present in proteins, is the thiol group of cysteine. It may be bound to PEG, yielding stable thioethers or disulfites. The lost bond may be cleaved *in vivo* by free thiol-containing molecules (e.g., glutathione), yielding the starting drug, as in the case of common prodrugs (Fig. 11.3).

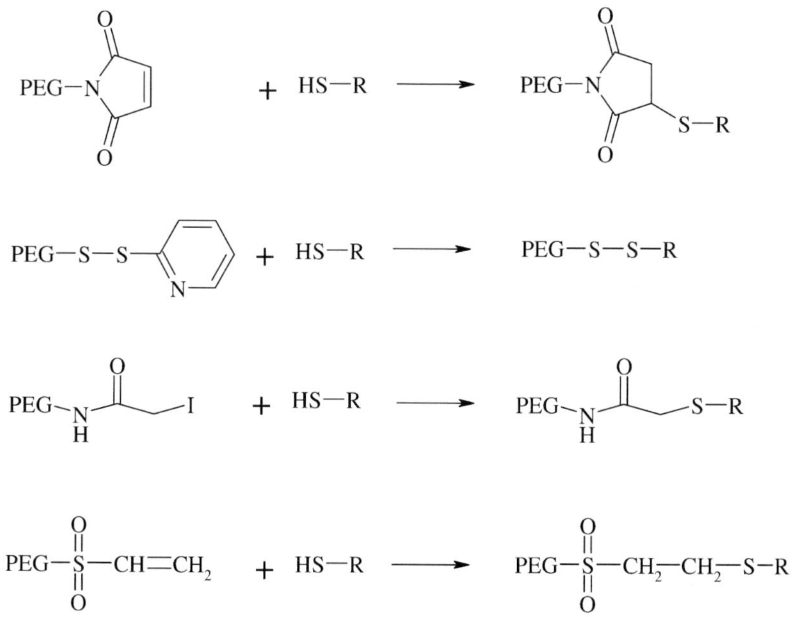


**Fig. 11.3** PEG conjugation to the thiol group of cysteine yields stable thioesters, or disulfites that can be cleaved by free thiols.

Other sites of PEG conjugation in peptides and proteins are the imidazole amino group of histidine, and the hydroxyl groups of tyrosine, serine, or threonine (Fig. 11.4). Hydroxyl groups are also the sites of conjugation in oligonucleotides. In all of these cases the drug–polymer conjugates are not stable, and *in vivo* may undergo simple hydrolysis by water without the need of enzymes. Hydroxyl groups are also usually employed for the preparation of long-lasting nonprotein drug–PEG conjugates.

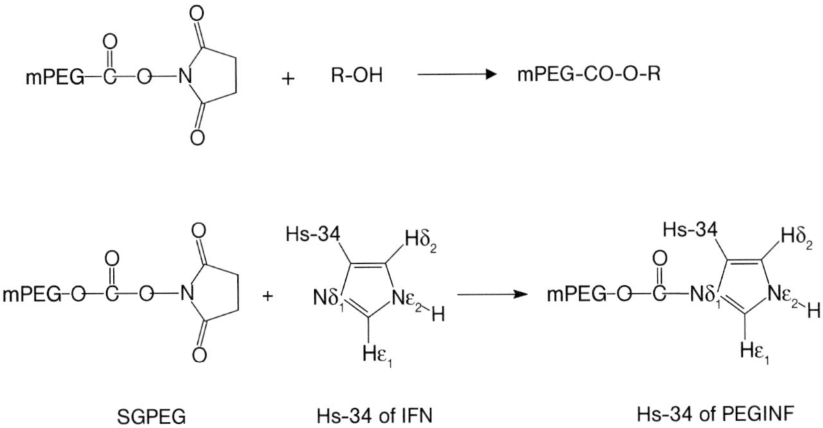


**Fig. 11.4** The binding of acylating PEGs to hydroxyl or imidazole nitrogen of histidine yields bonds easily cleavable by water, without the need for enzymes. IFN: interferon.

As an alternative means of obtaining labile prodrugs, advantage can be taken of specially designed, easy cleavable spacer arms between PEG and the drug. Examples of these strategies include:

1. A special amino acid sequence that is recognized and cleaved inside the cell by lysosomal enzymes (Fig. 11.5).
2. The presence of a labile ester bond between PEG and an alkyl acid linking the drug (Fig. 11.6).
3. The presence of a spacer arm that may facilitate, by anchimeric assistance, the hydrolysis of a bond between PEG and the drug (Fig. 11.7).

These strategies may be of value when drug release at only a specific body site is required (case 1), or if there is a need to modify the release rate (case 2) or to allow release from an otherwise water-stable bond (case 3).

The *in-vivo* residence time and localization in tissue is also strongly related to the PEG size and structure. A direct relationship was demonstrated in part between the PEG mass linked to a drug and the drug elimination rate. This relation-

PEG-OCO-$CH_2$-$CH_2$-$CO_2$-NHS
labile bond

**Fig. 11.5** Typical peptide sequences that are recognized and cleaved by lysosomal enzymes. They can be used as arm between PEG and drug for a targeted release.

PEG-**X**-$CO_2$-NHS

**X**= -Gly-Leu-Phe-Gly
-Gly-Phe-Leu-Gly-
-Gly-Phe-Gly-

**Fig. 11.6** An ester bond between PEG and drug is a potential site of easy cleavage by water under physiological conditions.

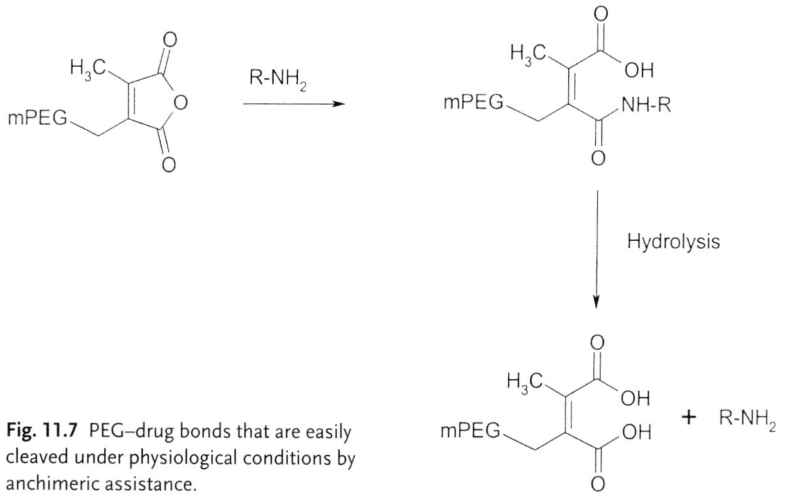


**Fig. 11.7** PEG–drug bonds that are easily cleaved under physiological conditions by anchimeric assistance.

ship is linear up to a value of about 50 kDa, a molecular weight value which, corresponding to the mass of albumin, is critical for the glomerular filtration of macromolecules. The hydrodynamic volume of the polymer, however, seems to be even more relevant to this relationship than the molecular weight. It was also found that the value of the mass (or better the hydrodynamic value) of linked PEG may be reached either by a single long molecule of polymer or by the cooperative effect of several small chains (Table 11.1).

The influence of the special structure of PEG on drug conjugate behavior has been well demonstrated, and is now exploited in the preparation of PEG-branched conjugates. This polymer structure is characterized by two PEG chains linked to the two nitrogens of a branching amino acid (generally lysine), leaving free the

**Table 11.1** Effect of number of PEG chains, PEG mass and apparent mass of the product, on the elimination half-life of recombinant interleukin-2 (rIL-2) from blood.[a]

| Product | No. of PEG chains | PEG mass [kDa] | Apparent size of product [kDa] | $t_{1/2}$ [min] |
|---|---|---|---|---|
| Unmodified rIL-2 | – | – | 19.5 | 44 |
| rIL-2–PEG 350 Da | 5 | 0.7 | 21 | 57 |
| rIL-2–PEG 4 kDa | 2 | 8 | 66 | 162 |
| rIL-2–PEG 5 kDa | 5 | 10 | 72 | 238 |
| rIL-2–PEG 10 kDa | 1 | 20 | 103 | 263 |
| rIL-2–PEG 20 kDa | 2 | 40 | 256 | 326 |

**a)** From data of M.J. Knauf, D.B. Bell, P. Hirtzer, Z.P. Luo, J.D. Joung, and N.V. Kartre, *J. Biol. Chem.* **1988**;*263*:15064–15070.

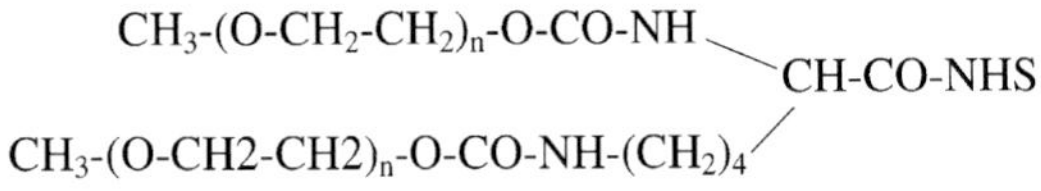


**Fig. 11.8** Structure of branched PEG which, because of its increased hindrance, prevents the approach of proteolytic enzyme and antibodies, increases the residence time in the systemic circulation, and sustains the biological activity longer as compared to linear PEG.

carboxylic group for the conjugation to drugs (Fig. 11.8). The resulting structure yields a greater protein surface covered by PEG at the level of the binding site, based on the repulsion of the two polymer chains. Furthermore, the two chains of PEG react only at the level of one residue in the protein, and this is reflected in a lower decrease in biological activity as compared to the linking of several chains with multiple residues. Peptides, proteins and oligonucleotides modified with branched PEG were thus found to exhibit a decreased loss of biological activity, increased stability towards proteolysis and antibodies, and prolonged residence time in the circulation. Successful examples of such conjugates with proteins or oligonucleotides have already reached the market place.

A further advancement of PEGylation that is expected to become an important application is based on the chemical structure of some PEGs that allow linking of a drug and a targeting agent. Examples of these reagents are presented in Fig. 11.9. Special features of these polymers are the presence of two different terminal ends, typically a carboxylic group reactive towards an amine or hydroxyl group of a drug on one side of the chain, and a thiol reactive group on the other side of the chain. In these cases, PEG acts as a spacer arm between the drug and the desired targeting moiety, as well as by changing the pharmacokinetics of the drug due to the increased mass of the conjugate.

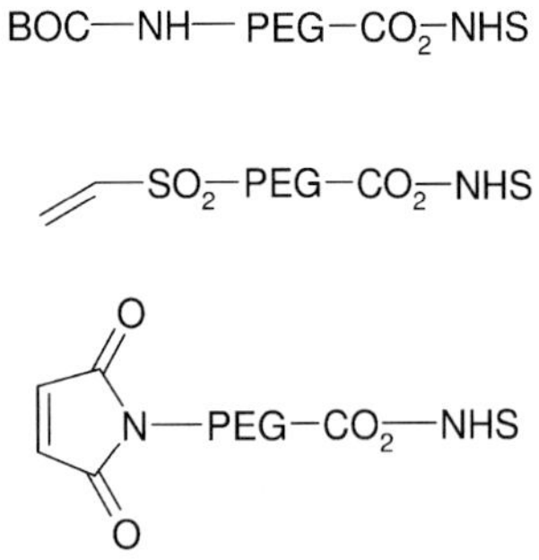


**Fig. 11.9** PEGs bearing two different reactive groups at the chain terminal ends, thus allowing the binding of a drug on one side and a targeting molecule on the other.

## 11.5
## Insulin

Since its discovery in 1921, insulin has rapidly become the standard pharmacotherapeutic intervention in the treatment of insulin-dependent diabetes mellitus (IDDM). However, throughout the years many attempts have been made to develop new insulin forms that could mimic as close as possible insulin secretion from the pancreas in the various IDDM situations. For example, fast-acting insulins are required in preprandial administrations, while long-acting insulins allow for the maintenance of the basal hormone level. In order to modulate the biopharmaceutical properties of insulin, the peptide structure was modified either by genetic manipulation or chemical modification. Both structural modifications were aimed at altering the formation of dimers or hexamers, the interaction with the cellular insulin receptor, the immunogenicity, the interaction with specific biological structures, and the hydrodynamic size. All of these parameters determine the pharmacokinetic/pharmacodynamic (PK/PD) behavior of this protein and, in turn, its therapeutic profile. The formation of multimers, for example, delays insulin absorption and may be responsible of inter-meal hypoglycemia. Changes in the receptor interaction are reflected in altered bioactivity and receptor-mediated hormone clearance. The immunogenic character promotes the processing and elimination by the immune system. Finally, the presence of specific moieties may alter the biodistribution profile and prolong the residence time in the bloodstream, while the size enlargement can delay absorption through biological membranes and slow down renal filtration of the drug.

Although amino acid sequence manipulation has been found useful in enhancing the pharmacokinetic properties of insulin. It can in fact generate mutagenic and carcinogenic derivatives [9]. For this reason, chemically modified insulins have been investigated as alternatives to genetically engineered insulins. Chemical modifications have been performed either by conjugation of macromolecules or by low molecular-weight derivatives such as fatty acids, polymers, and glycosylic functions. The modification site, extent of modification and chemical nature of the modifier have been found to play a crucial role in obtaining derivatives with the desired properties. Usually, the modification is carried out at the level of a few amino acids in order to avoid alteration of the receptor recognition, improve the protein's stability, and allow for interaction with specific biological structures. The most common conjugation sites of insulin are A1 Gly, B1 Phe and B29 Lys (Fig. 11.10). The modification of B29 Lys and B1 Phe, both of which are located on the hydrophobic surface involved in the dimerization, prevent aggregation, while modification of A1 Gly and B29 Lys reduces the interaction with the cellular receptor.

The conjugation of *p*-succinamidophenyl glucopyranoside, where the glucopyranoside function may be -galactoside, -fucoside and -mannoside, was found to strongly influence the PK/PD properties of the hormone. Glycosylation alters the receptor interaction, promotes binding with circulating lectins, and prevents the aggregation process. In particular, glycosylation stabilizes the monomeric form,

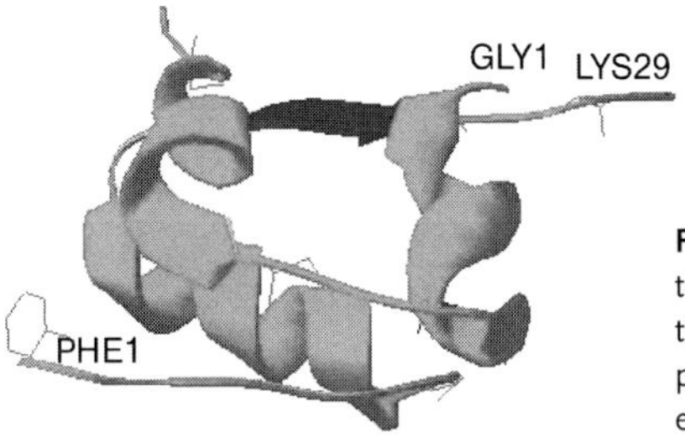


**Fig. 11.10** Insulin monomer structure with the three amino acids involved in polymer conjugation. A1 Gly is located in a hindered hydrophobic pocket, the ε-amino group of B29 Lys is the most exposed and available to conjugation.

allowing for rapid absorption while the interaction with lectins prolongs the residence time in the bloodstream. The reduced receptor affinity prevents rapid clearance by receptor-mediated endocytosis. These effects were found to be dependent on the modification site and extent. While glycosylation of the B1 Phe yields derivatives with prolonged absorption and delayed activity, glycosylation of the A1 Gly yields products with fast absorption, but low activity.

Fatty acid conjugation, investigated by Markussen and coworkers, was found to enhance insulin absorption and to prolong the hypoglycemic activity, though a general decrease in biological activity was observed. The conjugation of one palmitoyl molecule resulted in a derivative with activity similar to native insulin, but with a prolonged duration of action. In contrast, the conjugation of two palmitoyl moieties remarkably reduced both the maximal activity and the duration of action, though the maximal activity was delayed as compared to both the native and mono-palmitoyl forms [10].

Advances in polymer synthesis, characterization and conjugation chemistry have stimulated the development of drug macromolecularization technology. In this regard, a variety of studies on insulin modification with polysaccharides, polyvinyls and polyethylene glycols have been carried out.

Jain and coworkers showed that polysialylation significantly changes the PK/PD profile of insulin, prolonging the hypoglycemic activity after subcutaneous (SC) administration to T/O mice. Slight differences were observed between derivatives obtained with 22 kDa and 39 kDa polysialic acid; these were attributed to the slower hormone absorption obtained with the latter polymer, which favors hormone degradation at the injection site and hinders the interaction with the receptor more efficiently than the former polymer [11].

Synthetic polymers were generally preferred to the natural ones in protein conjugation because they are particularly efficient in modifying the biopharmaceutical, stability and immunological properties of insulin. Among the polymers used were polyvinylpyrrolidone, *N*-(2-hydroxypropyl) methacrylamides, and PEGs [12, 13].

Since the first insulin PEGylation studies of Ehrart and Luisi, several investigations have been carried to improve the physico-chemical and biological properties of this hormone by site-specific or random attachment of one or more PEG chains with different molecular weight and shape. Although the array of events which determines changes in the *in-vivo* behavior of insulin are often difficult to pinpoint, these studies showed that the biopharmaceutical properties of the hormone are correlated to the polymer conjugation: polymer size and shape, the site of conjugation, and the number of attached chains.

Hinds and Kim demonstrated that the site-specific conjugation of 750-Da and 2000-Da PEGs to the B1 Phe enhances and delays hormone absorption after SC administration, respectively. Furthermore, PEGylation slightly slows down the *in-vivo* elimination of insulin after intravenous (IV) injection. This effect, being proportional to the PEG molecular weight, was ascribed to the increased protein size, which in turn is reflected in reduced kidney filtration, increased interaction with circulating cells, and reduced interaction with the cellular receptor involved in insulin clearance [14, 15]. Studies carried out with 5- or 20-kDa PEG conjugation underlined the effect of polymer mass relative to protein surface on the biopharmaceutical and pharmacodynamic properties of insulin. The conjugation of one or two high molecular-weight chains was found to deeply affect insulin bioavailability, after either IV or SC administration [16, 17]. In particular, the attachment of one or two 5-kDa PEG slightly reduced the maximal hormone activity, which was partially counterbalanced by a prolonged duration of action of the PEGylated form. The conjugation with 20-kDa PEG induced a more dramatic alteration on the PK/PD behavior of insulin. The attachment of one 20-kDa PEG was found to create a derivative with slightly lower activity but longer-lasting action and six- to eight-fold higher pharmacological bioavailability compared to native insulin. In contrast, the attachment of two PEG chains was found to compromise the PK/PD behavior of insulin. A likely mechanism explaining this observation is the large hydrodynamic volume of the conjugate, which prevents renal filtration as well as receptor interaction.

Due to their interesting PK/PD properties, PEGylated insulins have been exploited for the preparation of new pharmaceutical formulations. Indeed, a proper PEGylation can promote insulin absorption through the mucosa, stabilize the peptide, and prevent its degradation by enzymatic proteolysis. On this basis, a new device for pulmonary delivery of insulin modified with 750-Da PEG is currently under development [18]. Insulin-PEG derivatives were also formulated in mucoadhesive tablets, and demonstrated a certain efficacy after peroral administration to diabetic mice [19]. Insulin modified with 5-kDa PEG was used in the preparation of PLGA microparticles. This study showed the successful combination of protein modification, which stabilizes the protein, and the microencapsulation technology that can provide for a formulation with prolonged drug release [20].

Based on results obtained after insulin PEGylation, NOBEX Corporation (Research Triangle Park, NC, USA) has recently developed a technology for the oral delivery of insulin centered around the hormone modification by attachment of one or more amphiphilic oligomers. Amphiphilic oligomers, obtained by alkyl groups or fatty acid radicals grafted onto PEGs, were attached to insulin to produce peptides that can penetrate the aqueous layer and the lipid portion of the epithelium, resist self-association, and resist excessive degradation of the hormone. The insulin modification yields an orally absorbable, bioactive conjugate, which is safe and rapidly absorbed, and which demonstrates dose-dependent glucose-lowering effects in animal models, healthy volunteers and type 1 diabetic patients. The results of recently completed and ongoing studies in both type 1 and type 2 diabetes suggest a promising role for conjugated insulin in the management of fasting and postprandial hyperglycemia. Orally administered, conjugated insulin is delivered first to the liver through the portal circulation, similar to the physiological route of insulin secretion in nondiabetic individuals. Potential benefits from this route of insulin delivery include an improved disease management and a reduction in the long-term complications of diabetes [21].

## 11.6 Interferons

Interferons are a class of cytokines effective in inducing the remission of chronic hepatitis C and other disease conditions. Due to the very short half-life of these proteins, interferons must be administered three times each week, which implies a number of problems in patient compliance. Furthermore, the frequent administration yields large fluctuations in peak-to-trough plasma drug concentrations, which compromises the anti-viral effectiveness. New formulations, as well as new bioconjugates, have been developed to improve the pharmacokinetic behavior of interferons. PEG conjugation, in particular, has been successful in overcoming the problems associated with interferon administration, though improvements in the biopharmaceutical properties of interferon require careful optimization of the PEGylation [22].

Interferon PEGylation has been performed by using PEGs with different molecular weight and shape, including linear 5-, 10-, 12- and 20-kDa PEG and branched 20- and 40-kDa PEG. The PEG conjugation has been carried out under different conditions in order to obtain various isomers, mono- or oligo-conjugates, site-specific conjugation and different protein–polymer linkages. Similarly to other proteins, the PK/PD properties of the interferon–PEG derivatives depend on the polymer size, the number of polymer chains, and the site of modification. Typically, the higher the total polymer mass on the interferon surface (which depends either on the PEG size or the number of conjugated PEG molecules), the lower is the protein interaction with the cellular membrane receptors. Furthermore, it was shown that the modification of few amino acids strictly involved in the receptor interaction provokes a dramatic reduction in interferon activity. This is the case of several lysines, the usual binding site in PEGylation, the modification of which hampers receptor interaction. Generally, lysine modification reduces the interferon activity by more than 80%, depending on the polymer structure and lysine position. In contrast, the derivatization of His34, which is located in a position far from the interaction area, was found to yield a derivative with high bioactivity. Therefore, the reduction of biological activity can be minimized by adopting proper PEGylation strategies that favor the formation of mono-substituted derivatives to specific amino acids. The choice of correct PEG size must also be considered in order to produce derivatives with the desired properties. PEGylation must be correctly manipulated in order to balance the positive and negative effects, namely reduction in biological activity versus enhanced pharmacokinetic, immunological and physico-chemical properties.

Linear 5- and 20-kDa PEGs and branched 20- and 40-kDa PEGs activated as succinimidyl esters were investigated in a study using SC administration of the conjugates to rats. This study showed that the conjugation of one PEG molecule increases the interferon-α2a half-life from 2 h in the case of native interferon to 3.4 h in the case of the 5-kDa PEG derivative and to 15 h for the 40-kDa PEG derivative. The di-PEGylated interferon obtained by conjugation of two 20-kDa PEGs showed a half-life of 23 h. Similarly, the plasma residence time was 2.0 h in the case of the 5-kDa PEG derivative and 32 h in the case of the di-20-kDa derivative [23].

The attachment of one branched 40-kDa PEG chain to interferon-α2a was found to yield a mixture of positional isomers in which PEG was covalently and irreversibly bound to Lys groups. Despite the unspecific site-coupling reaction, four main positional isomers were obtained: Lys31, Lys121, Lys131, and Lys134 [24]. The mixture was found to maintain only 7% of the pharmacologic activity of unmodified interferon. However, according to the consideration reported above, the loss of bioactivity was largely offset by the prolonged residence time in the bloodstream (Fig. 11.11).

After SC injection, the time of the maximal peak ($T_{max}$) of interferon shifted from about 8 h for the native protein to about 80 h for the PEGylated form. Furthermore, the PEGylated interferon displayed a 100-fold lower clearance than the native protein. The effective plasma concentration was maintained over 240 h after administration of the PEGylated interferon, while the native protein disappeared in about 20 h. The volume of distribution for the conjugate was found to be about 12 L, indicating that this product was mainly confined in the bloodstream. In a study of once-weekly administration of interferon-PEG over 48 weeks, the drug showed sustained absorption and a consistently lower total body clearance compared to native interferon; this resulted in a flat drug concentration–time profile throughout the dosing interval [25].

The conjugation of one 12-kDa linear PEG to interferon-α2b was carried out under controlled reaction conditions in order to modify preferentially His34. The PEGylation strategy adopted in this case has a triple effect:

- The lower PEG molecular weight has a limited hindering effect on the receptor recognition.
- The modification of His34, which is located far from the protein area involved in receptor binding, prevents impairment of the receptor interaction.
- Unlike Lys-PEG, the polymer linkage at His34 has a reversible nature, which yields to a pro-drug-like derivative [26].

As a consequence, interferon modified at the level of His34 with 12-kDa PEG maintains higher biological activity as compared to its counterpart prepared by 40-kDa PEG modification at the lysine groups, but has a much shorter duration of action. $T_{max}$ was ca. 20 h, the clearance was ten-fold higher than for interferon-PEG 40 kDa, and the volume of distribution was ca. 70 L, which is similar to that

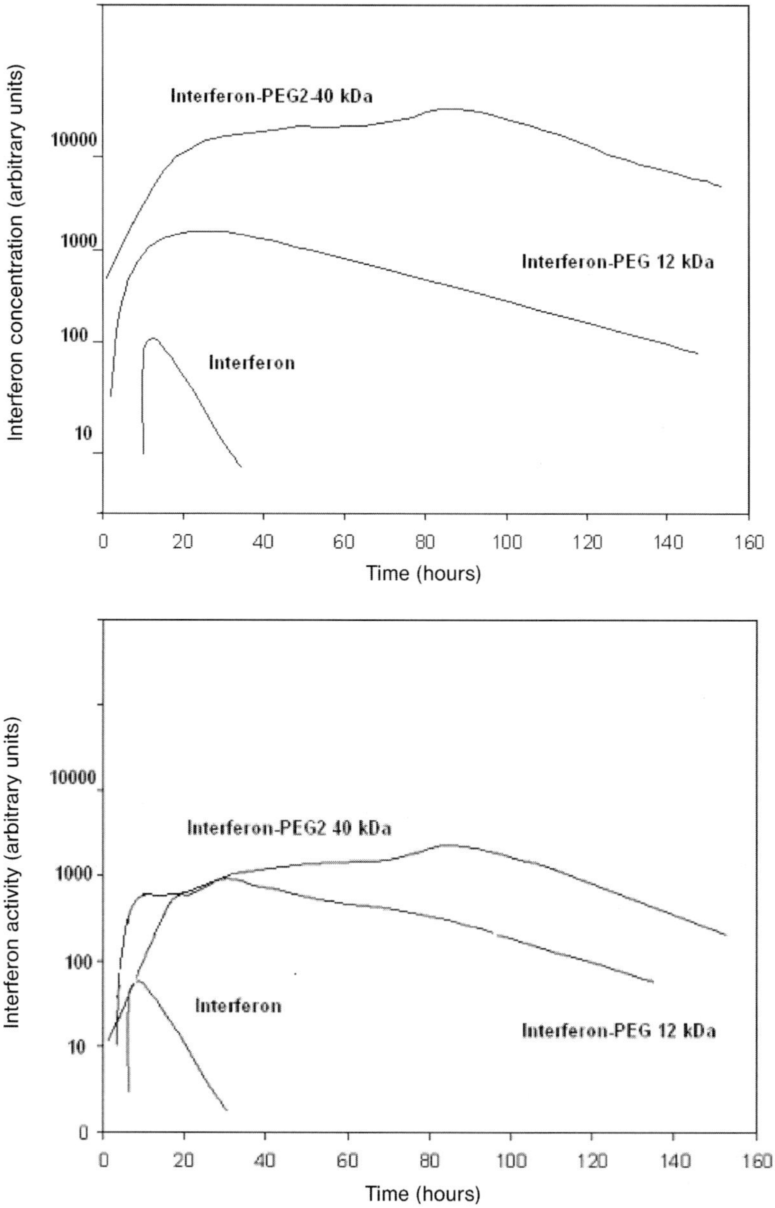


**Fig. 11.11** Pharmacokinetic (upper panel) and pharmacodynamic (lower panel) behaviors of native and PEGylated interferon administered subcutaneously to humans. Interferon–PEG 12 kDa and interferon–PEG2 40 kDa doses were 1.5 μg/kg and 180 μg, respectively. The native interferon dose was 3 MLN units. The interferon amounts reported here correspond to the recommended single doses adopted for therapy with the three cytokine forms. In the case of interferon–PEG 12 kDa the single dose is adjusted to body weight (1.5 μg/kg), while in the case of interferon–PEG2 40 kDa and native interferon a fixed dose is administered, independent of body weight (180 μg and 3 MLN units, respectively).

of the native protein. After SC administration, effective concentrations were maintained for approximately 180 h [27].

In conclusion, both mono-PEGylated interferons present superior properties for protein delivery, resulting in a longer-lasting activity as compared to the native protein, avoiding frequent administration, and improving the therapeutic performance of interferon.

## 11.7 Avidin

The high affinity that avidin shows for the low molecular-weight vitamin biotin forms the basis of several biological tools, either for *in-vitro* or *in-vivo* applications. Despite its high potential in diagnosis and therapy, the *in-vivo* use of avidin is limited by its poor immunological and pharmacokinetic properties. In aiming to increase the residence time of avidin in the bloodstream and to reduce its immunogenicity and antigenicity, several structural modifications have been investigated such as deglycosylation, acetylation, succinylation, and polymer conjugation. Avidin modifications have been mainly directed at modifying the isoelectric point and the glycosylic composition, both of which play roles in the distribution and elimination from the systemic circulation. The glycosylic functions are responsible for avidin's tendency to accumulate in the liver, while the high isoelectric point (10.5) is responsible for its accumulation in the kidneys. Polymer conjugation to the protein surface, which results in masking the protein structure and charges and increasing the protein size, has also been explored.

Recombinant avidin obtained by a synthetic cDNA encoding for the full and correct sequence of chicken avidin cloned into an *Escherichia coli* expression vector was found to maintain the tetrameric structure of the native protein, and to lack a carbohydrate moiety. Despite the absence of carbohydrates, which should promote protein disposition into the liver, the protein was cleared quite rapidly from the blood [28]. Similar results were obtained with Lite avidin, a partially deglycosylated avidin which was cleared from the circulation more rapidly than the native protein. Schechter and coworkers showed, in fact, that the concentration–time profile of nonglycosylated avidin was quite similar to that of the glycosylated form. In addition, the accumulation profile in liver and spleen did not change after deglycosylation, but the deglycosylated form is disposed to a higher extent in the kidneys [29]. These results seem to indicate that deglycosylation *per se* is not sufficient to prolong the residence time in the bloodstream.

It has been suggested that the modification of the protein's isoelectric point could result in an alteration of its pharmacokinetic profile. Avidin acylation was performed by lysine amino group derivatization with succinyl anhydride or other anhydrides, which allowed the isoelectric point to be shifted to more acidic values, depending on the level of modification. Indeed, the protein anionization induced a reduction of accumulation in the liver, but resulted only in a limited prolongation of residence time in the circulation [30, 31].

The simultaneous modification of the glycosylic composition and alteration of the isoelectric point had a remarkable effect on the residence time of avidin in the systemic circulation, an effect which results from a combination of reduced liver and renal uptake. A typical example of modified avidin with altered pharmacokinetic properties is represented by Neutrolite avidin, a recombinant nonglycosylated and acidic form of avidin. This avidin derivative was obtained by replacement of five out of the eight arginine residues with neutral amino acids, and two of the lysine residues by glutamic acid. In addition, the carbohydrate-bearing asparagine-17 residue was changed to an isoleucine [32]. Due to its acidic isoelectric point and lack of glycosylic functions, Neutrolite avidin presents a very low clearance (0.47 mL/min/kg) as compared to the native protein (24.6 mL/min/kg).

Polymer modification was also investigated as a means of obtaining new derivatives with enhanced pharmacokinetic properties as a consequence of protein size enlargement, protein structure masking, and isoelectric point alteration. Avidin was PEGylated using PEGs of different molecular weight and shape. Furthermore, the conjugation was performed under different reaction conditions in order to yield derivatives with different degrees of modification. The study results showed that the pharmacokinetic, biodistribution, immunological and biological properties of the conjugates are strictly related to the polymer mass on the protein surface, which in turn depends on the polymer molecular weight and number of attached polymer chains [33, 34].

The attachment of about four polymer chains, corresponding to the modification of 10% of the protein amino groups, yielded derivatives which maintained high affinity towards biotin and biotinylated antibodies. PEGylation was also found significantly to reduce the protein's antigenicity and immunogenicity which, together, are responsible for side effects and protein removal by activation of the immune system. The PEGylated avidins displayed slower elimination via the liver and kidneys, and a prolonged residence time in the bloodstream. PEGylation also promoted avidin distribution into solid tumors, which takes place by passive diffusion. The effect on avidin biodistribution and pharmacokinetic behavior was directly related to the polymer size used in the modification. In particular, as the PEG size increased, the avidin clearance decreased, the volume of distribution decreased, and diffusion into the peripheral compartment was slowed. PEGylated avidins also display a more favorable tumor-to-blood and tumor-to-organ concentration than nonPEGylated avidins, which makes them useful for *in-vivo* applications. These results were attributed to the size enlargement of PEGylated avidins and to the efficiency of the high molecular-weight PEGs in masking the glycosylic functions in the core of the construct. Conjugation of the polymer at the lysine amino groups also resulted in a lowering of the isoelectric point, which has already been identified as an important parameter in dictating the pharmacokinetic behavior of this protein [33].

In a recent study, avidin was extensively modified with linear 5- and 10-kDa PEGs and with a branched 20-kDa PEG. In order to maintain high biological activity, the polymer was conjugated in the presence of a macromolecular active site protective agent, which was used to avoid polymer attachment on the protein area close to the bioactive site. The adopted reaction conditions allowed for the modification of about 16 of 40 amino groups. The derivatives displayed high recognition properties towards both biotin and biotinylated antibodies. The polymer conjugation was found drastically to modify the pharmacokinetic properties of the protein. The prolonged residence time of avidin in the circulation was accompanied by a remarkable reduction in the volume of distribution (over 100-fold lower compared to native avidin), and by an absence of localization in peripheral organs. Although all the conjugates presented a similar distribution profile, the accumulation in a solid tumor implanted subcutaneously into mice was decreased as the PEG size was increased. This effect can be related to the reduced diffusion properties of the large conjugates, which prevents tissue uptake.

It was also observed that the pharmacokinetic properties of the conjugates cannot be explained by a simple reduction in renal filtration due to size enlargement. In fact, all of the derivatives had sizes larger than the glomerular filtration limit. In addition, extensive PEGylation was found to suppress the immunogenic and antigenic properties of avidin independently of the PEG molecular weight, and the protein was stable towards degradation. Thus, it is reasonable to assume that other mechanisms are involved in determining the *in-vivo* behavior of this protein, such as specific interactions with biological structures or circulating macromolecules [34].

Thermoresponsive acrylamide co-polymers were also used to alter the physicochemical and biopharmaceutical properties of avidin. Similar to PEG, the acrylamide co-polymers with a lower critical solution temperature (LCST) of about 37 °C were conjugated to the protein amino groups. The polymers were conjugated either by polymer multipoint attachment using polyfunctional polymers or by single chain attachment using end-chain monoactivated polymer. In both cases, the polymer conjugation was found to produce bioactive derivatives with reversible thermal character (Fig. 11.12).

The pharmacokinetic properties of avidin modified with mono-functional *N*-isopropylacrylamide-*co*-acrylamide (pNIPAAm; MW 6 kDa) showed that the polymer alters the *in-vivo* concentration–time course of the protein, though the effect was less pronounced than that obtained by PEGylation [35]. The smaller effect of pNIPAAm as compared to PEG may be due to the different physico-chemical properties of the two polymers. PEG is a highly hydrophilic compound which, in water, takes on an extended conformation. The PEG conformation, the coordination of several water molecules and its high flexibility, result in the large hydrodynamic size of these conjugates and in high protection of the protein in the core of the construct. In contrast, polymers with a lower hydrophilic character such as pNIPAAm can partially collapse on the protein surface, resulting in conjugates with smaller size and larger exposition of the protein structure to the environment. Therefore, pNIPAAm is less efficient in enlarging the protein size and masking the protein structure – two parameters which are known to determine the pharmacokinetic profile of the protein. It should be noted that other polymers such as polyvinyl pyrrolidone or polyacryloyl morpholine have been found to influence protein pharmacokinetics similarly to pNIPAAm [36].

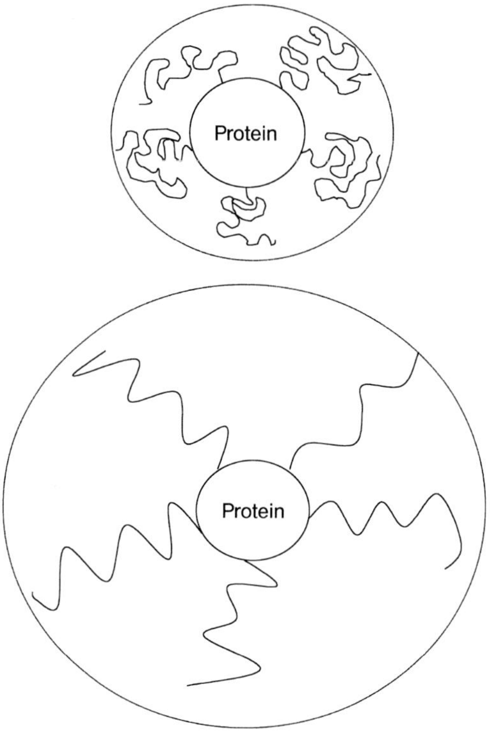


**Fig. 11.12** Structural representation of protein modified with polymers. The polymer dictates the hydrodynamic volume, the stability to proteolysis, and antigenicity of the conjugate.

## 11.8 Non-Peptide Drug Conjugation

The strategy for non-peptide drug conjugation may differ in many aspects from that employed for proteins, for four reasons:

- Usually, the linkage must be reversible in the plasma or inside the cell.
- The hydrolysis and release of the free product must take place at a defined rate in order to ensure therapeutic drug concentrations *in vivo*.
- The molecular weight of the polymer must be high enough to prevent too-rapid kidney filtration.
- A high drug payload is desired to avoid macromolecule overloading in the body.

These requirements have stimulated the research for new chemical strategies of coupling, and in particular for linkers that are easily cleaved. These include esters, lisosomotropic peptides that are stable in plasma but cleavable inside the cells by cathepsins [37], and spacers that can be directly cleaved under acidic conditions or by anchimeric assistance such as *cis*-citraconic acid amides or hydrazones derivatives [38]. Furthermore, pro-drug approaches or trimethyl lock systems have been developed where the carrier is cleaved in two steps, initially by water or enzymes and later by a chemical mechanism [39].

Non-peptide drug conjugation is simpler than that for proteins, as the products are easily characterized. On the other hand, this approach presents new problems such as the correct tuning of polymer weight and chemistry of linkage in order to reach and maintain the desired drug concentration in the body. One general result that is easily pursued by hydrophilic polymer conjugation is the increased solubility of water-insoluble drugs. Toxicity may also be reduced, provided that release of the free drug occurs at a rate slow enough to maintain plasma free drug concentrations below toxic levels.

Among the many non-peptide drugs studied to date, a few are presented here that appear to be typical examples of polymer–drug constructs.

### 11.8.1 Amphotericin B

Amphotericin B, characterized in its native form by extremely low solubility and toxicity, was conjugated to 40-kDa PEG following different strategies. The polymer is conjugated to the amino group present in the sugar moiety to yield a carbamate linkage through a suitable aromatic arm linked to PEG. For the preparation of this pro-drug, the 1,6-benzyl elimination strategy was exploited (Fig. 11.13) [40]. Correct substitution in the aromatic arm allows drug release to be controlled by slowing the hydrolysis rate via steric hindrance. Among the several products prepared, the one chosen for biological studies exhibited a 200-fold increased solubility and six-fold reduced toxicity compared to the free drug, and was stable in water as it did not release the native drug. The presented half-life for this compound was 3 h in plasma, and the AUC in rats was 3715 μg h/mL, while for the free drug the value was 0.68 μg h/mL. All these characteristics indicate that the conjugate is a promising candidate for improved therapeutic properties.

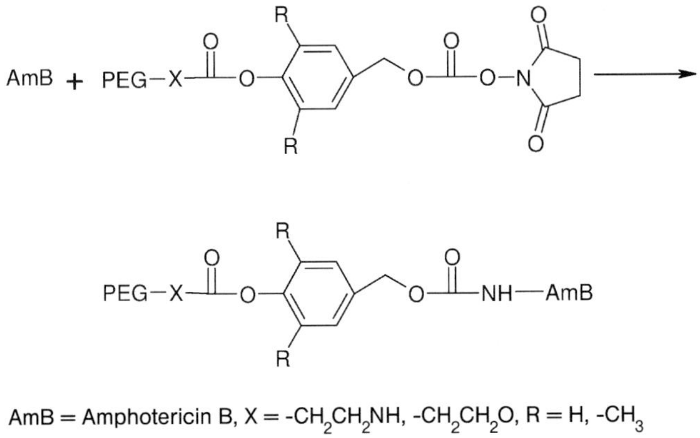


**Fig. 11.13** Scheme of PEG conjugation to amphotericin B.

## 11.8.2
## Camptothecins

The products belonging to this class of drugs differ in terms of the constituents at the ring. These drugs have important anti-tumor properties, but unfortunately also low solubility and high toxicity. Many PEG–camptothecin conjugates were synthesized by exploiting different groups for conjugation, a variety of polymer mass, and various chemistries of linkage.

In one case an amino group in position 10 of the aromatic ring A of the 10-amino-7-ethylcamptothecin was conjugated to 10-kDa PEG by an amide linkage [41]. A lysosomotropic tetrapeptide spacer between the polymer and the drug was added to achieve drug release inside the cell only by the lysosomal cathepsins (Fig. 11.14). As expected, the PEG conjugate was much more soluble than the free drug, and was also stable in plasma. *In vitro*, the drug was released by cathepsin B incubation. Following administration to rats, the PEG conjugate presented an eightfold lower maximal plasma concentration and a threefold longer mean residence time compared to the free drug. These data suggest that the PEG conjugate acts as a circulating reservoir of the anti-tumor agent. When tested biologically, it was found equally active as the reference compound towards leukemia, but more active towards fibrosarcoma.

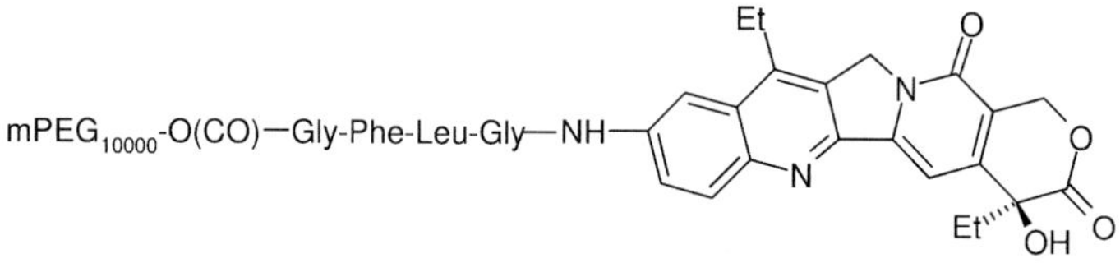


**Fig. 11.14** Chemical structure of the 1 : 1 PEG–camptothecin conjugate.

In a different study, camptothecin was linked at the level of the hydroxy group in position 20 of the lactone ring [42]. In order to increase the drug:polymer ratio, a PEG diol was used to allow the linkage of two drug molecules per PEG chain (Fig. 11.15). Also in this case, a high molecular-weight polymer (40-kDa PEG) was used as carrier to slow the renal filtration. An amino acid moiety, preferably glycine, was used as spacer between PEG and the drug. The polymer conjugation at the 10-hydroxy group of camptothecin stabilized the lactone form, thus contributing to reduced toxicity which is known to be higher for the open compared to the

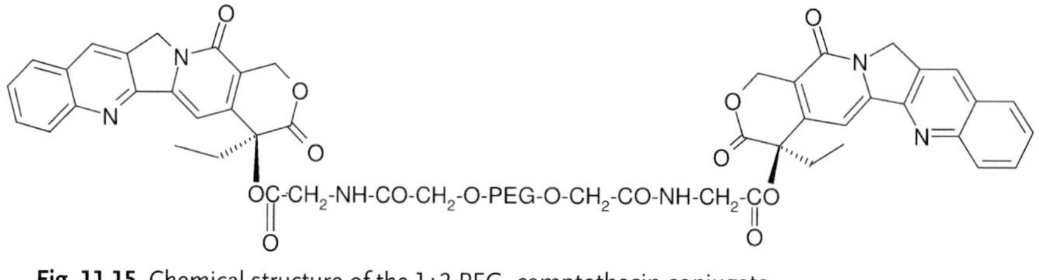


**Fig. 11.15** Chemical structure of the 1 : 2 PEG–camptothecin conjugate.

lactone form. In mice, the blood $t_{1/2}\alpha$ (6 min), $t_{1/2}\beta$ (10.2 h) and mean residence time (14.5 h) were higher than those of the unconjugated drug. Interestingly, the conjugate also displayed increased accumulation in tumors, as verified with tritium-labeled product and increased anti-tumor activity. The increased tumor accumulation and activity is probably related to a higher permeability through the neovasculature in tumors, accompanied by a lack of an effective lymphatic drainage system in this tissue. The combination of these effects results in an increased retention inside the tumor interstitium for macromolecular drugs.

## 11.8.3
## Cytosine Arabinoside (Ara-C)

Ara-C is a drug with low solubility in water, but it is usually administered at high doses. The low biological activity prevents the use of a mono- or bifunctional PEGs for conjugation. These two PEGs, in fact, allow only for a low drug loading with the result of a nonacceptable viscosity of the solution at the concentration needed for therapy. This stimulated the design of new forms of PEG with a branching at the extremes of the chain (Fig. 11.16).

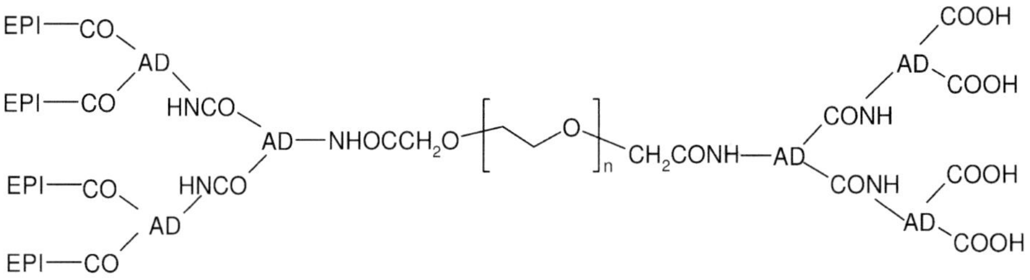


**Fig. 11.16** Chemical structure of the 1:2 PEG–Ara C conjugate.

Following different chemical strategies, PEGs with four of eight carboxylic groups at the extremes of the polymer were prepared, to which four or eight molecules of Ara-C were respectively bound [43, 44]. Conjugates carrying a spacer with the trimethyl lock system were also synthesized. All of these multi-loaded Ara-C pro-drugs exhibited greater water solubility and less viscosity than the mono- or disubstituted PEG analogues. The conjugation also increased plasma stability and residence time *in vivo*, and resulted in a release of free drug at different rates, depending upon the general architecture of the conjugate. The tetrameric and octameric Ara-C conjugates were found to be much more effective in the treatment of solid and ascites tumors as compared to the native drug.

## 11.9 Concluding Remarks

Due to their peculiar physico-chemical nature, biotech drugs often display poor pharmacokinetic behavior which prevents their optimal therapeutic performance. In order to improve the biopharmaceutical properties of this class of therapeutics, structural modifications are frequently undertaken. In the case of polypeptides and proteins, successful results have been obtained by genetic manipulation which allowed for creating muteins with high stability and activity as well as prolonged residence time in the systemic circulation. The chemical bioconjugation of moieties such as lipids, glycosides, and polymers has also been found to improve the pharmacokinetic profile of biotech as well as low molecular-weight drugs. In particular, polymer bioconjugation has been used to enhance the *in-vivo* behavior of proteins and small drugs by reducing kidney ultrafiltration, as well as degradation and elimination by the immune system. Moreover, the bioconjugation of polymers and specific moieties has been applied to yield drug targeting to the disease sites, either by passive or active mechanisms. Due to the versatility and wide array of possible chemical modifications, bioconjugation techniques can be considered the most promising method to expand the therapeutic application of either biotech or traditional bioactives, the use of which is limited by poor physico-chemical and biopharmaceutical properties and low therapeutic indices.

## 11.10 References


1 Creutzfeldt, W. **2001**. The entero-insular axis in type 2 diabetes – incretins as therapeutic agents. *Exp. Clin. Endocrinol. Diabetes* 109 : 288–303.

2 Lance V.A., W.A. Murphy, J. Sueiras-Diaz, and D.H. Coy. **1984**. Super-active analogs of growth hormone-releasing factor (1–29)-amide. *Biochem. Biophys. Res. Commun.* 119 : 265–272.

3 Brocchini, S., and R. Duncan. **1999**. In: Mathiowitz, E. (Ed.), *Encyclopedia of Controlled Drug Delivery*. Wiley, New York, pp. 786–816.

4 Wong K., L.G. Cleland, and M.J. Poznansky. **1980**. Enhanced anti-inflammatory effect and reduced immunogenicity of bovine liver superoxide dismutase by conjugation with homologous albumin. *Agents Actions* 10 : 231–239.

5 Maeda H., J. Takeshita, and R. Kanamaru. **1979**. A lipophilic derivative of neocarzinostatin. A polymer conjugation of an antitumor protein antibiotic. *Int. J. Peptide Protein Res.* 14 : 81–1487.

6 Caliceti P., O. Schiavon, M. Morpurgo, F.M. Veronese, L. Sartore, E. Ranucci, and P. Ferruti. **1995**. Physico-chemical and biological properties of monofunctional hydroxy terminating poly(*N*-vinylpyrrolidone) conjugated superoxide dismutase. *J. Bioact. Comp. Polym.* 10 : 103–120.

7 Abuchowski A., J.R. McCoy, N.C. Palczuk, T. van Es, and F.F. Davis. **1977**. Effect of covalent attachment of polyethylene glycol on immunogenicity and circulating life of bovine liver catalane. *Biol. Chem.* 252 : 3582–3586.

8 Tochilin V.P., I. Voronkov, and A.V. Mazaev. **1982**. Use of immobilized streptokinase (streptodecase) for treating thromboses. *Ter. Arkh.* 54 : 21–25

9 Kurtzahls P., L. Schaffer, A. Sorensen, C. Kristensen, I. Jonassen, C. Schmid and T. Trub. **2000**. Correlation of receptor bind-

ing and metabolic and mitogenic properties of insulin analogs designed for clinical use. *Diabetes* 49:999–1005.

**10** Hashimoto M., K. Takada, Y. Kiso, and S. Muranishi. **1986**. Synthesis of palmitoyl derivatives of insulin and their biological activities. *Pharm. Res.* 6:171–176

**11** Jain S., D. H. Hreczuk-Hirst, B. McCormack, M. Mital, A. Epenetos, P. Laing, and G. Gregoriadis. **2003**. Polysialylated insulin: synthesis, characterization and biological activity in vivo. *Biochim. Biophys. Acta* 1622:42–49.

**12** Chytry V., D. Letourneur, M. Baudys, and J. Jozefonvicz. **1996**. Conjugates of insulin with copolymers of *N*-(2-hydroxypropyl) methacrylamide: effects on smooth muscle cell proliferation. *J. Biomed. Mater. Res.* 31:265–272.

**13** Von Specht B.U., H.J. Kolb, R. Renner, and K.D. Hepp. **1978**. Preparation and physical-chemical characterization of poly-*N*-vinylpyrrolidone-insulin. *Hoppe Seyler's Z. Physiol. Chem.* 359:231–238

**14** Hinds K.D., and S.W. Kim. **2002**. Effects of PEG conjugation on insulin properties. *Adv. Drug Deliv. Rev.* 54:505–530.

**15** Hinds K., J.J. Koh, L. Joss, F. Liu, M. Baudys, and S.W. Kim. **2000**. Synthesis and characterization of poly(ethylene glycol)-insulin conjugates. *Bioconjug. Chem.* 11:195–201.

**16** Caliceti P., and F.M. Veronese. **1999**. Improvement of insulin physico-chemical and biopharmaceutical properties by poly(ethylene glycol) conjugation. *J. Drug Deliv. Sci. Technol.* 9:107–113.

**17** Caliceti P., and F.M. Veronese. Successful insulin delivery by PEG conjugation. 25th International Symposium on Controlled Release of Bioactive Materials, June 21–24, 1998, 347.

**18** Insulin inhalation – Pfizer/Nektar Therapeutics: HMR 4006, inhaled PEG-insulin – Nektar, PEGylated insulin – Nektar, Drugs R D **2004**. 5:166–170.

**19** Caliceti P., S. Salmaso, G. Walker, and A. Bernkop-Schnürch. **2004**. Development and in vivo evaluation of an oral insulin-PEG delivery system. *Eur. J. Pharm. Sci.* 22:315–323.

**20** Hinds K. D., K.M. Campbell, K.M. Holland, D. H. Lewis, C. A. Piché, and P. G. Schmidt. **2005**. PEGylated insulin in PLGA microparticles. In vivo and in vitro analysis. *J. Control. Release* 104:447–460.

**21** Still G. **2002**. Development of oral insulin: progress and current status. *J. Diabetes Metab. Res. Rev. Suppl.* 1:S29–S37

**22** Rajender Reddy K., M.W. Modi, and S. Pedder. **2002**. Use of peginterferon alfa-2a (40 KD) (Pegasys) for the treatment of hepatitis C. *Adv. Drug Deliv. Rev.* 54:571–586.

**23** Bailon P., C.L. Spence, C.A. Schaffer, K. Prinzo, S. Monkrash, J.E. Porter, W.-J. Fung, W.E. DePinto, A. Aglione, L. Stern, G. Truitt, A. Pulleroni, Z.-X. Xu, J. Hoffman, S. Pedder and M. Brunda. **1999**. Pharmacokinetic properties of five poly(ethylene glycol) conjugates of interferon alfa-2a. *Antivir. Ther.* 4:27–37.

**24** Bailon P., A. Pelleroni, C.A. Schaffer, C.L. Spence, W.-J. Fung, J.E. Porter, G.K. Erlich, W. Pan, Z.-X. Xu., M.W. Modi, A. Farid, and W. Berthold. **2001**. Rational design of a potent, long acting form of interferon's 40 kDa branched poly-ethylene glycol conjugated interferon α 2a for the treatment of hepatitis C. *Bioconjug. Chem.* 12:1995–2002.

**25** Modi M.W., M.W. Fried, R.W. Reindollar, V. R. Rustgi, R. Kenny, T.L. Wright, A. Gibas N.E. Martin, M.L. Shiffman, and L.S. Marsano. **2000**. The pharmacokinetic behaviour of pegylated (40 kDa) interferon alfa-2a (Pegasys) in chronic hepatitis C patients after multiple dosing. *Hepatology* 32:394A.

**26** Roisman L.C., J. Piehler, J.Y. Trosset, H.A. Scheraga, and G. Schreiber. **2001**. Structure of the interferon-receptor complex determined by distance constraints from double-mutant cycles and flexible docking. *Proc. Natl. Acad. Sci. USA* 98:13231–13236

**27** Glue P., J.W. Fang, R. Rouzier-Panis, C. Raffanel, R. Sabo, S.K. Gupta, M. Salfi, and S. Jacobs. **2000**. Pegylated interferon-α2b: pharmacokinetics, pharmacodynamics, safety, and preliminary efficacy data. Hepatitis C Intervention Therapy Group. *Clin. Pharmacol. Ther.* 68: 556–567.

**28** Chinol M., P. Casalini, M. Maggiolo, S. Canevari, E.S. Omodeo, P. Caliceti,

F.M. Veronese, M. Cremonesi, F. Chiolerio, E. Nardone, A.G. Siccardi, and G. Paganelli. **1998**. Biochemical modifications of avidin improve pharmacokinetics and biodistribution, and reduce immunogenicity. *Br. J. Cancer* 78 : 189–197.

**29** Schechter B., R. Silberman, R. Arnon, and M. Wilchek. **1990**. Tissue distribution of avidin and streptavidin injected into mice. Effect of avidin carbohydrate, streptavidin truncation and exogenous biotin. *Eur. J. Biochem.* 189 : 327–331.

**30** Kang Y.S., Y. Saito, and W.M. Pardrige. **1995**. Pharmacokinetics of [3H]biotin bound to different avidin analogues. *J. Drug Target.* 3 : 159–165.

**31** Rosebrough S.F., and D.F. Hartley. **1996**. Biochemical modification of streptavidin and avidin: in vitro and in vivo analysis. *J. Nucleic Med.* 37 : 1380–1384.

**32** Marttila A.T., O.H. Laitinen, K.J. Airenne, T. Kulik, E.A. Bayer, M. Wilchek, and M.S. Kulomaa. **2000**. Recombinant NeutraLite avidin: a non-glycosylated, acidic mutant of chicken avidin that exhibits high affinity for biotin and low non-specific binding properties. *FEBS Lett.* 467 : 31–36.

**33** Caliceti P., M. Chinol, M. Roldo, F.M. Veronese, A. Semenzato, S. Salmaso, and G. Paganelli. **2002**. Poly(ethylene glycol)-avidin bioconjugates: suitable candidates for tumor pre-targeting. *J. Control. Release* 83: 97–108.

**34** Caliceti P., M. Chinol, G. Paganelli, S. Salmaso, S. Bersani, and A. Semenzato. **2005**. Preparation and characterization of active site protected poly(ethylene glycol) avidin bioconjugates. *Biochim. Biophys. Acta* 1726 : 57–66.

**35** Salmaso S., S. Pennadam, D. Cunliffe, C. Alexander, D. Górecki, and P. Caliceti. Physico-Chemical and Pharmacokinetic Studies of Avidin Bioconjugates with Thermosensitive Polymers, 5th International Symposium on Controlled Release of Bioactive Materials, June 21–24, 1998, 347.

**36** Caliceti P., O. Schiavon, F.M. Veronese. **1999**. Biopharmaceutical properties of uricase conjugated to neutral amphiphilic polymers, *Bioconjug. Chem.* 10: 638–646.

**37** Duncan R. **2003**. The drawing era of polymer therapeutics. *Nature Rev. Drug Discov.* 2: 347–360

**38** Kratz, F., U. Beyer, and M. T. Schutte. **1999**. Drug–polymer conjugates containing acid-cleavable bonds carrier system. *Crit. Rev. Ther. Drug Carrier Syst.* 16, 245–288.

**39** Greenwald, R. B., Y. H. Choe, C. D. Conover, K. Shum, D. Wu, and M. Royzen. **2000**. Drug delivery systems based on trimethyl lock lactonization: poly(ethylene glycol) prodrugs of amino-containing compounds. *J. Med. Chem.* 2000, 43, 475–487.

**40** Conover C.D., H. Zhao, C.B. Longley, K.L. Shum, and R.B. Greenwald. **2003**. Utility of poly(ethylene glycol) conjugation to create prodrugs of amphotericin B. *Bioconjug. Chem.* 14 : 661–666.

**41** Guiotto A., M. Canevari, M. Pozzobon, S. Moro, P. Orsolini, and F.M. Veronese. **2004**. Anchimeric assistance effect on regioselective hydrolysis of branched PEGs: a mechanistic investigation. *Bioorg. Med. Chem.* 12 : 5031–5037.

**42** Conover C.D., R.B. Greenwald, A. Pendri, C.W. Gilbert, and K.L. Shum. **1998**. Camptothecin delivery systems: enhanced efficacy and tumor accumulation of camptothecin following its conjugation to polyethylene glycol via a glycine linker. *Cancer Chemother. Pharmacol.* 42 : 407–414.

**43** Schiavon O., G. Pasut, S. Moro, P. Orsolini, A. Guiotto, and F.M. Veronese. **2004**. PEG-Ara-C conjugates for controlled release. *Eur. J. Med. Chem.* 2 : 123–133.

**44** Choe Y.H., C.D. Conover, D. Wu, M. Royzen, Y. Gervacio, V. Borowski, M. Mehlig, and R.B. Greenwald. **2002**. Anticancer drug delivery systems: multiloaded N4-acyl poly(ethylene glycol) prodrugs of ara-C. II. Efficacy in ascites and solid tumors. *J. Control. Release* 79 : 55–70.

# 12
# Exposure–Response Relationships for Therapeutic Biologic Products

*Mohammad Tabrizi and Lorin K. Roskos*

## 12.1
## Introduction

Therapeutic biologic products exhibit a number of pharmacokinetic and pharmacodynamic features that distinguish them from small-molecule drugs [1, 2]. Unlike most small-molecule therapeutics, the disposition of protein drugs is often target-mediated. Elimination occurs by high-capacity, nonspecific processes (such as renal metabolism and reticuloendothelial clearance) and sometimes by highly specific, saturable processes such as receptor-mediated clearance. Inter-subject pharmacokinetic variability is often low, which facilitates analysis of complex pharmacokinetics and exposure–response relationships. Protein drugs generally exhibit high target specificity, which also facilitates mechanism-based, pharmacodynamic analyses. This chapter reviews the exposure–response relationships of biologic drug products, and presents case examples for several categories of these drugs.

## 12.2
## Overview of Pharmacokinetics and Pharmacodynamics

### 12.2.1
### Pharmacokinetics

The optimal administration of drugs in clinical practice is facilitated by effective application of the principles of clinical pharmacokinetics (PK) and pharmacodynamics (PD). Relationships between drug levels in the systemic circulation and various body compartments (e.g., tissues and biophase) following drug administration depend on factors governing drug absorption, distribution, elimination, and excretion (ADME). Collectively, the study of the factors that govern the ADME processes is termed pharmacokinetics.

*Pharmacokinetics and Pharmacodynamics of Biotech Drugs: Principles and Case Studies in Drug Development.* Edited by Bernd Meibohm

ISBN: 3-527-31408-3

#### 12.2.1.1 Absorption

A general summary of the pharmacokinetics of biologics in comparison to small-molecule drugs is provided in Table 12.1. Although oral administration of small-molecule drugs is usually preferred, the oral route is usually not a viable route for the administration of peptides or proteins because of gastrointestinal degradation. Most biologics must be administered parenterally. Drug absorption can be impacted by factors such as molecular charge, size, dissolution, gastric emptying, membrane permeability and diffusion, and can be highly variable for small-molecule drugs. For biologics, intravenous (IV) and subcutaneous (SC) administration are the two dosing routes most frequently used. Following SC administration, absorption can occur via the lymphatic system or blood capillaries. Lymphatic absorption has been associated with slower absorption profiles as compared to absorption via blood capillaries [3]. A linear and positive correlation between lymphatic absorption and molecular weight has been reported [4, 5]. The cumulative recovery of human recombinant interferon $\alpha$-2a (MW 19 kDa) via the lymphatic system was reported to be around 60%. Larger proteins, such as monoclonal antibodies (mAbs), are expected to undergo predominantly lymphatic absorption. The absorption kinetics through the vascular and lymphatic pathways can be modeled by two parallel, first-order absorption processes with different absorption rates [6].

#### 12.2.1.2 Distribution

The biodistribution of biologics is affected by factors such as molecular weight and the distribution and density of the target protein. Blood capillary structure and morphology offer different permeability profiles for the distribution of biologics. Capillary beds rich in tight junctions (continuous) and fenestrated capillaries that cover organs such as the blood–brain barrier, skin, muscle, the gastrointestinal tract, glands, and the kidneys are not permeable to macromolecules. On the other hand, more porous vascular beds such as sinusoidal capillaries (pore size 100 nm) found in organs such as the liver, spleen, and bone marrow are permeable to macromolecules [7]. The diffusion of high molecular-weight substances such as dextran and IgG (both 150 kDa) in normal tissues was reported using an *in-vivo* rabbit model [8]. These studies indicated an increase in diffusion (eight-fold) and microvascular permeability (33-fold) of neoplastic tissues relative to normal tissues for high molecular-weight substances.

#### 12.2.1.3 Elimination

In contrast to small-molecule drugs, the clearance of biologics is not influenced by the activity of xenobiotic-metabolizing enzyme systems such as the cytochrome P450 (CYP450) enzyme superfamily. Therefore, pharmacokinetic drug–drug interactions and polymorphic metabolism are rarely expected. Many factors related to the biophysical properties of the biologic (e.g., hydrodynamic radius and affinity for the target), or host factors such as receptor expression, receptor internalization rate, and patient status can affect the clearance of biologics.

**Table 12.1** Comparison of pharmacokinetics of small-molecule drugs and biologic products.

| Parameter | Small-molecule drugs | Biologic products |
|---|---|---|
| *Absorption* | | |
| Oral | Low to high bioavailability | Negligible to low bioavailability |
| Subcutaneous | Vascular absorption | Vascular and lymphatic Absorption |
| *Distribution* | | |
| Extravascular | Limited to extensive | Limited |
| Protein binding | Non-specific | Specific binding (e. g., shed receptors) |
| Target-dependent | Uncommon | Common |
| *Elimination* | | |
| Hepatic | Xenobiotic metabolizing enzymes; biliary excretion | RES (Kupffer cells) |
| Renal | Dependent on lipophilicity | Dependent on hydrodynamic size |
| Extrahepatic/renal | Variable (e. g., plasma carboxylesterases, intestinal CYP3A4) | RES (vascular endothelium and phagocytic blood cells) |
| Target-mediated | Rare | Common |
| Pharmacokinetic drug–drug interactions | Common | Uncommon |
| Immunogenicity | Rare | Common |

RES: Reticuloendothelial system.

For intact antibodies or Fc-fusion proteins, the nonspecific clearance of antibodies through the cells of the reticuloendothelial system (RES) can be regulated through interaction with specific Fc receptors [9]. These receptors are expressed by various phagocytic cells such as monocytes, macrophages, neutrophils, eosinophils and other cells of the immune system, and can have either a salvage effect (FcRn) or be involved in clearance (Fc$\gamma$R) of antibody–antigen complexes [9–12]. Kupffer cells (liver macrophages) in the liver and other phagocytic cells can internalize antigen–antibody complexes at a rapid rate [13]. Engineered mutations of IgG Fc residues (position 428, and 250 alone or in combination) that increased the binding affinity of the antibody for FcRn (four- to 27-fold better binding affinity to rhesus FcRn than the wild-type antibody at pH 6.0) resulted in a two-fold increase in serum half-lives of the mutant $IgG_2$ antibodies [11, 14].

The elimination of biologics can be affected to a great extent by their interaction with the target. When a biologic ligand binds to a cell membrane receptor, the ligand can induce receptor internalization, particularly if the ligand is an agonist. The ligand can also be internalized as the receptor undergoes its normal basal level of internalization, degradation, and recycling [15]. Once internalized, the ligand and receptor can undergo lysosomal or proteosomal degradation, or the ligand might be recycled to the cell surface if the ligand remains bound to a recycled receptor. When the target protein affects the clearance of a biologic, the effect is usually manifested as a dose-dependent clearance and half-life. At low drug doses that do not saturate the target, the half-life is shorter; but as the dose is increased and the binding to the target protein is saturated, an increase in half-life and decrease in clearance rate is observed. The target-dependent clearance pathway is often referred to as a "receptor sink" or an "antigen sink" for therapeutic antibodies. Target-related sinks are most commonly observed for biologics targeting internalizing cell membrane receptors with high normal tissue expression.

#### 12.2.1.4 **Immunogenicity**

Immunogenicity of biologic products can be a significant problem in therapeutic use, and can adversely affect the product pharmacokinetics, safety, and efficacy [16]. Many factors can influence the generation of anti-drug antibody responses following biologic therapy [16,17]. Immunogenicity can alter pharmacokinetics by affecting clearance and biodistribution, it can reduce efficacy, and also introduce serious safety concerns. Immune complex formation in the serum has been shown to accelerate the clearance of mAbs by the RES [18]. Serious adverse events and safety risks such as hypersensitivity and anaphylactoid reactions can be associated with immunogenicity [16].

The potential to develop neutralizing antibodies that crossreact with endogenous proteins is the greatest concern around immunogenicity. Rare cases of pure red cell aplasia have occurred in patients who developed neutralizing antibodies to epoetin-$\alpha$ [19]. Antibodies to thrombopoietin analogues that neutralized endogenous thrombopoietin and caused severe thrombocytopenia in some subjects in clinical trials led to the discontinuation of the clinical development of thrombopoietin analogues [20].

### 12.2.2 Pharmacodynamics

Pharmacodynamics is the study of drug effects on living organisms. The effect of most drugs results from their interaction with various macromolecules within the organism, which cause alterations of functional and biochemical pathways (i.e., inhibition or stimulation of the receptors and the relevant signaling pathways). These biochemical or functional changes characterize the specific biological response to the administered drug. Exposure–response relationships can be elucidated by studying pharmacodynamic data in conjunction with dosing and pharmacokinetic information.

The concentration of drug at the proximity of the biological receptor (biophase) determines the magnitude of the observed response. In simple cases, the relationships between drug dose or concentration and the observed pharmacological effect can be characterized by linear or log-linear pharmacodynamic models (Fig. 12.1B) in the form of $E = E_0 + m\ C$ (or ln C), where E and $E_0$ denote the observed and baseline (no-drug) effects, respectively, and m is the slope of the concentration–effect relationship. For some drugs, however, these simple models do not sufficiently capture the concentration–effect profile. In these instances, the maximum effect ($E_{max}$) model or the sigmoid $E_{max}$-model (Fig. 12.1A) can be a more suitable pharmacodynamic model [21–24]. Additionally, further complexities such as counterclockwise (Fig. 12.1C) or clockwise hysteresis loops can be observed in concentration–effect relationships for some drugs at non-steady-state serum/tissue concentrations. The underlying causes for these additional complexities in exposure–response relationships can be diverse, and related to factors such as active or inhibitory metabolites, indirect response characteristics, a transient time delay between concentration and effect, or an equilibration delay in the

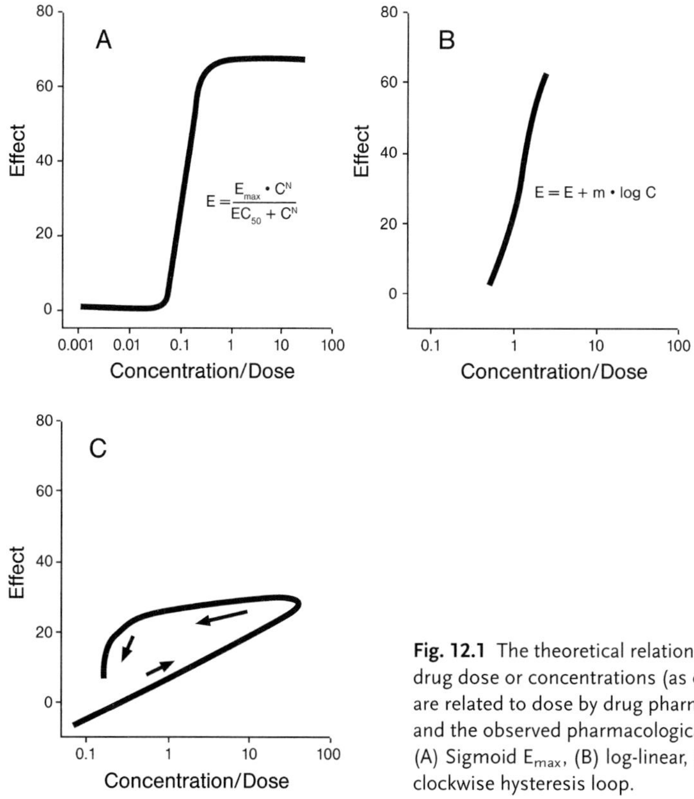


**Fig. 12.1** The theoretical relationships between drug dose or concentrations (as concentrations are related to dose by drug pharmacokinetics) and the observed pharmacological effect. (A) Sigmoid $E_{max}$, (B) log-linear, (C) counterclockwise hysteresis loop.

distribution of drug to the biophase compartment. Various modeling approaches have been employed to address these complexities [24–30].

Biologic products can exert their pharmacological effects via a number of different mechanisms:

- neutralizing target function,
- activating receptors by mimicking endogenous receptor ligands,
- delivering toxins to specific cells (targeted delivery),
- eliciting effector functions in conjunction with target modulation.

Under certain circumstances where target expression is high in critical organs (e.g., heart, lung, and vasculature), effector function might not be desirable and could be deleterious. In other instances (such as applications in hematological malignancies), effector functions can be a significant part of the mechanism of action, and maximizing effector functions can be highly desirable [10, 11].

Engineered IgG antibodies have been constructed with altered affinity to human Fc$\gamma$ receptors and altered potency *in vitro* and in animal models [14, 31–33]. Mutations of critical residues in the Fc region ($C_H2$ domain or the hinge region joining $C_H1$ and $C_H2$), have enhanced or decreased antibody-dependent cellular cytotoxicity (ADCC) and complement-dependent cytotoxicity (CDC) [33–37]. In addition, alterations in residues located at the $C_H2$ domain of $IgG_1$ involved in binding with C1q protein, a component of the complement activation cascade, resulted in a significant increase in CDC activity [34, 35].

## 12.3 Hormones

Hormones are secreted by specialized glands (adrenal, hypothalamus, ovary, pancreas, parathyroid, pineal, pituitary, testes, thyroid) or other tissues (e.g., heart, gut, and kidney), and regulate the cellular activities of distant tissues. Plasma levels of hormones are tightly regulated through homeostatic feedback systems. The peptide and protein hormones typically have short half-lives (minutes), which allow rapid changes in plasma levels and rapid enhancement or attenuation of their biological effects.

The therapeutic administration of biologic hormones should, ideally, mimic the endogenous secretion patterns of the hormone to achieve optimal effects. Hormone formulations and dosing regimens have been developed to reflect the normal time-course of exposure to the hormone. In some cases, the effects of the hormone can be highly dependent on the dose schedule. These important aspects of the exposure–response relationships for hormones are illustrated here for insulin and parathyroid hormone.

### 12.3.1 Insulin

The endogenous secretion of insulin maintains stable blood glucose concentrations (euglycemia) during periods of fasting and feeding. Basal insulin secretion maintains hepatic gluconeogenesis. Increases in blood glucose following food intake stimulate increased insulin secretion, and increased levels of insulin stimulate glucose uptake into cells. In nondiabetic subjects, insulin levels peak at about 15–30 min following the initiation of feeding [38]. Subsequently, as blood glucose levels decline due to increased cellular uptake, insulin secretion returns to the basal state. Insulin dosing in diabetic patients is intended to mimic this normal secretion pattern of insulin. Such a pattern is accomplished by the SC administration of insulin formulations with different absorption rates [39]. Since the half-life of insulin in plasma is very short (5–6 min), SC dosing results in flip-flop kinetics, which permits modulation of $C_{max}$, $T_{max}$, and the terminal half-life of insulin, depending on the absorption characteristics of the administered formulation.

Rapid (regular insulin, lispro, insulin aspart), intermediate (NPH, Lente), and slow (Ultralente, protamine zinc, and glargine) -acting formulations of insulin are available. Insulin self-associates into hexamers at the concentrations attained in pharmaceutical formulations. In order to be absorbed from the injection site, insulin must dissociate into dimers or monomers [40]. Different absorption rates have been attained by modifying either the insulin molecule or the formulation to change the rate of dissociation of the insulin hexamers. A normal exposure pattern to insulin can be mimicked through the convenient administration of a mixture of rapid-acting and intermediate-acting insulin just prior to the morning and evening meals [41].

The pharmacokinetics and glucodynamics of novel insulin formulations have been investigated through euglycemic clamp studies in healthy volunteers. In these studies, subjects receive a dose of insulin and a variable-rate glucose infusion to maintain stable blood glucose levels [42]. Typically, the time-course of the glucose infusion rate is presented graphically and used to represent the glucodynamic profile [39]. Recently, a PK/PD model has been developed to describe the pharmacokinetic and glucodynamic effects of different insulin products in euglycemic clamp studies, and has been used to differentiate insulin aspart and regular human insulin [43]. The model incorporates insulin pharmacokinetics, the kinetics of endogenous and infused glucose, and the effects of insulin on blood glucose levels. An example of PK/PD modeling of regular human insulin using this model is shown in Fig. 12.2. The model successfully described absorption differences between the insulin analogues, and detected no major differences in PD parameters after accounting for PK differences.

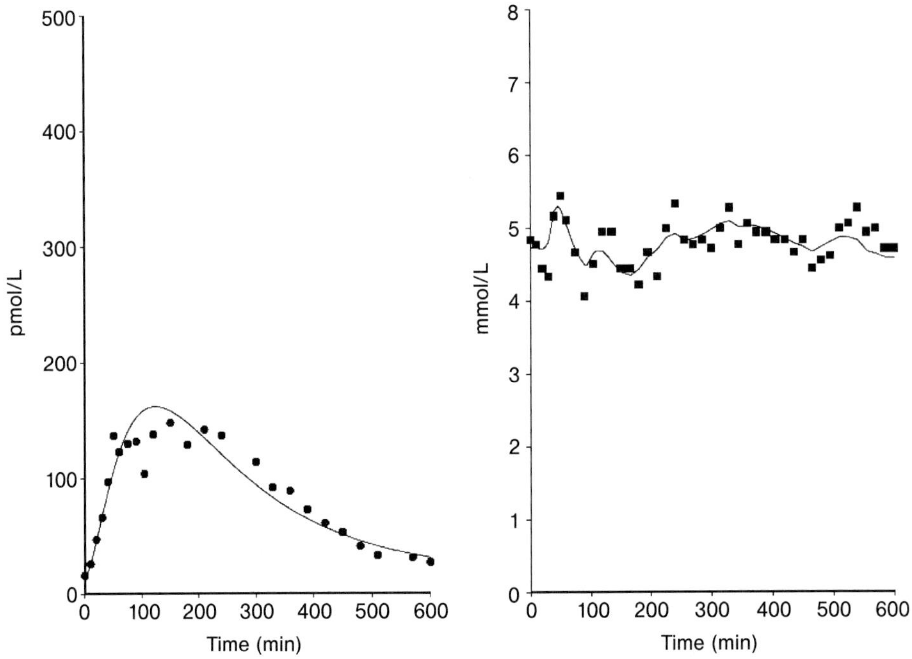


**Fig. 12.2** Pharmacokinetic-pharmacodynamic modeling of pharmacokinetics (left panel) and glucodynamics (right panel) in a healthy volunteer receiving regular human insulin in a euglycemic clamp study (from [43]).

### 12.3.2
### **Parathyroid Hormone**

Parathyroid hormone (PTH) is an 84-amino acid peptide secreted by the parathyroid glands, and is the principal regulator of extracellular calcium levels [44, 45]. The effects of PTH on extracellular calcium are mediated directly or indirectly through effects on bone, kidney, and intestine. A decrease in extracellular calcium causes an increase in PTH secretion. As a consequence, the rise in PTH levels causes increased bone resorption and the release of calcium from bone, decreased calcium excretion by the kidney, and increased intestinal calcium absorption. The therapeutic application of PTH has centered on the bone effects as an anabolic treatment for osteoporosis. PTH increases the activity of both osteoblasts (which form bone) and osteoclasts (which mediate bone resorption). The desirable anabolic effects of PTH on osteoblasts appear to be highly dependent on dose schedule and the duration of daily exposure.

Teriparatide, the N-terminal 34-mer of full-length PTH, is currently approved for the treatment of men or post-menopausal women who are at a high risk for fracture. Teriparatide is the only approved anabolic treatment for osteoporosis; other drugs for osteoporosis primarily affect bone resorption [46, 47]. Teriparatide is administered as a daily 20-μg SC injection, but because of its short half-life (ca. 1 h, reflecting the absorption half-life), therapeutic exposure occurs for only a few hours each day. Daily dosing leads to long-term increases in the biomarkers of bone formation [serum bone alkaline phosphatase (ALP) and carboxy-terminal propeptide of type I procollagen (PICP)] and bone resorption [urinary deoxypyridinolone (DPD) and N-telopeptide (NTx)] (Fig. 12.3) [48]. Early one-month changes in bone formation markers, but not bone resorption markers, were significantly correlated with improvements in bone structure after 22 months of treatment. The net anabolic effect is believed to be due to preferential stimulation of osteoblasts by intermittent daily exposure. While daily SC dosing is inconvenient, a need for intermittent daily exposure might preclude the development of sustained release or long half-life PTH analogues that could be dosed less frequently.

Rat models of the anabolic and catabolic effects of teriparatide on bone have supported the requirement for intermittent exposure [49]. Anabolic effects in rats required exposure to PTH for 1 h each day; an equivalent level of exposure for 6 h or more each day led to catabolic effects. Likewise, a DNA microarray analysis indicated a differential pattern of gene expression when rats were exposed to the catabolic and anabolic dosing schedules, respectively [50]. While the expression profiles, *per se*, did not clearly indicate a preferential shift between bone resorption and bone formation, some aspects of the gene expression changes following the catabolic regimen (e.g., up-regulation of several matrix metalloproteases) were more consistent with an effect on matrix degradation and bone resorption. Although the molecular mechanism of the schedule dependency is not fully understood, it is possibly related to down-regulation or desensitization of the receptor, PTH1R, during continuous exposure to PTH. Upon ligand binding, PTH1R is reversibly internalized, down-regulated, and inactivated by binding to β-arrestin [51]. A mathematical, two-state receptor sensitization model has been developed that potentially explains receptor desensitization on continuous PTH exposure and sensitization following intermittent exposure [52].

## 12.4 Cytokines

Cytokines are regulatory proteins that are secreted by a variety of cell types. These proteins exert pleiotropic effects on cells of the immune system and can enhance or attenuate immune responses. Cytokines are an integral part of the immune response to infections and to the malignant transformation of host cells. Cytokines have been successfully applied for a variety of therapeutic indications using pharmacological doses that yield concentrations above physiological levels and augment the normal immune response to infectious diseases and cancer. However, pharmacological doses can be associated with serious and life-threatening toxicities; therefore, exposure–response relationships for therapeutic cytokines have generally been based on tolerability and dose-limiting toxicity rather than therapeutic response.

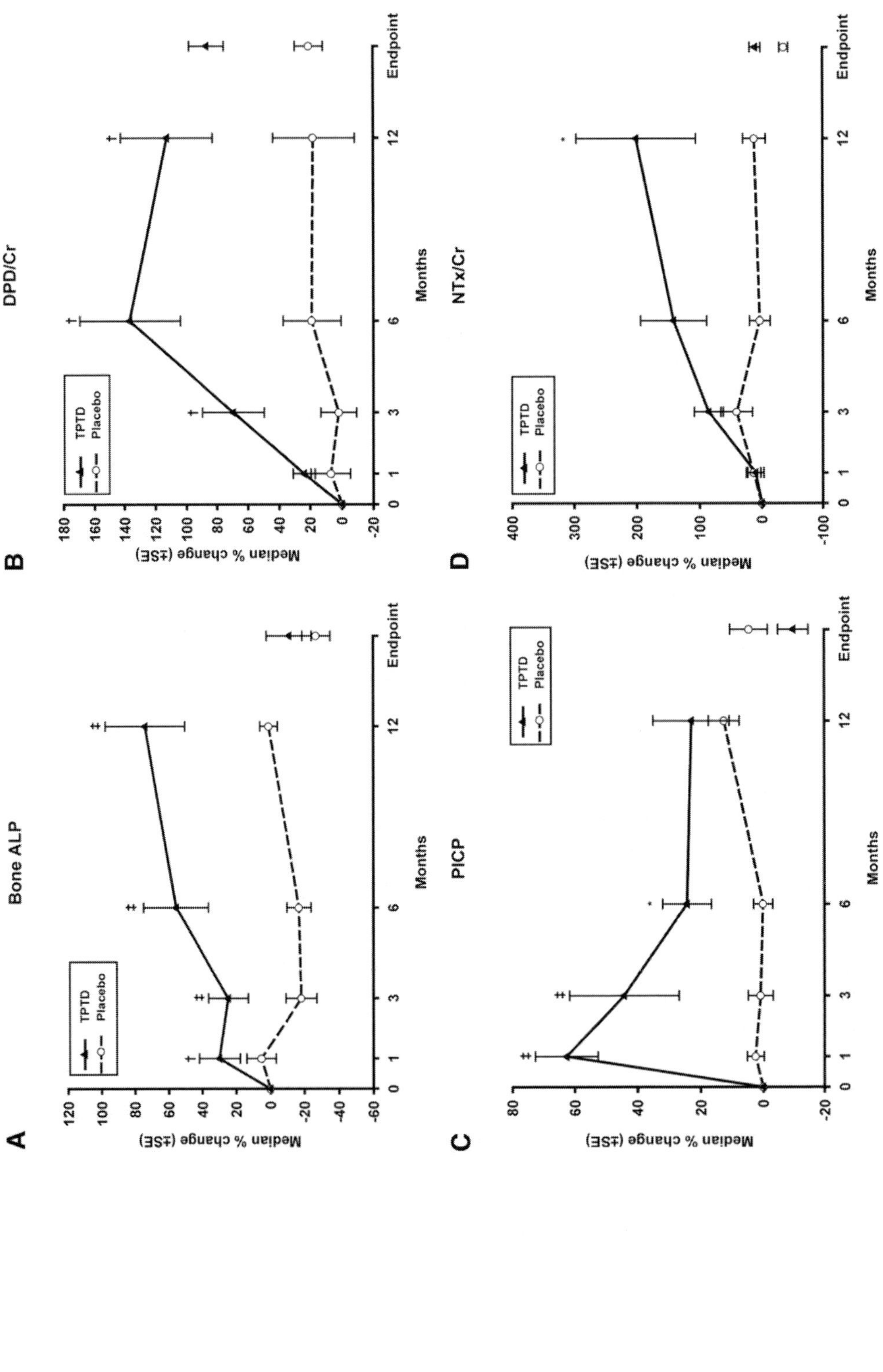
A
Bone ALP
B
DPD/Cr
C
PICP
D
NTx/Cr
TPTD
Placebo
Median % change (±SE)
Months
Endpoint

Several Type I and Type II interferons (IFN) and interleukin-2 (IL-2) have been approved to treat infectious diseases or malignancies by augmenting immune responses: IFN-α2a (chronic hepatitis C); IFN-α2b (hepatitis C, melanoma, chronic myelogenous leukemia, hairy-cell leukemia, Kaposi's sarcoma, cutaneous T-cell lymphoma, renal cell carcinoma); IFN-αcon1 (chronic hepatitis C); IFN-β1a (multiple sclerosis); IFN-β1b (multiple sclerosis); IFN-γ1b (malignant osteopetrosis); and IL-2 (renal cell carcinoma).

### 12.4.1 Interleukin-2

High-dose IL-2 is approved for the treatment of metastatic renal cell carcinoma. The tumor is intrinsically immunogenic, and can elicit a host immune response that infrequently results in spontaneous remission of disease; administration of IL-2 is believed to augment the normal immune response [53]. In a Phase II study that supported the approval of high-dose IL-2 for this indication, 15% of patients achieved objective responses (8% were partial responses and 7% complete responses). In this trial, patients received IL-2 at 600 000 or 720 000 U/kg per dose. The cumulative dose and the efficacy were similar between dose groups: patients receiving the higher dose tolerated fewer cycles of therapy due to toxicity.

Therapy with IL-2 can cause severe and life-threatening toxicities that include hypotension and capillary leak syndrome [54]. Approximately 50–75% of patients receiving high-dose IL-2 experience grade 3–4 hypotension requiring fluid resuscitation and pressor support. Capillary leak syndrome that can result in ascites, respiratory distress, and pleural effusions has been observed in 10–20% of treated patients. Due to the toxicity profile, studies have evaluated the comparative efficacy of low-dose and high-dose IL-2 for metastatic renal cell carcinoma.

A randomized Phase III study conducted by the National Cancer Institute compared high-dose (720 000 U/kg) IL-2 to a low-dose (72 000 U/kg) regimen [55]. Although the low-dose regimen was better tolerated, high-dose IL-2 was superior in overall survival and complete response rate. As with many toxic cancer chemotherapy regimens, the greatest efficacy of IL-2 is achieved at the maximum tolerated dose. While randomized trials have been conducted to evaluate lower doses and subcutaneous outpatient dosing of IL-2 in combination with other therapies such as IFN-α2b, these trials to date have supported high-dose IL-2 as the preferred dosing regimen for the treatment of renal cell carcinoma [56].

◀ **Fig. 12.3** Effects of daily teriparatide (TPTD) administration on biomarkers of bone formation (bone ALP and PICP) and bone resorption (DPD/Cr and NTx/Cr) in postmenopausal women with osteoporosis. Changes in bone formation markers at 1 month were significantly correlated with improvements in bone structure after 22 months of treatment (from [48]).

## 12.5
## Growth Factors

Growth factors stimulate the lineage-selective proliferation of mitotic cell populations, and also stimulate the mobilization and maturation of post-mitotic cells. Ultimately, the growth factors produce an increase in the number of terminally differentiated cells of a particular lineage. Approved growth factors include palifermin for the treatment of chemotherapy- and radiation-induced mucositis, epoetin-α and darbepoetin-α for the treatment of anemia, and filgrastim and pegfilgrastim for the treatment of neutropenia. PK/PD relationships for pegfilgrastim are discussed in detail in Chapter 15.

The drug development of hematopoietic growth factors (HGFs) has provided a unique opportunity for understanding the exposure–response relationships of these biologic products. Increases in precursor cells and terminally differentiated blood cells in the circulation provide a convenient PD endpoint that permits complex modeling of PK/PD relationships. The HGFs are subject to receptor-mediated clearance as part of the homeostatic mechanisms that regulate blood cell numbers in the circulation. In modeling the nonlinear pharmacokinetics of the HGFs, the peripheral blood cell count can be used as a proportionality factor to modulate the maximum velocity ($V_{max}$) of the receptor-mediated elimination rate. Thus, the nonlinearity of HGF pharmacokinetics can be modeled according to drug level and time.

In cytokinetic models, the lifespan of the cell must be considered. Cells that enter the circulation at the same time do not have identical lifespans; therefore, the variability of cell lifespan should also be considered. Cells can be eliminated from the circulation by apoptosis, senescence, or random destruction. Erythrocytes and platelets at normal levels are primarily eliminated by senescence. For platelets, a random component of elimination is evident at low platelet counts. Neutrophils are eliminated randomly from the circulation, and the elimination can be modeled appropriately as a first-order loss. A general model that can be applied to the cytokinetics of HGF-stimulated hematopoiesis is illustrated in Fig. 12.4 [57]. This model, which was applied to peripheral platelet counts and the kinetics of autologous platelet tracers following administration of a thrombopoietin analogue to healthy volunteers, accounts for the concentration–response relationship for stimulation of precursor cells, delays in the emergence of new platelets from marrow, random destruction of platelets, and the intra-subject variability of platelet lifespan. Modeling the cell count as the sum of a series of transit compartments, where the cell lifespan is represented by the mean transit time through the compartments, imparts a gamma distribution to variance of the intra-subject cell lifespan; the variance can be decreased by increasing the number of transit compartments. The same model architecture can be applied readily to growth factor-stimulated proliferation of other cell types.

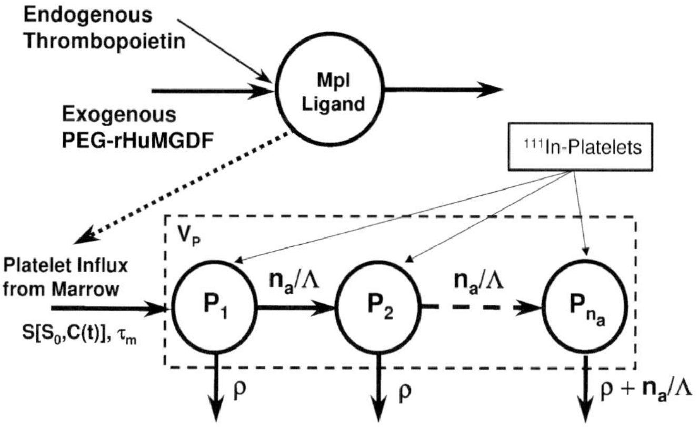


**Fig. 12.4** Pharmacokinetic-pharmacodynamic model of the thrombopoietic effects of a thrombopoietin analogue (PEG-rHuMGDF) in healthy volunteers. The intrinsic longevity of platelets ($\Lambda$), nonlinear random destruction of platelets ($\rho$), and the intra-subject variability of intrinsic platelet longevity (controlled by n, the number of catenary-linked transit compartments) are represented by the model. Simultaneous modeling of autologous radiolabeled platelet tracers allowed analysis of the effect of drug on platelet survival (from [57]).

### 12.5.1
### Epoetin-α

Epoetin-α (recombinant human erythropoietin) is approved for the treatment of anemia in hemodialysis patients and those receiving chemotherapy. A hyperglycosylated analogue of erythropoietin (darbepoetin-α) that has a lower clearance rate and can be dosed less frequently has also been approved. Erythropoietin stimulates the proliferation of erythrocyte progenitor cells in bone marrow and increases peripheral red blood cell (RBC) counts.

The first example of modeling the exposure–response relationship for an HGF was conducted for hemodialysis patients receiving erythropoietin, and was conducted as a population dose–response analysis [58]. The modeling included parameterization of the lifespan of the erythrocyte. Although the model assumed a zero variance for intra-subject erythrocyte lifespan, this simplification is usually valid for the modeling of erythrocyte, hematocrit, or hemoglobin kinetics. The model fit to the time-course of a patient's hematocrit is shown in Fig. 12.5. As is seen in cases where the cell lifespan is longer than the mean residence time of the HGF, the time to steady-state pharmacodynamic response is determined by the cell lifespan. In modeling the dose–response relationship, high variability in the pharmacodynamics was observed. One factor contributing to this variability was inter-subject variation in erythrocyte lifespan. On average, erythrocyte lifespan in hemodialysis patients was about one-half the typical lifespan in healthy volunteers.

Recently, more complex PK/PD analyses of erythropoietin and darbepoetin-α in normal volunteers and patients have been conducted, and have contributed to the understanding of the exposure–response relationships of these drugs [59–61].

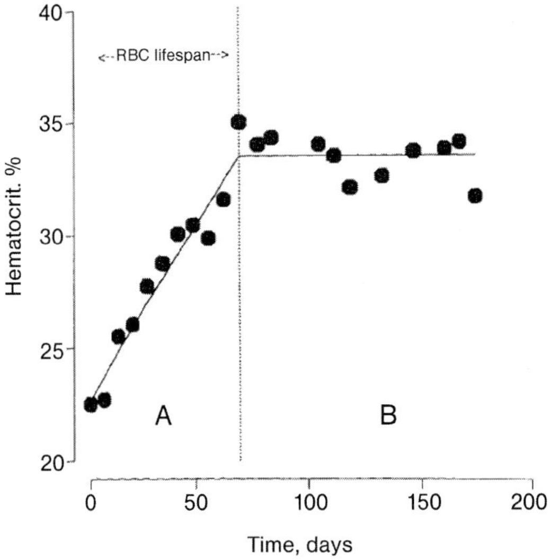


**Fig. 12.5** Dose–response modeling of the effects of epoetin-α on the hematocrit of a hemodialysis patient. The time to reach a new steady-state hematocrit is determined by the erythrocyte lifespan. *Phase A:* Epoetin-α causes an increase in RBC production rate; consequently the hematocrit increases as newly produced erythrocytes do not die at the early stage. *Phase B:* After reaching one RBC lifespan, erythrocytes die at the current production rate (from [58]).

## 12.6 Soluble Receptors

The extracellular domain of cell membrane receptors can be produced recombinantly and used as a therapeutic product. The soluble receptor can be used to bind and neutralize the receptor's endogenous ligands (e.g., etanercept), or it can be used for stimulation of a co-receptor (e.g., abatacept). Soluble receptors are typically produced as IgG Fc fusion proteins to extend the half-life of the receptor in circulation.

### 12.6.1 Etanercept

Etanercept is dimeric fusion protein that consists of the extracellular ligand-binding protein of the human 75-kDa tumor necrosis factor (TNF) receptor linked to the Fc portion of human $IgG_1$ immunoglobulin. Etanercept has a short half-life (3–6 days) and is administered at doses of 25 mg SC twice weekly [62]. The population PK/PD of etanercept were described recently [63]. Following SC administration of the protein at 25 mg and 50 mg twice weekly doses in rheumatoid arthritis (RA) patients, linear pharmacokinetic behavior was observed. Predicted serum concentration–time profiles for etanercept using the population pharmacokinetic model are shown in Fig. 12.6A. Dose-proportional increases in average serum concentrations following multiple dose administration were observed. Mean steady-state serum concentrations ranged between 4 and 5 μg/mL following administration of the 50-mg dose twice weekly. The mean steady-state serum concentrations following administration of 25-mg twice weekly were reported as 2–3 μg/mL. The efficacy of etanercept, as a monotherapy, in psoriasis patients following SC administration of low (25 mg, once weekly), medium (25 mg, twice weekly) and high (50 mg, twice weekly) doses was also reported recently [64]. The relationship between predicted mean serum concentrations and the number of patients (%) with ≥90% improvements in PASI (Psoriasis Area and Severity Index) score on study week 24 is shown in Fig. 12.6B. The mean concentration–effect (PASI score) relationship for etanercept in psoriasis patients appeared to be best described by a sigmoid $E_{max}$-model with predicted $EC_{50}$ (serum concentrations corresponding to 50% effect) of around 2 μg/mL. These predictions are comparable to previous published data on the efficacy of etanercept in RA patients following administration of 10 and 20 mg twice weekly for 12 months [65].

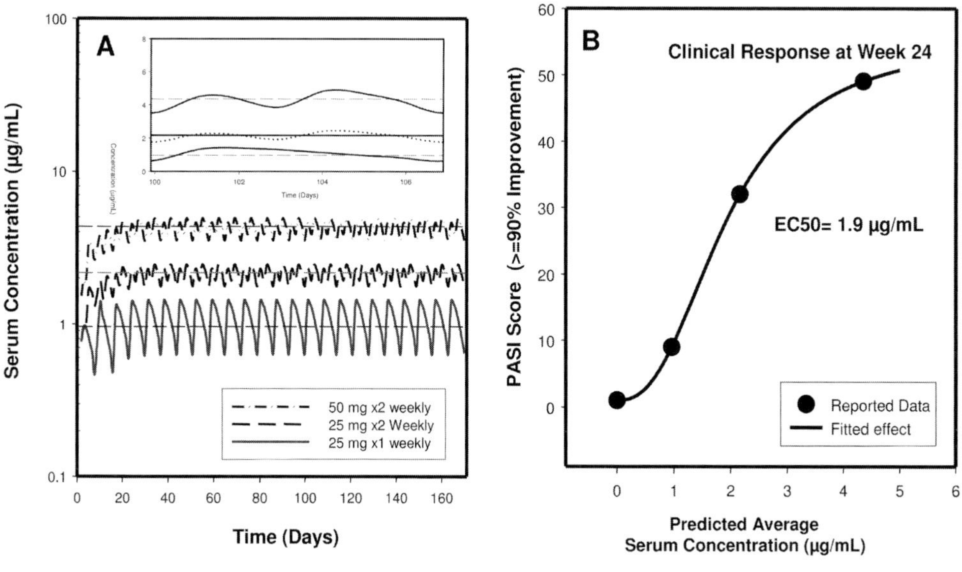


**Fig. 12.6** (A) Predicted serum concentration–time profiles for etanercept following subcutaneous administration of the low (25 mg once weekly), medium (25 mg twice weekly), and high (50 mg twice weekly) dose in rheumatoid arthritis patients using the population pharmacokinetic model described previously (from [63]). (B) Relationship between predicted populations mean serum concentrations and the number of patients (%) with ≥90% improvements in PASI (Psoriasis Area and Severity Index) score on Study Week 24. (Figure produced from data reported in [63, 64]).

## 12.7 Monoclonal Antibodies (mAbs)

Antibodies serve two important functions: (1) to bind and modulate antigens; and (2) to bind complement and immune effector cells such as natural killer cells and monocytes (see Chapter 3). Each IgG molecule contains two identical heavy chains and two identical light chains (Fig. 12.7). Antibody structure has evolved to accommodate the diverse antigen binding specificities through variations in the variable region sequence. The antigen-binding site is formed by the intertwining of the light chain variable domain ($V_L$) and the heavy chain variable domain ($V_H$). Each V domain contains three short stretches of peptide known as the complementarity-determining regions (CDRs); the CDRs are the major determinants of antigen binding affinity and specificity. The light chain contains one constant domain, $C_L$. The heavy chain contains three constant domains, $C_H1$, $C_H2$, and $C_H3$. The $C_H2$ and $C_H3$ domains can allow interactions of the IgG molecule with various components of the immune system by either binding C1q, which can activate the complement cascade and elicit complement-dependent cytotoxicity, or by binding to Fcγ receptors on immune effector cells, which can elicit antibody-dependent cellular cytotoxicity. These same variable and constant domains of the molecule also affect IgG catabolism and elimination.

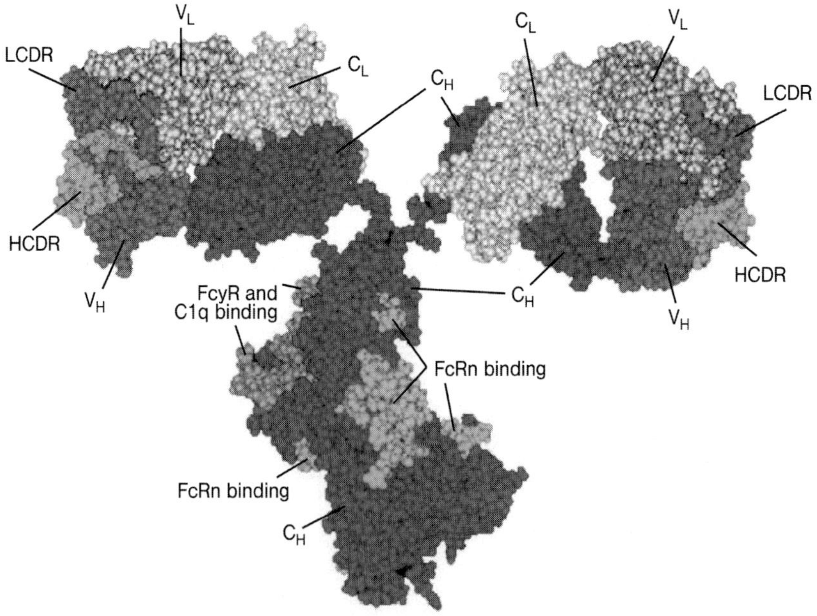


**Fig. 12.7** Space-filling model of an IgG molecule. $V_L$ and $V_H$: variable domain light and heavy chain; $C_L$ and $C_H$: constant domain light and heavy chain; LCDR and HCDR: complementarity-determining regions on $V_L$ and $V_H$ domains.

A unique characteristic of antibody function is the exquisite specificity of the interaction with antigens. The immune system has evolved to generate specific antibody molecules that allow recognition of almost any foreign protein. In living organisms, antibodies are generated via complex processes that require random arrangements of gene segments into functional genes, thus allowing the generation of diverse sets of specificities. Once the immune system is activated, a particular set of antibody-generating B cells are selected and the antibody is refined by undergoing affinity maturation by somatic mutation [37, 66, 67]. This process allows the immune system to generate antibodies for optimal performance against various antigens.

Several technologies have been employed for the generation of therapeutic antibodies. Although the early marketed antibodies were fully murine (containing 100% mouse protein), the evolution of mAb technology over the past two decades has led to generation of fully human (containing 100% human protein) antibodies using display libraries [68, 69], or transgenic mice that have been genetically engineered to produce human antibodies [70]. The advances in mAb technology have been crucial in reducing immunogenicity and improving the safety profiles of therapeutic antibodies (see Chapter 3, Section 3.4.2).

Currently, 17 mAbs have been approved in the United States (US) for therapeutic use in organ transplant, percutaneous coronary intervention, prophylaxis of respiratory syncytial virus, RA, Crohn's disease, asthma, chronic lymphocytic leukemia (CLL), acute myeloid leukemia, non-Hodgkin's lymphoma, breast cancer, and colorectal cancer. All approved mAbs are of the IgG class: 13 human $IgG_1$, one $IgG_4$, and two murine $IgG_{2a}$. Thirteen are intact mAbs, three are conjugated, and one is a Fab fragment (Tables 12.2 to 12.4).

## 12.7.1 Therapeutic Antibodies in Inflammatory Diseases

Monoclonal antibodies have been utilized successfully in the management of various inflammatory diseases such as RA, psoriasis, Crohn's disease, organ transplant rejection, and asthma (Table 12.2). The focal role of TNF-$\alpha$ in the underlying pathology of psoriasis, RA and Crohn's disease is now well established [65, 71, 72]. A chimeric antibody (infliximab), a fully human antibody (adalimumab), and a fusion protein (etanercept), have been successfully used in the management of RA, psoriasis, or Crohn's disease by targeting TNF-$\alpha$. Efalizumab, a humanized $IgG_1$ antibody against CD11a, was approved in 2003 for the treatment of adult patients with chronic moderate to severe plaque psoriasis [73]. Due to the central role of IgE in allergic responsiveness, a selective anti-IgE therapy has proven efficacious in the treatment of asthma. Omalizumab is a marketed-humanized $IgG_1$ mAb that inhibits the binding of IgE to the high-affinity IgE receptor (Fc$\gamma$RI) on the surface of mast cells and basophils [74]. Three antibodies have been marketed in the US as immunosuppressive therapies for the prophylaxis of acute organ rejection following transplantation: muromonab-CD3, a murine $IgG_{2a}$ antibody, binds to the lymphocyte-CD3 complex; basiliximab, a chimeric $IgG_1$ antibody, and

**Table 12.2** Approved therapeutic monoclonal antibodies for use in inflammatory diseases.

| Trade name | INN | Technology | mAb isotype | Target | Indication | Route | Dose | Schedule |
|---|---|---|---|---|---|---|---|---|
| Humira | Adalimumab | Human (Phage display) | $IgG_1$ | TNF-α | RA | SC | 40 mg | Biweekly |
| Orthoclone-OKT3 | Muromonab-CD3 | Murine | $IgG_{2a}$ | CD3 | Organ Transplant | IV | 5 mg | Daily (10–14 days) |
| Remicade | Infliximab | Chimeric | $IgG_1$ | TNF-α | RA, Crohn's disease | IV | 3–10 mg/kg | Biweekly-bimonthly |
| Raptiva | Efalizumab | Humanized | $IgG_1$ | CD11a | Plaque psoriasis | SC | 1 mg/kg | Weekly |
| Simulect | Basiliximab | Chimeric | $IgG_1$ | CD25 | Organ transplant | IV | 20 mg | 2 doses (days 1 and 5) |
| Xolair | Omalizumab | Humanized | $IgG_1$ | IgE | Asthma | SC | 150–375 mg | Biweekly-monthly |
| Zenapax | Daclizumab | Humanized | $IgG_1$ | CD25 | Organ transplant | IV | 1 mg/kg | Biweekly (5 doses) |

**Table 12.3** Approved unconjugated therapeutic monoclonal antibodies for treatment of cancer.

| Tradename | INN | Technology | mAb isotype | Target | Indication | Route | Dose | Schedule |
|---|---|---|---|---|---|---|---|---|
| Avastin | Bevacizumab | Humanized | $IgG_1$ | VEGF | Colorectal cancer | IV | 5 mg/kg | Biweekly |
| Campath | Alemtuzumab | Humanized | $IgG_1$ | CD52 | B-cell CLL | IV | 3–30 mg | 3 doses per week |
| Erbitux | Cetuximab | Chimeric | $IgG_1$ | EGFR | Colorectal cancer | IV | 250 mg/m$^2$ | Weekly |
| Herceptin | Trastuzumab | Humanized | $IgG_1$ | HER2 | Breast cancer | IV | 2 mg/kg; 4 mg/kg load | Weekly |
| Rituxan | Rituximab | Chimeric | $IgG_1$ | CD20 | B-cell NHL | IV | 375 mg/m$^2$ | Weekly |

NHL: Non-Hodgkin's lymphoma.

**Table 12.4** Approved conjugated therapeutic monoclonal antibodies for treatment of cancer.

| Tradename | USAN | Technology | mAb isotype | Target | Indication | Route | Dose | Schedule |
|---|---|---|---|---|---|---|---|---|
| Bexxar | Tositumomab | Murine | $IgG_{2a}$ | CD20 | B-cell NHL | IV | 450 mg tositumomab and $^{131}I$-tositumomab by dosimetry | 1 Dose (65–75 cGy) |
| Mylotarg | Gemtuzumab Ozogamicin | Humanized | $IgG_4$ | CD33 | CD33+ AML | IV | 9 $mg/m^2$ | 2 doses biweekly |
| Zevalin | Ibritumomab Tiuxetan | Murine | $IgG_1$ | CD20 | B-cell NHL | IV | 0.4 mCi/kg | 1 dose $\leq 30$ mCi |

NHL: Non-Hodgkin's lymphoma.

daclizumab, a humanized $IgG_1$ antibody, inhibit the actions of IL-2 on its receptor on the activated T cells [75, 76].

#### 12.7.1.1 **Anti-TNF-α Antibodies**

TNF-α plays a central role in the pathogenesis of RA and other immune-mediated inflammatory disorders. Inhibition of TNF-α allows rapid control of the inflammatory manifestations of RA and retards cartilage and bone destruction [77]. Two biologically active forms of TNF-α have been characterized: a membrane-bound and a soluble form. The soluble cytokine is a 17-kDa molecule that aggregates and forms a trimer in biological fluids (serum concentrations 50 pg/mL in RA patients) and acts through two receptors, p55 TNFR1 and p75 TNFR2, which are ubiquitously expressed.

Three anti-TNF-α biologics are currently marketed, including two monoclonal antibodies and one fusion protein. Although these anti-TNF-α biologics bind to and neutralize TNF-α, they share few similarities with respect to their pharmacokinetics, recommended dose, and dosing frequency. The pharmacokinetic differences have influenced the dose and dosing frequency for these anti-TNF-α biologics. Etanercept, a fusion protein with the shortest pharmacokinetic half-life, is administered twice weekly. The recommended dosing frequency for adalimumab, a fully human $IgG_1$ antibody with an average half-life of 12 days, is every one to two weeks. Higher doses and more frequent administration are recommended for infliximab, a chimeric $IgG_1$ antibody, for the treatment of RA and Crohn's disease: 3–10 mg/kg twice weekly to twice monthly.

Anti-TNF-α biologics are highly effective for the treatment of RA. Following a single dose administration of adalimumab (0.5–10 mg/kg) in RA patients, dose-dependent improvements in ACR20 and EULAR (European League Against Rheumatism) responses were observed [78]. (ACR20 is an American College of Rheumatology standard assessment used to measure patients' responses to anti-rheumatic therapies, and indicates a 20% improvement in symptoms and other predefined disease measures.) The ACR20 dose–response curve for adalimumab between 24 h and 29 days post antibody administration in this patient population in shown in Fig. 12.8A. The dose–response curve reached a plateau at a dose of 1 mg/kg. Therapeutic effects of the single dose became evident within 24 h to one week after antibody administration, and peaked after one to two weeks.

To evaluate the theoretical suppression of serum TNF-α and to predict unbound TNF-α time-course profiles following administration of single doses of adalimumab in RA patients, computer simulations were conducted by the present authors using a bimolecular pharmacokinetic model that simulated the antibody–antigen interaction in the serum of RA patients. The results of simulations are shown in Figure 12.8B. These simulations predict dose-dependent suppression of serum TNF-α following administration of 0.5 to 10 mg/kg antibody doses. Additionally, >95% suppression of serum TNF-α was predicted immediately following administration of the antibody doses; a longer duration of antigen suppression was predicted with higher doses (Fig. 12.8B). The recommended dose of adalimumab for

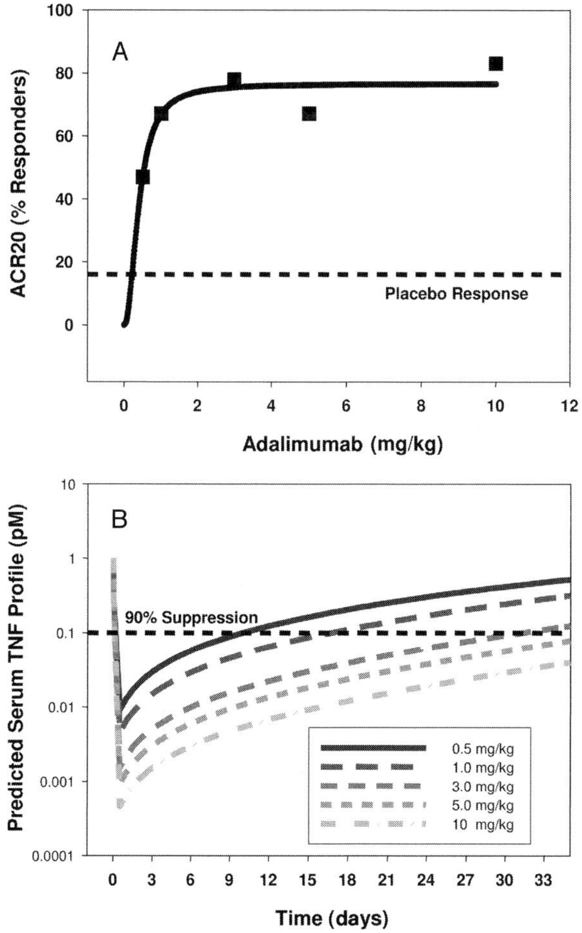


**Fig. 12.8** (A) Dose–response relationship for adalimumab following single-dose administration in patients with rheumatoid arthritis. (Figure produced from data reported in [78]). (B) Simulated suppression of serum TNF-α following single-dose administration of adalimumab.

adult patients in RA is a fixed dose of 40 mg (~0.6 mg/kg for a 70-kg patient) administered every other week as a SC injection. According to the model predictions, this dosing regimen should result in >90% serum antigen suppression at steady-state antibody serum trough concentrations of 8–9 μg/mL.

Relationships between serum infliximab concentrations and clinical improvements in RA were also examined recently [79]. Following the administration of 3 and 10 mg/kg doses every eight to 10 weeks, dose-proportional increases in maximum steady-state serum concentrations ($C_{max}$) of the antibody were evident. Considerable variability in the trough serum concentrations ($C_{min}$) of the antibody was observed. The highest proportion of response (ACR50 and ACR70) was reported in patients with serum trough levels of >1 to 10 μg/mL, which should correspond to >90% serum antigen suppression based on theoretical predictions.

In general, a review of the available literature for anti-TNF-α biologics reflects that, at steady state, the highest proportion of the observed response (maximum therapeutic effect) appears to occur in patients with steady-state serum concentrations ranging between 1 to 10 μg/mL, independent of the compound. This is anticipated since the reported affinities of the anti-TNF-α biologics to TNF-α are comparable (etanercept, 65 pM; adalimumab, 78 pM; and infliximab, 100 pM). Comparable steady-state serum trough concentrations of the drugs are therefore necessary for a >90% suppression of serum TNF-α.

Treatment with available anti-TNF-α inhibitors can be associated with the development of antibodies to the administered biologics [10]. The incidence is reported to be higher in patients receiving infliximab (13 to 60%), the chimeric monoclonal antibody containing a murine variable region, compared with the incidences reported for the fusion protein etanercept (<5%) or the fully human antibody, adalimumab (~12% as monotherapy). The observed incidence of antibody formation is reduced by concomitant immunosuppressive therapies, such as methotrexate. Lower efficacy and higher incidences of infusion-related reactions have been reported in antibody-positive patients receiving infliximab [80].

#### 12.7.1.2 **Efalizumab**

Efalizumab binds to the α-subunit of the leukocyte function-associated antigen 1 (LFA-1), which is expressed on all leukocytes. Intracellular adhesion molecules (ICAMs), which are expressed on a variety of cells such as lymphocytes, endothelial cells and epidermal keratinocytes, facilitate the binding of antigen-presenting cells to T cells through their interaction with LFA-1. Efalizumab inhibits adhesion of leukocytes to other cells by blocking the binding of LFA-1 to ICAMs.

The population PK/PD of efalizumab were recently evaluated in patients with moderate to severe plaque psoriasis following SC administration of 1.0 and 2.0 mg/kg for 12 weeks [81–83]. Steady-state serum concentrations were achieved by four to eight weeks following administration of 1 and 2 mg/kg doses, respectively. At both doses, CD11a expression on T lymphocytes was reported to be maximally down-modulated. In addition, at doses of 1 and 2 mg/kg, >95% of CD11a binding-sites were reported to be saturated at steady-state serum trough concentrations of 9 and 24 μg/mL, respectively. The improvement in PASI scores was observed quickly, and efalizumab administration was reported to result in 60–70% improvement in PASI scores when compared to baseline after 12 weeks of treatment. The current recommended dose for efalizumab is a single 0.7 mg SC conditioning dose which is followed by weekly SC doses of 1 mg/kg.

The pharmacokinetics of efalizumab are highly influenced by the target expression, indicating the presence of a receptor-mediated clearance pathway [81–83]. Using purified mouse and human T cells, internalization of anti-CD11a antibodies was observed following interaction with CD11a. Internalized antibodies moved in endosomes to lysosomes and were catabolized within the cells [84, 85].

The target-mediated clearance of efalizumab following administration of single IV doses in human (0.1–10 mg/kg) and chimpanzee (0.5–10 mg/kg) was also reported [83]. An approximately 50-fold decrease in the clearance of efalizumab was observed in humans, with an approximately 100-fold increase in the dose (from 0.1 to 10 mg/kg). The clearance of efalizumab was saturable at serum concentrations $\geq 10\ \mu g/mL$, and this concentration was reported to be in agreement with the *in-vitro* binding data. *In vitro*, the half-maximal binding of efalizumab to lymphocytes was achieved at an $EC_{50}$ of 0.1 μg/mL, with saturation requiring concentrations around 10 μg/mL [84, 85]. Among various covariates, a population PK analysis identified body weight as the covariate with largest impact on clearance, and supported the body weight-adjusted dosing strategy [82, 83]. Covariates with a modest impact on clearance were baseline lymphocyte counts, PASI score, and the age of the patient.

#### 12.7.1.3 **Omalizumab**

Asthma is a chronic respiratory condition characterized by episodes of airway hyper-responsiveness. Inflammation of the airways contributes to the underlying pathology, airflow limitation, and bronchoconstriction [86,87]. Due to the central role of IgE in asthma, omalizumab has proven efficacious in adults and adolescents (aged $\geq 12$ years) with moderate to severe persistent asthma. Omalizumab is a humanized $IgG_1$ anti-IgE antibody that inhibits binding of IgE to its high-affinity receptor [88]. The pharmacokinetics of omalizumab are similar in adults and adolescents following administration of an initial IV dose of 2 mg/kg followed by six doses of 1.0 mg/kg each over 77 days [88,89]. Due to target-mediated disposition of omalizumab (formation of large complexes with IgE), nonlinear pharmacokinetics have been reported at doses below 0.5 mg/kg. The efficacy of omalizumab in asthma is related to the magnitude of IgE reduction. Dose-dependent reductions in unbound IgE levels were reported following omalizumab administration. Reductions in unbound IgE levels were correlated with a concomitant decrease in mast cell degranulation and significant reductions in inflammatory cells and mediators. The recommended dosing for omalizumab is based on the baseline serum IgE levels before initiation of treatment, and on the patient's body weight.

### 12.7.2
### Therapeutic Antibodies in Oncology

Currently, a total of five naked (unconjugated) humanized or chimeric $IgG_1$ antibodies has been marketed in the US, with applications in hematological malignancies, colorectal, and breast cancer (Table 12.3). Two radioimmunoconjugates, ibritumomab tiuxetan and tositumomab, are approved for treatment of relapsed or refractory, low-grade, follicular, or transformed B-cell non-Hodgkin's lymphoma. One antibody–drug conjugate, gemtuzumab ozogamicin, is approved for treatment of CD33-positive acute myeloid leukemia (Table 12.4).

Rituximab, the first mAb approved for the treatment of cancer, binds to the CD20 antigen found on the surface of normal and the malignant B lymphocytes, and is currently approved for the treatment of patients with relapsed or refractory low-grade or follicular, CD20-positive, B-cell non-Hodgkin's lymphoma [90]. Binding of the antibody to the CD20 positive B cells has been shown to result in cell lysis and *in-vivo* B-cell depletion.

Alemtuzumab is a humanized antibody that is directed against the cell-surface glycoprotein expressed on the surface of the normal and malignant B and T lymphocytes, and is indicated for the treatment of B-cell CLL [91].

Bevacizumab, a humanized $IgG_1$, and cetuximab, a chimeric $IgG_1$, are currently marketed in the US for treatment of metastatic colorectal cancer [92, 93]. Bevacizumab neutralizes the biological activity of vascular endothelial growth factor (VEGF), while cetuximab binds specifically to the extracellular domain of the human epidermal growth factor receptor (EGFR). Bevacizumab, in combination with IV 5-fluorouracil (5-FU) -based chemotherapy, is indicated for first-line treatment of metastatic colorectal cancer, whereas cetuximab is used in patients refractory to or intolerant to irinotecan-based chemotherapy. The clinical pharmacokinetics of cetuximab are discussed in detail in Chapter 14.

HER2 is an internalizing transmembrane receptor that belongs to the epidermal growth factor receptor family and is expressed in human breast cancer. Trastuzumab is a humanized monoclonal antibody against the HER2 receptor and is currently indicated as a single agent or in combination with chemotherapy regimens for treatment of patients with metastatic breast cancer whose tumors overexpress the HER2 protein [94, 95].

#### 12.7.2.1 **Rituximab**

Rituximab is a chimeric antibody that contains the human $IgG_1$ constant region and murine variable region, and binds the membrane-associated human CD20 antigen. Evidence for multiple mechanisms of rituximab action has been reported [96, 97]. Rituximab can deplete CD20-positive B cells via ADCC, CDC, or apoptosis; however, it is not clear which is the most important mechanism in humans. Resistance to rituximab's *in-vivo* effects has been reported, but the underlying resistance mechanisms are not well understood.

The pharmacokinetics of rituximab following administration of multiple doses in non-Hodgkin's lymphoma patients were reported to be nonlinear and influenced by the density of the target antigen [98]. Upon administration of multiple doses of 375 mg/m$^2$, mean maximum steady-state serum concentrations of 486 μg/mL were achieved. In addition, a 2.5-fold increase in the terminal half-life was observed after the fourth weekly dose. The change in half-life following administration of anti-CD20 antibody was partially attributed to reductions in tumor burden (CD20-positive cells) in lymphoma patients following rituximab therapy. Administration of rituximab resulted in a rapid and sustained depletion of circulating and tissue B cells in humans and monkeys [99, 100].

In patients with CLL, much higher doses are required for efficacy compared with the dose used for non-Hodgkin's lymphoma patients. The higher dose requirement is presumably due to a greater tumor burden and antigen sink [101, 102]. One group has attributed the higher dose requirement to very high levels of soluble CD20 in the serum of CLL patients that compete with the binding of rituximab to tumor cells [103].

#### 12.7.2.2 **Bevacizumab**

Angiogenesis is the process of formation of new blood vessels from the pre-existing blood vessels, and is of crucial importance in embryonic development, wound healing, as well as tumor growth and metastasis [104]. VEGF is a potent stimulator of angiogenesis, and expression of VEGF is up-regulated in a broad array of tumor types [104, 105]. Bevacizumab is a humanized $IgG_1$ monoclonal antibody that binds and inhibits VEGF interaction with its receptors Flt-1 and KDR in both *in-vitro* and *in-vivo* systems. Bevacizumab is currently recommended for first-line treatment of patients with metastatic carcinoma of the colon or rectum, in combination with 5-FU-based chemotherapy, at 5 mg/kg given once every 14 days as an IV infusion until disease progression is detected [92].

Anti-tumor activity of bevacizumab has been reported in various preclinical animal models (primary and metastatic) with a broad array of tumor types [106, 107]. Clinical studies have further validated the focal role of VEGF in cancer. A single infusion of bevacizumab at 5 mg/kg in patients with primary and locally advanced adenocarcinoma of the rectum demonstrated direct and rapid antivascular effect in human tumors, with decreases in tumor perfusion, vascular volume, microvascular density, and interstitial pressure [108]. Clinical efficacy of bevacizumab in combination with 5-FU- and irinotecan-based regimens has been demonstrated in patients with metastatic colorectal cancer: a significant improvement in overall survival time was observed compared with chemotherapy alone (20.3 versus 15.6 months for chemotherapy plus bevacizumab versus chemotherapy alone) [109].

Following administration of multiple doses ranging from 0.3 to 10 mg/kg in patients with advanced cancer, linear pharmacokinetics were observed with clearance values ranging from 2.7 to 5 mL/kg per day [110]. Mean maximum serum concentrations ($C_{max}$) of 280 μg/mL were achieved following administration of a 10 mg/kg dose. Serum total VEGF concentrations (unbound plus antibody-bound VEGF) increased two- to four-fold the baseline value by 72 days following the initiation of therapy. Unbound serum VEGF was not detectable following administration of a 0.3 mg/kg dose [110, 111]. Similar results were observed in a Phase Ib clinical trial when bevacizumab was administered in combination with various chemotherapeutic regimens. The pharmacokinetics of bevacizumab (mean clearance: 3.2 mL/kg per day) in patients with recurrent non-small-cell lung cancer following multiple dosing in combination with the EGFR tyrosine kinase inhibitor, erlotinib, were similar to that observed following administration of the antibody alone [112]. The clearance of bevacizumab was reported to vary with body weight and gender. In addition, patients with a higher tumor burden had an approximately 20% higher clearance. To date, the relationship between exposure and clinical outcomes for bevacizumab has not been reported.

#### 12.7.2.3 **Trastuzumab**

HER2 is an internalizing transmembrane receptor that belongs to the EGFR family and is overexpressed in human breast cancer [95, 113]. Trastuzumab is a humanized monoclonal antibody that was developed to target the HER2 receptor. In clinical studies, patient selection was determined by testing tumor specimens for overexpression of the HER2 protein, and the beneficial treatments effects were largely limited to patients with highest levels of HER2 protein overexpression [95, 113]. Following administration of trastuzumab at the currently recommended standard dose (4 mg/kg loading infusions and weekly maintenance doses of 2 mg/kg), steady-state serum concentrations achieved mean peak and trough concentrations of 123 μg/mL and 79 μg/mL, respectively [114].

The pharmacokinetics of trastuzumab are influenced by HER2 expression, which indicates the presence of a receptor-mediated clearance pathway [114–117]. Following single-dose administration of trastuzumab in patients with HER2-overexpressing metastatic breast cancer, a 2.5-fold decrease in clearance of the antibody was reported with increases in antibody dose from 1 to 8 mg/kg (Fig. 12.9). The population pharmacokinetics of trastuzumab following administration of the recommended standard dose (4 mg/kg loading infusions and weekly maintenance doses of 2 mg/kg) to patients with HER2-positive metastatic breast cancer from Phase II and III clinical studies have also been reported [114]. A relatively large inter-patient variability in clearance and volume of distribution (43% and 29%, respectively) was reported. Among the covariates examined, the number of meta-

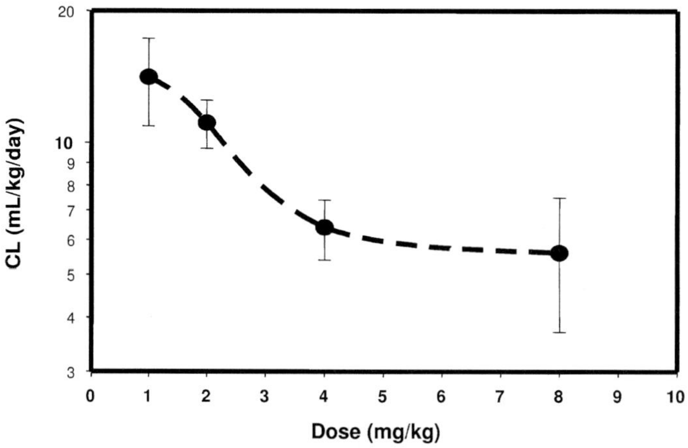


**Fig. 12.9** Nonlinear relationship between dose and antibody clearance following administration of single doses of trastuzumab in patients with HER2/overexpressing metastatic breast cancer. (Figure produced from data reported in [117]).

static sites, plasma levels of the soluble antigen (HER2 receptor extracellular domain), and patient body weight had a significant effect on clearance, volume of distribution, or both. Patients with four or more metastatic sites had 18% lower steady-state exposures. In patients with shed HER2 extracelluar domain concentrations >200 ng/mL, a 14% higher clearance and 40% greater volume of distribution were reported. Concomitant chemotherapy (paclitaxel or anthracycline plus cyclophosphamide) did not appear to influence the antibody clearance. The rationale for selection of the recommended maintenance dose of 2 mg/kg per week is not clear from the published literature; however, the clearance rate derived from population modeling (0.225 L/day) is similar to the clearance rate for $IgG_1$ antibodies with no antigen sink, suggesting that the systemic HER2 pool was, on average, saturated by trastuzumab when administered at the recommended dose.

## 12.8 Conclusions

Well-defined exposure–response relationships have been established for biologic products using conventional approaches to pharmacokinetic and pharmacodynamic analysis. Characteristics of the exposure–response relationships that are generally unique to biologic products, such as target-mediated disposition and homeostatic regulation of pharmacokinetics, have been successfully incorporated into analyses. Mechanistic pharmacodynamic models representing complex biological effects have been developed and have contributed to the understanding of the biology and clinical activity of these therapeutic products. The routine application of PK/PD modeling techniques should continue to facilitate the development of novel biologic therapeutics.

## 12.9 References


1 Galluppi, G.R., M.C. Rogge, L.K. Roskos, L.J. Lesko, M.D. Green, D.W. Feigal, Jr., and C.C. Peck. **2001**. Integration of pharmacokinetic and pharmacodynamic studies in the discovery, development, and review of protein therapeutic agents: a conference report. *Clin. Pharmacol. Ther.* 69:387–399.

2 Tang, L., A.M. Persky, G. Hochhaus, and B. Meibohm. **2004**. Pharmacokinetic aspects of biotechnology products. *J. Pharm. Sci.* 93:2184–2204.

3 Toon, S. **1996**. The relevance of pharmacokinetics in the development of biotechnology products. *Eur. J. Drug Metab. Pharmacokinet.* 21:93–103.

4 Supersaxo, A., W. Hein, H. Gallati, and H. Steffen. **1988**. Recombinant human interferon alpha-2a: delivery to lymphoid tissue by selected modes of application. *Pharm. Res.* 5:472–476.

5 Supersaxo, A., W.R. Hein, and H. Steffen. **1990**. Effect of molecular weight on the lymphatic absorption of water-soluble compounds following subcutaneous administration. *Pharm. Res.* 7: 167–169.

6 Wang, B., T.M. Ludden, E.N. Cheung, G.G. Schwab, and L.K. Roskos. **2001**. Population pharmacokinetic-pharmacodynamic modeling of filgrastim (r-metHuG-CSF) in healthy volunteers.

*J. Pharmacokinet. Pharmacodyn.* 28:321–342.

7 Weinstein, J.N. and W. van Osdol. **1992**. The macroscopic and microscopic pharmacology of monoclonal antibodies. *Int. J. Immunopharmacol.* 14:457–463.

8 Gerlowski, L.E. and R.K. Jain. **1986**. Microvascular permeability of normal and neoplastic tissues. *Microvasc. Res.* 31:288–305.

9 Ghetie,V., J.G. Hubbard, J.K. Kim, M.F. Tsen,Y. Lee, and E.S. Ward. **1996**. Abnormally short serum half-lives of IgG in beta 2-microglobulin-deficient mice. *Eur. J. Immunol.* 26:690–696.

10 Roskos, L.K., G.C. Davis, and G.M. Schwab. **2004**. The clinical pharmacology of therapeutic monoclonal antibodies. *Drug Dev. Res.* 61:108–120.

11 Tabrizi, M.A., C.-M.L. Tseng, and L.K. Roskos. **2006**. Elimination mechanisms of therapeutic monoclonal antibodies. *Drug Discovery Today* 11:81–88.

12 Cohen-Solal, J.F., L. Cassard,W.H. Fridman, and C. Sautes-Fridman. **2004**. Fc gamma receptors. *Immunol. Lett.* 92:199–205.

13 Bogers,W.M., R.K. Stad, D.J. Janssen, N. van Rooijen, L.A. van Es, and M.R. Daha. **1991**. Kupffer cell depletion in vivo results in preferential elimination of IgG aggregates and immune complexes via specific Fc receptors on rat liver endothelial cells. *Clin. Exp. Immunol.* 86:328–333.

14 Hinton, P.R., M.G. Johlfs, J.M. Xiong, K. Hanestad, K.C. Ong, C. Bullock, S. Keller, M.T. Tang, J.Y. Tso, M. Vasquez, and N. Tsurushita. **2004**. Engineered human IgG antibodies with longer serum half-lives in primates. *J. Biol. Chem.* 279:6213–6216.

15 Neel, N.F., E. Schutyser, J. Sai, G.H. Fan, and A. Richmond. **2005**. Chemokine receptor internalization and intracellular trafficking. *Cytokine Growth Factor Rev.* 16:637–658.

16 Schellekens, H. **2005**. Factors influencing the immunogenicity of therapeutic proteins. *Nephrol. Dial. Transplant.* 20 (Suppl. 6):vi3–vi9.

17 Roskos, L.K., S.A. Kellermann, and K.A. Foon. **2005**. Human Antiglobulin Responses. In: M.T. Lotze and A.W. Thomson (Eds.), *Measuring Immunity: Basic Science and Clinical Practice.* Elsevier Academic Press, London, pp. 172–186.

18 Johansson, A., A. Erlandsson, D. Eriksson, A. Ullen, P. Holm, B.E. Sundstrom, K.H. Roux, and T. Stigbrand. **2002**. Idiotypic-anti-idiotypic complexes and their in vivo metabolism. *Cancer* 94:1306–1313.

19 Kharagjitsingh, A.V., J.C. Korevaar, J.P. Vandenbroucke, E.W. Boeschoten, R.T. Krediet, M.R. Daha, F.W. Dekker. **2005**. Incidence of recombinant erythropoietin (EPO) hyporesponse, EPO-associated antibodies, and pure red cell aplasia in dialysis patients. *Kidney Int.* 68:1215–1222.

20 Li, J., C. Yang,Y. Xia, A. Bertino, J. Glaspy, M. Roberts, and D.J. Kuter. **2001**. Thrombocytopenia caused by the development of antibodies to thrombopoietin. *Blood* 98:3241–3248.

21 Holford, N.H. and L.B. Sheiner. **1981**. Understanding the dose-effect relationship: clinical application of pharmacokinetic-pharmacodynamic models. *Clin. Pharmacokinet.* 6:429–453.

22 Holford, N.H. and L.B. Sheiner. **1981**. Pharmacokinetic and pharmacodynamic modeling in vivo. *Crit. Rev. Bioeng.* 5:273–322.

23 Holford, N.H. **1991**. Relevance of pharmacodynamic principles in therapeutics. *Ann. Acad. Med. Singapore* 20:26–30.

24 Sheiner, L.B. and J.L. Steimer. **2000**. Pharmacokinetic/pharmacodynamic modeling in drug development. *Annu. Rev. Pharmacol. Toxicol.* 40:67–95.

25 Mager, D.E., E. Wyska, and W.J. Jusko. **2003**. Diversity of mechanism-based pharmacodynamic models. *Drug Metab. Dispos.* 31:510–518.

26 Mager, D.E. and W.J. Jusko. **2001**. General pharmacokinetic model for drugs exhibiting target-mediated drug disposition. *J. Pharmacokinet. Pharmacodyn.* 28:507–532.

27 Mager, D.E. and W.J. Jusko. **2001**. Pharmacodynamic modeling of time-dependent transduction systems. *Clin. Pharmacol. Ther.* 70:210–216.

**28** Sharma, A. and W.J. Jusko. **1998**. Characteristics of indirect pharmacodynamic models and applications to clinical drug responses. *Br. J. Clin. Pharmacol.* 45:229–239.

**29** Sun, Y.N. and W.J. Jusko. **1998**. Transit compartments versus gamma distribution function to model signal transduction processes in pharmacodynamics. *J. Pharm. Sci.* 87:732–737.

**30** Tabrizi-Fard, M.A. and H.L. Fung. **1998**. Effects of nitro-L-arginine on blood pressure and cardiac index in anesthetized rats: a pharmacokinetic-pharmacodynamic analysis. *Pharm. Res.* 15:1063–1068.

**31** Dall'Acqua, W.F., R.M. Woods, E.S. Ward, S.R. Palaszynski, N.K. Patel, Y.A. Brewah, H. Wu, P.A. Kiener, and S. Langermann. **2002**. Increasing the affinity of a human $IgG_1$ for the neonatal Fc receptor: biological consequences. *J. Immunol.* 169:5171–5180.

**32** Raghavan, M., V.R. Bonagura, S.L. Morrison, and P.J. Bjorkman. **1995**. Analysis of the pH dependence of the neonatal Fc receptor/immunoglobulin G interaction using antibody and receptor variants. *Biochemistry* 34:14649–14657.

**33** Shields, R.L., A.K. Namenuk, K. Hong, Y.G. Meng, J. Rae, J. Briggs, D. Xie, J. Lai, A. Stadlen, B. Li, J.A. Fox, and L.G. Presta. **2001**. High resolution mapping of the binding site on human $IgG_1$ for Fc gamma RI, Fc gamma RII, Fc gamma RIII, and FcRn. and design of $IgG_1$ variants with improved binding to the Fc gamma R. *J. Biol. Chem.* 276: 6591–6604.

**34** Presta, L.G., R.L. Shields, A.K. Namenuk, K. Hong, and Y.G. Meng. **2002**. Engineering therapeutic antibodies for improved function. *Biochem. Soc. Trans.* 30: 487–490.

**35** Idusogie, E.E., P.Y. Wong, L.G. Presta, H. Gazzano-Santoro, K. Totpal, M. Ultsch, and M.G. Mulkerrin. **2001**. Engineered antibodies with increased activity to recruit complement. *J. Immunol.* 166:2571–2575.

**36** Ward, E.S. and V. Ghetie. **1995**. The effector functions of immunoglobulins: implications for therapy. *Ther. Immunol.* 2:77–94.

**37** Abbas, A.K. and C.A. Janeway, Jr. **2000**. Immunology: improving on nature in the twenty-first century. *Cell* 100:129–138.

**38** Pfeifer, M.A., R.J. Graf, J.B. Halter, and D. Porte, Jr. **1981**. The regulation of glucose-induced insulin secretion by pre-stimulus glucose level and tolbutamide in normal man. *Diabetologia* 21:198–205.

**39** Roach, P. and J.R. Woodworth. **2002**. Clinical pharmacokinetics and pharmacodynamics of insulin lispro mixtures. *Clin. Pharmacokinet.* 41:1043–1057.

**40** Brange, J., D.R. Owens, S. Kang, and A. Volund. **1990**. Monomeric insulins and their experimental and clinical implications. *Diabetes Care* 13:923–954.

**41** Vignati, L., J.H. Anderson, Jr., and P.W. Iversen. **1997**. Efficacy of insulin lispro in combination with NPH human insulin twice per day in patients with insulin-dependent or non-insulin-dependent diabetes mellitus. Multicenter Insulin Lispro Study Group. *Clin. Ther.* 19: 1408–1421.

**42** DeFronzo, R.A., J.D. Tobin, and R. Andres. **1979**. Glucose clamp technique: a method for quantifying insulin secretion and resistance. *Am. J. Physiol.* 237:E214–E223.

**43** Osterberg, O., L. Erichsen, S.H. Ingwersen, A. Plum, H.E. Poulsen, and P. Vicini. **2003**. Pharmacokinetic and pharmacodynamic properties of insulin aspart and human insulin. *J. Pharmacokinet. Pharmacodyn.* 30:221–235.

**44** Chorev, M. **2002**. Parathyroid hormone 1 receptor: insights into structure and function. *Receptors Channels* 8:219–242.

**45** Fukugawa, M. and K. Kurokawa. **2002**. Calcium homeostasis and imbalance. *Nephron* 92 (Suppl. 1):41–45.

**46** Berg, C., K. Neumeyer, and P. Kirkpatrick. **2003**. Teriparatide. *Nat. Rev. Drug Discov.* 2:257–258.

**47** Deal, C. **2004**. The use of intermittent human parathyroid hormone as a treatment for osteoporosis. *Curr. Rheumatol. Rep.* 6:49–58.

**48** Dobnig, H., A. Sipos, Y. Jiang, A. Fahrleitner-Pammer, L.G. Ste-Marie,

J.C. Gallagher, I. Pavo, J. Wang, and E.F. Eriksen. **2005**. Early changes in biochemical markers of bone formation correlate with improvements in bone structure during teriparatide therapy. *J. Clin. Endocrinol. Metab.* 90 : 3970–3977.
**49** Frolik, C.A., E.C. Black, R.L. Cain, J.H. Satterwhite, P.L. Brown-Augsburger, M. Sato, and J.M. Hock. **2003**. Anabolic and catabolic bone effects of human parathyroid hormone (1–34) are predicted by duration of hormone exposure. *Bone* 33 : 372–379.
**50** Onyia, J.E., L.M. Helvering, L. Gelbert, T. Wei, S. Huang, P. Chen, E.R. Dow, A. Maran, M. Zhang, S. Lotinun, X. Lin, D.L. Halladay, R.R. Miles, N.H. Kulkarni, E.M. Ambrose, Y.L. Ma, C.A. Frolik, M. Sato, H.U. Bryant, and R.T. Turner. **2005**. Molecular profile of catabolic versus anabolic treatment regimens of parathyroid hormone (PTH) in rat bone: an analysis by DNA microarray. *J. Cell Biochem.* 95 : 403–418.
**51** Chauvin, S., M. Bencsik, T. Bambino, and R.A. Nissenson. **2002**. Parathyroid hormone receptor recycling: role of receptor dephosphorylation and beta-arrestin. *Mol. Endocrinol.* 16 : 2720–2732.
**52** Potter, L.K., L.D. Greller, C.R. Cho, M.E. Nuttall, G.B. Stroup, L.J. Suva, and F.L. Tobin. **2005**. Response to continuous and pulsatile PTH dosing: a mathematical model for parathyroid hormone receptor kinetics. *Bone* 37 : 159–169.
**53** McDermott, D.F. and M.B. Atkins. **2004**. Application of IL-2 and other cytokines in renal cancer. *Expert Opin. Biol. Ther.* 4 : 455–468.
**54** Rosenberg, S.A., J.C. Yang, S.L. Topalian, D.J. Schwartzentruber, J.S. Weber, D.R. Parkinson, C.A. Seipp, J.H. Einhorn, and D.E. White. **1994**. Treatment of 283 consecutive patients with metastatic melanoma or renal cell cancer using high-dose bolus interleukin 2. *JAMA* 271 : 907–913.
**55** Yang, J.C., R.M. Sherry, S.M. Steinberg, S.L. Topalian, D.J. Schwartzentruber, P. Hwu, C.A. Seipp, L. Rogers-Freezer, K.E. Morton, D.E. White, D.J. Liewehr, M.J. Merino, and S.A. Rosenberg. **2003**. Randomized study of high-dose and low-dose interleukin-2 in patients with metastatic renal cancer. *J. Clin. Oncol.* 21 : 3127–3132.
**56** McDermott, D.F., M.M. Regan, J.I. Clark, L.E. Flaherty, G.R. Weiss, T.F. Logan, J.M. Kirkwood, M.S. Gordon, J.A. Sosman, M.S. Ernstoff, C.P. Tretter, W.J. Urba, J.W. Smith, K.A. Margolin, J.W. Mier, J.A. Gollob, J.P. Dutcher, and M.B. Atkins. **2005**. Randomized phase III trial of high-dose interleukin-2 versus subcutaneous interleukin-2 and interferon in patients with metastatic renal cell carcinoma. *J. Clin. Oncol.* 23 : 133–141.
**57** Harker, L.A., L.K. Roskos, U.M. Marzec, R.A. Carter, J.K. Cherry, B. Sundell, E.N. Cheung, D. Terry, and W. Sheridan. **2000**. Effects of megakaryocyte growth and development factor on platelet production, platelet life span, and platelet function in healthy human volunteers. *Blood* 95 : 2514–2522.
**58** Uehlinger, D.E., F.A. Gotch, and L.B. Sheiner. **1992**. A pharmacodynamic model of erythropoietin therapy for uremic anemia. *Clin. Pharmacol. Ther.* 51 : 76–89.
**59** Krzyzanski, W., W.J. Jusko, M.C. Wacholtz, N. Minton, and W.K. Cheung. **2005**. Pharmacokinetic and pharmacodynamic modeling of recombinant human erythropoietin after multiple subcutaneous doses in healthy subjects. *Eur. J. Pharm. Sci.* 26 : 295–306.
**60** Ramakrishnan, R., W.K. Cheung, M.C. Wacholtz, N. Minton, and W.J. Jusko. **2004**. Pharmacokinetic and pharmacodynamic modeling of recombinant human erythropoietin after single and multiple doses in healthy volunteers. *J. Clin. Pharmacol.* 44 : 991–1002.
**61** Jumbe, N., B. Yao, R. Rovetti, G. Rossi, and A.C. Heatherington. **2002**. Clinical trial simulation of a 200-microg fixed dose of darbepoetin-α in chemotherapy-induced anemia. *Oncology (Williston. Park)* 16 : 37–44.
**62** Immunex Corp. **2005**. Enbrel prescribing information.
**63** Lee, H., H.C. Kimko, M. Rogge, D. Wang, I. Nestorov, and C.C. Peck. **2003**. Population pharmacokinetic and pharmacody-

namic modeling of etanercept using logistic regression analysis. *Clin. Pharmacol. Ther.* 73 : 348–365.

**64** Leonardi, C.L., J.L. Powers, R.T. Matheson, B.S. Goffe, R. Zitnik, A. Wang, and A.B. Gottlieb. **2003**. Etanercept as monotherapy in patients with psoriasis. *N. Engl. J. Med.* 349 : 2014–2022.

**65** Bathon, J.M., R.W. Martin, R.M. Fleischmann, J.R. Tesser, M.H. Schiff, E.C. Keystone, M.C. Genovese, M.C. Wasko, L.W. Moreland, A.L. Weaver, J. Markenson, and B.K. Finck. **2000**. A comparison of etanercept and methotrexate in patients with early rheumatoid arthritis. *N. Engl. J. Med.* 343 : 1586–1593.

**66** Abbas, A.K. **1982**. Immunoglobulin idiotypes; experimental and clinical applications. *Indian J. Pediatr.* 49 : 641–648.

**67** Abbas, A.K. **1989**. Antigen presentation by B lymphocytes: mechanisms and functional significance. *Semin. Immunol.* 1 : 5–12.

**68** Huse, W.D., L. Sastry, S.A. Iverson, A.S. Kang, M. Alting-Mees, D.R. Burton, S.J. Benkovic, and R.A. Lerner. **1989**. Generation of a large combinatorial library of the immunoglobulin repertoire in phage lambda. *Science* 246 : 1275–1281.

**69** McCafferty, J., A.D. Griffiths, G. Winter, and D.J. Chiswell. **1990**. Phage antibodies: filamentous phage displaying antibody variable domains. *Nature* 348 : 552–554.

**70** Mendez, M.J., L.L. Green, J.R. Corvalan, X.C. Jia, C.E. Maynard-Currie, X.D. Yang, M.L. Gallo, D.M. Louie, D.V. Lee, K.L. Erickson, J. Luna, C.M. Roy, H. Abderrahim, F. Kirschenbaum, M. Noguchi, D.H. Smith, A. Fukushima, J.F. Hales, S. Klapholz, M.H. Finer, C.G. Davis, K.M. Zsebo, and A. Jakobovits. **1997**. Functional transplant of megabase human immunoglobulin loci recapitulates human antibody response in mice. *Nat. Genet.* 15 : 146–156.

**71** Anderson, P.J. **2005**. Tumor necrosis factor inhibitors: clinical implications of their different immunogenicity profiles. *Semin. Arthritis Rheum.* 34 : 19–22.

**72** Nahar, I.K., K. Shojania, C.A. Marra, A.H. Alamgir, and A.H. Anis. **2003**. Infliximab treatment of rheumatoid arthritis and Crohn's disease. *Ann. Pharmacother.* 37 : 1256–1265.

**73** Genentech Inc. **2005**. Raptiva prescribing information.

**74** Genentech Inc. **2005**. Xolair prescribing information.

**75** Chapman, T.M. and G.M. Keating. **2003**. Basiliximab: a review of its use as induction therapy in renal transplantation. *Drugs* 63 : 2803–2835.

**76** Wiland, A.M. and B. Philosophe. **2004**. Daclizumab induction in solid organ transplantation. *Expert Opin. Biol. Ther.* 4 : 729–740.

**77** Feldmann, M. and R.N. Maini. **2003**. Lasker Clinical Medical Research Award. TNF defined as a therapeutic target for rheumatoid arthritis and other auto immune diseases. *Nat. Med.* 9 : 1245–1250.

**78** den Broeder, A., L. van de Putte, R. Rau, M. Schattenkirchner, P. Van Riel, O. Sander, C. Binder, H. Fenner, Y. Bankmann, R. Velagapudi, J. Kempeni, and H. Kupper. **2002**. A single dose, placebo controlled study of the fully human antitumor necrosis factor-alpha antibody adalimumab (D2E7) in patients with rheumatoid arthritis. *J. Rheumatol.* 29 : 2288–2298.

**79** St Clair, E.W., C.L. Wagner, A.A. Fasanmade, B. Wang, T. Schaible, A. Kavanaugh, and E.C. Keystone. **2002**. The relationship of serum infliximab concentrations to clinical improvement in rheumatoid arthritis: results from ATTRACT, a multicenter, randomized, double-blind, placebo-controlled trial. *Arthritis Rheum.* 46 : 1451–1459.

**80** Baert, F., M. Noman, S. Vermeire, G. Van Assche, G. D'Haens, A. Carbonez, and P. Rutgeerts. **2003**. Influence of immunogenicity on the long-term efficacy of infliximab in Crohn's disease. *N. Engl. J. Med.* 348 : 601–608.

**81** Bauer, R.J., R.L. Dedrick, M.L. White, M.J. Murray, and M.R. Garovoy. **1999**. Population pharmacokinetics and pharmacodynamics of the anti-CD11a antibody hu1124 in human subjects with psoriasis. *J. Pharmacokinet. Biopharm.* 27 : 397–420.

**82** Mortensen, D.L., P.A. Walicke, X. Wang, P. Kwon, P. Kuebler, A.B. Gottlieb, J.G. Krueger, C. Leonardi, B. Miller, and A. Joshi. **2005**. Pharmacokinetics and pharmacodynamics of multiple weekly subcutaneous efalizumab doses in patients with plaque psoriasis. *J. Clin. Pharmacol.* 45:286–298.

**83** Sun, Y.N., J.F. Lu, A. Joshi, P. Compton, P. Kwon, and R.A. Bruno. **2005**. Population pharmacokinetics of efalizumab (humanized monoclonal anti-CD11a antibody) following long-term subcutaneous weekly dosing in psoriasis subjects. *J. Clin. Pharmacol.* 45:468–476.

**84** Coffey, G.P., E. Stefanich, S. Palmieri, R. Eckert, J. Padilla-Eagar, P.J. Fielder, and S. Pippig. **2004**. In vitro internalization, intracellular transport, and clearance of an anti-CD11a antibody (Raptiva) by human T-cells. *J. Pharmacol. Exp. Ther.* 310:896–904.

**85** Coffey, G.P., J.A. Fox, S. Pippig, S. Palmieri, B. Reitz, M. Gonzales, A. Bakshi, J. Padilla-Eagar, and P.J. Fielder. **2005**. Tissue distribution and receptor-mediated clearance of anti-CD11a antibody in mice. *Drug Metab. Dispos.* 33:623–629.

**86** Babu, K.S., S.H. Arshad, and S.T. Holgate. **2001**. Anti-IgE treatment: an update. *Allergy* 56:1121–1128.

**87** Milgrom, H., R.B. Fick, Jr., J.Q. Su, J.D. Reimann, R.K. Bush, M.L. Watrous, and W.J. Metzger. **1999**. Treatment of allergic asthma with monoclonal anti-IgE antibody. rhuMAb-E25 Study Group. *N. Engl. J. Med.* 341:1966–1973.

**88** Easthope, S. and B. Jarvis. **2001**. Omalizumab. *Drugs* 61:253–260.

**89** Ruffin, C.G. and B.E. Busch. **2004**. Omalizumab: a recombinant humanized anti-IgE antibody for allergic asthma. *Am. J. Health Syst. Pharm.* 61:1449–1459.

**90** Biogen Idec Inc. and Genentech Inc. **2005**. Rituxan prescribing information.

**91** Genzyme Inc. **2005**. Campath prescribing information.

**92** Genentech Inc. **2005**. Avastin prescribing information.

**93** Imclone Systems Inc. and Bristol-Myers Squibb Corp. **2004**. Erbitux prescribing information.

**94** Genentech Inc. **2005**. Herceptin prescribing information.

**95** McKeage, K. and C.M. Perry. **2002**. Trastuzumab: a review of its use in the treatment of metastatic breast cancer overexpressing HER2. *Drugs* 62:209–243.

**96** Smith, M.R. **2003**. Rituximab (monoclonal anti-CD20 antibody): mechanisms of action and resistance. *Oncogene* 22:7359–7368.

**97** Dall'Ozzo, S., S. Tartas, G. Paintaud, G. Cartron, P. Colombat, P. Bardos, H. Watier, and G. Thibault. **2004**. Rituximab-dependent cytotoxicity by natural killer cells: influence of FCGR3A polymorphism on the concentration-effect relationship. *Cancer Res.* 64:4664–4669.

**98** Mangel, J., R. Buckstein, K. Imrie, D. Spaner, E. Franssen, P. Pavlin, A. Boudreau, N. Pennell, D. Combs, and N.L. Berinstein. **2003**. Pharmacokinetic study of patients with follicular or mantle cell lymphoma treated with rituximab as 'in vivo purge' and consolidative immunotherapy following autologous stem cell transplantation. *Ann. Oncol.* 14:758–765.

**99** Reff, M.E., K. Carner, K.S. Chambers, P.C. Chinn, J.E. Leonard, R. Raab, R.A. Newman, N. Hanna, and D.R. Anderson. **1994**. Depletion of B cells in vivo by a chimeric mouse human monoclonal antibody to CD20. *Blood* 83:435–445.

**100** Schroder, C., A.M. Azimzadeh, G. Wu, J.O. Price, J.B. Atkinson, and R.N. Pierson. **2003**. Anti-CD20 treatment depletes B-cells in blood and lymphatic tissue of cynomolgus monkeys. *Transpl. Immunol.* 12:19–28.

**101** Keating, M. and S. O'Brien. **2000**. High-dose rituximab therapy in chronic lymphocytic leukemia. *Semin. Oncol.* 27:86–90.

**102** O'Brien, S.M., H. Kantarjian, D.A. Thomas, F.J. Giles, E.J. Freireich, J. Cortes, S. Lerner, and M.J. Keating. **2001**. Rituximab dose-escalation trial in chronic lymphocytic leukemia. *J. Clin. Oncol.* 19:2165–2170.

**103** Manshouri, T., K.A. Do, X. Wang, F.J. Giles, S.M. O'Brien, H. Saffer, D. Thomas, I. Jilani, H.M. Kantarjian, M.J. Keating, and M. Albitar. **2003**. Circulating CD20 is detectable in the plasma of patients with chronic lymphocytic leukemia and is of prognostic significance. *Blood* 101 : 2507–2513.

**104** Verheul, H.M. and H.M. Pinedo. **2005**. Inhibition of angiogenesis in cancer patients. *Expert Opin. Emerg. Drugs* 10 : 403–412.

**105** Rini, B.I. and E.J. Small. **2005**. Biology and clinical development of vascular endothelial growth factor-targeted therapy in renal cell carcinoma. *J. Clin. Oncol.* 23 : 1028–1043.

**106** Motl, S. **2005**. Bevacizumab in combination chemotherapy for colorectal and other cancers. *Am. J. Health Syst. Pharm.* 62 : 1021–1032.

**107** Gerber, H.P. and N. Ferrara. **2005**. Pharmacology and pharmacodynamics of bevacizumab as monotherapy or in combination with cytotoxic therapy in preclinical studies. *Cancer Res.* 65 : 671–680.

**108** Willett, C.G., Y. Boucher, E. di Tomaso, D.G. Duda, L.L. Munn, R.T. Tong, D.C. Chung, D.V. Sahani, S.P. Kalva, S.V. Kozin, M. Mino, K.S. Cohen, D.T. Scadden, A.C. Hartford, A.J. Fischman, J.W. Clark, D.P. Ryan, A.X. Zhu, L.S. Blaszkowsky, H.X. Chen, P.C. Shellito, G.Y. Lauwers, and R.K. Jain. **2004**. Direct evidence that the VEGF-specific antibody bevacizumab has antivascular effects in human rectal cancer. *Nat. Med.* 10 : 145–147.

**109** Aggarwal, S. and E. Chu. **2005**. Current therapies for advanced colorectal cancer. *Oncology (Williston. Park)* 19 : 589–595.

**110** Gordon, M.S., K. Margolin, M. Talpaz, G.W. Sledge, Jr., E. Holmgren, R. Benjamin, S. Stalter, S. Shak, and D. Adelman. **2001**. Phase I safety and pharmacokinetic study of recombinant human antivascular endothelial growth factor in patients with advanced cancer. *J. Clin. Oncol.* 19 : 843–850.

**111** Margolin, K., M.S. Gordon, E. Holmgren, J. Gaudreault, W. Novotny, G. Fyfe, D. Adelman, S. Stalter, and J. Breed. **2001**. Phase Ib trial of intravenous recombinant humanized monoclonal antibody to vascular endothelial growth factor in combination with chemotherapy in patients with advanced cancer: pharmacologic and long-term safety data. *J. Clin. Oncol.* 19 : 851–856.

**112** Herbst, R.S., D.H. Johnson, E. Mininberg, D.P. Carbone, T. Henderson, E.S. Kim, G. Blumenschein, Jr., J.J. Lee, D.D. Liu, M.T. Truong, W.K. Hong, H. Tran, A. Tsao, D. Xie, D.A. Ramies, R. Mass, S. Seshagiri, D.A. Eberhard, S.K. Kelley, and A. Sandler. **2005**. Phase I/II trial evaluating the anti-vascular endothelial growth factor monoclonal antibody bevacizumab in combination with the HER-1/epidermal growth factor receptor tyrosine kinase inhibitor erlotinib for patients with recurrent non-small-cell lung cancer. *J. Clin. Oncol.* 23 : 2544–2555.

**113** Harari, P.M. **2004**. Epidermal growth factor receptor inhibition strategies in oncology. *Endocr. Relat. Cancer* 11 : 689–708.

**114** Bruno, R., C.B. Washington, J.F. Lu, G. Lieberman, L. Banken, and P. Klein. **2005**. Population pharmacokinetics of trastuzumab in patients with HER2+ metastatic breast cancer. *Cancer Chemother. Pharmacol.* 56 : 361–369.

**115** Pegram, M.D., A. Lipton, D.F. Hayes, B.L. Weber, J.M. Baselga, D. Tripathy, D. Baly, S.A. Baughman, T. Twaddell, J.A. Glaspy, and D.J. Slamon. **1998**. Phase II study of receptor-enhanced chemosensitivity using recombinant humanized anti-p185HER2/neu monoclonal antibody plus cisplatin in patients with HER2/neu-overexpressing metastatic breast cancer refractory to chemotherapy treatment. *J. Clin. Oncol.* 16 : 2659–2671.

**116** Leyland-Jones, B., K. Gelmon, J.P. Ayoub, A. Arnold, S. Verma, R. Dias, and P. Ghahramani. **2003**. Pharmacokinetics, safety, and efficacy of trastuzumab administered every three weeks in combination with paclitaxel. *J. Clin. Oncol.* 21 : 3965–3971.

**117** Tokuda, Y., T. Watanabe, Y. Omuro, M. Ando, N. Katsumata, A. Okumura, M. Ohta, H. Fujii, Y. Sasaki, T. Niwa, and T. Tajima. **1999**. Dose escalation and pharmacokinetic study of a humanized anti-HER2 monoclonal antibody in patients with HER2/neu-overexpressing metastatic breast cancer. *Br. J. Cancer* 81 : 1419–1425.

# Part IV
# Examples for the Integration of Pharmacokinetic and Pharmacodynamic Concepts Into the Biotech Drug Development Plan

*Pharmacokinetics and Pharmacodynamics of Biotech Drugs: Principles and Case Studies in Drug Development.* Edited by Bernd Meibohm

ISBN: 3-527-31408-3

# 13
# Preclinical and Clinical Drug Development of Tasidotin, a Depsi-Pentapeptide Oncolytic Agent

*Peter L. Bonate, Larry Arthaud, and Katherine Stephenson*

## 13.1
## Introduction

Peptides, which are any member of a class of low molecular-weight compounds that yield two or more amino acids when hydrolyzed [1], exist pharmacokinetically between small molecules and proteins. For example, peptides may differ in their degree of oral bioavailability, in their specificity towards traditional drug-metabolizing enzymes such as cytochrome P450, and in their degree of protein binding. Few reviews have been published concerning the pharmacokinetics of peptides, notably Humphrey and Ringrose [2] and Tang et al. [3], although the latter review combines proteins and peptides into the same grouping. Peptides must be examined on a case-by-case basis because their size and structural properties render them unique, and generalities are few.

The role of pharmacokinetics in the clinical development of oncology drugs is varied. Since Phase I studies for oncology drugs are usually performed in patients rather than in healthy subjects, and most cytotoxic drugs are dosed to the maximum tolerated dose (MTD), the development of exposure–response relationships is used not so much in driving drug development as in confirming clinical activity. This chapter aims to review the role that preclinical pharmacology, toxicology, and pharmacokinetics has played in the development of tasidotin, a cytotoxic, naturally occurring pentapeptide that is currently being studied for the treatment of solid tumors.

## 13.2
## The Dolastatins

Many anticancer agents have their origins in pharmacognosy, including paclitaxel, etoposide, and the camptothecin analogues, topotecan and irinotecan, to name just a few. The dolastatins are a unique class of compounds isolated from the Indian Ocean sea-hare *Dolabella auricularia* that are referred to as depsipeptides,

*Pharmacokinetics and Pharmacodynamics of Biotech Drugs: Principles and Case Studies in Drug Development.* Edited by Bernd Meibohm

ISBN: 3-527-31408-3

which are peptides that are naturally secreted by bacteria, fungi, or other organisms and which can be synthesized in the laboratory. (NB. This is not to be confused with depsipeptide, a drug currently being developed by the National Cancer Institute and Fujisawa Company; NSC 630176, FR901228.)

Dolastatin 10 (NSC 376128; Fig. 13.1), developed by the National Cancer Institute, was the first dolastatin isolated and shown to be a potent antimitotic agent through its effects on tubulin [4, 5]. In Phase I clinical trials in patients with advanced solid tumors, dolastatin 10 had a MTD of 400 $\mu g/m^2$ administered once every three weeks (q3w) in minimally pretreated patients, and 325 $\mu g/m^2$ q3w in heavily pretreated patients, with neutropenia and mild sensory neuropathy as the dose-limiting toxicities (DLT). Dolastatin 10 was completely cleared from plasma within 24 h [6]. A strong correlation between absolute neutrophil count nadir and total area under the curve (AUC) was observed. A second Phase I study was conducted in patients with advanced solid tumors under a q3w schedule [7]. The MTD was 300 $\mu g/m^2$ with granulocytopenia as the DLT. Dolastatin 10 pharmacokinetics were consistent with a three-compartment model having $\alpha$-, $\beta$-, and $\gamma$-half-lives of 0.1, 1.0, and 18.9 h, respectively. The absolute neutrophil count nadir was predicted using an $E_{max}$ model with AUC as the dependent variable [8]. Dolastatin 10 has completed Phase II trials in patients with hormone refractory metastatic prostate cancer [9], metastatic melanoma [10], and advanced breast cancer [11] using the MTD identified in the Phase I studies. Although acceptable toxicity was observed in these investigations, poor efficacy was also identified, leading to the conclusion that dolastatin 10 should not be pursued further as single-agent therapy.

Dolastatin 15 (Fig. 13.1) a seven-subunit depsipeptide also derived from *Dolabella auricularia*, was shown to have similar activity *in vitro* compared to dolastatin 10. Dolastatin 15 was never pursued in the clinic (it is unclear why this was so from the literature, but may have been due to an inability to secure intellectual property), but it was used as a backbone by scientists at BASF Bioresearch Corporation (Worcester, MA, USA) to form more stable synthetic derivatives. One of these derivatives is cemadotin (LU103793; Fig. 13.1), in which the C-terminus of dolastatin 15, which is apparently not a site on the molecule needed for activity, was replaced by a benzylamide subunit. Cemadotin has *in-vitro* and *in-vivo* activity similar to that of dolastatin 10, and was taken into Phase I trials. The first trial was conducted in patients with advanced solid tumors as a 5-min bolus intravenous (IV) administration q3w schedule [12]. The MTD was 20 $mg/m^2$, with serious cardiac events as the DLT (i. e., some patients experienced hypertension and cardiac infarction). A second trial was initiated using a longer infusion time, 24 h q3w, under the belief that the adverse cardiac effects could be ablated by diminishing the maximal concentration [13]. The MTD was 15 $mg/m^2$, with similar adverse events observed as in the first study. Many patients experienced hypertension, and three experienced cardiac ischemia. Patients also experienced neutropenia, asthenia, tumor pain, and elevated liver enzymes. A third study was initiated using a 72-h infusion q3w. The MTD was 12.5 $mg/m^2$, with the primary DLT being neutropenia. Importantly, no cardiac events were noted using this dosing schedule. Pharmacokinetically, cemadotin was shown to be metabolized to a

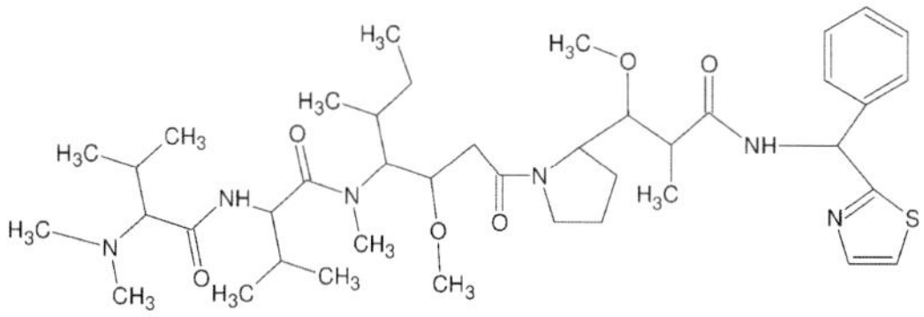

NSC 376128

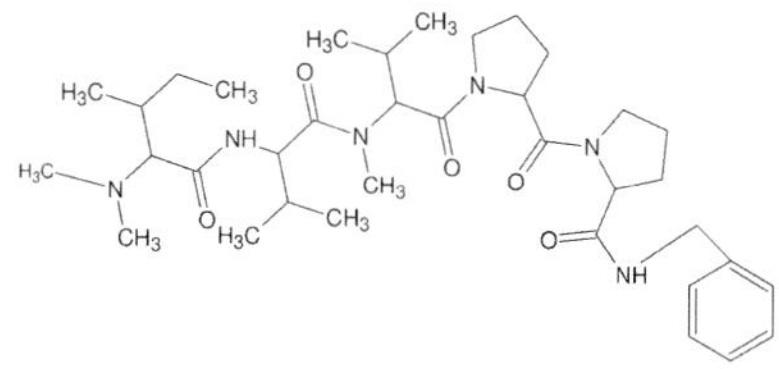

Cemadotin

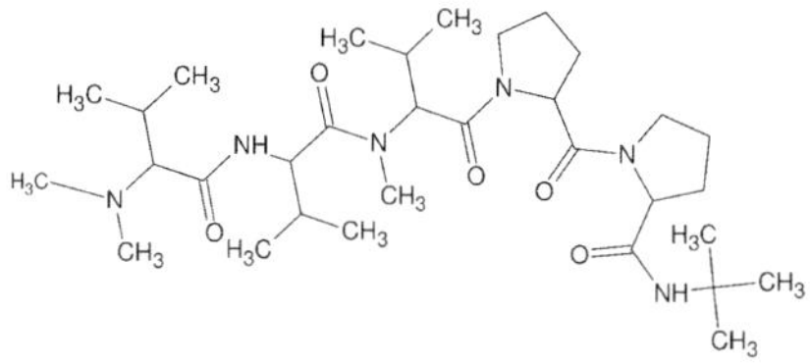

Tasidotin

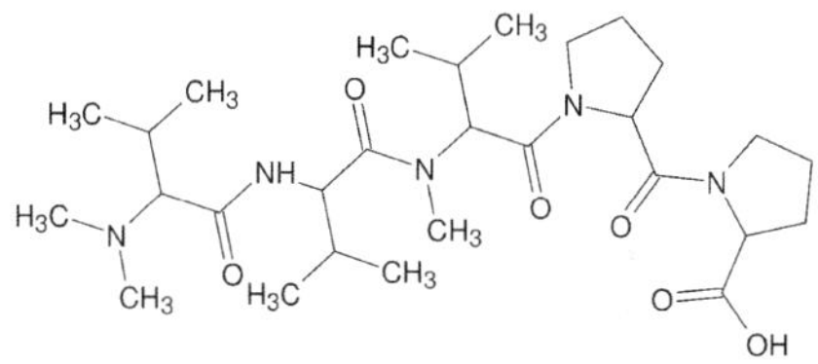

ILX651-C-carboxylate

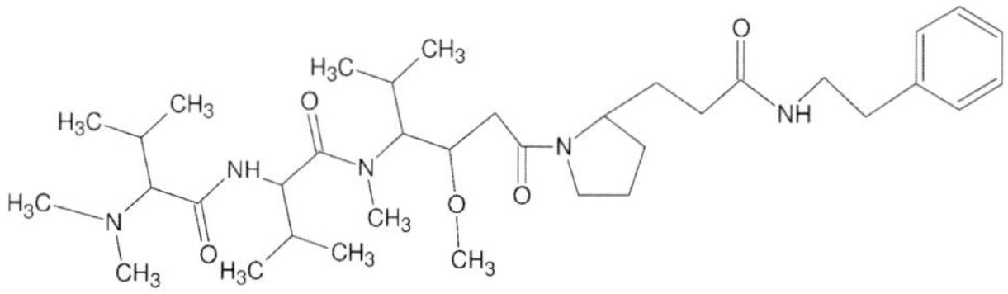

TZT-1027

**Fig. 13.1** Structures of dolastatin analogues.

C-carboxylate metabolite, had a half-life of approximately 10–13 h, and a small volume of distribution (ca. 9–10 L/m$^2$), indicating minimal tissue penetration. Even if this drug had shown activity, because a long infusion time was needed to avoid toxicity, such a regimen would not be commercially viable. Thus, the scientists returned to the laboratory to create a better analogue that did not cause the cardiac events seen with cemadotin.

## 13.3 Discovery and Preclinical Pharmacokinetics of Tasidotin

In response to the safety issues seen with cemadotin, scientists at BASF Bioresearch Corporation synthesized LU223651, a 607-Da depsi-pentapeptide, in which the benzylamide group at the C-terminus of cemadotin is replaced by a *t*-butylamine group (Fig. 13.1). Preclinical studies showed that cemadotin was 90–95% metabolized by prolyl oligopeptidase (POP), with the remainder being metabolized by cytochrome P450. When LU223651 was incubated with purified recombinant POP isolated from *Flavobacterium meningosepticum* for 4 h, less than 20% of cemadotin remained compared to more than 95% of LU223651. After 24 h, more than 90% of LU223651 remained intact. The primary LU223651 metabolite identified was the same metabolite formed from the biotransformation of cemadotin by POP, which was a carboxylic acid (Fig. 13.1). Further, when 80 mg/kg of LU223651 or cemadotin was administered to mice, plasma concentrations were still detectable at 1 h after IV administration, but cemadotin concentrations were not [14]. LU223651 was also shown to be 100% unbound in rat, mouse, and human plasma, as was its carboxylate metabolite.

Before these preclinical findings could be translated into the clinic, BASF Bioresearch Corporation was acquired by Abbott Laboratories (Abbott Park, IL, USA). During portfolio review, Abbott decided to not pursue further development of LU223651. In 2000, ILEX Oncology (San Antonio, TX, USA) licensed LU223651 and renamed the drug ILX651. In 2004, ILX651 was given the generic name Tasidotin by the United States Adopted Names Council. Also, in 2004, ILEX Oncology was acquired by Genzyme Corporation (Cambridge, MA). Hence, tasidotin development is currently being conducted by Genzyme Corporation.

## 13.4 Preclinical Pharmacology of Tasidotin and ILX651-C-Carboxylate

In-vitro studies in human cell lines of colon, mammary, and ovarian carcinoma, melanoma, and leukemia showed that the cytotoxicity of tasidotin was comparable with that of doxorubicin, but less than that of cemadotin. $IC_{50}$ values ranged from $1 \times 10^{-6}$ to $2 \times 10^{-8}$ M. In mouse xenograft studies, tasidotin showed good-to-excellent antitumor activity after IV administration against the following tumor types: P388 (leukemia), MX-1 (breast carcinoma), LOX (melanoma), PC-3 (prostate), LX-1 (small cell lung), CX-1 (colon) MiaPaCa and PANC-1 (pancreas), RL (lymphoma), H-MESO-1 (mesothelioma), and A-673 (sarcoma) [15]. Tasidotin administered perorally showed similar antitumor activity against these tumor types whenever tested.

The carboxylate metabolite of tasidotin was also studied using *in-vitro* cell models which showed the metabolite to have pharmacologic activity 10- to 30-fold less potent than the parent tasidotin ($IC_{50}$ values ranged from $10^{-7}$ M to $10^{-6}$ M). When the metabolite was studied *in vivo* using an MX-1 (breast cancer) xenograft model in mice, the metabolite showed no activity.

## 13.5 Toxicology of Tasidotin

A standard toxicology package was developed for tasidotin for filing the Investigational New Drug Application with the Food and Drug Administration (FDA). Single-dose studies under the proposed route of administration (IV) were initiated in rats and mice. Multiple-dose studies were completed in two species (rodent and nonrodent), rats and dogs, using IV as the proposed route of administration. Additional studies using an oral route of administration were also performed. The primary target tissues were the rapidly proliferating tissues: testes, bone marrow, skin, intestinal epithelium, and hematopoietic tissues. Toxicologic effects were reversible and consistent across species. The no observable adverse effect level (NOAEL) was determined to be 1 mg/kg per day (6 mg/m$^2$) in rats, with an MTD of 3.9 mg/kg per day (23.4 mg/m$^2$) IV for five days, repeated every 28 days. The NOAEL was 0.225 mg/kg per day (4.5 mg/m$^2$), with an MTD of 0.45 mg/kg per day (9 mg/m$^2$) IV for five days, repeated every 28 days in dogs [15].

In addition, because of the cardiovascular effects observed with cemadotin, safety pharmacology studies were initiated in rats and dogs administered tasidotin once daily for five days. No changes were observed in blood pressure, locomotor activity, body temperature, or body weight, and although the heart rate increased by 25% during the treatment period, it quickly returned to normal (this effect that may have been due to the stress of animal handling rather than to any drug-related effect). No changes in any cardiovascular parameters were seen in conscious, normotensive dogs. In anesthetized, normotensive dogs administered a single IV dose of tasidotin, slight dose-dependent decreases in cardiac output were observed concurrent with slight increases in coronary, femoral, and total peripheral resistance. However, these changes were only one-tenth the degree of those observed with cemadotin.

## 13.6 Clinical Pharmacology and Studies of Tasidotin in Patients with Solid Tumors

Preclinical studies showed that tasidotin was equally active as cemadotin against tumor cell lines, was more metabolically stable than cemadotin, and did not have the cardiac effects seen with cemadotin. The findings were promising enough that a decision was made to proceed into clinical development. In late 2000 and early 2001, two open-label, multicenter Phase I studies were initiated in patients with advanced solid tumors using different schedules: every day for five days once every three weeks (Study 101); and every other day for five days once every three - weeks (Study 102). The inclusion criteria were standard for such studies: otherwise healthy adult (aged $\geq$18 years) males or females having a confirmed diagnosis of malignant solid tumor for which no standard treatment existed, or that had progressed or recurred with prior therapy and having the following: a life expectancy of at least 12 weeks, a Eastern Cooperative Oncology Group performance status $\leq$2, no chemotherapy (including investigational agents) or major surgery within four weeks of enrollment and adequately recovered from any prior therapy, and adequate organ and immune function. Females of childbearing potential were required to show a negative serum pregnancy test within one week of enrollment. The following exclusion criteria applied: known hypersensitivity to the drug or its analogues; presence of any psychiatric disorder or chemical abuse that would interfere with consent; pregnant or lactating females; presence of any clinically apparent central nervous system malignancies or carcinomatous meningitis; and HIV-positive or AIDS-related illness. In Study 102, patients could also not have had prior radiation therapy to $\geq 25\%$ of the bone marrow (and must have recovered from prior radiation therapy) or prior bone marrow/stem cell transplantation. All studies were conducted in accordance with the Declaration of Helsinki, and patients provided their informed consent.

These studies were multiple-dose, ascending-dose tolerance studies with three patients per cohort, except at the MTD where usually six or more patients were enrolled. In the event of a DLT, the cohort was to be expanded up to six patients. Dose escalation was to proceed until the MTD was identified, with MTD being defined as the highest dose at which less than two of six patients experienced DLT during, or as a consequence of, treatment with tasidotin. Toxicity was analyzed according to the National Cancer Institute Common Toxicity Criteria, version 2.0 [16]. Serial blood samples were collected for pharmacokinetic analysis, as were urine collections on days 1 and 5.

The starting dose in each of these studies was based on one-tenth the MTD in rats, or one-third the MTD in dogs, whichever was smaller. In this case, the starting dose was based on the MTD in rats (23.4 mg/m$^2$ administered every day for five days, repeated every 28 days, equal to 117 mg/m$^2$ per cycle). Since in Study 101, five doses per cycle were to be administered, the starting dose was [(1/10 $\times$ 117 mg/m$^2$)/5 days] or 2.3 mg/m$^2$ for the every- day schedule. Performing the same calculations for the every-other-day dosing regimen, the starting dose was 3.9 mg/m$^2$. In late 2001, a third Phase I study (Study 103) was initiated in the same population using a once-weekly dosing schedule for three weeks, repeated every four weeks. The starting dose for that study was based on the cohorts that had already been observed. In Study 102, three patients had already been dosed with 7.8 mg/m$^2$ (equal to 23.4 mg/m$^2$ per cycle), with no toxicity or greater than Grade 1 adverse events. Hence, the starting dose for Study 103 was 7.8 mg/m$^2$.

In Study 101, 36 patients were enrolled and treated at doses up to 36.3 mg/m$^2$ [17]. The MTD was 27.3 mg/m$^2$, with the principal DLT being neutropenia (four of 14 patients experienced Grade 4 toxicity), which was consistent with preclinical toxicology studies. The other principal toxicities consisted of mild-to-moderate transaminitis, alopecia, fatigue, and nausea. One patient with melanoma metastatic to liver and bone who was treated at 15.4 mg/m$^2$ experienced a complete response and received 20 courses of tasidotin. Two other patients with melanoma had mixed responses of cutaneous metastases at 27.3 mg/m$^2$ per day associated with either stable or progressive visceral disease. In addition, nine patients with various other malignancies had stable disease.

In Study 102, 32 patients were enrolled and treated with doses up to 45.7 mg/m$^2$ [18]. The MTD was 34.4 mg/m$^2$, with the DLT being neutropenia. Other common, drug-related toxicities included mild to moderate fatigue, anemia, nausea, anorexia, emesis, alopecia, and diarrhea. The best observed antitumor response consisted of stable disease, and this was noted in 10 patients; the median duration on study for those patients with stable disease was 99.5 days compared to 37.5 days for those with progressive disease.

In Study 103, 30 patients were enrolled and treated with doses up to 62.2 $mg/m^2$ [19]. The MTD was 46.8 $mg/m^2$, with the DLT being neutropenia. Other non-hematologic toxicities included nausea and vomiting, diarrhea and fatigue, all of which were mild to moderate in severity, and manageable. In terms of response, one patient with metastatic non-small lung carcinoma experienced a minor response, and one patient with hepatocellular carcinoma had stable disease lasting 11 months.

Pharmacokinetic data from all studies were pooled and analyzed using noncompartmental methods [20] and WinNonlin Professional software (Version 4, Pharsight Corp., Mountain View, CA, USA). Because of the repeated-measures nature of the data, pharmacokinetic data were analyzed with linear mixed-effects models [21] using the MIXED procedure in SAS (version 8.0, SAS Institute, Cary, NC, USA). Clearance and volume of distribution were log-transformed prior to analysis, whereas half-life was analyzed untransformed. Dose proportionality was assessed using a power model [22]. Log-transformed $AUC_{0-\infty}$ and maximal concentrations ($C_{max}$; for ILX651-C-carboxylate only) were used as dependent variables. Initially, a full model was examined. For all models, patients were treated as random effects. Log-transformed dose, day of administration, and body surface area (BSA) were treated as fixed effects. Model terms not statistically significant at the 0.05 level were removed from the model. The reduced model was then re-fitted.

Pharmacokinetic data were available for 97 patients. The tasidotin plasma concentrations declined rapidly, and were less than 1% of maximal concentrations within about 8 h after dosing (Fig. 13.2). With such a short effective half-life, accumulation with daily tasidotin administration was not likely – once-daily multiple doses resemble a series of single doses. Concentrations appeared to decline in a biphasic manner (Fig. 13.2). The presence of a third, gamma phase was observed in some patients, but was not consistently detected, and for this reason the effective half-life was calculated instead of the terminal elimination phase half-life.

Dose proportionality was assessed using a power model. Less than 0.3% of the total AUC was extrapolated on average, with the largest percentage area extrapolated being 4.0%. $AUC_{0-\infty}$ did not change over day of administration, but was affected by dose ($p < 0.0001$) and BSA ($p < 0.0001$). $AUC_{0-\infty}$ increased disproportionally with increasing dose (Fig. 13.3). Under the power law model of dose proportionality, if a pharmacokinetic parameter is dose-linear, then the 90% confidence interval (CI) associated with dose for that parameter should contain the value 1.0. For $AUC_{0-\infty}$ the 90% CI for the slope related to dose was {1.15, 1.29} with an estimate of 1.22. Hence, the 90% CI did not contain the value 1.0. For a two-fold increase in dose, $AUC_{0-\infty}$ would increase by 2.3-fold ($= 2^{1.22}$), a difference of 16% higher than expected based on linear pharmacokinetics. $C_{max}$ did not change over day of administration (due to lack of accumulation), but was affected by dose ($p < 0.0001$) and BSA ($p = 0.0061$). The 90% CI for $C_{max}$ for the slope related to $C_{max}$ was {1.00, 1.17} with an estimate of 1.08. Hence, tasidotin $AUC_{0-}$ showed mild, probably clinically insignificant non-linear pharmacokinetics, but $C_{max}$ showed linear pharmacokinetics.

Tasidotin total systemic clearance did not change over day of administration, but did decrease with increasing dose ($p < 0.0001$) and increased with increasing

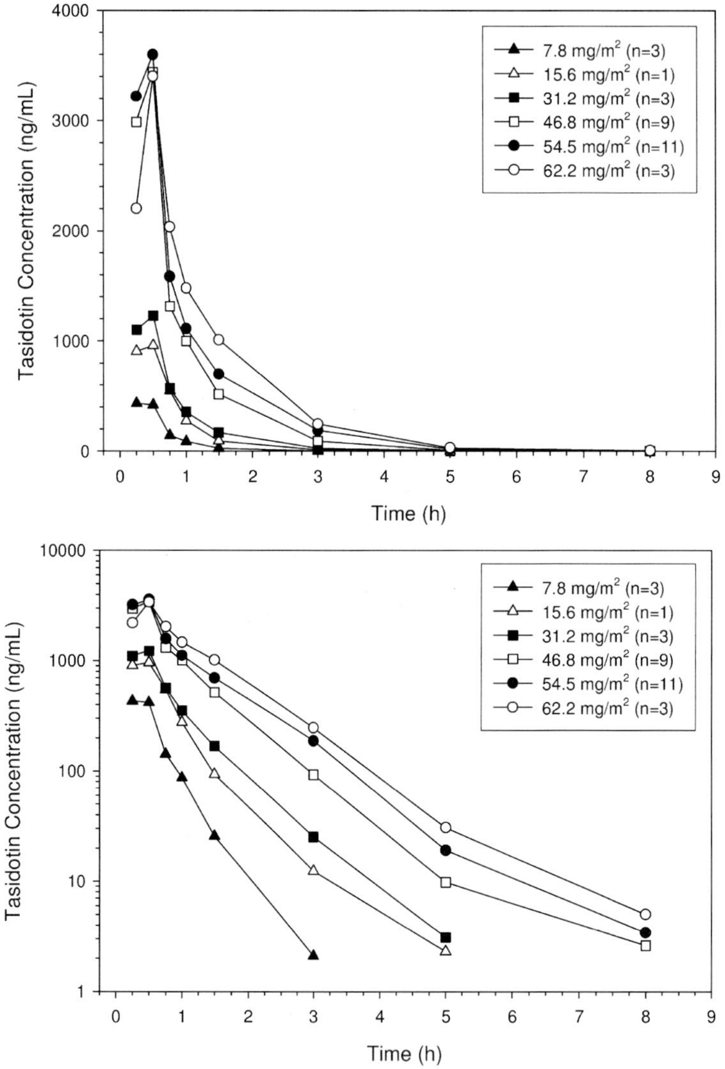


**Fig. 13.2** Mean concentration–time profiles for tasidotin after single-dose administration for patients enrolled in Study 103.

BSA ($p < 0.0001$). The least-squares mean clearance for a patient with a BSA of 1.83 $m^2$ at 2.3 $mg/m^2$ was 62 L/h, but was 30 L/h at 62.2 $mg/m^2$. Decreasing clearance with increasing dose is consistent with Michaelis–Menten elimination kinetics. Between-subject variability was moderate at approximately 30%. Tasidotin did not show any major renal elimination, with only ca. 13% of the dose being found in the urine as unchanged drug. In Study 103, the least-squares mean tasidotin renal clearance was approximately 4.3 L/h (about 13% of total systemic clearance), with a between-subject variability of approximately 51%. Given a glo-

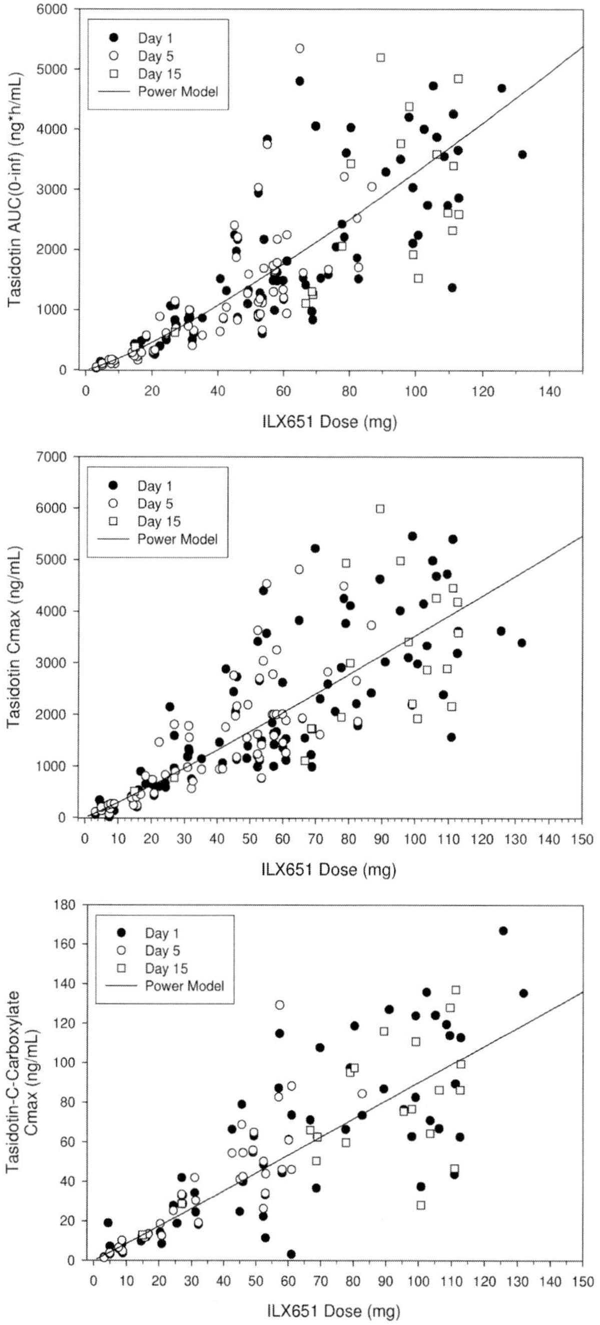


**Fig. 13.3** Dose proportionality assessment. The solid line is the predicted exposure based on a linear mixed effects power model having a reference subject. For the tasidotin plots, the predicted line is for a reference patient having a BSA of 1.83 $m^2$.

merular filtration rate in humans of 7.5 L/h, this suggests that tubular reabsorption or renal metabolism play a role in the urinary excretion of tasidotin. As renal clearance constituted a small part of total systemic clearance, the most likely cause of the nonlinearity was saturation of metabolic enzymes.

The least-squares mean tasidotin total volume of distribution at steady state was 10 L, and was not affected by dose or day of administration. Between-subject variability was estimated at 39%. The tasidotin half-life did not change over day of administration, but increased with increasing dose ($p < 0.0001$) and decreased with increasing BSA ($p = 0.0144$). The least squares mean half-life was 26 min at 2.3 mg/m$^2$, but was 46 min at 62.2 mg/m$^2$. Between-subject variability was estimated at approximately 20%. The finding that the half-life was dose-dependent was not surprising, as total systemic clearance was affected by both dose and BSA, whereas volume of distribution at steady state was unaffected by dose or BSA. Under these conditions, the half-life would be expected to change inversely proportional to clearance.

Controversy currently exists regarding the dogmatic dosing of oncolytic drugs per body surface area in the absence of any evidence for a relationship between BSA and pharmacokinetic parameters such as clearance [23]. The theory is that, in the absence of evidence, dosing per BSA might actually increase between-subject variability. Linear mixed effects model analysis indicated that BSA ($p < 0.0001$) was a significant predictor of tasidotin clearance (Fig. 13.4), but that neither dose nor BSA were significant predictors of volume of distribution at steady state. This analysis supports the continued dosing of tasidotin on a BSA-basis.

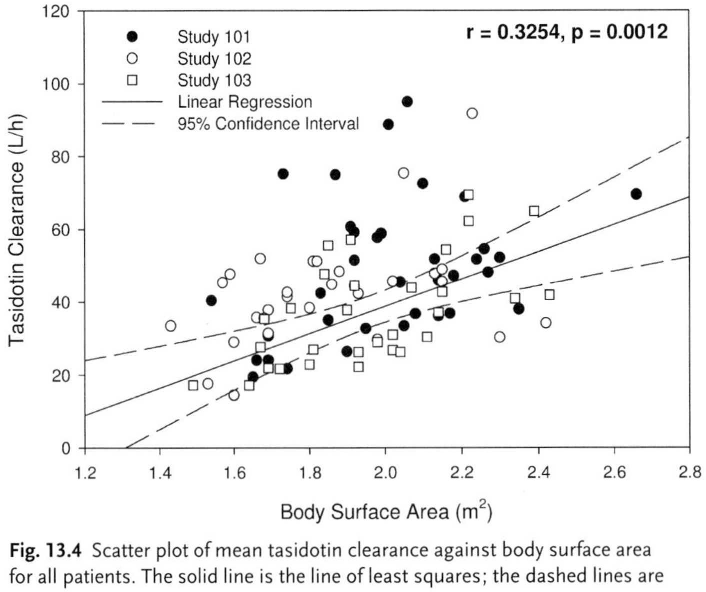


**Fig. 13.4** Scatter plot of mean tasidotin clearance against body surface area for all patients. The solid line is the line of least squares; the dashed lines are the 95% CI for the slope of the line.

## 13.7 Clinical Pharmacology of ILX651-C-Carboxylate

Metabolite data were available from 66 patients. Metabolite concentrations increased to a maximum and then declined in a monophasic manner (Fig. 13.5). In general, ILX651-C-carboxylate concentrations were approximately one-tenth of the parent drug concentrations, but declined much more slowly. Maximal concentra-

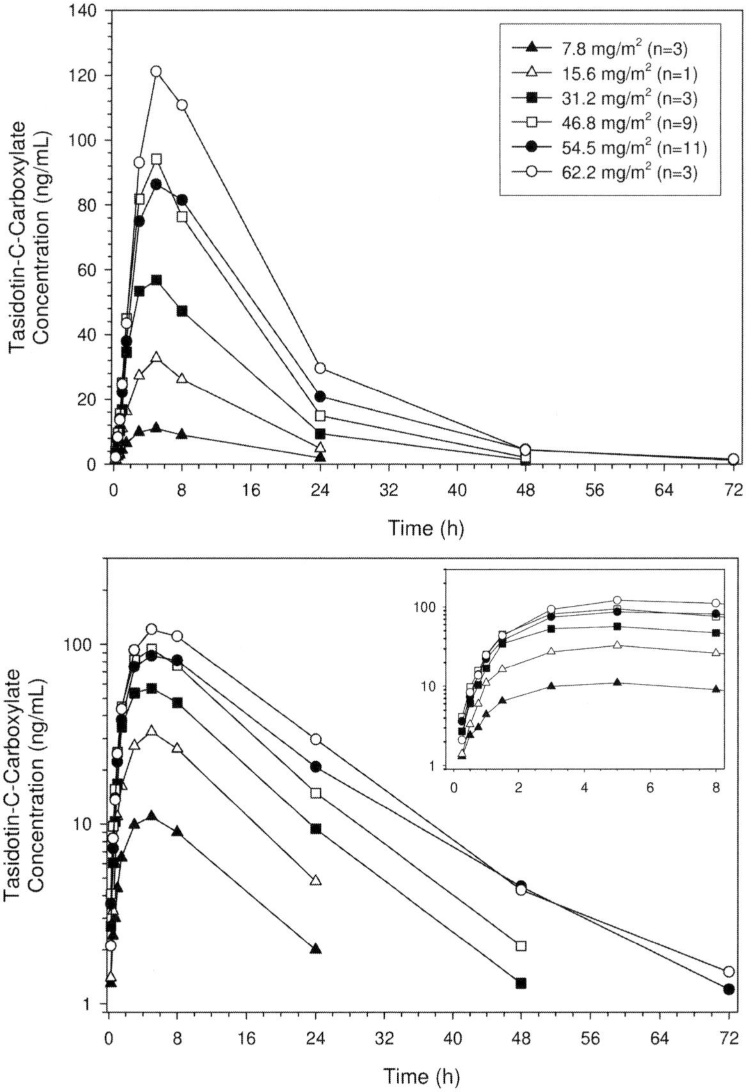


**Fig. 13.5** Mean concentration–time profiles for ILX651-C-carboxylate after single-dose tasidotin administration in patients enrolled in Study 103.

tions were reached within 3–8 h after dosing, with most patients achieving maximal metabolite concentrations at 5 h after starting the infusion. There was no apparent effect of dose on time to maximal concentration ($T_{max}$).

Dose proportionality was assessed using the power model. $AUC_{0-24}$ and $C_{max}$ did not change over day of administration, but did increase with increasing dose ($p < 0.0001$; see Fig. 13.3). For $AUC_{0-24}$, the 90% CI for the slope related to dose was {0.96, 1.18} with a point estimate of 1.07. The 90% CI for the slope related to $C_{max}$ was 1.02, with a 90% CI of {0.92, 1.13}. Hence, the 90% CI for both $AUC_{0-24}$ and $C_{max}$ contained the value 1.0 and both parameters were dose-proportional. The between-subject variability for $AUC_{0-24}$ and $C_{max}$ was 32% and 40%, respectively.

Most subjects provided insufficient data to estimate the half-life of the carboxylate metabolite. Nevertheless, sufficient data were available from 14 patients, from which an estimate of ILX651-C-carboxylate half-life could be determined. Neither day of administration, BSA, nor dose had any influence on ILX651-C-carboxylate half-life. The least-squares mean ILX651-C-carboxylate half-life was 8 h, which was considerably longer than the half-life of tasidotin. Between-subject variability in metabolite half-life was 17%. Given a parent half-life of less than 1 h, metabolite concentrations were not formation rate-limited. Given the half-life of the metabolite, little accumulation of the metabolite would be expected.

The metabolite ratio, defined as the ratio of $AUC_{0-\infty}$ of the metabolite to that of the parent, is a measure of the relative exposure of the metabolite to that of the parent, and could only be computed on 19 observations. The metabolite ratio for ILX651-C-carboxylate was not affected by dose, BSA, or day of administration, and was estimated at 0.59, indicating that the metabolite had about 59% the exposure of the parent. Hence, about one-third the total exposure (parent + metabolite) was due to the metabolite. This conclusion must be tempered, however, as few observations were made at the lower doses due to an inability to estimate $AUC_{0-\infty}$.

## 13.8 Exposure–Response Relationships

$AUC_{0-\infty}$ and $C_{max}$ data for parent tasidotin and $C_{max}$ data for the carboxylate metabolite were pooled across studies and averaged across days by subject in order to obtain average summary estimates of exposure for parent and metabolite. Because of the sampling design, an appropriate AUC measure for the metabolite was unavailable except for Study 103, wherein samples were collected for a sufficiently long period to permit accurate estimation of the metabolite $AUC_{0-24}$. The highest grade of neutropenia was determined for each subject without regard to cycle, as was the highest grade of nausea or vomiting. (When using Case Report Forms, a patient may have grade 1 nausea and grade 2 vomiting; however, for this analysis the response was modeled as grade 2 nausea and vomiting.) Hence, the following measures of exposure were available: dose, dose per $m^2$, total dose per cycle, parent $AUC_{0-\infty}$, parent $C_{max}$, metabolite $AUC_{0-24}$ for Study 103 only, and metabolite $C_{max}$. Note that metabolite concentrations were not measured in Study 102; consequently, no exposure data for the metabolite were available for that study. The following measures of response were examined: highest grade neutropenia; highest grade nausea and vomiting; and time to disease progression.

Ordinal logistic regression was used to relate grade of response to measures of exposure for the variables neutropenia and nausea/vomiting. Patient gender was also used as a covariate in the analysis of the nausea and vomiting data. Linear regression was used to relate time to disease progression to response measures. Missing data were deleted from the analysis.

Exposure–response data were available from 98 patients (66 patients not including Study 102, with the missing metabolite exposure data). All measures of exposure were predictive of degree of neutropenia, with the best predictor being $AUC_{0-24}$ of the metabolite. Also, all measures of exposure related to parent tasidotin were related to nausea and vomiting, with increasing exposure to parent tasidotin increasing the probability of nausea and vomiting. The best predictor of nausea and vomiting was daily dose per $m^2$. Patient gender was also an important predictor of nausea and vomiting, with females being 4.6-fold more likely to experience nausea and vomiting compared to males. None of the measures of metabolite exposure was related to nausea and vomiting; this was not unexpected as metabolite formation does not peak until 5–8 h after dosing, which is several hours after the nausea and vomiting typically occurs. All measures of exposure, including parent and metabolite, were predictive of time to disease progression, with the best predictor being $AUC_{0-24}$ of the metabolite, followed by metabolite $C_{max}$ (Figs. 13.6 and 13.7).

Combining the neutropenia results with time to disease progression produced the plot shown in Fig. 13.8, which shows that neutropenia and time to disease progression go hand-in-hand. Patients who become neutropenic tend to have the longest times until disease progression. Further, patients tend to have a linear increase in the probability of experiencing Grade 3 or higher nausea and vomiting with increasing dose intensity, and with females having a greater chance than males of experiencing Grade 3 or higher nausea and vomiting across all doses.

## 13.9 Discussion

Tasidotin displayed mild and clinically insignificant nonlinear elimination kinetics, which was not surprising for a cytotoxic chemotherapeutic agent. Nevertheless, tasidotin was rapidly cleared from the plasma with concentrations decreasing more than 99% by 8 h post dose. One question that arose early in development was the difference in half-life between tasidotin and cemadotin. Cemadotin has a reported half-life of 10 h [12], compared to 45 min or less for tasidotin. This was a large discrepancy for such a small structural difference between the two drugs. A careful review of the literature revealed that the analytical assay for cemadotin used a radioimmunoassay that was not specific for parent cemadotin, but showed crossreactivity with the metabolite. Hence, an aggregate of cemadotin and its metabolites was measured rather than parent cemadotin. Tasidotin was analyzed by

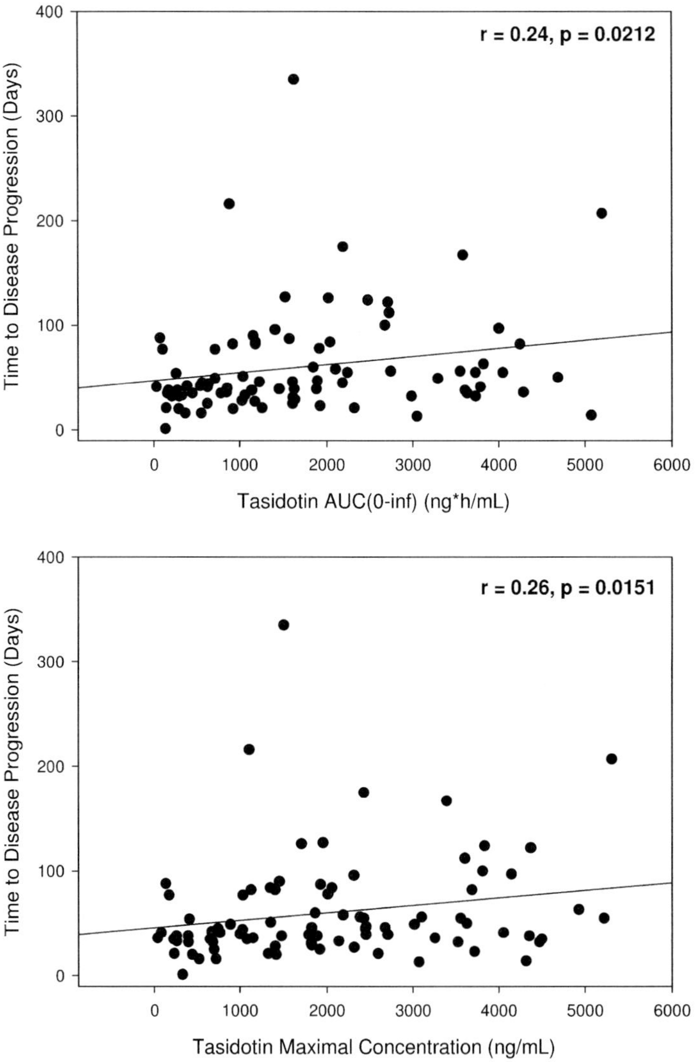


**Fig. 13.6** Scatter plot of time to disease progression as a function of tasidotin exposure. The solid line is the line of linear least squares.

LC-MS/MS, which was specific for parent tasidotin and did not show any cross-reactivity with metabolites. Hence, the half-life reported for cemadotin was not the true half-life, but probably that of the metabolite.

In contrast to tasidotin, plasma concentrations of the metabolite, ILX651-C-carboxylate, though much lower than those of the parent drug at early time periods, remained in the plasma longer and had a much longer half-life of 8 h, though this value was based on a relatively small number of subjects. Because of its longer half-life, the metabolite contributed about one-third to the total circulating

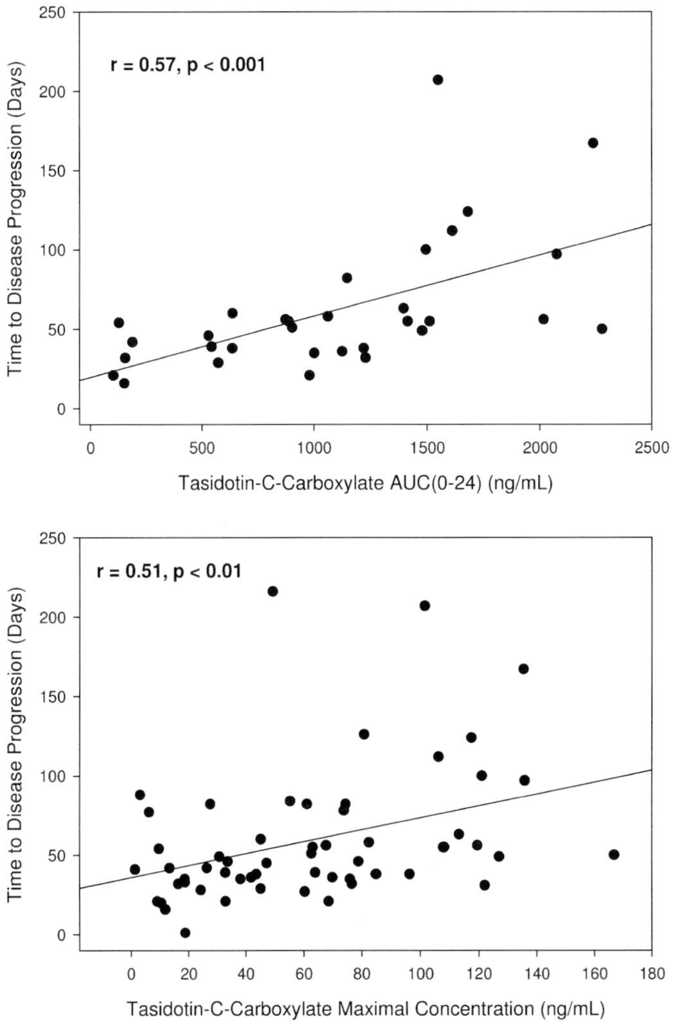


**Fig. 13.7** Scatter plot of time to disease progression as a function of ILX651-C-carboxylate metabolite $AUC_{0-24}$ and $C_{max}$. The solid line is the line of linear least squares.

exposure. Hence, preclinical studies were initiated to further understand the activity of the metabolite. These studies indicated that the metabolite was less active than tasidotin *in vitro*, and showed no activity when administered to mice implanted with tumors. One explanation of this observation is that, being a free carboxylic acid and highly polar, the metabolite does not penetrate the cell membrane to reach intracellular targets *in vivo*. Nonetheless, it is interesting that metabolite concentrations do not peak until tasidotin concentrations are essentially cleared from the plasma (3–8 h after administration). This would suggest that the meta-

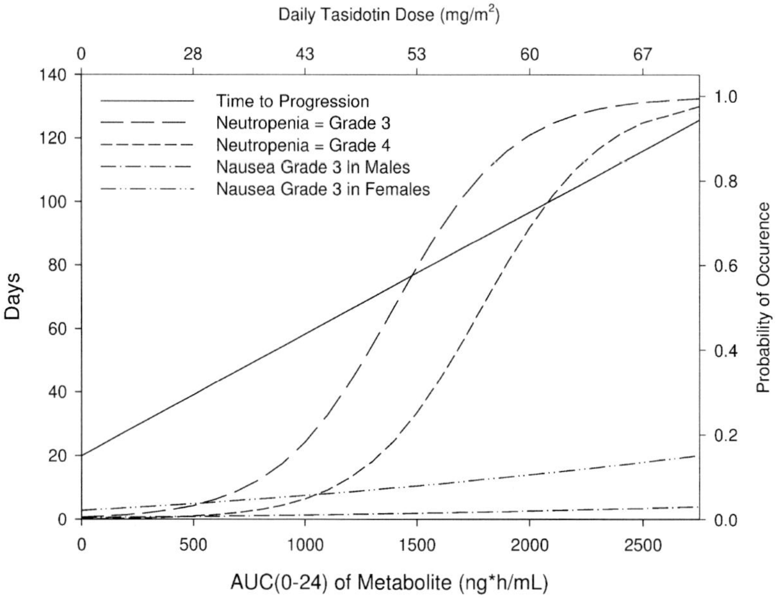


**Fig. 13.8** Probability of neutropenia and time to disease progression as a function of $AUC_{0-24}$ of the metabolite ILX651-C-carboxylate.

bolite is formed at a site exterior to the sampling compartment (e.g., a tissue), and then diffuses or leaches from the intracellular domain back into the plasma. If this were true, then the metabolite may indeed be active. Ongoing studies with the metabolite using Raman spectroscopy [24] and classic transport techniques in A549 (non-small cell lung cancer) cell lines [25] are currently being conducted to further determine the role of the metabolite in tasidotin's activity.

Given its small volume of distribution, tasidotin showed minimal tissue penetration, with the drug being largely confined to the extravascular and plasma compartments. Since tasidotin is 100% unbound, the small volume of distribution was likely a property of its high molecular weight. Tasidotin was not excreted into the urine to any significant extent as the parent drug (only ca. 13%). Given tasidotin's molecular weight (607 Da), biliary secretion would be expected to play the dominant role in its clearance. These results, together with those from preclinical studies using isolated perfused rat livers, are consistent with that hypothesis.

One interesting pharmacokinetic phenomenon observed with tasidotin was that parent concentrations prior to ending the infusion were higher than at the end of the infusion (Fig. 13.9). For a drug with constant clearance, this should not be possible. Although unusual, this phenomenon is not unheard of, having been reported for suberoylanilide hydroxamic acid, a histone deacetylase inhibitor [26]. In that report, the authors could not postulate a reason for such pharmacokinetic behavior. The same effect has also been reported for furosemide, perhaps due to the drug's behavior within the renal tubule [27].

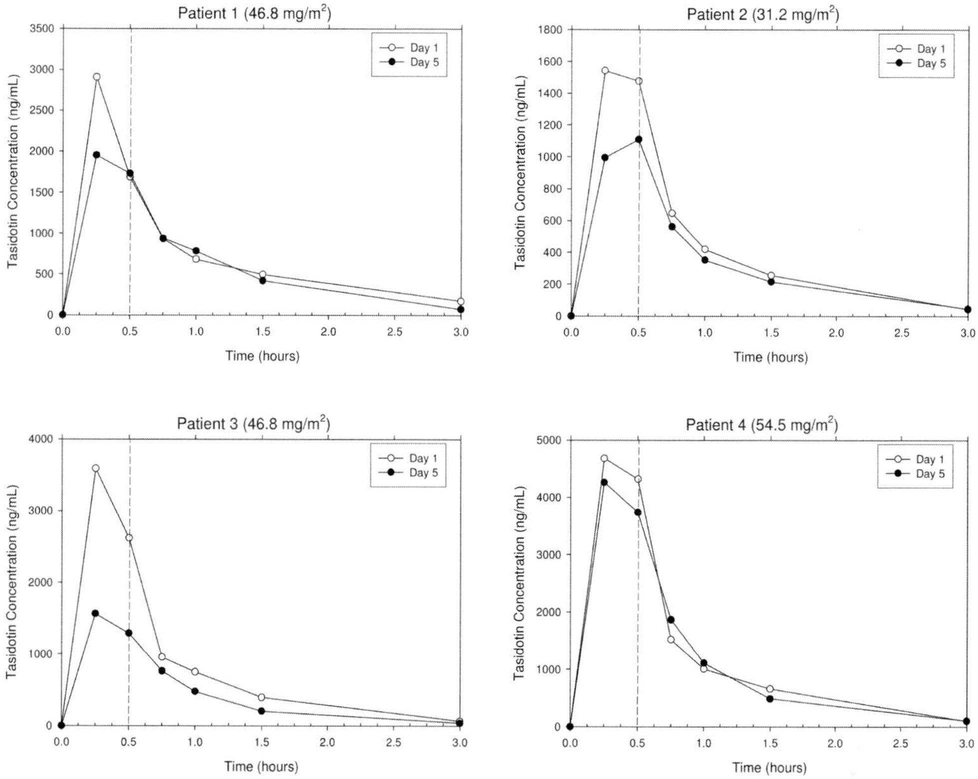


**Fig. 13.9** Scatter plots of representative tasidotin concentration–time profiles. The end of infusion is denoted with a vertical dashed line.

Several reasons might explain this observation, both physiological and study-related. First, the infusion may have been terminated (in error) prior to the scheduled end of infusion, such that the concentration data were artifactual. However, this seems unlikely as the same effect was observed in all clinical studies. In addition, the end of infusion time and pharmacokinetic sampling times were confirmed in the study Case Report Forms with, in all cases, the sampling times and end of infusion times being accurate. A second possibility is that there may have been an analytical error, though this too seems unlikely as the assay was validated according to FDA guidelines and the phenomenon was observed across multiple bioanalytical batches from several studies. While these two possibilities seem the most obvious, they can be eliminated, however.

The concentration of drug in plasma is due to three factors: the rate of drug input (rate of infusion); the rate of elimination; and the volume of distribution. Since the rate of input is a constant, only a change in the rate of elimination or volume of distribution emerge as possible reasons for such an unusual finding. It is possible that the volume of distribution changes over time. It is also possible that tasidotin is distributed by a transporter that must be activated. Basically, with this explanation, tasidotin distributes into the blood, after which there is a lag period before it distributes to other tissues – much like inverting a plugged bottle and then pulling the plug – though this seems unlikely. Alternatively, the altered distribution could be the result of altered protein binding, though as tasidotin is 100% unbound in plasma this too seems unlikely.

A more likely explanation relates to a nonconstant rate of elimination. It is possible that the protease enzymes which metabolize tasidotin need to be activated for maximal activity – that is, they are inactive initially but are activated in the presence of tasidotin. Tasidotin is metabolized at the proline end of the molecule to a carboxylic acid. The enzyme responsible for this conversion is largely POP (EC 3.4.21.26), a large intracellular enzyme of the serine protease family (distinct from trypsin) that preferentially hydrolyzes proline-containing peptides 30 amino acids in length or less at the carboxyl end of prolyl-residues to carboxylic acids [28, 29]. Protease activation is relatively common in eukaryotes, and there is evidence to suggest that serine proteases in particular require activation for maximal activity [30]. Under this scenario, tasidotin distributes into the body and tissues, but is metabolized slowly for a small period of time (based on the data in the present study this appears to be ca. 15 min) until the enzyme is activated. Thereafter, the metabolism is rapid and sustainable until tasidotin is cleared from the blood. The enzyme then resets itself until the next administration of tasidotin. This explanation appears plausible, and preclinical studies are being initiated using purified recombinant POP to determine the kinetic mechanism of catalysis of tasidotin, though these results are not yet available. Interestingly, TZT-1027, another dolastatin 10 analogue (see Fig. 13.1), does not show this phenomenon [31]. However, TZT-1027 differs from tasidotin in one important aspect – it is metabolized predominantly by cytochrome P450 3A4, and not by POP. CYP3A4 does not show activation kinetics as POP might. Hence, while both tasidotin and TZT-1027 are both dolastatins they are metabolized completely differently. As an aside, TZT-1027 is also more than 95% protein bound to plasma proteins, primarily to $\alpha_1$-acid-glycoprotein, whereas tasidotin is 100% unbound. So, despite small structural differences between the two drugs, they demonstrate major differences in their pharmacokinetics.

Based on a surrogate marker for clinical activity, time to disease progression [32] was prolonged as exposure to either the parent drug or metabolite was increased. However, coupled to this was the fact that the probability of dose-limiting neutropenia and severe nausea and vomiting also increased. Hence, an ideal dose regimen would be one where efficacy occurred prior to the adverse events. One suggestion would be to increase the time of infusion in order to reduce the parent drug concentrations and hence decrease the probability of nausea and vomiting. In addition, a growth factor might be added to the treatment to reduce the incidence of neutropenia. These two factors might allow clinicians to administer higher doses in order to achieve a longer time to disease progression.

## 13.10
## Summary

At present, tasidotin is still in development, and questions regarding its pharmacokinetic and mechanism of action remain unanswered, but that is the nature of drug development. As new data become available, new questions arise such that, at the time of regulatory approval, the sponsor should have a clear understanding of the mechanism of action, relationship between toxicology, preclinical and clinical pharmacokinetics, and the safety and efficacy of the drug.

## Acknowledgments

Drug development is a team sport, and these analyses could not have been carried out without the help and support of all investigators involved in the studies.

## 13.11
## References


1 www.biology-online.org. 2005. Peptides, Definition.

2 Humphrey, M.J., and P.S. Ringrose. **1986**. Peptides and related drugs: a review of their absorption, metabolism, and excretion. *Drug Metab. Rev.* 17:283–310.

3 Tang, L., A.M. Persky, G. Hochhaus, and B. Meibohm. **2004**. Pharmacokinetic aspects of biotechnology products. *J. Pharm. Sci.* 93:2184–2204.

4 Bai, R.L., G.R. Pettit, and E. Hamel. **1990**. Binding of dolastatin 10 to tubulin at a distinct site for peptide antimitotic agents near the exchangeable nucleotide and Vinca alkaloid sites. *J. Biol. Chem.* 265:17141–17149.

5 Pettit, G.R., Y. Kamano, C.L. Herald, A.A. Tuinman, F.E. Boettner, H. Kizu, J.M. Schmidt, L. Baczynsky, K.B. Tomer, and R.J. Bontems. **1987**. The isolation and structure of a remarkable marine animal antineoplastic constituent: dolastatin 10. *J. Am. Chem. Soc.* 109: 6883–6885.

6 Pitot, H.C., E.A. McElroy, J.M. Reid, A.J. Windebank, J.A. Sloan, C. Erlichman, P.G. Bagniewski, D.L. Walker, J. Rubin, R.M. Goldberg, A.A. Adjei, M.M. Ames. **1999**. Phase I trial of dolastatin-10 (NSC 376128) in patients with advanced solid tumors. *Clin. Cancer Res.* 5 525–531.

7 Madden, T., H.T. Tran, D. Beck, R. Huie, R.A. Newman, L. Pusztai, J.J. Wright, and J.L. Abbruzzese. **2000**. Novel marine-derived anticancer agents: A Phase I clinical, pharmacological, and pharmacodynamic study of dolastatin 10 (NSC 376128) in patients with advanced solid tumors. *Clin. Cancer Res.* 6:1293–1301.

8 Wagner, J.G. **1993**. *Pharmacokinetics for the Pharmaceutical Scientist.* Technomic Publishing Company Inc., Lancaster, PA.

9 Vaishampayan, U., M. Glode, W. Du, A. Kraft, G. Hudes, J. Wright, and M. Hussain. **2000**. Phase II study of dolastatin-10 in patients with hormone-refractory metastatic prostate adenocarcinoma. *Clin. Cancer Res.* 6:4205–4208.

10 Margolin, K., J. Longmate, T.W. Synold, D.R. Gandara, J. Weber, R. Gonazalez, M.J. Johansen, R. Newman, T. Baratta, and J.H. Doroshow. **2001**. Dolastatin-10 in metastatic melanoma: a phase II and pharmokinetic trial of the California Cancer Consortium. *Invest. New Drugs* 19:335–340.

11 Perez, E.A., D.W. Hillman, P.A. Fishkin, J.E. Krook, W.W. Tan, P.A. Kuriakose,

S.R. Alberts, and S.R. Dakhil. **2005**. Phase II trial of dolastatin-10 in patients with advanced breast cancer. *Invest. New Drugs* 23:257–261.

**12** Mross, K., K. Herbst, W.E. Berdel, A. Korfel, I.M. von Broen, Y. Bankmann, and D.K. Hossfeld. **1996**. Phase I clinical and pharmacokinetic study of LU103793 (cemadotin hydrochloride) as an intravenous bolus injection in patients with metastatic solid tumors. *Onkologie* 19:490–495.

**13** Supko, J.G., T.J. Lynch, J.W. Clark, R. Fram, L.F. Allen, R. Velagapudi, D.W. Kufe, and J.P. Eder, Jr. **2000**. A phase I clinical and pharmacokinetic study of the dolastatin analogue cemadotin administered as a 5-day continuous intravenous infusion. *Cancer Chemother. Pharmacol.* 46: 319–328.

**14** Nelson, C.M., D. Conlon, P.A. Smith, A.A. Admed, A. Haupt, D. Choquette, and T. Barlozzari. **1999**. Preclinical pharmacology of LU223651, orally available analog of cemadotin HCl. *Proc. Am. Assoc. Cancer Res.* 40:1908.

**15** Roth, S., R. Krumbholz, L. Arthaud, S. Weitman, and K. Stephenson. **2005**. In vivo and in vitro antitumor effects of ILX651, a pentapeptide with a novel mechanism of action. *Proc. Am. Assoc. Cancer Res.* 45:2121.

**16** National Cancer Institute. **1999**. Common Toxicity Criteria, version 2.0.

**17** Ebbinghaus, S., E. Rubin, E. Hersh, L.D. Cranmer, P.L. Bonate, R.J. Fram, A. Jekunen, S. Weitman, and L.A. Hammond. **2005**. A Phase I study of the dolastatin-15 analog tasidotin (ILX651) administered intravenously daily for five consecutive days every three weeks in patients with advanced solid tumors. *Clin. Cancer Res.* 7807–7816.

**18** Ebbinhaus, S., E. Rubin, E. Hersch, L.D. Cranmer, P.L. Bonate, R.J. Fram, A. Jejuken, S. Weitman, and L.A. Hammond. **2005**. A Phase I study of the dolastatin-15 analogue tasidotin (ILX651) administered intravenously daily for 5 consecutive days every 3 weeks in patients with advanced solid tumors. *Clin. Cancer Res.* 11:7807–7816.

**19** Cunningham, C., L.J. Appleman, M. Kirvan-Visovatti, D.P. Ryan, E. Regan, S. Vukelja, P.L. Bonate, F. Ruvuna, R.J. Fram, A. Jekunen, S. Weitman, L.A. Hammond, and J.P. Eder, Jr. **2005**. Phase I and pharmacokinetic study of the dolastatin-15 analogue tasidotin (ILX651) administered intravenously on days 1, 3, and 5 every 3 weeks in patients with advanced solid tumors. *Clin. Cancer Res.* 11:7825–7833.

**20** Cawello, W. et al. **1999**. *Parameters For Compartment-Free Pharmacokinetics: Standardisation of Study Design, Data Analysis, and Reporting*. Shaker Verlag, Aachen, Germany.

**21** Verbeke, G., and G. Molenberghs. **1997**. *Linear Mixed Models in Practice: A SAS-Oriented Approach*. Springer-Verlag, New York

**22** Gough, K., M. Hutcheson, O. Keene, B. Byrom, S. Ellis, L. Lacey, and J. McKellar. **1995**. Assessment of dose proportionality: Report from the Statisticians in the Pharmaceutical Industry/Pharmacokinetics UK Joint Working Party. *Drug Information J.* 29:1039–1048.

**23** Sawyer, M., and M.J. Ratain. **2001**. Body surface area as a determinant of pharmacokinetics and drug dosing. *Invest. New Drugs* 19:171–177.

**24** Ling, J., S.D. Weitman, M.A. Miller, R.V. Moore, and A.C. Bovik. **2002**. Direct Raman imaging techniques for study of the subcellular distribution of a drug. *Appl. Optics* 41:6006–6017.

**25** Kuh, H.-J., S.H.J. Jang, G. Wientjes, and J. Au. **2000**. Computational model of intracellular pharmacokinetics of paclitaxel. *J. Pharmacol. Exp. Ther.* 293:761–770.

**26** Kelly, W.K., V.M. Richon, O. O'Connor, T. Curley, B. MacGregor-Curtelli, W. Tong, M. Klang, L. Schwartz, S. Richardson, E. Rosa, M. Drobnjak, C. Cordon-Cordo, J.H. Chiao, R. Rifkind, P.A. Marks, and H. Scher. **2003**. Phase I clinical trial of histone deacetylase inhibitor: suberoylanilide hydroxamic acid administered intravenously. *Clin. Cancer Res.* 9: 3578–3488.

**27** van Meyel, J.J.M., P. Smits, F.G.M. Russel, P.G.G. Gerlag, Y. Tan, and F.W.J. Gribnau. **1992**. Diuretic efficiency of furosemide during continuous administration versus bolus injection in healthy volunteers. *Clin. Pharmacol. Ther.* 51 : 440–444.

**28** Polgar, L. **2002**. The prolyl oligopeptidase family. *Cell Molec. Life Sci.* 59 : 349–362.

**29** Wilk, S. **1983**. Prolyl endopeptidase. *Life Sci.* 33 : 2149–2157.

**30** Dixon, M., and E.C. Webb. **1979**. *Enzymes.* Academic Press, New York.

**31** de Jonge, M.J.A., A. van der Gaast, A.S. Planting, L. van Doorn, A. Lems, I. Boot, J. Wanders, M. Satomi, and J. Verweij. **2005**. Phase I and pharmacokinetic study of the dolastatin 10 analogue TZT-1027, given on days 1 and 8 of a 3-week cycle in patients with advanced solid tumors. *Clin. Cancer Res.* 11 : 3806–3813.

**32** DHHS, U.S. Food and Drug Administration, CBER, CDER. **2005**. *Guidance for Industry: Clinical Trial Endpoints for the Approval of Cancer Drugs and Biologics.* Rockville, MD.

# 14
# Clinical Drug Development of Cetuximab, a Monoclonal Antibody

*Arno Nolting, Floyd E. Fox, and Andreas Kovar*

## 14.1
## Introduction

Cetuximab is a chimeric monoclonal antibody (mAb) of the immunoglobulin G1 class (IgG1), which was obtained by attaching the Fv regions of the murine mAb M225 against epidermal growth factor receptor (EGFR) to constant regions of human IgG1 (i. e., the κ light chain and the γ-1 heavy chain). Cetuximab binds specifically and with high affinity to the extracellular domain of the human EGFR, a transmembrane glycoprotein that belongs to the Erb-B tyrosine kinase growth factor receptor family [1]. EGFR is also expressed in many normal human epithelial tissues, including the skin follicle. EGFR is expressed in a variety of tumor types. The expression is frequently associated with poor clinical prognosis [2, 3]. By binding to EGFR, cetuximab antagonizes the binding of endogenous EGFR ligands such as EGF and transforming growth factor-α (TGF-α). Because of the diversity of the EGFR-dependent intracellular signal pathways [4], the biological effects of the blockade of ligand receptor binding by cetuximab are highly pleiotropic. They comprise most cellular functions implicated in tumor growth and metastasis such as cell proliferation, cell survival, cell motility, cell invasion, tumor angiogenesis, and desoxyribonucleic acid (DNA) repair. Cetuximab also induces the internalization of EGFRs, which is thought to lead to EGFR down-regulation and a reduction in EGFR signaling.

Currently, cetuximab (Erbitux®) is indicated for the treatment of patients with EGFR-expressing metastatic colorectal cancer after failure of irinotecan-including cytotoxic therapy. Depending on the country, cetuximab is approved in combination with irinotecan or in addition as monotherapy. The approved dosing regimen consists of an initial dose of 400 mg/m$^2$ body surface area (BSA), followed by weekly doses of 250 mg/m$^2$. The (intravenous; IV) infusion durations are 2 h for the initial infusion, and 1 h for the subsequent weekly infusions.

First approval worldwide was granted for cetuximab in Switzerland in December 2003 for the combination therapy with irinotecan. Subsequently, approvals were granted in the US (mono- and combination therapy) and all 25 membership

*Pharmacokinetics and Pharmacodynamics of Biotech Drugs: Principles and Case Studies in Drug Development.* Edited by Bernd Meibohm

ISBN: 3-527-31408-3

countries of the European Union, as well as Iceland and Norway (combination therapy). As of March 2006, cetuximab has been approved by regulatory agencies of 53 countries.

Colorectal cancer (CRC) is the third most commonly diagnosed cancer worldwide, and the second most common cause of cancer mortality in Europe and North America. Worldwide, approximately one million cases were diagnosed in 2000, accounting for 500 000 deaths. Age represents the most important risk factor, since about 90% of CRC patients are aged 50 years or more [5]. The EGFR expression rate in CRC is reported to be between 75 and 89% [6–8]. EGFR-expressing CRC tumors are associated with a poor prognosis in terms of survival [9–11].

The rationale for using cetuximab in combination with irinotecan in metastatic CRC is based on the fact that inhibitors of growth factor receptor signaling and genotoxic agents (chemotherapeutics, radiation) interfere with cell cycle progression and promote apoptosis. They also modulate expression and activity of common intermediates of the signaling pathways controlling cell cycle and apoptosis in the same direction.

Merck KGaA, Darmstadt, Germany, licensed the right to market and develop Erbitux outside the USA and Canada from ImClone Systems Inc. of New York in 1998. In Japan, Merck KGaA has co-exclusive marketing and development rights with ImClone Systems. Cetuximab is being further developed in this program, which was joined in 2001 by Bristol Myers Squibb Company, for the treatment of several tumor types that express EGFR such as squamous cell carcinoma of the head-and-neck, non-small cell lung cancer and pancreatic cancer, including early lines of treatment as well as combinations with various chemotherapeutic agents or radiotherapy.

## 14.2
## Specific Considerations in Oncologic Drug Development

Compared to other therapeutic areas, the preclinical and clinical development of a biotechnology drug for use in oncology requires consideration of several additional unique issues [12]:

- Early clinical studies are performed in cancer patients in hospitals instead of healthy volunteers, and in specialized Phase I units. Selection criteria for patients entered into cetuximab Phase I studies included various important factors such as disease state, life expectancy (>3 months), prior treatment, organ function, age, tumor type and target (i.e., EGFR expression). Therefore, pharmacokinetic (PK) data obtained from these individuals is confounded by numerous factors, a fact usually absent in conventional studies with tightly controlled, well-selected healthy subjects performed for non-oncologic drugs.
- Prolonged timelines are required to complete a single Phase I study. Oncologic Phase I studies are typically conducted in cancer patients for whom all other existing therapy has failed. Thus, competition exists between various sponsors

of oncology agents, since such patients are relatively scarce. In dose-escalation studies, enrollment is usually slow since the maximum tolerated dose (MTD), the identification of which is an important objective in oncologic Phase I studies, may not be known and few patients are willing to enroll at a potentially subtherapeutic dose.

- Trial design and objectives. The major purpose of the early cetuximab Phase I studies was to define a safe and tolerable dose and dosing schedule either for cetuximab alone or combined with chemotherapeutic agents to be investigated in the later stages of clinical development. In general, the objectives were to determine the MTD, to characterize side effects and identify dose-limiting toxicity, to define the PK profile, to measure response to treatment, and to define the potential immunologic response to cetuximab treatment. Patients were typically treated in small cohorts of three to six per dose group, and dose-escalation steps were performed according to a common methodology, the modified Fibonacci Scheme [13, 14].
- Dosing. In oncology, the administration of a drug in relation to BSA (in $mg/m^2$) tends to be standard practice. The rationale for the use of BSA originates from the observation that there is a proportional relationship between renal function and BSA [15]. Over several decades, and for many cytotoxic drugs, an additional relationship between specific toxicity and dosing per BSA could be established. Dosing for cetuximab per BSA followed this historic path.
- Concomitant medication. Patients involved in Phase I oncologic studies have typically undergone several rounds of previous anti-cancer therapies, which had to be discontinued due to progression of the underlying disease or due to unacceptable toxicity. The general health status of cancer patients enrolled in these studies may, therefore, be comparatively poor. Hence, in contrast to Phase I studies in healthy volunteers, palliative and supportive care for disease-related symptoms and for toxicity associated with treatment must be given to all patients in oncologic trials. Such intervention may include the administration of medications such as antibiotics, analgesics, antihistamines, antiemetics, steroids, growth factors (including erythropoietin), or procedures such as paracentesis or thoracentesis or the administration of blood products such as red blood cells, platelets or fresh-frozen plasma transfusions.
- Data quality. Oncologic Phase I studies are often performed as multicentric studies because the recruitment of cancer patients tends to be slow. Since these studies are performed in the hospital setting at multiple sites, and not at dedicated Phase I units, the ensurance of data quality is more challenging. Not all hospitals and their staff are equally well experienced in adhering to the demands of a rigorous PK sampling schedule, and the requirements regarding sampling preparation, handling, storage and shipment. More often than not, especially in Phase II studies, hospitals and their staff are more mindful of the need to ensure the quality of the safety, tolerability and efficacy data rather than of PK data. This fact may add to the overall variability seen in oncologic trials regarding PK data.

## 14.3
## Introduction to the Clinical Pharmacokinetics of Cetuximab

PK data for cetuximab were derived from 19 clinical studies performed up to 2003 [16]. Ten of these studies were conducted as dose-escalation/dose-finding studies, while the remaining nine were conducted at the approved dosing regimen of 400 mg/m$^2$ initially, followed by weekly doses of 250 mg/m$^2$.

The PK of cetuximab were investigated after single IV doses ranging from 5 to 500 mg/m$^2$. PK data obtained after administration of cetuximab monotherapy were available from 82 patients in five of the dose-escalation studies, and from 214 patients in four of the studies with the approved dosing regimen. In two studies, cetuximab was administered both as a monotherapy and in combination with irinotecan; single- as well as multiple-dose cetuximab concentration data were available from 300 patients in these two studies.

In 10 studies the corresponding clinical study protocol stipulated that cetuximab was to be given in combination with a chemotherapeutic agent (irinotecan, paclitaxel, gemcitabine, cisplatin, carboplatin, or doxorubicin) or in combination with radiation therapy.

In the dose-escalation studies, PK data were obtained for patients with a variety of EGFR-expressing solid cancers, while the studies at the approved dosing regimen were conducted in the indications of colorectal carcinoma, squamous cell cancer of the head-and-neck, pancreatic cancer, and renal cell cancer.

Cetuximab serum concentrations were measured using validated enzyme-linked immunosorbent assay (ELISA) methods or a validated surface plasmon resonance assay. These bioanalytical assays were crossvalidated to allow pooling of PK data across studies.

PK data were collected by the traditional approach of frequent sampling in all studies, except for one Phase II study [17] in which a sparse population sampling design (four to six PK samples per patient) was chosen. Two studies investigated the single-dose PK of cetuximab for up to three weeks prior to the administration of subsequent weekly cetuximab doses of 250 mg/m$^2$. In all other studies, PK sampling was carried out within the dosing interval of one week. Good agreement was observed between PK data obtained from three-week and one-week sampling schedules. Results from individual studies were analyzed by non-compartmental PK analysis and are summarized in Sections 14.4 and 14.5. Additionally, cetuximab concentration data from all studies were analyzed by a population PK approach within the framework of an integrated PK database analysis. The results of the integrated PK database analysis are summarized in Section 14.6.

## 14.4
## Early Attempts to Characterize the PK of Cetuximab

Initially, PK results of the first three Phase I studies in patients were evaluated after pooling [18]. This was carried out in an attempt to identify possible dose-related patterns in the PK parameters across a dose range wider than that which could be investigated in one study alone.

A total of 52 patients with either head-and-neck cancer or with non-small cell lung cancer received either a single cetuximab dose (n = 13), multiple weekly cetuximab doses (n = 17) and multiple weekly cetuximab doses with cisplatin (n = 22). Cetuximab IV doses of 5, 20, 50, 100, 200 and 400 $mg/m^2$ were administered. Following the administration of 5 $mg/m^2$, serum cetuximab concentrations were generally below the limit of quantification. For all other doses, cetuximab serum concentration data were analyzed using non-compartmental approaches. Results from these preliminary analyses indicated that the volume of distribution was independent of dose, while clearance (CL) appeared to decrease with dose, with little difference in CL between the 200 $mg/m^2$ and 400 $mg/m^2$ dose levels. A mean CL value of 0.127 was reported for the 20 $mg/m^2$ dose, but this decreased to 0.034 $L/h/m^2$ for the 100 $mg/m^2$ dose. At doses of 200 $mg/m^2$ and higher, a mean CL value of 0.018 $L/h/m^2$ was determined (these values were converted from the original unit, assuming a BSA of 1.7 $m^2$ and a body weight of 70 kg). In addition, the mean CL value at the 100 $mg/m^2$ dose level after monotherapy was similar to that obtained after cetuximab was coadministered with cisplatin, suggesting a lack of an effect by cisplatin on the clearance of cetuximab.

## 14.5 PK of Cetuximab Following Pooling of Data Across All Studies

### 14.5.1 Comparison of Single-Dose PK Parameters at Various Dose Levels

Subsequently, with the accumulation of more single-dose cetuximab PK data, and in order to obtain a sufficiently high number of observations at each dose level, PK data for the same dose or dose level were pooled across studies [19]. Thus, the PK characteristics could be described across a dose range of 5 to 500 $mg/m^2$. Mean single-dose cetuximab PK parameters are displayed by dose in Table 14.1, across all studies.

#### 14.5.1.1 Maximum Serum Concentration

Mean maximum observed cetuximab ($C_{max}$) values after first cetuximab administration are displayed as a function of dose in Fig. 14.1. Serum concentrations of cetuximab typically reached maximum levels approximately 1–2 h following the end of the cetuximab infusion, independent of dose. Mean cetuximab $C_{max}$ values ranged from 8.7 to 283.8 μg/mL after single cetuximab doses of 20 to 500 $mg/m^2$, respectively. A linear relationship (coefficient of correlation, $r^2 = 0.98$) between cetuximab dose and mean $C_{max}$ value was observed, indicating that maximum serum concentrations following infusion are predictable for each dose used.

**Table 14.1** Single-dose pharmacokinetic parameters for cetuximab across all studies.

| **Dose [mg/m²]**<br>**No. of studies** | | **20**<br>**4** | **50**<br>**7** | **100**<br>**7** | **200**<br>**4** | **250**<br>**2** | **300**<br>**1** | **400**<br>**8** | **500**<br>**5** |
|---|---|---|---|---|---|---|---|---|---|
| $C_{max}$ | N | 13 | 23 | 52 | 14 | 8 | 4 | 56 | 20 |
| [µg/mL] | Mean | 8.69 | 22.19 | 46.77 | 102.36 | 140.20 | 133.25 | 184.51 | 283.80 |
| | SD | 4.19 | 4.74 | 11.59 | 29.37 | 19.63 | 47.66 | 54.62 | 84.08 |
| | Range | 3–19 | 12.58– 32.63 | 17–73 | 53–167 | 119.7–170.17 | 79–190 | 0–327.03 | 155.37–498 |
| $AUC_{0\text{-}\infty}$ | N | 10 | 22 | 48 | 14 | 8 | 4 | 53 | 18 |
| [µg h/mL] | Mean | 343 | 1031 | 2912 | 9923 | 12414 | 16311 | 21142 | 32448 |
| | SD | 228 | 440 | 1060 | 3226 | 3332 | 3786 | 8657 | 12880 |
| | Range | 133–866 | 394–1929 | 1136–7522 | 4683–17 015 | 9290–19 159 | 11 265–19 388 | 9412–48 039 | 14 659–57 700 |
| $t_{1/2}$ | N | 10 | 22 | 48 | 14 | 8 | 4 | 53 | 18 |
| [h] | Mean | 33.3 | 32.3 | 44.8 | 79.8 | 65.9 | 90.5 | 97.2 | 119.4 |
| | SD | 29.2 | 9.8 | 12.8 | 19.6 | 18.8 | 13.8 | 37.4 | 76.9 |
| | Range | 14.74–106.66 | 15.35–57.91 | 20.85–95.26 | 51.19–117.53 | 45.48–104.11 | 77.28–108.07 | 41.39–213.38 | 33.58–338.07 |
| CL | N | 10 | 22 | 48 | 14 | 8 | 4 | 53 | 18 |
| [L/h/m²] | Mean | 0.079 | 0.059 | 0.039 | 0.020 | 0.022 | 0.019 | 0.022 | 0.018 |
| | SD | 0.039 | 0.028 | 0.015 | 0.010 | 0.005 | 0.005 | 0.009 | 0.008 |
| | Range | 0.023–0.15 | 0.026–0.127 | 0.013–0.088 | 0–0.042 | 0.013–0.027 | 0.015–0.027 | 0.008–0.042 | 0.009–0.034 |
| $V_{ss}$ | N | 10 | 22 | 48 | 14 | 8 | 4 | 53 | 18 |
| [L/m²] | Mean | 2.81 | 2.49 | 2.48 | 2.48 | 1.99 | 2.52 | 2.88 | 2.68 |
| | SD | 1.09 | 0.70 | 0.91 | 1.05 | 0.29 | 0.49 | 0.93 | 0.75 |
| | Range | 1.87–5.09 | 1.69–4.31 | 1.51–5.95 | 1.71–5.05 | 1.32–2.25 | 1.82–2.93 | 1.51–6.19 | 1.52–4.23 |

*Abbreviations:* $C_{max}$: maximum cetuximab serum concentration; $AUC_{0\text{-}\infty}$: area under the concentration–time curve extrapolated to infinity; CL: total body clearance; $V_{ss}$: volume of distribution at steady state; $t_{1/2}$: terminal elimination half-life; SD: standard deviation.
Note: At the 5 mg/m² dose level, only $C_{max}$ was determined: Mean: 2.38 ± 2.77 µg/mL; range: 1.00–9.00 (n = 8).

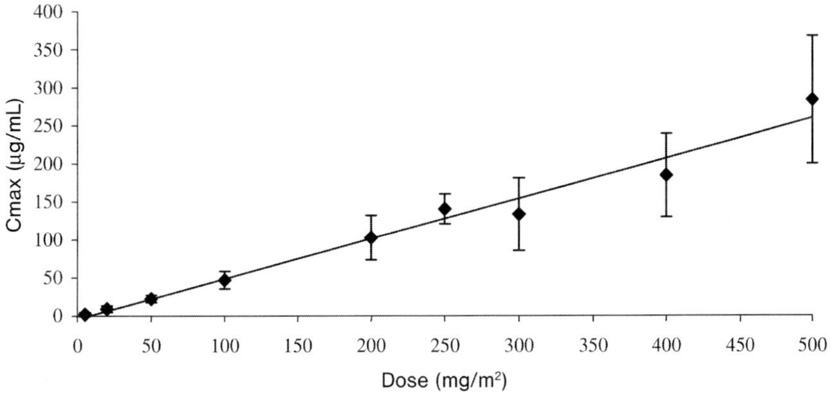


**Fig. 14.1** Mean (± SD) maximum serum concentration ($C_{max}$) for cetuximab versus dose across all studies.

#### 14.5.1.2 Area Under the Concentration-Time Curve

Mean area under the concentration–time curve ($AUC_{0\text{-}\infty}$) values after the first cetuximab administration are displayed as a function of dose in Fig. 14.2. Mean cetuximab $AUC_{0\text{-}\infty}$ values ranged from 343 to 32448 µg·h/mL after single cetuximab doses of 20 to 500 mg/m$^2$, respectively. Overall, a linear relationship ($r^2 = 0.98$) between cetuximab dose and $AUC_{0\text{-}\infty}$ value was observed, indicating that the exposure to cetuximab, as measured by AUC, is predictable at dose levels of 200 mg/m$^2$ and higher. However, at lower doses (20 to 100 mg/m$^2$), deviations from this linear relationship were observed which are thought to be due the specific elimination characteristics of cetuximab at lower doses, as described in the following paragraph.

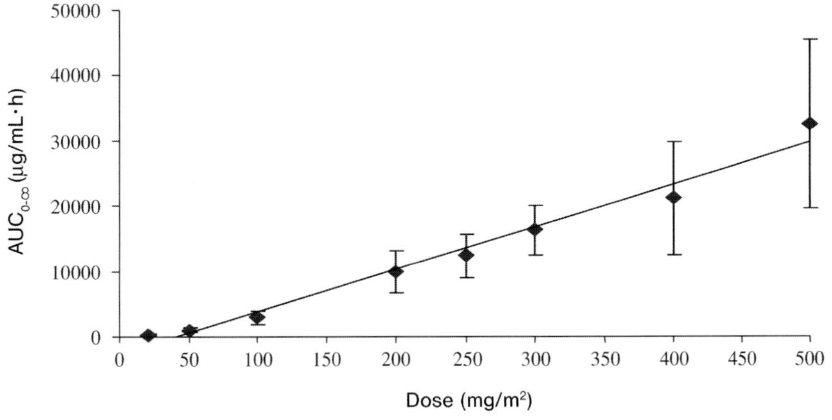


**Fig. 14.2** Mean (± SD) area under the concentration–time curve ($AUC_{0\text{-}\infty}$) for cetuximab versus dose across all studies.

#### 14.5.1.3 Clearance

Mean clearance (CL) values for cetuximab are displayed as a function of dose in Fig. 14.3. Mean CL values decreased from 0.079 to 0.018 $L/h/m^2$ after single cetuximab doses of 20 to 500 $mg/m^2$, respectively. In the dose range 20 to 200 $mg/m^2$, CL values decreased with dose. At doses of 200 $mg/m^2$ and greater, CL values leveled off at a value of approximately 0.02 $L/h/m^2$. This biphasic behavior suggests the existence of two elimination pathways. The elimination of cetuximab apparently involves a specific, capacity-limited elimination process that is saturable at therapeutic concentrations, in parallel with a nonspecific first-order elimination process that is non-saturable at therapeutic concentrations. Increasing doses of cetuximab will therefore ultimately lead to the saturation of the elimination process that is capacity-limited and that follows Michaelis–Menten kinetics, whereas the first-order process will become the dominant mechanism of elimination beyond a particular dose range.

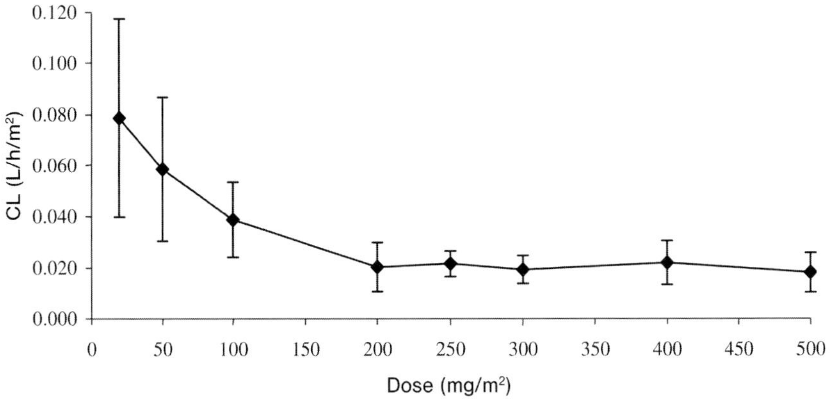


**Fig. 14.3** Mean (± SD) clearance (CL) for cetuximab versus dose across all studies.

For cetuximab, it is assumed that the specific, saturable elimination mechanism acts via the complementarity-determining regions (CDRs) of the antibody, which is reflective of the target-specific elimination process. Specifically, cetuximab binds to EGFR via its CDRs, and the resulting cetuximab/EGFR complex is either recycled with recurrence of the receptor at the cell surface or degraded in lysosomes. Due to the large number of EGFRs in the body, internalization markedly contributes to the elimination of cetuximab from the systemic circulation.

At a certain serum concentration range (i.e., at a certain corresponding cetuximab dose), EGFRs will become saturated with cetuximab molecules. When concentrations increase beyond this saturation range, the nonspecific, nonsaturable, elimination pathway becomes the predominant elimination pathway, as indicated by the observation that CL and $t_{1/2}$ values remain constant. This nonspecific elimination pathway is common for all antibodies, independent of their origin, and is nonsaturable at therapeutic antibody concentrations. Antibodies are usually recognized by several receptors that have binding affinities to either the protein and carbohydrate moieties on the Fc region. Liver cells or cells of the reticuloendothelial system, for example, are known to express Fc-recognizing receptors. Binding of antibodies to these receptors is usually followed by internalization and further catabolism [20, 21]. Due to the very large number of receptors in the body it is believed that these receptors are not saturable at cetuximab doses that were applied in clinical studies.

#### 14.5.1.4 Elimination Half-Life

Mean values for the elimination half-life ($t_{1/2}$) of cetuximab increased from 33.3 to 119.4 h after single cetuximab doses of 20 to 500 mg/m$^2$, respectively (Fig. 14.4). At the approved dosing regimen of 400/250 mg/m$^2$, $t_{1/2}$ values were approximately 66 to 97 h.

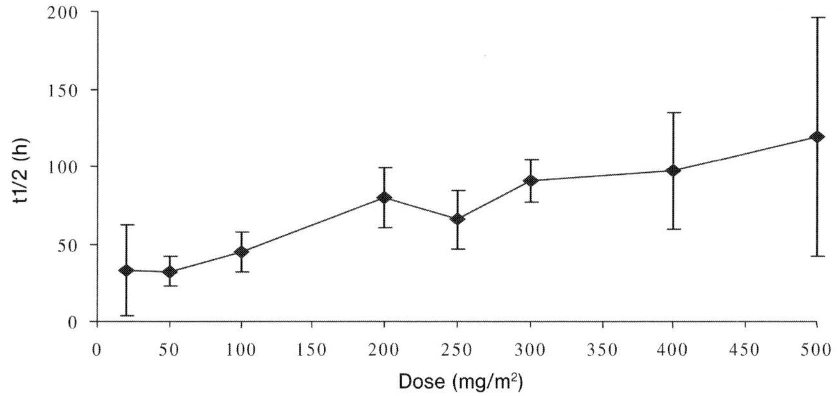


**Fig. 14.4** Mean (± SD) terminal half-life ($t_{1/2}$) for cetuximab versus dose across all studies.

#### 14.5.1.5 Volume of Distribution

Values for mean volume of distribution at steady state ($V_{SS}$) appeared to be independent of dose, and ranged from 1.99 to 2.88 L/m$^2$ after the first dose of 20 to 500 mg/m$^2$ of cetuximab (Fig. 14.5). These volumes are consistent with that of the vascular space, and in agreement with values of other IgG1-derived monoclonal antibodies.

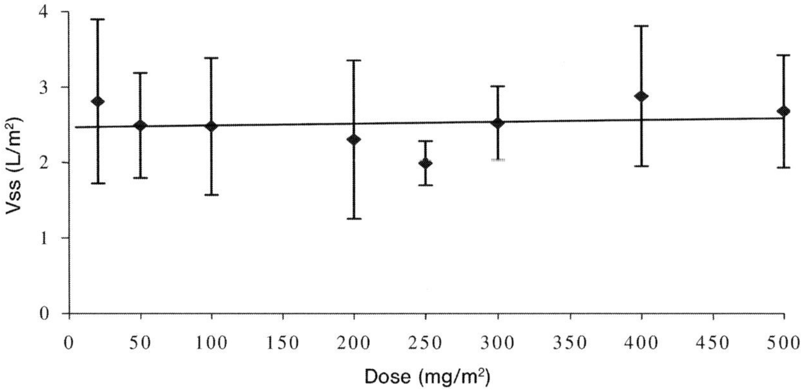


**Fig. 14.5** Mean (± SD) volume of distribution ($V_{ss}$) for cetuximab versus dose across all studies.

### 14.5.2
### Drug Metabolism and *in-vitro* Drug–Drug Interaction Studies

To date, no studies on the metabolism of cetuximab have been performed in humans or in animals. Indeed, metabolism studies are not generally performed for mAbs. Several pathways have been described that may contribute to antibody metabolism, all of which involve biodegradation of the antibody to smaller molecules (i. e., small peptides or amino acids). This fact has been recognized in the International Conference on Harmonization (ICH) guidance document "Preclinical Safety Evaluation of Biotechnology-Derived Pharmaceuticals" [22], where it is stated in Section 4.2.3 that "... the expected consequence of metabolism of biotechnology-derived pharmaceuticals is the degradation to small peptides and individual amino acids ..." and that therefore classical biotransformation studies as performed for traditional small molecule pharmaceuticals are not needed.

### 14.5.3
### Comparison of Single- and Multiple-Dose PK at the Approved Dosing Regimen

Mean values of CL, AUC, $t_{1/2}$, and $V_{ss}$ following the administration of up to four doses of cetuximab are displayed in Table 14.2. These data suggest that PK parameters remained constant for up to four cetuximab doses at the approved dosing regimen (400/250 mg/m$^2$).

In a recently completed Phase I study [23], cetuximab was administered to 49 patients with metastatic CRC. Treatment regimens included the approved dosing regimen (400 mg/m$^2$, followed by weekly doses of 250 mg/m$^2$), as well as two experimental regimens of 250 mg/m$^2$ weekly and 350 mg/m$^2$ weekly. The main PK parameters in week 4 of the study are listed in Table 14.3. The PK parameters observed in this study were in good agreement with previous data. In addition, values for CL, $V_{ss}$ and $t_{1/2}$ were in good agreement across the three treatment groups. Dose-related increases in exposure, as expected for linear PK characteris-

**Table 14.2** Pharmacokinetic parameters for cetuximab after multiple doses at the approved dosing regimen (400/250 mg/m²).

| Week | No. of studies | Statistic | CL [L/h/m²] | AUC [µg h/mL] | $t_{1/2}$ [h] | $V_{ss}$ [L/m²] |
|---|---|---|---|---|---|---|
| 1 | 8 | N | 53 | 53 | 53 | 53 |
| | | Mean | 0.022 | 21 142 | 97.24 | 2.88 |
| | | SD | 0.009 | 8657 | 37.38 | 0.93 |
| | | Range | 0.008–0.042 | 9412–48 039 | 41.39–213.38 | 1.51–6.19 |
| 3 | 2 | N | 8 | 8 | 8 | 8 |
| | | Mean | 0.020 | 22 723 | 123.25 | 2.30 |
| | | SD | 0.006 | 10 313 | 41.39 | 0.83 |
| | | Range | 0.002–0.019 | 12 291–40 850 | 81.92–187.87 | 1.33–3.91 |
| 4 | 3 | N | 13 | 11 | 11 | 11 |
| | | Mean | 0.017 | 24 329 | 108.09 | 2.00 |
| | | SD | 0.006 | 11 202 | 29.32 | 0.59 |
| | | Range | 0.008–0.028 | 11 704–47 180 | 74.97–173.16 | 1.11–2.68 |

**Table 14.3** Main pharmacokinetic parameters for cetuximab during week 4 of dosing.

| Dosing regimen (initial/weekly) | Statistic | CL [L/h/m²] | AUC [µg h/mL] | AUC dose-normalized [µg h/mL/mg] | $t_{1/2}$ [h] | $V_{ss}$ [L/m²] |
|---|---|---|---|---|---|---|
| 400/250 mg/m² | N | 11 | 11 | 11 | 11 | 11 |
| | Mean | 0.020 | 13 181 | 30.1 | 97 | 2.69 |
| | SD | 0.005 | 2560 | 6.9 | 24 | 0.47 |
| | Range | 0.014–0.030 | 8165–17 474 | 17.0–41.6 | 56–125 | 2.12–3.73 |
| 250/250 mg/m² | N | 17 | 17 | 17 | 17 | 17 |
| | Mean | 0.020 | 14 968 | 34.8 | 89 | 2.23 |
| | SD | 0.009 | 5792 | 15.1 | 32 | 0.70 |
| | Range | 0.010–0.043 | 5791–25 144 | 11.6–64.1 | 30–164 | 1.59–4.33 |
| 350/350 mg/m² | N | 17 | 17 | 17 | 17 | 17 |
| | Mean | 0.019 | 22 524 | 37.3 | 109 | 2.62 |
| | SD | 0.010 | 9044 | 16.1 | 32 | 0.77 |
| | Range | 0.008–0.051 | 6852–42 460 | 10.9–73.2 | 40–163 | 1.7–3.8 |

tics, were observed following the dosing regimens of 250/250 $mg/m^2$ weekly compared to 350/350 $mg/m^2$ weekly.

## 14.6 Characterization of Cetuximab PK by a Population PK Approach

Cetuximab concentration data from the 19 studies were combined into a common PK database for an integrated PK analysis [16]. This analysis was performed for the following purposes:

- to identify predictors of exposure to the drug (e.g., demographic factors, laboratory values, concomitant therapy, renal and hepatic status);
- to identify if any patient subpopulation exhibits altered PK;
- to estimate the inter-patient variability of cetuximab PK.

A total of 8388 observations (i.e., concentration values) from 906 patients was used for this analysis. The following covariates were investigated for a possible impact on the PK of cetuximab: age, gender, race, BSA, creatinine clearance, bilirubin, alanine aminotransferase, aspartate aminotransferase, concomitant therapy, hepatic status and renal status.

Population PK models were built using a nonlinear mixed-effect modeling approach with the NONMEM software (double precision, Version V, Level 1.1) and NMTRAN pre-processor [24]. Models were run using the Digital Visual Fortran Compiler (Version 5.0D) on a personal computer under the Microsoft Windows NT 4.0 operating system. The interface software PDx-Pop® was used to run NONMEM. Goodness-of-fit diagnostic plots were prepared within S-Plus 2000 Professional Release 3. Screening of potential covariates was conducted using the General Additive Modeling (GAM) feature of Xpose, version 3 [25].

In the initial model building, an initial dataset consisting of 5357 concentration values from 13 studies was used for the development of the structural base model. The following alternative models were assessed:

- a one-compartment model with Michaelis–Menten saturable elimination;
- a two-compartment model with first-order elimination;
- a two-compartment model with Michaelis–Menten saturable elimination;
- a two-compartment model with two parallel elimination pathways (one saturable and one nonsaturable pathway).

The selection of the appropriate population pharmacokinetic base model was guided by the following criteria: a significant reduction in the objective function value ($p < 0.01$, 6.64 points) as assessed by the Likelihood Ratio Test; the Akaike Information Criterion (AIC); a decrease in the residual error; a decrease in the standard error of the model parameters; randomness of the distribution of individual weighted residuals versus the predicted concentration and versus time post start of cetuximab administration; randomness of the distribution of the observed concentration versus individual predicted concentration values around the line of identity in a respective plot.

The first model tested was a one-compartment model with Michaelis–Menten elimination, which served as a base against which to compare more complex models. A marked decrease in the objective function value (OFV) was obtained when a two-compartment model with Michaelis–Menten elimination was fitted to the data ($\delta = -743$). The addition of a linear component to the elimination from the central compartment further reduced the OFV by an additional 245 points. The two-compartment models incorporating saturable elimination with and without the linear component were then further evaluated. Addition of the linear elimination pathway from the central compartment resulted in a decrease of 56 points in OFV compared to the model with saturable elimination only.

Both candidate models were then re-evaluated using the final dataset of 8388 observations from 906 patients of the 19 studies. With the final dataset, the addition of the linear elimination pathway resulted in only a single point reduction in OFV compared to the model with saturable elimination only. Therefore, a two-compartment model with saturable elimination was considered the final structural model and was used for the development of the covariate model (see Table 14.4).

**Table 14.4** Cetuximab population pharmacokinetic analysis: final model parameter estimates.

| Parameter | Typical value (%RSE)[a] |
|---|---|
| $V_1$ [L] | 4.54 (3.5) |
| Q [L] | 0.0493 (16.3) |
| $V_2$ [L] | 4.49 (11.7) |
| $V_{max}$ [mg/h] | 5.4 (19.1) |
| $K_m$ [μg/mL] | 91.5 (38.4) |

**a)** RSE: percent relative standard error of the estimate = SE/parameter estimate × 100.
*Abbreviations:* $V_1$: volume of central compartment; Q: inter-compartment clearance; $V_2$: volume of peripheral compartment; $V_{max}$: maximum elimination capacity; $K_m$: concentration at which half maximum capacity is achieved.

None of the tested covariates was identified as having a clinically relevant impact on the PK of cetuximab. Thus, changes in dose or dosing regimen do not appear to be necessary in any of the subpopulations defined by these covariates. However, it should be noted that the majority of the studied patients had adequate hepatic and renal function. Overall, more than 90% of the patients included in the PK database had normal hepatic function, more than 60% had normal renal function, and a further 32% had only mild impairment (creatinine clearance 50–80 mL/min). Hence, the effect of more severe renal or hepatic impairment on cetuximab PK remains to be elucidated.

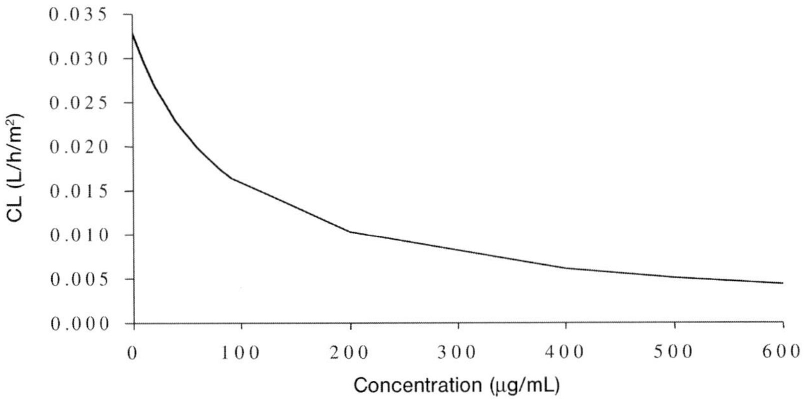


**Fig. 14.6** Simulation of clearance (CL) for cetuximab versus concentration, based on the results of the integrated population pharmacokinetic analysis.

Based on the integrated PK database analysis, CL values were estimated and are displayed in Fig. 14.6 as a function of the cetuximab serum concentrations. A concentration-dependent decrease in CL was observed in the integrated PK database analysis, similar to that observed in the non-compartmental analysis.

In summary, the pharmacostatistical model developed in the population PK analysis provided an excellent basis for the description of the cetuximab concentration data, and demonstrated that the PK of cetuximab are not likely to be influenced by intrinsic or extrinsic factors, as indicated by the results of the covariate analysis.

## 14.7
## Drug–Drug Interaction Studies

One PK interaction study was performed to investigate potential interactions of cetuximab with irinotecan and its active metabolites SN-38 and SN-38 glucuronide (SN-38G) [26]. In this study, 14 patients with advanced, EGFR-expressing, solid tumors were assigned consecutively to one of two treatment groups. Group A (n = 6) received a single dose of irinotecan in week 1, followed by three weeks of cetuximab at the approved dosing regimen. In week 4, another irinotecan dose was administered (i.e., irinotecan doses of 350 mg/m$^2$ at weeks 1 and 4, plus the approved cetuximab dosing regimen starting in week 2). PK data on irinotecan were determined from blood samples collected after administration of the drug at weeks 1 (irinotecan alone) and 4 (irinotecan plus cetuximab). Results from this group were used to investigate the impact of cetuximab on the PK of irinotecan. Group B (n = 8) received three weeks of cetuximab monotherapy at the approved dosing regimen, followed by one week of combination therapy (initial cetuximab dose of 400 mg/m$^2$ at week 1, followed by weekly doses of 250 mg/m$^2$ at weeks 2,

3, and 4, as well as a single dose of 350 mg/m$^2$ irinotecan at week 4). PK data on cetuximab were determined from blood samples collected at weeks 3 (cetuximab alone) and 4 (cetuximab plus irinotecan). Results from this group were used to investigate the possible impact of a single dose of irinotecan on the PK of cetuximab. The two treatment groups were similar with regard to demographic and baseline characteristics. Group A included three male and three female patients with a mean (± SD) age of 51.7 ± 15.2 years. Group B included three male and five female patients with a mean age of 52.1 ± 8.4 years.

The main PK parameters for irinotecan are summarized in Table 14.5. The PK parameters for irinotecan at week 1 were similar to those at week 4, suggesting the absence of a PK interaction with cetuximab. For the metabolite SN-38 the measured concentrations were in good agreement with those cited in the literature [27]; however, the number of samples with quantifiable concentrations was not sufficient to allow a meaningful analysis.

**Table 14.5** Pharmacokinetic parameters of irinotecan before and after cetuximab co-administration.

| Parameter | | Week 1 (irinotecan alone) | Week 4 (irinotecan+cetuximab) | Week 4 as % of Week 1[a)] |
|---|---|---|---|---|
| $C_{max}$ | N | 6 | 6 | 6 |
| [ng/mL] | Mean | 8129 | 6783 | 90 |
| | SD | 2882 | 1293 | 29 |
| | Range | 5150–13 407 | 5095–8799 | 48–127 |
| $AUC_{0-\infty}$ | N | 6 | 6 | 6 |
| [ng h/mL] | Mean | 44 243 | 40 394 | 96 |
| | SD | 23 683 | 18 365 | 21 |
| | Range | 24 820–80 875 | 20 515–74 046 | 64–123 |
| $t_{1/2}$ | N | 6 | 6 | 6 |
| [h] | Mean | 9.8 | 9.8 | 102 |
| | SD | 2.6 | 2.0 | 16 |
| | Range | 6.8–12.9 | 6.9–12.4 | 75–124 |
| CL | N | 6 | 6 | 6 |
| [L/h/m$^2$] | Mean | 9.7 | 10.0 | 107 |
| | SD | 4.2 | 4.3 | 26 |
| | Range | 4.3–14.1 | 4.4–17.1 | 81–151 |
| $V_{ss}$ | N | 6 | 6 | 6 |
| [L/m$^2$] | Mean | 83 | 85 | 106 |
| | SD | 21 | 15 | 21 |
| | Range | 57–117 | 68–105 | 71–130 |

**a)** Mean of individual ratios.

The main PK parameters for cetuximab are summarized in Table 14.6. The PK parameters for cetuximab in week 3 were similar to those in week 4, suggesting the absence of a PK interaction with irinotecan.

The possible impact of co-administered chemotherapies and radiation therapy on the PK of cetuximab was furthermore assessed using the population PK approach, as described in Section 14.6. The co-administered chemotherapies included cisplatin, carboplatin, paclitaxel, doxorubicin, irinotecan, and gemcitabine. The results of the analysis indicate that neither the co-administered chemotherapies nor radiation therapy had a significant impact on the PK of cetuximab. This finding suggests that the potential for PK-based drug–drug interactions with cetuximab is low.

**Table 14.6** Pharmacokinetic parameters of cetuximab before and after irinotecan co-administration.

| **Parameter** | | **Week 3 (cetuximab alone)** | **Week 4 (cetuximab+irinotecan)** | **Week 4 as % of week 3[a)]** |
|---|---|---|---|---|
| $C_{max}$ | N | 7 | 7 | 7 |
| [μg/mL] | Mean | 153 | 162 | 106 |
| | SD | 38 | 43 | 11 |
| | Range | 112–225 | 115–225 | 87–122 |
| $AUC_{0-\infty}$ | N | 7 | 7 | 7 |
| [μg h/mL] | Mean | 13 039 | 14 923 | 117 |
| | SD | 4783 | 5029 | 14 |
| | Range | 6234–19 019 | 8918–22 386 | 102–143 |
| $t_{1/2}$ | N | 7 | 7 | 7 |
| [h] | Mean | 119 | 117 | 107 |
| | SD | 42 | 32 | 38 |
| | Range | 82–188 | 85–173 | 51–167 |
| CL | N | 7 | 7 | 7 |
| [$L/h/m^2$] | Mean | 0.020 | 0.018 | 91 |
| | SD | 0.006 | 0.007 | 8 |
| | Range | 0.013–0.027 | 0.011–0.028 | 80–103 |
| $V_{ss}$ | N | 7 | 7 | 7 |
| [$L/m^2$] | Mean | 2.07 | 1.89 | 92 |
| | SD | 0.55 | 0.55 | 13 |
| | Range | 1.33–2.82 | 1.11–2.48 | 67–106 |

**a)** Mean of individual ratios.

## 14.8 Conclusions

The PK of cetuximab after doses ranging from 5 to 500 $mg/m^2$ are well described in a broad range of studies and tumor types. Largely linear relationships with dose were observed for $C_{max}$ and $AUC_{0-\infty}$ after single-dose administration, indicating that the exposure to cetuximab is predictable at doses of 200 $mg/m^2$ and higher. The volume of distribution is observed to be independent of dose and consistent with a distribution of cetuximab in the vascular space. Dose-dependent relationships were observed for $t_{1/2}$ and CL, predominantly at lower doses. At concentrations following the approved dosing regimen for cetuximab, CL values are constant (0.02 $L/h/m^2$), indicating predictable PK in this dose range. PK parameters obtained after a three-week sampling period were in good agreement to PK data obtained after a one-week sampling schedule. The possibility of PK-based drug–drug interactions with cetuximab is low, which increases the potential of cetuximab to be combined with a variety of chemotherapeutic agents, without the need for a change in its approved dosing regimen. Furthermore, based on the PK characteristics of cetuximab, dose adjustments in special subpopulations do not appear necessary, as indicated by the results of the population PK analysis.

It should be noted that, as a protein, cetuximab has the potential to induce an immune response in individuals who receive it as a therapeutic agent. During the cetuximab clinical development program, patient sera were monitored for induction of an anti-cetuximab or human anti-chimeric antibody (HACA) response. Although anti-cetuximab antibodies were identified in some patients, they were found not to interfere with the biological properties of cetuximab, such as its ability to bind EGFR (i. e., they are non-neutralizing).

Currently, clinical studies are ongoing in order to investigate modifications to the approved dosing regimen for cetuximab. Of particular interest are changes in dosing frequency, from a weekly to a bi-weekly regimen, in order to harmonize dosing with that of other treatment regimens involving chemotherapeutic agents such as 5-fluorouracil, folinic acid, irinotecan, or oxaliplatin. These initial studies are designed to identify bi-weekly cetuximab doses that result in similar efficacy, safety and PK profiles, compared to those observed for the approved cetuximab dosing regimen.

## 14.9 References


1 Baselga, J. **2001**. The EGFR as a target for anticancer therapy – focus on cetuximab. *Eur. J. Cancer* 37:S16–S22.

2 Brandt, R., R. Eisenbrandt, Leenders, W. Zschiesche, B. Binas, C. Juergensen, and F. Theuring. **2000**. Mammary gland specific hEGF receptor transgene expression induces neoplasia and inhibits differentiation. *Oncogene* 19:2129–2137.

3 Dassonville, O., J.L. Formento, M. Francoual, A. Ramaioli, J. Santini, M. Schneider, F. Demard, and G. Milano. **1993**. Expression of epidermal growth factor receptor and survival in upper aerodigestive tract cancer. *J. Clin. Oncol.* 11:1873–1878.

4 Yarden Y., and M.X. Sliwkowski. **2001**. Untangling the ErbB signalling network. *Nat. Rev. Mol. Cell Biol.* 2: 127–137.

5 Chau, I., and D. Cunningham. **2002**. In: M.C. Perry (Ed.), *Adjuvant chemotherapy in colon cancer: state of the art.* ASCO educational book, 38th Annual Meeting, ASCO Alexandria,VA, USA, pp. 228–239.

6 Goldstein, N.S. and M. Armin. **2001**. Epidermal growth factor receptor immunohistochemical reactivity in patients with American Joint Committee on stage IV colon adenocarcinoma: implications for a standardized scoring. *Cancer* 92:1331–1346.

7 Saltz, L. B., N.J. Meropol, P.J. Loehrer, M.N. Needle, J. Kopit, and R.J. Mayer. **2004**. Phase II trial of cetuximab in patients with refractory colorectal cancer that expresses the epidermal growth factor receptor. *J. Clin. Oncol.* 22:1201–1208.

8 Cunningham, D.,Y. Humblet, S. Siena, D. Khayat, H. Bleiberg, A. Santoro, D. Bets, M. Mueser, A. Harstrick, C. Verslype, I. Chau, and E. van Cutsem. **2004**. Cetuximab monotherapy and cetuximab plus irinotecan in irinotecan-refractory metastatic colorectal cancer. *N. Engl. J. Med.* 351:337–345.

9 Hemming, A.W., N.L. Davis, A. Kluftinger, B. Robinson, N.F. Quenville, B. Liseman, and J. LeRiche. **1992**. Prognostic markers of colorectal cancer: An evaluation of DNA content, epidermal growth factor receptor, and Ki-67. *J. Surg. Oncol.* 51:147–152.

10 Mayer, A., M. Takimoto, E. Fritz, G. Schellander, K. Kopfler, H. Ludwig. **1993**. The prognostic significance of proliferating cell nuclear antigen, epidermal growth factor receptor, and mdr gene expression in colorectal cancer. *Cancer* 71: 2454–2460.

11 Messa, C., F. Russo, M.G. Caruso, and A. DiLeo. **1998**. EGF,TGF-, and EGF-R in human colorectal carcinoma. *Acta Oncol.* 37:285–289.

12 Kovar, A., and B. Meibohm. **2004**. Drug development in oncology. In: P. Bonate and D. Howard (Eds.), *Pharmacokinetics in Drug Development. Regulatory and Development Paradigms.* Volume 2. American Association of Pharmaceutical Scientists Press, Arlington,VA, pp. 281–304.

13 Hansen, H. **1970**. Clinical experience with 1-(2-chloroethyl)3-cyclohexyl-1-nitrosurea. *Proc. Am. Assoc. Cancer Res.* 1:43.

14 Von Hoff, D.D., J. Kuhn, and G. Clark. **1983**. Design and conduct of Phase 1 trials. In: M. Buyse, M. Staquet, and R. Sylvester (Eds.), *Cancer Clinical Trials.* Oxford University Press, Oxford, UK, pp. 210–220.

15 Pinkel, D. **1958**. The use of body surface area as a criterion of drug dosage in cancer chemotherapy. *Cancer Res.* 18:853–856.

16 Fox, F.E, D. Mauro, S. Bai, R. Raymond, C. Farrell, H. Pentikis, A. Nolting, M. Birkhofer, and M. Needle. **2004**. A population pharmacokinetic (PPK) analysis of the anti-EGFR specific IgG1 monoclonal antibody cetuximab. Poster at the Annual ASCO Gastrointestinal Cancers Symposium, January 22–24, A290.

17 Cunningham, D.,Y. Humblet, S. Siena, D. Khayat, H. Bleiberg, A. Santoro, D. Bets, M. Mueser, A. Harstrick, C. Verslype, I. Chau, and E. van Cutsem. **2004**. Cetuximab monotherapy and cetuximab plus irinotecan in irinotecan-refractory metastatic colorectal cancer. *N. Engl. J. Med.* 351:337–345.

**18** Baselga, J., D. Pfister, M.R. Cooper, R. Cohen, B. Burtness, M. Bos, G. D'Andrea, A. Seidman, L. Norton, K. Gunnet, J. Falcey, V. Anderson, H. Waksal, and J. Mendelsohn. **2000**. Phase 1 Studies of anti-epidermal growth factor receptor chimeric antibody C225 alone and in combination with cisplatin. *J. Clin. Oncol.* 18 : 904–914.

**19** Nolting, A., F.E. Fox, D. Mauro, and A. Kovar. **2004**. Pharmacokinetics of cetuximab (Erbitux®) after single and multiple intravenous doses in cancer patients. *AAPS Journal* 6 : 4046.

**20** Makiya, R. and T. Stigbrand. **1992**. Placental alkaline phosphatase is related to human IgG internalization in HEp2 cells. *Biochem. Biophys. Res. Commun.* 182 : 624–630.

**21** Stone D.L., Y. Suzuke, and G.W. Wood. **1987**. Human amnion as a model for IgG transport. *Am. J. Reprod. Immunol. Microbiol.* 13(2):36–43.

**22** International Conference on Harmonization (ICH). **1997**. *Topic S6: Note for Guidance on Preclinical Safety Evaluation of Biotechnology-Derived Pharmaceuticals.* Dated July 16. www.ich.org.

**23** Humblet, Y., M. Peeters, H. Bleiberg, R. Stupp, C. Sessa, A. Roth, J. Nippgen, A. Nolting, P. Stuart, and G. Giaccone. **2005**. An open-label, phase I study of cetuximab to assess the safety, efficacy and pharmacokinetics (PK) of different cetuximab regimens in patients with epidermal growth factor receptor (EGFR)-expressing metastatic colorectal cancer (mCRC). *Proc. ASCO* 23 : 3632.

**24** Beal, S.L. and Sheiner, L.B., NONMEM Users Guides – Part I-VIII. NONMEM ProSject Group C255, University of California at San Francisco, San Francisco, 1988–1998.

**25** Jonsson, E.N. and M.O. Karlsson. **1999**. Xpose – an S-PLUS-based population pharmacokinetic/pharmacodynamic model building aid for NONMEM. *Comp. Methods Prog. Biomed.* 58 : 51–64.

**26** Delbaldo, C., J.-Y. Pierga, V. Dieras, S. Faivre, V. Laurence, J.-C. Vedovato, M. Bonnay, M. Mueser, A. Nolting, A. Kovar, and E. Raymond. **2005**. Pharmacokinetic profile of cetuximab (Erbitux™) alone and in combination with irinotecan in patients with advanced EGFR-positive adenocarcinoma. *Eur. J. Cancer* 41 : 1739–1745.

**27** Rivory, L.P., M.C. Haaz, P. Canal, F. Lokiec, J.P. Armand, and J. Robert. **1997**. Pharmacokinetic interrelationships of irinotecan (CPT-11) and its three major plasma metabolites in patients enrolled in phase I/II trials. *Clin. Cancer Res.* 3 : 1261–1266.

# 15
# Integration of Pharmacokinetics and Pharmacodynamics Into the Drug Development of Pegfilgrastim, a Pegylated Protein

*Bing-Bing Yang*

## 15.1
## Introduction

Cytotoxic chemotherapy suppresses the growth and proliferation of rapidly dividing neoplastic cells. However, traditional chemotherapy does not discriminate between cancerous and rapidly dividing noncancerous cells. Noncancerous cells include white blood cells and their progenitor cells in the bone marrow, the body's primary defense against infection. As such, one of the most serious side effects of many types of chemotherapy is a low white blood cell count. Chemotherapy agents that cause this side effect are termed "myelosuppressive", as they suppress the production of white blood cells in the bone marrow. Neutrophils account for 50–70 % of all white blood cells, and a low neutrophil count is referred to as "neutropenia". Chemotherapy-induced neutropenia is associated with several adverse consequences such as febrile neutropenia (absolute neutrophil count (ANC) $<0.5 \times 10^9$/L and temperature $\geq 38.2\,^\circ$C for >1 h) that requires hospitalization, infection, and infection-related mortality; it can also compromise the delivery of the planned chemotherapy dose on schedule and thus the efficacy of the treatment [1, 2]. These complications adversely influence the quality of life of affected patients. Following the advent of recombinant biotechnology, the use of hematopoietic growth factors is now integral to managing neutropenia that can lead to complications from infection, which are the most common dose-limiting adverse events during cancer chemotherapy.

Granulocyte colony-stimulating factor (G-CSF) is an endogenous hematopoietic growth factor that selectively stimulates granulopoietic cells of the neutrophil lineage [3]. It acts at all stages of neutrophil development, and is a very potent late-acting growth factor, increasing the proliferation and differentiation of neutrophils from committed progenitor cells. G-CSF also enhances the survival and function of mature neutrophils by increasing phagocytic activity, antimicrobial killing, and antibody-dependent cell-mediated cytotoxicity [4].

Filgrastim (Neupogen®; Amgen Inc.) is a recombinant human methionyl G-CSF that is produced in genetically modified *Escherichia coli* bacteria [5–7]. It is

*Pharmacokinetics and Pharmacodynamics of Biotech Drugs: Principles and Case Studies in Drug Development.* Edited by Bernd Meibohm

ISBN: 3-527-31408-3

composed of 175 amino acids in a sequence that is identical to that of endogenous G-CSF, a glycoprotein with 204 amino acids, 30 of which are a signal sequence and are removed from the secreted form. Because it is produced in *E. coli*, filgrastim is nonglycosylated; however, it has a methionine molecule added to the N-terminus to add stability in the bacterial expression system. The biologic activities of filgrastim are identical to those of the endogenous molecule. Since its approval for marketing in the United States in 1991, filgrastim has been widely used by patients with cancer who receive myelosuppressive chemotherapy or bone marrow transplantation, patients with acute myeloid leukemia receiving induction or consolidation chemotherapy, patients undergoing peripheral blood progenitor cell collection and therapy, and patients with severe chronic neutropenia [8, 9].

Filgrastim is safe and effective in reducing the severity and duration of neutropenia and its complications, thus making possible the delivery of full-dose and high-dose chemotherapy; however, its short circulating half-life necessitates that it be given daily. Amgen Inc. therefore had an opportunity to improve the product by reducing the frequency of injections while maintaining the therapeutic effect. A more specific goal was to develop a molecule that had filgrastim as the backbone because of its history of safety and efficacy. Further, the molecule had to have no greater toxicity but would provide similar efficacy with just one dose per chemotherapy cycle, offering the benefits of greater patient compliance, uninterrupted therapy when patients cannot visit the clinic, convenience, and potential cost savings – in general, simplifying the management of chemotherapy-induced neutropenia for both patients and healthcare providers.

After many second-generation candidates had been screened, pegfilgrastim was chosen for continued development [10–12], and its sustained activity has been demonstrated in preclinical and clinical studies. Pivotal trials showed that the reduction in the duration of severe neutropenia with a single injection of pegfilgrastim was clinically and statistically equivalent to that with approximately 11 daily injections of filgrastim per chemotherapy cycle. Pegfilgrastim has been approved and is now marketed worldwide under the trade names Neulasta® and Neupopeg®. This chapter describes how pharmacokinetic (PK) and pharmacodynamic (PD) data were used in the development strategy that brought this molecule to the market.

## 15.2
## Overview of Filgrastim Pharmacokinetics

The pharmaceutical industry commonly utilizes two main strategies in designing a long-acting form of a drug:

- slowing its absorption (e. g., by sustained-release formulations), and
- reducing its clearance (e. g., by conjugating it to inert polymers).

As the latter approach has been the most common and practical approach for modifying recombinant therapeutic proteins, Amgen Inc. elected this strategy to engineer a long-acting filgrastim. Therefore, an understanding how filgrastim is cleared from the body was essential.

The PK of filgrastim have been extensively reviewed [13, 14]. As the dose of filgrastim given by intravenous (IV) or subcutaneous (SC) administration increases, its clearance in animals and humans decreases. The saturable clearance pathway for filgrastim has been attributed to neutrophil-mediated clearance, presumably involving G-CSF receptors on neutrophils and neutrophil precursors. Internalization and degradation of G-CSF derivatives by neutrophils and bone marrow *in vitro* have been suggested for a mutein G-CSF derivative, nartograstim [15–17]. Several studies have shown that the serum concentrations of filgrastim or lenograstim, another recombinant human G-CSF, are inversely correlated with the number of circulating neutrophils [18–22]. The clearance of lenograstim was shown to be closely related not only to the number of circulating neutrophils but also to the percentage of G-CSF receptor-positive neutrophils [23]. Therefore, it was suggested that the neutrophil-mediated clearance of recombinant human G-CSF is mediated by G-CSF receptors on neutrophils and neutrophil precursors rather than by nonspecific endocytosis by neutrophils. G-CSF receptors are normally present on myeloid progenitor cells in the bone marrow and on peripheral neutrophils [24].

Compartmental analyses of filgrastim PK data collected from bilateral nephrectomized and sham-operated rats also suggested that, in addition to neutrophil-mediated clearance, two linear clearance pathways for filgrastim are present [25], one of which has been attributed to renal clearance. Filgrastim, which has a molecular weight of 18.8 kDa, is readily excreted by the kidney. The clearance of filgrastim is lower in nephrectomized rats than in sham-operated rats [25, 26]. It is not clear what constitutes the second linear clearance pathway, though proteases and soluble G-CSF receptors might play a role. Nevertheless, the contribution of the second linear pathway to the total clearance is very minor.

Based on the clearance mechanisms for filgrastim, three options seemed to be available for modifications to engineer a long-acting filgrastim. However, the pharmacologic effect of filgrastim is to stimulate the production of neutrophils and neutrophil precursors; as such, neutrophil-mediated clearance was not considered a viable pathway to modulate. Next, as stated previously, the mechanism of the nonspecific second linear pathway is not known and plays a very minor role in the clearance of filgrastim. This left altering the renal clearance the most practical option for developing a long-acting form of filgrastim. Interestingly, the impactful ramifications resulting from a reduction in the renal clearance of the drug were unrealized at that time.

## 15.3
## The Making of Pegfilgrastim

The covalent attachment of polyethylene glycol (PEG) polymers to therapeutic drugs (pegylation) is one of the most widely used and well-established engineering technologies to enhance the performance of biotech drugs (see Chapter 11) [27, 28]. The unique physico-chemical properties of PEG polymers result in several benefits observed in pegylated therapeutic proteins, including lower toxicity, increased drug stability and solubility, and, under certain instances, increased therapeutic efficacy through reduction or elimination of immunogenicity. Another benefit – and one which is most relevant to the development of pegfilgrastim – is a reduced drug clearance that allows for a less-frequent dosing schedule. The attachment of PEG moieties to a protein will result in a molecule with a reduced renal clearance. Furthermore, the shielding effect of the PEG moieties leads not only to decreased immunogenicity but also to reduced proteolysis for these bioconjugates. Together, these properties increase the circulation residence time of pegylated therapeutic proteins.

The reduction in renal clearance stems from the ability of the PEG moiety to become highly hydrated in aqueous solution (partially because of chain flexibility and extensive hydration) [11]. It has been shown that the molecular size increases more than proportionally with increasing numbers of 5-kDa PEG polymers attached to the native protein [29]. Studies of PEG polymers in solution have shown that each ethylene oxide unit is tightly associated with two to three water molecules.

Various pegylated derivatives of filgrastim were produced by modifying filgrastim with PEGs of different molecular weights, different numbers of PEGs per molecule, and with linear or branched PEGs [11]. The candidates were first screened in a battery of *in-vitro* assays and then ranked *in vivo* on the basis of the elicited PD data (ANC) in the normal mouse model. Because the responses in normal mice could not distinguish many candidates, and because the pegylated filgrastim was to be developed first for the treatment of chemotherapy-induced neutropenia, the fluorouracil-treated mouse model was used to further rank the candidates.

Pegfilgrastim had the most promising profile, and was chosen for subsequent development. It is produced by covalently attaching a single 20-kDa linear, monofunctional monomethoxy-PEG molecule to the N-terminus of filgrastim (Fig. 15.1). It was expected that targeting the amino end with pegylation would minimize any potential interference with G-CSF receptor interaction, not unduly upset the tertiary structure of the molecule, and produce a consistent and defined product [12]. *In-vitro* studies have shown that the biologic activity and mechanism of action of pegfilgrastim are identical to those of filgrastim; however, *in- vivo* testing in mice revealed that pegfilgrastim increases ANC for a substantially longer period of time than filgrastim [30].

## 15.4
## Preclinical Pharmacokinetics and Pharmacodynamics of Pegfilgrastim

Dose-ranging PK and PD studies of pegfilgrastim have been conducted in mice, rats, and monkeys. Similar to filgrastim, the PK of pegfilgrastim are nonlinear after SC administration; the clearance of pegfilgrastim decreases with increasing

**Fig. 15.1** Structure of pegfilgrastim. A relatively compact protein (filgrastim) is shown at the top of this molecule model. Despite having a similar molecular weight to filgrastim (18.8 kDa), the PEG moiety (20 kDa), shown at the bottom, is loosely hydrated and occupies a relatively large volume. (Adapted with permission from [12]).

doses. The saturable clearance of pegfilgrastim is also assumed to be attributed to neutrophil-mediated clearance.

When pegfilgrastim was given every other day to rats, or once weekly to cynomolgus monkeys, the plasma concentrations of pegfilgrastim were lower after the last dose than after the first (Fig. 15.2), indicating that clearance increased with repeated dosing. In addition, the ANC was higher after the last dose than after the first. The parallel changes in pegfilgrastim clearance and ANC suggested that the expansion of neutrophil and neutrophil precursor mass resulted in greater neutrophil-mediated clearance with time. The time-dependent nonlinearity was more apparent with lower doses (50–100 μg/kg), when neutrophil-mediated clearance predominates, than with higher doses (500–1000 μg/kg), when it is saturated. As a result of the increase in neutrophil-mediated clearance with repeated dosing, no accumulation of pegfilgrastim was observed.

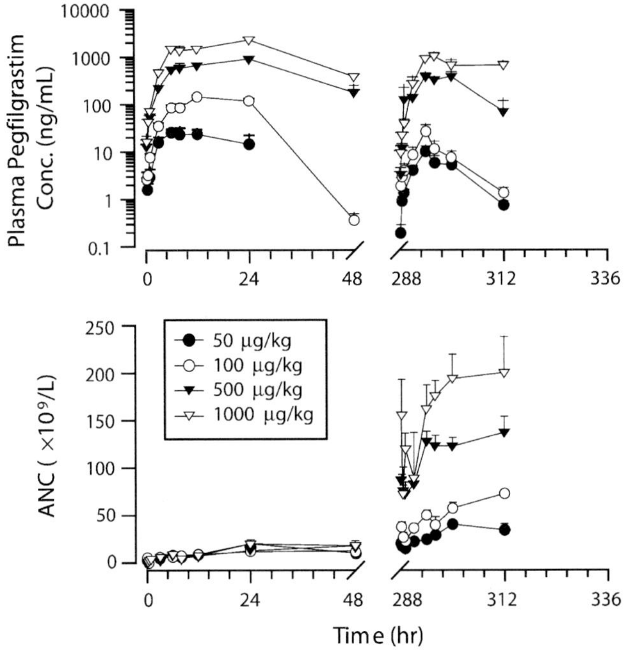


**Fig. 15.2** Mean (+ SE) plasma pegfilgrastim concentration–time and ANC–time profiles in rats after subcutaneous administration of pegfilgrastim every other day for two weeks.

The role of neutrophil-mediated clearance in pegfilgrastim clearance was further investigated in a study in G-CSF receptor-knockout mice. The murine G-CSF receptor gene consists of 17 exons, and G-CSF receptor-knockout mice lack exons 3 to 8, which results in a nonfunctional G-CSF receptor protein. Mice were given a single dose of filgrastim or pegfilgrastim 10 or 100 µg/kg by the IV route; this route was used to avoid complications of the absorption process in interpreting the PK and PD data subsequently collected. Drug exposure was significantly higher in the G-CSF receptor-knockout mice than in wild-type mice [31], regardless of which dose level was administered, or whether the mice received filgrastim or pegfilgrastim. These findings suggested the importance of functional G-CSF receptors in the clearance of pegfilgrastim and filgrastim.

Because the strategy for increasing the residence time of filgrastim was reducing or eliminating its renal clearance, the role of the kidney in the clearance of pegfilgrastim was evaluated in bilateral nephrectomized and sham-operated rats; filgrastim was used as a comparator [25]. Bilateral nephrectomy resulted in decreased clearance of filgrastim by 60–75%; however, exposure to pegfilgrastim was similar in sham-operated and bilateral nephrectomized rats, suggesting that the kidney has a very minor role in the clearance of pegfilgrastim (Fig. 15.3). The almost negligible contribution of the kidney to the elimination of pegfilgrastim is most likely related to its hydrodynamic radius. The molecular weight of

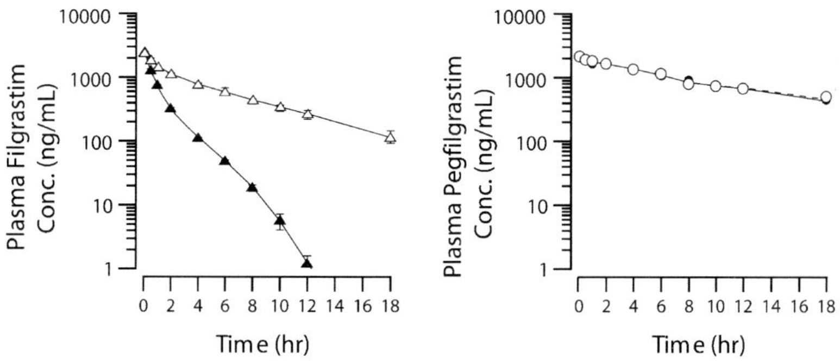


**Fig. 15.3** Plasma concentrations of filgrastim (left) and pegfilgrastim (right), both given at 100 µg/kg in sham-operated rats (closed symbols) and bilateral nephrectomized rats (open symbols). (Adapted with permission from [25]).

pegfilgrastim (38.8 kDa) is only approximately twice that of filgrastim, but its hydrodynamic radius is approximately 4.5 times as large (data on file at Amgen Inc.).

The results from the nephrectomy study in rats suggested that the goal of eliminating the renal clearance of pegfilgrastim had been achieved. What also began to become clear was the importance of leaving the neutrophil-mediated clearance pathway to predominate with little to no renal clearance. These characteristics would allow the molecule to have a very efficient self-regulating clearance mechanism; that is, the molecule could increase the neutrophil count, which in turn would clear the drug from the circulation when the body no longer needed it. To confirm this supposition for the second-generation filgrastim molecule selected for development, clinical trials as well as PK and PD modeling were conducted.

## 15.5 Pharmacokinetic and Pharmacodynamic Modeling

The PK and ANC data obtained after the administration of pegfilgrastim to cynomolgus and rhesus monkeys were modeled to characterize the concentration–effect relationships [32]. Normal cynomolgus monkeys were given a single SC bolus dose of pegfilgrastim (100, 300, or 1000 µg/kg), while normal rhesus monkeys were given a single dose of pegfilgrastim (10 or 100 µg/kg) by either the IV or SC route.

Pegfilgrastim plasma concentrations were modeled using a two-compartment model with a delayed, first-order absorption process. The clearance of pegfilgrastim was described by parallel linear and neutrophil-mediated clearance pathways:

$$\mathrm{CL} = \frac{k_{\mathrm{cat}} \cdot \dfrac{\mathrm{ANC}}{\mathrm{ANC}_{\mathrm{baseline}}}}{k_{\mathrm{m}} + C} + \mathrm{CL}_2$$

The neutrophil-mediated clearance was expressed as a Michaelis–Menten function, because G-CSF receptor binding is subject to saturation. The product of $k_{cat}$ and the ratio of ANC and ANC at baseline is the maximum velocity of drug elimination from this pathway, $k_m$ is the Michaelis constant, and $C$ is the pegfilgrastim concentration. $CL_2$ denotes the linear clearance pathway.

A maturation-structured cytokinetic model was developed to describe the granulopoietic effects of pegfilgrastim and the feedback regulation of pegfilgrastim clearance (Fig. 15.4). This model consisted of an influx function into the postmitotic precursor pool that was expanded by plasma pegfilgrastim. It was also assumed that pegfilgrastim increased the rate of influx of marrow band cells and mature neutrophils into the circulation. The baseline residence times of postmitotic cells in the marrow (metamyelocytes, band cells, and segmented neutrophils) were fixed in the model based on published data [33]. The effects of pegfilgrastim on neutrophil margination (adhesion of peripheral neutrophils to blood vessels) were described by changes in the neutrophil dilution volume. Concentration–effect relationships were described by a saturable, simple $E_{max}$ function:

$$\text{Effect} = \frac{E_{max} \cdot C}{EC_{50} + C}$$

where $E_{max}$ is the maximum granulopoietic effect, and $EC_{50}$ is the concentration of pegfilgrastim that produced half-maximum stimulation of granulopoiesis. The $E_{max}$ model was applied to the mitotic and maturational effects of pegfilgrastim, with the assumption of the same $EC_{50}$ and a different $E_{max}$ for each process.

This PK/PD model was fit to the individual and naive-averaged PK profiles from monkeys. The $k_m$ for the nonlinear pathway was 5.3 ng/mL for cynomolgus

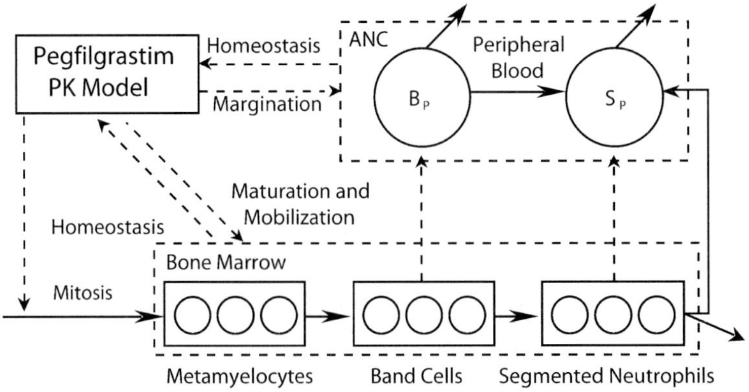


**Fig. 15.4** Pharmacodynamic model describing the granulopoietic effects of pegfilgrastim. Concentrations of pegfilgrastim stimulate mitosis and mobilization of band cells and segmented neutrophils in bone marrow and affect margination of the peripheral blood band cell ($B_P$) and segmented neutrophil ($S_P$) populations, which comprise the total absolute neutrophil count (ANC). Changes in neutrophil counts in peripheral blood provide feedback regulation of pegfilgrastim clearance. (Model provided courtesy of Dr. Lorin Roskos).

monkeys, and 5.7 ng/mL for rhesus monkeys. Pharmacodynamic modeling of concentration–effect relationships produced $EC_{50}$ values of 2.3 ng/mL for cynomolgus monkeys and 3.6 ng/mL for rhesus monkeys. The $k_m$ and $EC_{50}$ estimates were comparable, suggesting that nonlinear clearance and granulopoietic effect of pegfilgrastim are both mediated by the same receptor (the G-CSF receptor). It was then hypothesized that the plasma concentration of pegfilgrastim relative to the $k_m$ and $EC_{50}$ may be used as an indicator of receptor occupancy and a predictor of the PD response.

In normal cynomolgus monkeys given pegfilgrastim 100 or 300 µg/kg SC, the drug plasma levels fell below the targeted range (2–5 ng/mL) within 3–5 days after dosing. Since a neutropenic monkey model was not readily available during the early development of pegfilgrastim, simulations were conducted to predict whether the concentrations of pegfilgrastim could be maintained above the targeted range for a longer time during chemotherapy-induced neutropenia. The assumptions of the simulations were that: (1) the baseline PK parameters were identical in neutropenic and normal monkeys; (2) the ANC profile in neutropenic monkeys was similar to that in the placebo cohort in a clinical study of filgrastim; and (3) the antiproliferative effects of myelosuppressive chemotherapy were incorporated in the model at the level of mitosis. The simulations predicted that effective concentrations of pegfilgrastim might be maintained for 7–9 days during severe neutropenia in monkeys given a single 100- or 300-µg/kg SC dose (Fig. 15.5).

Most importantly, these simulations suggested that the clearance of a single dose of pegfilgrastim given after the chemotherapy would be negligible during the period of neutropenia, and that the plasma levels would be sustained until the onset of ANC recovery. It was expected therefore that, because of its highly efficient self-regulating clearance, a single dose of pegfilgrastim could be given once per chemotherapy cycle. The results from these simulations were provided as a product rationale in the pegfilgrastim Investigational New Drug Application submitted to the Food and Drug Administration (BB IND 7701). The prospective simulation was later validated with data from neutropenic monkeys [34] and from clinical studies in which patients received myelosuppressive chemotherapy.

## 15.6 Clinical Pharmacokinetics and Pharmacodynamics of Pegfilgrastim

The first clinical study of pegfilgrastim was conducted in patients with non-small cell lung cancer (NSCLC) who were given pegfilgrastim before chemotherapy (cycle 0) and after chemotherapy (cycle 1) to determine the safety and PK/PD properties of pegfilgrastim over a range of doses [35]. In cycle 0, chemotherapy-naive patients were randomized in a 3:1 ratio within a pegfilgrastim dose cohort to either a single SC dose of pegfilgrastim or daily SC doses of filgrastim 5 µg/kg for 5 days or until the ANC was $\geq 75 \times 10^9$/L, whichever occurred first. Three pegfilgrastim cohorts (30, 100, and 300 µg/kg) were sequentially enrolled; the doses of pegfilgrastim were chosen on the basis of the simulation results described in the pre-

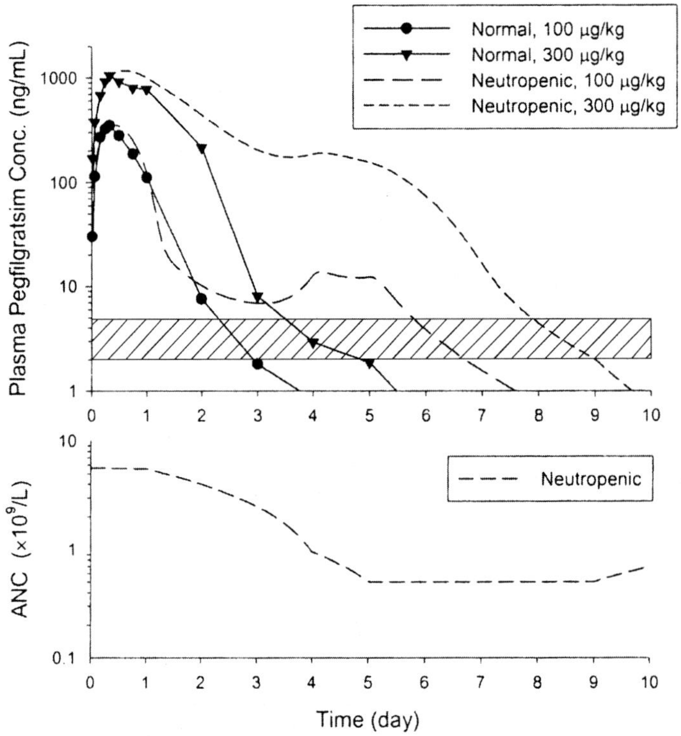


**Fig. 15.5** Upper: Prospective simulation of sustained plasma levels of pegfilgrastim in simulated neutropenic monkeys. Symbols indicate observed values in normal monkeys; dashed lines indicate simulated values during chemotherapy-induced neutropenia. The shaded area represents the targeted concentrations for granulopoietic response to pegfilgrastim. Lower: Assumed neutropenic ANC profile used in simulation to modulate neutrophil-mediated clearance of pegfilgrastim. (Data from Amgen Inc. BB IND 7701).

vious section. After a 14-day washout period (on day 15, the start of cycle 1), patients were given carboplatin (AUC 6) by a 30-min IV infusion and paclitaxel (225 mg/m$^2$) by a 24-h IV infusion. At 24 h after completion of the chemotherapy (day 17), patients were given the same dose of the study drug as in cycle 0; daily filgrastim was given until ANC was $\geq 10 \times 10^9$/L after the expected ANC nadir.

No dose-limiting adverse events were noted in this study. The results demonstrated the temporal PK/PD relationship of pegfilgrastim before and after chemotherapy (Fig. 15.6). The PK of pegfilgrastim appeared nonlinear in both cycles; the apparent time-averaged clearance of pegfilgrastim (estimated by the dose divided by the area under the pegfilgrastim concentration–time curve, AUC) decreased with increasing doses. Before chemotherapy, the median ANC in patients receiving pegfilgrastim increased in a dose-dependent fashion. In addition, the ANC response after a single dose of pegfilgrastim was sustained longer than that after five daily doses of filgrastim. After chemotherapy, neutropenia occurred as expected. The median ANC profiles were similar in the pegfilgrastim 30-µg/kg

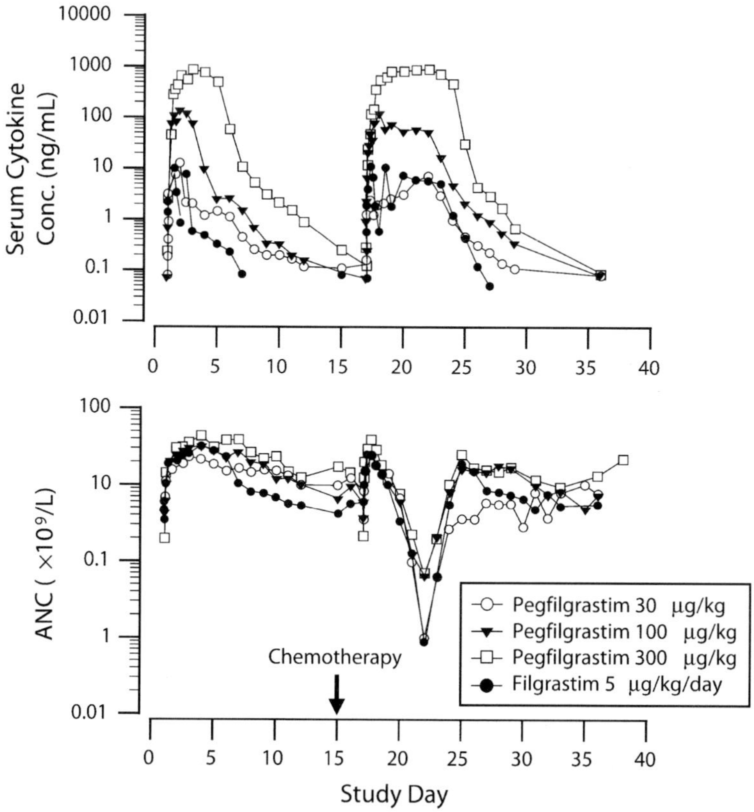


**Fig. 15.6** Median cytokine (pegfilgrastim or filgrastim) concentration–time profiles and ANC–time profiles in patients with non-small cell lung cancer before and after chemotherapy (n = 3–4 per cohort). (Adapted with permission from [35]).

and filgrastim cohorts; the ANC nadirs for these two groups were lower than those for the pegfilgrastim 100- and 300-μg/kg cohorts. The ANC profiles after chemotherapy for the 100- and 300-μg/kg cohorts were similar, suggesting that maximum response had been achieved by the 100-μg/kg dose. An interesting observation was contained in the PK profiles of pegfilgrastim. The maximum concentrations of pegfilgrastim after chemotherapy in each cohort were similar to those respectively before chemotherapy. However, prolonged plateaus in the concentrations were observed after chemotherapy; these concentrations began to decline only at the onset of neutrophil recovery. The clearance of pegfilgrastim after chemotherapy was lower than that before chemotherapy in each dose cohort, possibly because of the lower neutrophil-mediated clearance.

This first clinical study of pegfilgrastim in patients with NSCLC provided preliminary and encouraging evidence that a single dose of pegfilgrastim stimulates ANC recovery after chemotherapy in a manner similar to that with daily doses of filgrastim, and without greater toxicity. The data also suggested that the self-regulating clearance mechanism of pegfilgrastim could provide a potential therapeutic advantage in a variety of clinical settings associated with neutropenia. Subsequently, additional clinical studies were conducted in patients who had different types of cancer and who were treated with up to four cycles of chemotherapy [36–38].

One of these studies was a Phase II trial in women with high-risk stage II, III, or IV breast cancer conducted to evaluate the efficacy, safety, and PK of pegfilgrastim during four cycles of myelosuppressive chemotherapy [36]. At 24 h after completion of the chemotherapy (doxorubicin 60 $mg/m^2$ by IV bolus followed by docetaxel 75 $mg/m^2$ by 1-h IV infusion), patients either received a single SC dose of pegfilgrastim (30, 60, or 100 μg/kg) or began daily SC doses of filgrastim (5 μg/kg) that continued for 14 days or until ANC reached $10 \times 10^9$/L, whichever occurred first; treatment was repeated every 21 days for up to four cycles, as long as full hematopoietic recovery occurred, defined as ANC $>1 \times 10^9$/L and platelet count $>100 \times 10^9$/L. Similar to the observations in patients with NSCLC, the PK of pegfilgrastim was nonlinear in women with breast cancer; the apparent time-averaged clearance of pegfilgrastim decreased with increasing doses. The exposure to pegfilgrastim was lower in chemotherapy cycle 3 than in cycle 1 (Fig. 15.7). This observation is consistent with an increase in drug clearance, possibly because of an expansion of neutrophil and neutrophil precursor mass in the later cycles, and further supports the self-regulating clearance mechanism of pegfilgrastim. The importance of the results from this study was that for chemotherapy given over multiple cycles, a single SC injection of pegfilgrastim 100 μg/kg per chemotherapy cycle provided similar neutrophil support with respect to duration of grade 4 neutropenia (ANC $<0.5 \times 10^9$/L) compared with daily SC injections of filgrastim 5 μg/kg. In addition, pegfilgrastim was found to be as safe as filgrastim in this setting.

Subsequently, a randomized, double-blind, noninferiority Phase III study was conducted to determine whether a single SC injection of pegfilgrastim 100 μg/kg was as safe and effective in neutrophil support as daily filgrastim 5 mg/kg. Patients with breast cancer (n = 310) were randomized in a 1 : 1 ratio to receive a single dose of pegfilgrastim 100 μg/kg or daily doses of filgrastim 5 μg/kg per chemotherapy cycle; the chemotherapy treatment was the same as the above-mentioned Phase II study. Results from this Phase III study confirmed that a single injection of pegfilgrastim 100 μg/kg per chemotherapy cycle is as safe and effective as 11 daily injections of filgrastim 5 μg/kg in reducing neutropenia and its complications in patients with breast cancer [39].

Based on the results from these clinical studies, patients can be treated with pegfilgrastim once per chemotherapy cycle, with the same therapeutic benefit that daily filgrastim provides, thereby simplifying the management of chemotherapy-induced neutropenia. To further simplify treatment, giving pegfilgrastim at a fixed dose rather than a weight-adjusted dose (100 μg/kg) was considered. Instead of conducting a Phase II study to evaluate several dose levels, PK/PD modeling of clinical data was applied to facilitate selection of the optimal fixed dose, saving the expense, time, and effort typically associated with dose-finding studies [40].

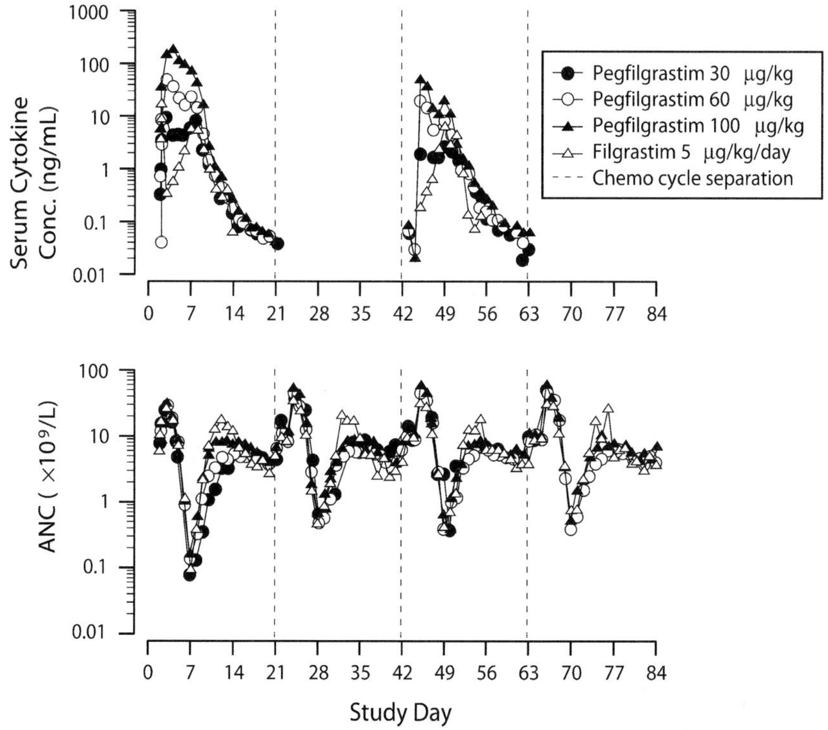


**Fig. 15.7** Median cytokine (pegfilgrastim or filgrastim) concentration–time profiles and median ANC–time profiles in patients with breast cancer. In cycle 1, n = 18, 59, and 44 for pegfilgrastim 30-, 60-, and 100-μg/kg cohorts, respectively, and n = 25 for the filgrastim cohort. In cycle 3, n = 4, 68, and 44 for pegfilgrastim 30-, 60-, and 100-μg/kg cohorts, respectively, and n = 25 for the filgrastim cohort. (Adapted with permission from [36]).

## 15.7
## Basis for the Fixed-Dose Rationale

The PK/PD model that was fit to the monkey PK and ANC data described previously was subsequently validated with data from the first clinical study of pegfilgrastim in patients with NSCLC [35]. Briefly, the model describes the PK of pegfilgrastim, the effects of the serum concentrations of pegfilgrastim on granulopoiesis, the feedback regulation of the clearance of pegfilgrastim by changing neutrophil mass, and the suppression of mitosis by chemotherapy [13].

The modeled PK and ANC profiles suggested that the PK/PD model adequately described the relationship between pegfilgrastim concentrations and neutrophil response before and after chemotherapy in patients with NSCLC (Fig. 15.8). In this analysis, the linear, neutrophil-independent clearance pathway for pegfilgrastim was significantly reduced compared with that for filgrastim; the predominant

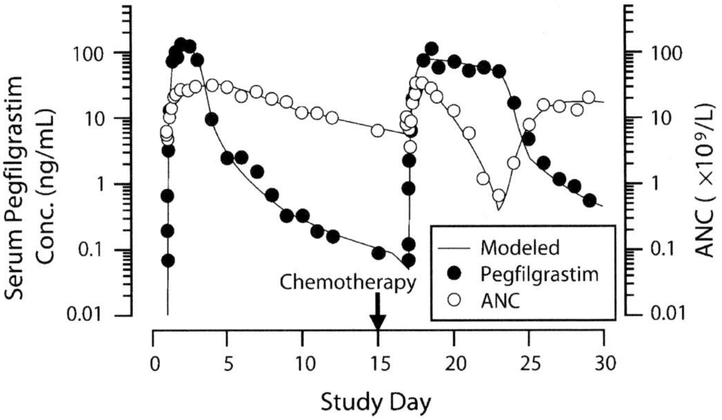


**Fig. 15.8** Pegfilgrastim concentration–time and ANC–time profiles in patients with non-small cell lung cancer after single subcutaneous administration of pegfilgrastim 100 μg/kg before and after chemotherapy. Symbols represent median observed serum pegfilgrastim concentrations (closed circles) and ANC values (open circles) that are presented in Fig. 15.6. The solid lines (generated by the PK/PD model presented in Fig. 15.4) suggest that the model provides a good description of the data.

dependency on neutrophil and neutrophil precursors for pegfilgrastim clearance causes pegfilgrastim PK to be under highly efficient, homeostatic control. The $EC_{50}$ for the effect of pegfilgrastim on mitosis and maturation was estimated to be approximately 8 ng/mL. This parameter estimate allowed for prediction of the optimum clinical dose of the drug.

In the clinical setting, pegfilgrastim is given once per chemotherapy cycle, and it is not feasible to titrate the dose for each patient to ensure optimal administration. Therefore, all patients should be given a dose sufficiently high to ensure adequate exposure to the drug. The $EC_{50}$ is considered a "threshold" concentration for response to pegfilgrastim, as this concentration falls in the middle of the steep region of the concentration–effect curve. Individuals exposed to drug concentrations below this threshold will have a low probability of responding adequately to cytokine therapy. The optimum population average response will be achieved when drug levels are near the $EC_{90}$ (72 ng/mL) of pegfilgrastim. Beyond the $EC_{90}$, there is limited incremental gain in effect when the drug level increases. Due to heterogeneity in response, an optimal dose of pegfilgrastim should thus provide a population average serum concentration of drug greater than or equal to the $EC_{90}$, with all individuals achieving serum concentration greater than or equal to the $EC_{50}$.

In order to relate drug exposure to the EC value, an unconventional PK parameter was estimated; the average concentration ($C_{avg}$) of pegfilgrastim was calculated as AUC from the time of pegfilgrastim dosing to the time of the ANC nadir divided by the time to the ANC nadir (Fig. 15.9). Because of the time required for progenitor cells (precursors) to mature, it is thought that the concentrations of pegfilgrastim before and at the time of the ANC nadir elicit the initial ANC recovery. The concentrations of pegfilgrastim after the ANC nadir, while important in

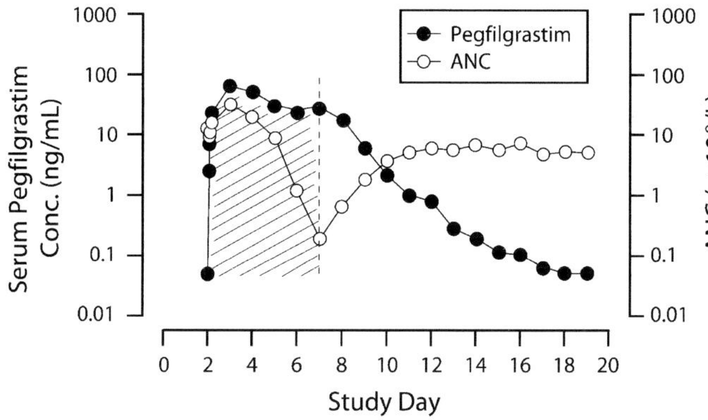


**Fig. 15.9** Pegfilgrastim average concentration was calculated by dividing the area under the pegfilgrastim concentration–time profile from time of pegfilgrastim administration to time of ANC nadir (shaded area) by the same time interval.

maintaining ANC recovery, are influenced by ANC variability, which contributes to variability in clearance. The primary objective of determining the $C_{avg}$ was to make it possible to relate drug exposure ($C_{avg}$ in ng/mL) to the $EC_{90}$ (ng/mL), which is not feasible with an AUC (ng·h/mL) representation.

Pharmacokinetic data from a Phase II study [39] in patients with breast cancer were reevaluated using the $C_{avg}$ approach. The median $C_{avg}$ values in patients treated with pegfilgrastim 30, 60, and 100 µg/kg were 6.99, 36.1, and 123 ng/mL, respectively. These drug levels correspond to $EC_{47}$, $EC_{82}$, and $EC_{94}$, respectively (Fig. 15.10). The $C_{avg}$ value in the majority of the patients in the 30-µg/kg group not only was lower than the $EC_{90}$ value (72 ng/mL) but was also lower than the $EC_{50}$ value (8 ng/mL). This confirms the observation that a dose of 30 µg/kg does not

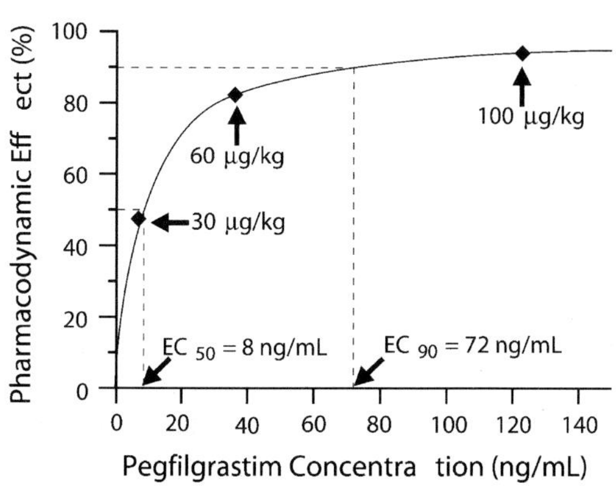


**Fig. 15.10** Concentration–effect relationship of pegfilgrastim based on a simple $E_{max}$ model. The average concentrations of pegfilgrastim from patients with breast cancer at doses of 30, 60, and 100 µg/kg are indicated (◆).

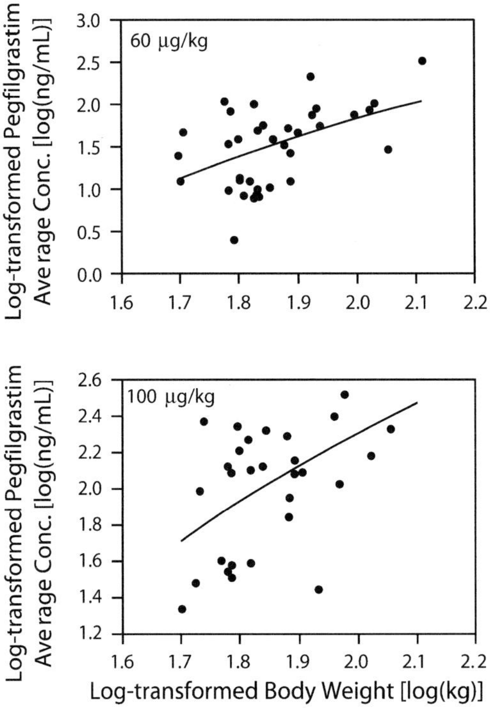


**Fig. 15.11** Relationship between average concentration of pegfilgrastim and body weight. A general trend of increasing average concentration of pegfilgrastim with increasing body weight was observed.

provide sufficient systemic drug exposure over time to support prompt and consistent ANC recovery. The median $C_{avg}$ value ($EC_{82}$) in the 60-µg/kg group was lower than the $EC_{90}$ (72 ng/mL) and, according to a log-normal distribution of $C_{avg}$ values, a $C_{avg}$ value below the $EC_{50}$ value would be expected in approximately 20% of patients treated with 60 µg/kg pegfilgrastim. At the 100-µg/kg dose level, not only was the median $C_{avg}$ value above the $EC_{90}$ in the 100-µg/kg group, but the $C_{avg}$ values were also expected to be above the $EC_{50}$ value in all patients. This retrospective analysis also confirmed that 100 µg/kg was an adequate dose on a weight basis.

Covariate analysis determined that body weight had the strongest correlation to the $C_{avg}$. The $C_{avg}$ values in the 60- and 100-µg/kg groups were higher in patients with higher body weight than in those with lower body weight (Fig. 15.11); this relationship was not observed with the subtherapeutic (30 µg/kg) dose, with which only low $C_{avg}$ values were achieved. A possible explanation for this phenomenon in the 60- and 100-µg/kg groups is that patients with a high body weight have a low volume of distribution relative to their body weight. Proteins with a size similar to that of pegfilgrastim are known to distribute to the extracellular water space, and it is proposed that the extracellular water volume does not increase proportionally with body weight, especially in overweight or obese subjects.

On the basis of the above observation, it was hypothesized that a fixed dose should provide more consistent exposure to pegfilgrastim. A suitable fixed dose (6 mg) was estimated on the basis of the average total dose given in the 60- and 100-µg/kg cohorts in a Phase II study in patients with breast cancer [39]. Additional PK analyses predicted that a fixed dose of 6 mg would provide adequate exposure with a median $C_{avg}$ value of 73.6 ng/mL, which approximates the $EC_{90}$, and that a $C_{avg}$ value above the $EC_{50}$ would be achieved in >99.99% of patients. To support this supposition, the simulated concentrations corresponding to a 6-mg dose were compared with the observed concentrations in breast cancer patients who had been given a total dose of pegfilgrastim of 5–7 mg. Both datasets matched well, and this confirmation of the predicted data by the observed clinical data provided evidence that a fixed dose of pegfilgrastim 6 mg would produce drug concentrations within the targeted range. Results from the PK analysis allowed advancement of the choice of 6 mg to the clinic for further evaluation in patients.

## 15.8 Clinical Evaluation of the Fixed Dose

A Phase III study in patients with breast cancer was designed to compare the safety and efficacy of the 6-mg fixed dose of pegfilgrastim given once per chemotherapy cycle with those of daily filgrastim (5 µg/kg) [41]. The overall study design was similar to that of the Phase III trial [36] in patients with breast cancer previously described, differing only in that a fixed dose of pegfilgrastim was given instead of a body weight-adjusted dose. The primary end point of the study was noninferiority (mean difference <1 day) in the duration of grade 4 neutropenia in cycle 1.

Results from this study showed that a single 6-mg dose of pegfilgrastim was as effective as daily injections of filgrastim in all efficacy measures in all four chemotherapy cycles; the median number of daily injections of filgrastim was 11 in cycles 1 through 3 and 10 in cycle 4. The mean (± SD) duration of grade 4 neutropenia in cycle 1 was 1.8 ± 1.4 days in the pegfilgrastim cohort, and 1.6 ± 1.1 days in the filgrastim cohort. A concern with the fixed dose was that the clinical benefit might not be as great in heavier patients, owing to their being given a lower dose per unit of body weight. Subset analysis by baseline body weight was therefore performed. Comparison of the duration of grade 4 neutropenia in cycle 1 for all subsets suggested that patients of all weights were adequately supported (Fig. 15.12). At the other extreme, another concern was that a fixed dose might result in an altered safety profile in lighter patients. When evaluated both between treatment groups and within weight groups, however, the incidence and severity of adverse events, including musculoskeletal bone pain, with a fixed dose of pegfilgrastim 6 mg were no different from those with filgrastim.

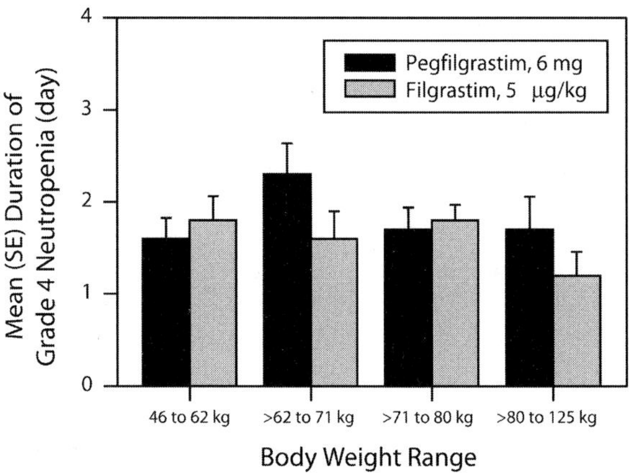


**Fig. 15.12** Mean (SE) duration of grade 4 neutropenia in patients with breast cancer during the first chemotherapy cycle. The responses for the different body weight subsets were similar after a single fixed dose of pegfilgrastim 6 mg and after daily doses of filgrastim 5 μg/kg. (Data from [41]).

To further support the use of a 6-mg fixed dose of pegfilgrastim, a Phase II study was conducted in 29 patients with non-Hodgkin's lymphoma who received a single SC 6-mg dose of pegfilgrastim approximately 24 h after the start of CHOP (cyclophosphamide, doxorubicin, vincristine, and prednisone) chemotherapy [42]. No relationship between the duration of grade 4 neutropenia and body weight was seen, and no unexpected adverse events were reported.

These two clinical trials validated that a fixed dose of pegfilgrastim 6 mg is as safe as, and provides the clinical benefits of, filgrastim. Subsequently, clinical efficacy data from two pivotal studies in which patients received either a single dose of pegfilgrastim 100 μg/kg or 6 mg or filgrastim 5 μg/kg/day were pooled for a retrospective analysis [43]. Collectively, the data showed that the relative decrease in risk of febrile neutropenia was significantly more pronounced (42%) in patients who received pegfilgrastim compared to those who received filgrastim. Furthermore, patients who received pegfilgrastim were also at a lower risk of requiring hospitalization and IV anti-infective medications than those administered filgrastim. These observations were noted regardless of age, performance status, disease stage, and prior treatment. Because of the advantage of once-per-cycle dosing, pegfilgrastim is expected to simplify the management of chemotherapy-induced neutropenia and to have significant quality-of-life benefits.

## 15.9 Summary

The development of pegfilgrastim is an example of the use of PK and PD principles in drug development. The PK/PD modeling of preclinical and clinical pegfilgrastim data has been important for describing the elimination pathways of the drug and selecting the optimal schedule and dose. The granulopoietic model takes into account the maturation and development of neutrophil precursor cells, which is an essential component for the elimination of pegfilgrastim and is consistent with feedback-regulated clearance by G-CSF receptors on neutrophils and neutrophil precursors. Both preclinical and clinical trials have verified the predictions in the PK/PD models. Pegfilgrastim can be administered as a fixed dose once per cycle, and is as safe and effective as filgrastim in decreasing the severity and duration of neutropenia and its complications in patients treated with myelosuppressive chemotherapy.

### Acknowledgments

The author thanks Lorin Roskos, PhD, for his PK/PD modeling and simulation efforts, including those prospectively provided in the Investigational New Drug Application for pegfilgrastim, and for his insights on data analyses and interpretation. Thanks are also expressed to the Amgen Pegfilgrastim Global Development Team for conducting many preclinical and clinical studies, which provided the PK and PD data described herein, and to Anna Kido for her assistance in writing this chapter.

## 15.10 References

1 Lyman GH, Kuderer NM. Filgrastim in patients with neutropenia: potential effects on quality of life. *Drugs* **2002**; *62* (suppl 1):65–78.

2 Dale D. Current management of chemotherapy-induced neutropenia: the role of colony-stimulating factors. *Semin. Oncol.* **2003**:*30*:3–9.

3 Demetri GD, Griffin JD. Granulocyte colony-stimulating factor and its receptor. *Blood* **1991**;*78*:2791–2808.

4 Bociek RG, Armitage JO. Hematopoietic growth factors. *CA Cancer J. Clin.* **1996**; *46*:165–184.

5 Souza LM, Boone TC, Gabrilove J, et al. Recombinant human granulocyte colony-stimulating factor: effects on normal and leukemic myeloid cells. *Science* **1986**;*232*:61–65.

6 Zsebo KM, Cohen AM, Murdock DC, et al. Recombinant human granulocyte colony stimulating factor: molecular and biological characterization. *Immunobiology* **1986**;*172*:175–184.

7 Herman AC, Boone TC, Lu HS. Characterization, formulation, and stability of Neupogen (filgrastim), a recombinant human granulocyte-colony stimulating factor. In: Pearlman R, Wang YJ (Eds.), *Formulation, Characterization, and Stability of Protein Drugs.* New York, NY: Plenum Press; **1996**:303–328.

8 Welte K, Gabrilove J, Bronchud MH, et al. Filgrastim (r-metHuG-CSF): the first 10 years. *Blood* **1996**;*88*:1907–1929.
9 Neupogen [package insert]. Thousand Oaks, CA: Amgen Inc., **2004**.
10 Molineux G, Kinstler O, Briddell B, et al. A new form of filgrastim with sustained duration in vivo and enhanced ability to mobilize PBPC in both mice and humans. *Exp. Hematol.* **1999**;*27*:1724–1734.
11 Kinstler O, Molineux G, Treuheit M, Ladd D, Gregg C. Mono-N-terminal poly(ethylene glycol)-protein conjugates. *Adv. Drug Deliv. Rev.* **2002**;*54*:477–485.
12 Molineux G. The design and development of pegfilgrastim (PEG-rmetHuG-CSF, Neulasta). *Curr. Pharm. Des.* **2004**;*10*: 1235–1244.
13 Roskos, LK, Yang B, Schwab G, et al. A cytokinetic model of r-metHuG-CSF-SD/01 (SD/01) mediated granulopoiesis and the "self-regulation" of SD/01 elimination in non-small cell lung cancer (NSCLC) patients. *Blood* **1998**;*92*:507a. Abstract 2085.
14 Kuwabara T, Kobayashi S, Sugiyama Y. Pharmacokinetics and pharmacodynamics of a recombinant human granulocyte colony-stimulating factor. *Drug Metab. Rev.* **1996**;*28*:625–658.
15 Kuwabara T, Uchimura T, Kobayashi H, Kobayashi S, Sugiyama Y. Receptor-mediated clearance of G-CSF derivative nartograstim in bone marrow of rats. *Am. J. Physiol.* **1995**;*269*:E1–E9.
16 Kuwabara T, Uchimura T, Takai K, Kobayashi H, Kobayashi S, Sugiyama Y. Saturable uptake of a recombinant human granulocyte colony-stimulating factor derivative, nartograstim, by the bone marrow and spleen of rats in vivo. *J. Pharmacol. Exp. Ther.* **1995**;*273*:1114–1122.
17 Kuwabara T, Kobayashi S, Sugiyama Y. Kinetic analysis of receptor-mediated endocytosis of G-CSF derivative, nartograstim, in rat bone marrow cells. *Am. J. Physiol.* **1996**;*271*:E73–E84.
18 Layton JE, Hockman H, Sheridan WP, Morstyn G. Evidence for a novel in vivo control mechanism of granulopoiesis: mature cell-related control of a regulatory growth factor. *Blood* **1989**;*74*:1303–1307.
19 Stute N, Santana VM, Rodman JH, Schell MJ, Ihle JN, Evans WE. Pharmacokinetics of subcutaneous recombinant human granulocyte colony-stimulating factor in children. *Blood* **1992**;*79*:2849–2854.
20 Kearns CM, Wang WC, Stute N, Ihle JN, Evans WE. Disposition of recombinant human granulocyte colony-stimulating factor in children with severe chronic neutropenia. *J. Pediatr.* **1993**;*123*:471–479.
21 Sturgill MG, Huhn RD, Drachtman RA, Ettinger AG, Ettinger LJ. Pharmacokinetics of intravenous recombinant human granulocyte colony-stimulating factor (rhG-CSF) in children receiving myelosuppressive cancer chemotherapy: clearance increases in relation to absolute neutrophil count with repeated dosing. *Am. J. Hematol.* **1997**;*54*:124–130.
22 Takatani H, Soda H, Fukuda M, et al. Levels of recombinant human granulocyte colony-stimulating factor in serum are inversely correlated with circulating neutrophil counts. *Antimicrob. Agents Chemother.* **1996**;*40*:988–991.
23 Terashi K, Oka M, Ohdo S, et al. Close association between clearance of recombinant human granulocyte colony-stimulating factor (G-CSF) and G-CSF receptor on neutrophils in cancer patients. *Antimicrob. Agents Chemother.* **1999**;*43*:21–24.
24 Avalos B. Molecular analysis of the granulocyte colony-stimulating factor receptor. *Blood* **1996**;*88*:761–777.
25 Yang B-B, Lum PK, Hayashi MM, Roskos LK. Polyethylene glycol modification of filgrastim results in decreased renal clearance of the protein in rats. *J. Pharm. Sci.* **2004**;*93*:1367–1373.
26 Tanaka H, Tokiwa T. Influence of renal and hepatic failure on the pharmacokinetics of recombinant human granulocyte colony-stimulating factor (KRN8601) in the rat. *Cancer Res.* **1990**;*50*:6615–6619.
27 Harris JM, Martin NA, Modi M. Pegylation: a novel process for modifying pharmacokinetics. *Clin. Pharmacokinet.* **2001**;*40*:539–551.
28 Roberts MJ, Bentley MD, Harris JM. Chemistry for peptide and protein Pegylation. *Adv. Drug Deliv. Rev.* **2002**;*54*:459–476.
29 Yamasaki M, Asano M, Okabe M, Morimoto M, Yokoo Y. Modification of recombinant human granulocyte colony-stimulat-

ing factor (rhG-CSF) and its derivative ND 28 with polyethylene glycol. *J. Biochem.* **1994**;*115*:814–819.

**30** Lord BI, Woolford LB, Molineux G. Kinetics of neutrophil production in normal and neutropenic animals during the response to filgrastim (r-metHu G-CSF) or filgrastim SD/01 (PEG-r-metHu G-CSF). *Clin. Cancer Res.* **2001**;*7*: 2085–2090.

**31** Kotto-Kome AC, Fox SE, Lu W, Yang B-B, Christensen RD, Calhoun DA. Evidence that the granulocyte colony-stimulating factor (G-CSF) receptor plays a role in the pharmacokinetics of G-CSF and PegG-CSF using a G-CSF-R KO model. *Pharm. Res.* **2004**;*50*:55–58.

**32** Cheung EN, Cosenza M, Lopez O, et al. Modeling of r-metHuG-CSF-SD/01 (SD/01) mediated granulopoiesis in normal animals with mathematical extrapolation to neutropenic settings. *Blood* **1998**;*92* (suppl 1):379a. Abstract.

**33** Athens JW. Granulocytes-neutrophils. In: Lee GR, Bithell TC, Foerster J, Athens JW, Lukens JN (Eds.), *Wintrobe's Clinical Hematology*. Vol. 1. 9th edn. Philadelphia, PA; Lea & Febiger; 1993:223–266.

**34** Farese AM, Yang B-B, Roskos L, Stead RB, MacVittie TJ: Pegfilgrastim, a sustained-duration form of filgrastim, significantly improves neutrophil recovery after autologous marrow transplantation in rhesus macaques. *Bone Marrow Transplant.* **2003**;*32*:399–404.

**35** Johnston E, Crawford J, Blackwell S, et al. Randomized, dose-escalation study of SD/01 compared with daily filgrastim in patients receiving chemotherapy. *J. Clin. Oncol.* **2000**;*18*:2522–2528.

**36** Holmes FA, Jones SE, O'Shaughnessy J, et al. Comparable efficacy and safety profiles of once-per-cycle pegfilgrastim and daily injection filgrastim in chemotherapy-induced neutropenia: a multicenter dose-finding study in women with breast cancer. *Ann. Oncol.* **2002**;*13*:903–909.

**37** Vose JM, Crump M, Lazarus H, et al. Randomized, multicenter, open-label study of pegfilgrastim compared with daily filgrastim after chemotherapy for lymphoma. *J. Clin. Oncol.* **2003**;*21*:514–519.

**38** Grigg A, Solal-Celigny P, Hoskin P, et al. Open-label, randomized study of pegfilgrastim vs. daily filgrastim as an adjunct to chemotherapy in elderly patients with non-Hodgkin's lymphoma. *Leuk. Lymphoma* **2003**;*44*:1503–1508.

**39** Holmes FA, O'Shaughnessy JA, Vukelja S, et al. Blinded, randomized, multicenter study to evaluate single administration pegfilgrastim once per cycle versus daily filgrastim as an adjunct to chemotherapy in patients with high-risk stage II or stage III/IV breast cancer. *J. Clin. Oncol.* **2002**; *20*:727–731.

**40** Yang B-B, Lum P, Renwick J, et al. Pharmacokinetic rationale for a fixed-dose regimen of a sustained-duration form of filgrastim in cancer patients. *Blood* **2000**;*96*:157b. Abstract 4384.

**41** Green MD, Koelbl H, Baselga J, et al. A randomized double-blind multicenter phase III study of fixed-dose single-administration pegfilgrastim versus daily filgrastim in patients receiving myelosuppression chemotherapy. *Ann. Oncol.* **2003**;*14*:29–35.

**42** George S, Yunus F, Case D, et al. Fixed-dose pegfilgrastim is safe and allows neutrophil recovery in patients with non-Hodgkin's lymphoma. *Leuk. Lymphoma* **2003**;*10*:1691–1696.

**43** Siena S, Piccart MJ, Holmes FA, Glaspy J, Hackett J, Renwick JJ. A combined analysis of two pivotal randomized trials of a single dose of pegfilgrastim per chemotherapy cycle and daily filgrastim in patients with stage II-IV breast cancer. *Oncol. Rep.* **2003**;*10*:715–724.

# Subject Index



*Pharmacokinetics and Pharmacodynamics of Biotech Drugs: Principles and Case Studies in Drug Development.* Edited by Bernd Meibohm

ISBN: 3-527-31408-3